Environmentally Friendly Polymers and Polymer Composites

Environmentally Friendly Polymers and Polymer Composites

Editors

Rafael Balart
Nestor Montanes
Franco Dominici
Sergio Torres-Giner
Teodomiro Boronat

MDPI • Basel • Beijing • Wuhan • Barcelona • Belgrade • Manchester • Tokyo • Cluj • Tianjin

MDPI

Editors
Rafael Balart
Universitat Politècnica de València (UPV)
Spain

Nestor Montanes
Universitat Politècnica de València (UPV)
Spain

Franco Dominici
Università di Perugia
Italy

Sergio Torres-Giner
Universitat Politècnica de València (UPV)
Spain

Teodomiro Boronat
Universitat Politècnica de València (UPV)
Spain

Editorial Office
MDPI
St. Alban-Anlage 66
4052 Basel, Switzerland

This is a reprint of articles from the Special Issue published online in the open access journal *Materials* (ISSN 1996-1944) (available at: https://www.mdpi.com/journal/materials/special_issues/green_polymers).

For citation purposes, cite each article independently as indicated on the article page online and as indicated below:

LastName, A.A.; LastName, B.B.; LastName, C.C. Article Title. *Journal Name* **Year**, *Volume Number*, Page Range.

ISBN 978-3-0365-0036-2 (Hbk)
ISBN 978-3-0365-0037-9 (PDF)

Cover image courtesy of Sergio Torres-Giner.

Contents

About the Editors vii

Preface to "Environmentally Friendly Polymers and Polymer Composites" ix

Rafael Balart, Nestor Montanes, Franco Dominici, Teodomiro Boronat and Sergio Torres-Giner
Environmentally Friendly Polymers and Polymer Composites
Reprinted from: *Materials* **2020**, *13*, 4892, doi:10.3390/ma13214892 1

Cristina Mellinas, Marina Ramos, Alfonso Jiménez and María Carmen Garrigós
Recent Trends in the Use of Pectin from Agro-Waste Residues as a Natural-Based Biopolymer for Food Packaging Applications
Reprinted from: *Materials* **2020**, *13*, 673, doi:10.3390/ma13030673 7

Siriporn Taokaew, Mitsumasa Ofuchi and Takaomi Kobayashi
Size Distribution and Characteristics of Chitin Microgels Prepared via Emulsified Reverse-Micelles
Reprinted from: *Materials* **2019**, *12*, 1160, doi:10.3390/ma12071160 25

Jyoti Ahlawat, Eva M. Deemer and Mahesh Narayan
Chitosan Nanoparticles Rescue Rotenone-Mediated Cell Death
Reprinted from: *Materials* **2019**, *12*, 1176, doi:10.3390/ma12071176 41

Kornkamol Potivara and Muenduen Phisalaphong
Development and Characterization of Bacterial Cellulose Reinforced with Natural Rubber
Reprinted from: *Materials* **2019**, *12*, 2323, doi:10.3390/ma12142323 55

Diego Lascano, Luis Quiles-Carrillo, Rafael Balart, Teodomiro Boronat and Nestor Montanes
Toughened Poly (Lactic Acid)—PLA Formulations by Binary Blends with Poly(Butylene Succinate-*co*-Adipate)—PBSA and Their Shape Memory Behaviour
Reprinted from: *Materials* **2019**, *12*, 622, doi:10.3390/ma12040622 73

Luis Quiles-Carrillo, Nestor Montanes, Fede Pineiro, Amparo Jorda-Vilaplana and Sergio Torres-Giner
Ductility and Toughness Improvement of Injection-Molded Compostable Pieces of Polylactide by Melt Blending with Poly(ε-caprolactone) and Thermoplastic Starch
Reprinted from: *Materials* **2018**, *11*, 2138, doi:10.3390/ma11112138 87

Beatriz Melendez-Rodriguez, Sergio Torres-Giner, Abdulaziz Aldureid, Luis Cabedo and Jose M. Lagaron
Reactive Melt Mixing of Poly(3-Hydroxybutyrate)/Rice Husk Flour Composites with Purified Biosustainably Produced Poly(3-Hydroxybutyrate-*co*-3-Hydroxyvalerate)
Reprinted from: *Materials* **2019**, *12*, 2152, doi:10.3390/ma12132152 107

Dumitru Bolcu and Marius Marinel Stănescu
The Influence of Non-Uniformities on the Mechanical Behavior ofHemp-Reinforced Composite Materials with a Dammar Matrix
Reprinted from: *Materials* **2019**, *12*, 1232, doi:10.3390/ma12081232 129

Pooria Khalili, Xiaoling LIU, Zirui ZHAO and Brina Blinzler
Fully Biodegradable Composites: Thermal, Flammability, Moisture Absorption and Mechanical Properties of Natural Fibre-Reinforced Composites with Nano-Hydroxyapatite
Reprinted from: *Materials* **2019**, *12*, 1145, doi:10.3390/ma12071145 . . . . **145**

Izaskun Larraza, Lorena Ugarte, Aintzane Fayanas, Nagore Gabilondo, Aitor Arbelaiz, Maria Angeles Corcuera and Arantxa Eceiza
Influence of Process Parameters in Graphene Oxide Obtention on the Properties of Mechanically Strong Alginate Nanocomposites
Reprinted from: *Materials* **2020**, *13*, 1081, doi:10.3390/ma13051081 . . . . **159**

Franco Dominici, María Dolores Samper, Alfredo Carbonell-Verdu, Francesca Luzi, Juan López-Martínez, Luigi Torre and Debora Puglia
Improved Toughness in Lignin/Natural Fiber Composites Plasticized with Epoxidized and Maleinized Linseed Oils
Reprinted from: *Materials* **2020**, *13*, 600, doi:10.3390/ma13030600 . . . . **177**

Lucia Gonzalez, Angel Agüero, Luis Quiles-Carrillo, Diego Lascano and Nestor Montanes
Optimization of the Loading of an Environmentally Friendly Compatibilizer Derived from Linseed Oil in Poly(Lactic Acid)/Diatomaceous Earth Composites
Reprinted from: *Materials* **2019**, *12*, 1627, doi:10.3390/ma12101627 . . . . **193**

Patricia Liminana, Luis Quiles-Carrillo, Teodomiro Boronat, Rafael Balart and Nestor Montanes
The Effect of Varying Almond Shell Flour (ASF) Loading in Composites with Poly(Butylene Succinate (PBS) Matrix Compatibilized with Maleinized Linseed Oil (MLO)
Reprinted from: *Materials* **2018**, *11*, 2179, doi:10.3390/ma11112179 . . . . **209**

Patricia Liminana, David Garcia-Sanoguera, Luis Quiles-Carrillo, Rafael Balart and Nestor Montanes
Optimization of Maleinized Linseed Oil Loading as a Biobased Compatibilizer in Poly(Butylene Succinate) Composites with Almond Shell Flour
Reprinted from: *Materials* **2019**, *12*, 685, doi:10.3390/ma12050685 . . . . **227**

Mirna Nunes Araújo, Leila Lea Yuan Visconte, Daniel Weingart Barreto, Viviane Alves Escócio, Ana Lucia Nazareth da Silva, Ana Maria Furtado de Sousa and Elen Beatriz Acordi Vasques Pacheco
The Use of Cashew Nut Shell Liquid (CNSL) in PP/HIPS Blends: Morphological, Thermal, Mechanical and Rheological Properties
Reprinted from: *Materials* **2019**, *12*, 1904, doi:10.3390/ma12121904 . . . . **241**

Vu Minh Thanh, Le Minh Bui, Long Giang Bach, Ngoc Tung Nguyen, Hoa Le Thi and Thai Thanh Hoang Thi
Origanum majorana L. Essential Oil-Associated Polymeric Nano Dendrimer for Antifungal Activity against *Phytophthora infestans*
Reprinted from: *Materials* **2019**, *12*, 1446, doi:10.3390/ma12091446 . . . . **267**

About the Editors

Rafael Balart received his Ph.D. from the Polytechnic University of Valencia (UPV), Spain, in 2003. Currently, he is Full Professor of Materials Science and Engineering at the Department of Mechanical and Materials Engineering. With more than 25 years of experience in the field of polymer and composite materials, he currently leads the Research Group on Environmentally Friendly Polymers and Composites (GiPCEco), which focuses its activity on the development and formulation of sustainable polymers and the manufacture of composite materials for industrial applications. In the last years, he has focused his research on upgrading industrial, agricultural, and agroforestry wastes with a biorefinery approach to obtain highly value-added products with applications in polymers and composites.

Nestor Montanes received his Ph.D. in Engineering from the Polytechnic University of Valencia (UPV), Spain, in 2017. Currently, he is Associate Professor of Materials Science and Engineering at the Department of Mechanical and Materials Engineering (UPV Campus d'Alcoi). With more than 10 years of experience in the university and 10 other years of experience in the packaging industry, he has actively participated in the implementation of new teaching methodologies in the classroom: reverse teaching, gamification, and so on. He joined the Technological Institute of Materials (ITM) in 2014 and he has specialized in polymer analysis and characterization, mainly biopolymers and polymer-based composites. His areas of interest include polymer manufacturing, compounding, formulation, additives, polymer blends, and upgrading industrial wastes in the polymer industry.

Franco Dominici received a degree in Industrial Engineering from the University of Perugia in 2014. Then, he specialized in the field of high-performance polymers and received his Ph.D. within the framework of an International Doctorate in Civil and Environmental Engineering in 2018. His research is focused on: processing and characterization of thermoplastic and thermosetting polymers, high-performance polymer matrix composites/nanocomposites, biopolymers, and green composites. Currently, he is working as technical staff in the Materials Science and Technology group, which operates at the Faculty of Engineering of the University of Perugia, in the Polo Scientifico e Didattico of Terni, aiming to develop knowledge on advanced and traditional materials technologies for their design and production and their engineering applications.

Sergio Torres-Giner received a Dipl-Ing in Chemical Engineering from the Polytechnic University of Valencia (UPV), Spain, in 2003. In 2004, he achieved a M.Sc. in Process Systems Technology at Cranfield University, England, followed by an MBA in Industrial Management in 2005 at Catholic University of Valencia 'San Vincente Martir', Spain. He completed his Ph.D. in 2010 in Food Science at the University of Valencia, Spain. He currently works as a scientist at the Research Institute of Food Engineering for Development (IIAD) in the field of macromolecular science of application interest in food packaging technology. He has more than 15 years of experience in both public research agencies and industrial R&D organizations. He has published over 75 peer-reviewed scientific papers indexed in JCR, 10 book and encyclopedia chapters, and 4 patents. His research activity has strongly contributed to advancing the knowledge of biopolymers and to transferring it into applications and products for food-related applications.

Teodomiro Boronat is a Mechanical Engineer and received his Ph.D. in Engineering from the Polytechnic University of Valencia (UPV), Spain, in 2009. Currently, he is Associate Professor of Manufacturing Engineering at the Department of Mechanical and Materials Engineering (UPV Campus d'Alcoi). He is also a researcher at the Technological Institute of Materials (ITM) and specializes in polymer manufacturing processes and additive manufacturing. His main research focus is studying the relationship between processing parameters and the obtained final properties. Other areas of interest include thermoplastic manufacturing processes, thermosetting systems, rheology, and additive manufacturing.

Preface to "Environmentally Friendly Polymers and Polymer Composites"

In the last few years, environmental concerns about the disposal and accumulation of plastic wastes have increased. Conventional plastics are produced from fossil fuels and, in general, these materials are not biodegradable in nature or even disintegrable in controlled compost soil. As a result, our society is becoming increasingly sensitive to pollution derived from plastic materials and aware of these environmental issues. Thus, relatively new topics related to petroleum depletion, climate change, sustainable development, Circular Economy, and waste management have recently arisen.

Within this context, important research has been conducted in the field of environmentally friendly polymers and polymer composites in the last decade. These investigations include a wide range of biopolymers that show improved environmental efficiency at different stages of their life cycle, spanning from polymers that are obtained from renewable resources—for example, bio-based polyethylenes (bio-PEs), bio-based polyethylene terephthalate (bio-PET), or bio-based polyamides (bio-PAs)—to biodegradable petroleum derived polymers such as poly(ϵ-caprolactone) (PCL), polyvinyl alcohol (PVA), polyglycolide (PGA), polybutylene succinate (PBS), and the poly(butylene adipate-co-terephthalate) (PBAT) and poly(butylene succinate-co-adipate) (PBSA) copolymers. In addition, novel research is being conducted on bio-based and biodegradable polymers, which include polysaccharide derived polymers (e.g., starch, cellulose, pectin, chitin, and its derivative chitosan), natural proteins (e.g., gluten, casein, ovalbumin, and collagen), and biopolyesters of more industrial significance, for instance, polylactide (PLA) and microbial polyhydroxyalkanoates (PHAs), such as poly(3-hydroxybutyrate) (PHB) or poly(3-hydroxybutyrate-co-3-valerate) (PHBV). Original research is being performed in the field of green composites, in which the matrix and the reinforcement phase are both obtained from renewable resources such as natural fiber reinforced plastics (NFRPs) and wood plastic composites (WPCs). These novel polymer composites can reduce the overall current cost of biopolyesters, improve their physical properties, and be used to develop products with a wood-like aspect. These environmentally friendly materials can be combined with vegetable oils, waste derived liquids (1), and essential oils to yield more sustainable and high-performance polymer-based materials.

The book is divided into 16 chapters that compile novel research works dealing with sustainable polymers and polymer composites to advance sustainable development. The first three chapters cover the potential use of different types of carbohydrates, which are promising in food packaging as active systems and in pharmaceutical and biomedical applications. This part of the book includes a review on pectin and two original research articles focused on chitin and chitosan. The next four chapters discuss strategies to improve the thermal resistance and mechanical ductility of naturally occurring polymers such as bacterial cellulose (BC), which habitually shows strong hydrophilicity, and PLA and PHB biopolyesters. The book continues with three more chapters demonstrating that biopolymers are excellent candidates for manufacturing WPCs and NFRPs in combination with plant fibers or for developing advanced composites using mineral and organic nanoparticles. Four subsequent chapters display the potential of multi-functionalized vegetable oils as novel renewable additives for the plasticization and compatibilization of biopolymer blends and green composites. All these research articles describe the improvements attained in thermal stability and mechanical strength due to the formation of ester bonds between the multiple reactive groups present in the oils and the terminal hydroxyl groups of the biopolyester matrices. The book ends with two more chapters discussing the application of renewable and waste derived liquids (2) for green composite formulations. This section is composed of two novel research articles dealing with the use of a by-product of the cashew agricultural industry and an essential oil as plasticizer and fungicidal alternatives to conventional chemical additives.

This book will serve to guide a diverse audience of material scientists since it provides an update of the state-of-the-art knowledge on environmentally friendly polymers and polymer composites. The book provides a valuable reference for chemical engineers to transfer new materials for industrial purposes, particularly in the food packaging sector due to the high volume of plastic waste generated by this industry.

Rafael Balart, Nestor Montanes, Franco Dominici, Sergio Torres-Giner, Teodomiro Boronat

Editors

materials

Editorial

Environmentally Friendly Polymers and Polymer Composites

Rafael Balart [1,*], Nestor Montanes [1], Franco Dominici [2], Teodomiro Boronat [1] and Sergio Torres-Giner [3,*]

1 Technological Institute of Materials (ITM), Universitat Politècnica de València (UPV), Plaza Ferrándiz y Carbonell 1, 03801 Alcoy, Spain; nesmonmu@upvnet.upv.es (N.M.); tboronat@dimm.upv.es (T.B.)

2 Civil and Environmental Engineering Department, University of Perugia, UdR INSTM, Strada di Pentima 4, 05100 Terni, Italy; franco.dominici@unipg.it

3 Research Institute of Food Engineering for Development (IIAD), Universitat Politècnica de València (UPV), Camino de Vera s/n, 46022 Valencia, Spain

* Correspondence: rbalart@mcm.upv.es (R.B.); storresginer@upv.es (S.T.-G.)

Received: 14 October 2020; Accepted: 28 October 2020; Published: 31 October 2020

Abstract: In the last decade, continuous research advances have been observed in the field of environmentally friendly polymers and polymer composites due to the dependence of polymers on fossil fuels and the sustainability issues related to plastic wastes. This research activity has become much more intense in the food packaging industry due to the high volume of waste it generates. Biopolymers are nowadays considered as among the most promising materials to solve these environmental problems. However, they still show inferior performance regarding both processability and end-use application. Blending currently represents a very cost-effective strategy to increase the ductility and impact resistance of biopolymers. Furthermore, different lignocellulosic materials are being explored to be used as reinforcing fillers in polymer matrices for improving the overall properties, lower the environmental impact, and also reduce cost. Moreover, the use of vegetable oils, waste derived liquids, and essential oils opens up novel opportunities as natural plasticizers, reactive compatibilizers or even active additives for the development of new polymer formulations with enhanced performance and improved sustainability profile.

Keywords: bio-based polymers; biodegradable polyesters; green composites; wood plastic composites; natural additives and fillers; composites characterization; bioplastics manufacturing

---

The demand for plastics has remarkably increased in recent decades and pollution deriving from these materials has become one of the most prominent environmental concerns of recent years. Moreover, conventional polymers are obtained from fossil fuels and show high persistence in the environment, which results in sustainability issues related to petroleum depletion and waste management. Recent regulations on the recyclability of materials and environmental requirements have compelled manufacturers and research institutions to develop environmentally friendly polymers and polymer composites in the last years. This Special Issue compiles one review and fifteen articles written by researchers and technologists that reflects the growing interest in the development of sustainable materials that offer the possibility of rendering interesting applications in food packaging as well as in other sectors such as pharmaceutics, agriculture or biomedicine.

In this context, the biopolymers obtained from renewable resources currently represent a sustainable alternative to petroleum derived polymers and they can also contribute to decrease the product carbon footprint. The next generation of biopolymers will be synthesized from non-edible and highly available plants and also from agro-food and industrial wastes or by-products, which contribute to the progress of the Circular Bioeconomy. Furthermore, some biopolymers are also biodegradable and their resultant articles can be disintegrable in controlled compost soil. These materials include different types of carbohydrates, for example thermoplastic starch (TPS), cellulose and its derivatives, alginates, pectin, chitin or chitosan, animal-based proteins, such as silk, gelatin or collagen, and plant-based proteins as well as lipids. In this regard, the review performed by Mellinas et al. [1] gathered the most recent studies regarding the sources, different types, structure, and potential uses of pectin. This natural-based polysaccharide is currently extracted from plants and thereafter isolated as a bioplastic material. However, by using innovative methods, pectin can also be obtained from different kinds of waste biomass, ranging from the by-products of juice manufacturing to the peels and seeds of orange, mango, banana, lime or pomegranate. Authors also showed that the final applications of pectin can be very diverse due to the variability in its structure, being very promising in food packaging in the development of active systems.

Chitin and chitosan are also relevant examples of polysaccharides showing a great deal of potential in pharmaceutical and biomedical applications. Chitin is a β-(1,4)-*N*-acetyl-D-glucosamine found in the shells of crabs, lobsters, and shrimps. Chitosan, which is obtained from the alkaline *N*-deacetylation of chitin, is a cationic polysaccharide composed of a linear chain of D-glucosamine and *N*-acetyl-D-glucosamine linked via a β-(1,4) bond. Both carbohydrates are largely present in marine fishery by-products and are also known to exhibit several active and bioactive properties. For instance, in the research article of Taokaew et al. [2], chitin was extracted from local snow crab shell waste and used to produce microgels with sizes ranging from 5 to 200 μm using a batch process of emulsification and gelation. Chitin microgels with narrow size distribution with an average size as low as 7 μm and porous spherical morphology were successfully achieved at chitin contents of 3 wt %, exhibiting pH-dependent swelling-shrinking behavior for pH values between 2 and 10. In another study, Ahlawat et al. [3] reported the antioxidant effect on a human SH-SY5Y neuroblastoma cell line of chitosan nanoparticles synthesized using an ionotropic gelation method with tripolyphosphate (TPP) as the cross-linking agent. Chitosan particles with an average size of ~200 nm showed reduced rotenone-initiated cytotoxicity and apoptotic cell death. According to the authors, these novel chitosan nanoparticles might be a neuroprotective agent for the prevention of Parkinson's disease.

The main drawbacks of most biopolymers are their inferior thermal resistance and poorer mechanical properties when compared to conventional polymers such as polyolefins and styrene-based polymers. In the case of naturally occurring polymers, they habitually show strong hydrophilic character, being highly affected by water. All these characteristics limit the expansion of biopolymers to food packaging and other commodity areas. Blending represents a very cost-effective solution to overcome or, at least, minimize the low ductility and toughness of biopolymers. For instance, bacterial cellulose (BC), which is mainly produced by the bacteria of genera Acetobacter, such as *Acetobacter xylinum*, can result in strong and high-barrier films composed of interconnected networks of cellulose nanofibers. Since it shows low-breaking elongation, Potivara and Phisalaphong [4] achieved a ~4-fold improvement in the tensile strength and elongation at break of BC films by the incorporation of natural rubber (NR). When BC pellicles were immersed in a diluted NR latex (NRL) suspension at 2.5–5.0% dry rubber content (DRC) in the presence of ethanol aqueous solution at 50–60 °C, the resultant NR/BC films showed improved resistance to water and also high resistance to non-polar solvents such as toluene, biodisintegrating completely in soil after 5–6 weeks. Among biopolyesters, polylactide (PLA) is already well positioned in the bioplastics market due to its good processability, balanced properties, relatively low price, and industrial compostability. Despite this, PLA materials are very brittle and for this reason Lascano et al. [5] developed binary blends of PLA with the petrochemical biodegradable copolyester poly(butylene succinate-*co*-adipate) (PBSA) and an epoxy styrene-acrylic oligomer (ESAO) as a reactive compatibilizer. The compatibilized blend containing 20 wt % of PBSA showed elongation-at-break values of approximately 121%, whereas the addition of 30 wt % of PBSA also improved the impact strength in V-notched samples from 2.48 to 5.75 kJ/m^2. Furthermore, the addition of 10 wt % PBSA slightly improved the shape memory recovery of PLA. With the same scope, ternary blends of PLA with 40 wt % of different poly(ε-caprolactone) (PCL) and TPS combinations were reported by Quiles-Carrillo et al. [6]. Although all the biopolymer blends were immiscible, the combination of PLA with 30 wt % of PCL and 10 wt % of TPS showed an elongation at break as high as 196.7%, which was approximately 40 times higher than that of neat PLA. The resultant improvement in toughness was ascribed to the "island-and-sea" morphology of the PLA-based blends, where the rubber-like PCL phase was finely dispersed in the form of micro-sized spherical domains favored by the co-presence of TPS.

Despite the high suitability of biopolymers as candidates for sustainable applications, they are still not cost-effective. This is particularly true for polyhydroxyalkanoates (PHAs), a family of linear polyesters produced in the nature by the action of bacteria during the fermentation of sugar or lipids in famine conditions. The use of biomass derived from food processing by-products and agro-food wastes does not only represent a possible strategy to reduce price but also allows to achieve a more sustainable material concept since they valorize large amounts of residues. Moreover, natural fillers offer several advantages such as improved biodegradability, low cost, low abrasion, high specific strength, low density, etc. Furthermore, when a bio-based and biodegradable polymer is combined with natural fillers, the resultant material was named "green composite", meaning that the whole material was obtained from renewable resources and is biodegradable. In this context, Melendez-Rodriguez et al. [7] developed green composites based on different blends of commercial poly(3-hydroxybutyrate) (PHB) and purified poly(3-hydroxybutyrate-*co*-3-hydroxyvalerate) (PHBV), which was produced by mixed microbial cultures (MMCs) using biomass derived from fruit pulp waste, filled with 10 wt % rice husk flour (RHF). The green composites were prepared using triglycidyl isocyanurate (TGIC) as a compatibilizer and dicumyl peroxide (DCP) as an initiator. The incorporation of the biosustainably produced PHBV at contents of 5–10 wt % counteracted the stiffness and fragility induced by RHF, yielding films with a balanced performance in terms of strength and ductility. The resultant films also showed improved thermal stability, thus having a larger processing window, and also a medium and high barrier to water vapor and aroma, respectively, being potential candidates for rigid packaging applications.

Biopolymers are also excellent candidates for manufacturing wood plastic composites (WPCs) and natural fiber-reinforced plastics (NFRPs) in combination with plant fibers, for instance flax, kenaf, ramie, jute, hemp or sisal. The mechanical behavior of composite materials can be, however, influenced by several environmental factors, for example temperature, humidity, radiation or chemical agents, as well as the type, variation in time, loading speed, direction or duration of the mechanical stresses to which they are subjected. As reported by Bolcu and Stănescu [8], an additional important aspect concerning the mechanical properties of polymer composite materials is given by the non-uniformities that appear in the technological process of fabrication. In the case of NFRPs, the main factor influencing their mechanical behavior is the uneven distribution of the fibers in matrix. In this research study, it was studied the influence of material irregularity on the mechanical behavior of hemp-reinforced composite materials produced with natural Dammar, a gum resin obtained from trees of the family *Dipterocarpaceae*, mixed with an epoxy resin. Authors concluded that the analyzed material was uniform and, in the case of the composites, the analyzed material showed reinforcing material proportions, resin specifically, different from those of the reference materials. According to this, when considering the non-uniformity degree, the resin transfer, the structural reactions, and the interface effects are phenomena that should be taken into account.

Inorganic fillers, such as nanoclays, or different types of nanoparticles, can also be used to enhance the flame retardancy and thermal stability of biopolymers and their NFRPs. In this context, Khalili et al. [9] developed natural fiber-reinforced PLA laminates by a conventional film stacking method using biopolymer films and natural fabrics with different cross-ply layups followed by hot compression. Natural fiber composites of PLA filled with varying concentrations of nanoparticles of hydroxyapatite (nHA) were produced by the same manufacturing technique. The flame behavior of the PLA/nHA composites was evaluated by the UL-94 test, demonstrating that only the composite containing the highest quantity of nHA, that is, 40 wt %, achieved an FH-1 rating and exhibited no recorded burn rate, whereas other composites obtained only an FH-3. Moreover, upon the addition of nHA, the thermal analysis indicated that the mass residue was improved by 279% whereas the thermal decomposition and the mass loss rate was also enhanced slightly. In another study, Larraza et al. [10] produced graphene oxide (GO), which was thereafter subjected to different levels of exfoliation by sonication and centrifugation and finally used to reinforce sodium alginate (SA). The latter biopolymer is a hydrophilic linear polysaccharide composed of (1→4)-β-D-mannuronic acid and (1→4)-α-L-guluronic acid units that can be extracted from brown algae (*Macrocystis pyrifera*). Due to restrictions in SA chains motion as a consequence of interactions with GO, the resistance to thermal degradation and mechanical properties of the biopolymer nanocomposites were increased. In particular, the loading of 8 wt % of GO led to an improvement of 65.3% and 83.3% for the tensile strength and Young's modulus, respectively. According to the authors, the resultant biopolymer nanocomposites show potential applications in pharmaceutics, biomedicine, food industry, etc.

Natural oils and, in particular, vegetable oils, represent a new generation of renewable materials that can positively contribute to progress in the preparation and commercialization of sustainable plasticizers. These are polymer additives that act as internal lubricants, favoring polymer chain mobility and enhancing processability and ductility. Some vegetable oils are also interesting from a chemical point of view due to their triglyceride structure based on a glycerol basic structure that is bonded through esters to fatty acids. They present a different number of unsaturations, which constitute the base for a chemical modification to provide the desired multi-functionality. Thus, selectively modified vegetable oils have been developed as novel renewable materials for the compatibilization of polymer blends and green composites. On this topic, Dominici et al. [11] evaluated the use of maleinized (MLO) and epoxidized (ELO) linseed oils in contents of up to 15 wt % as potential bio-based plasticizers for improving the toughness of Arboform®, a lignin/natural fiber commercial composite. It was observed that the addition of ELO at 2.5 wt % improved the impact-absorb energy from 5.4 to 11.1 kJ/m^2, while a similar improvement of 118% was obtained by the addition of 5 wt % of MLO. Furthermore, both MLO and ELO improved the thermal stability and tensile strength due to the formation of ester bonds between the multiple maleic and epoxy groups present in the oils and the hydroxyl groups of the matrix. In another study, Gonzalez et al. [12] reported the efficiency of MLO as compatibilizer in PLA/diatomaceous earth (DE) at a filler content of 10 wt %. Above five parts per hundred resin (phr) of MLO, the ductile properties were remarkably improved and the impact strength increased to nearly 22 kJ/m^2, which was almost double the value of the uncompatibilized PLA/DE composite. Similarly, in a first study, Liminana et al. [13] prepared green composite pieces of poly(butylene succinate) (PBS), a petrochemical biodegradable homopolyester, with varying loadings of almond shell flour (ASF) by extrusion with MLO and subsequent injection molding. The MLO content was kept constant at 10:1.5 (wt/wt) in relation to ASF and the optimal formulation in terms of the mechanical properties and filler content was attained for pieces filled with 30 wt % of ASF and containing 4.5 wt % of MLO. In a second work, Liminana et al. [14] used the previous green composites to study the effect of varying the MLO content for a constant ASF loading of 30 wt %. In the composition range from 2.5 to 10 wt %, MLO successfully plasticized the green composites and also acted as a compatibilizer by the reaction of the maleic anhydride pendant groups with the hydroxyl groups of both PBS end chains and cellulose and hemicelluloses of ASF. This compatibilizing effect was observed by a reduction of the gap between the ASF particles and the surrounding PBS matrix and also by the increase in the glass transition temperature (T_g) of PBS from −28 to −12 °C after the addition of 10 wt % of MLO. According to the authors, the developed green composites can reduce the overall current cost of biopolyesters by using large amounts of waste coming from the almond industry, leading to WPCs with a wood-like color.

Other studies have also been focused on the use of renewable and waste derived liquids for green composite formulations. For example, Araújo et al. [15] employed a by-product of the cashew agricultural industry, termed cashew nut shell liquid (CNSL), as a third component to polypropylene (PP) and high-impact polystyrene (HIPS) blends and studied its effect as a compatibilizer. Technical grade CNSL is indeed a natural lipid that consists of a mixture of phenols in which its major constituent is cardanol. The addition of 2 and 5 phr of CNSL in PP/HIPS blends successfully led to a size reduction of the HIPS domains, which ultimately facilitated the tension transfer from one phase to the other as well as the stabilization of the blended morphology. As a result, the industrial use of CNSL can be regarded as a cost-effective and sustainable compatibilizer in polymer blends since it is derived from waste and is also abundant. Alternatively, essential oils are natural mixtures of compounds that, being completely safe for both humans and the environment, show antimicrobial and antioxidant properties. In the study of Thanh et al. [16], marjoram (*Origanum majorana* L.) essential oil was incorporated into a polyamidoamine dendrimer generation 4.0 (PAMAM G4.0) for the development of a novel and sustainable nanocide against the plant disease *Phytophthora infestans*. The resulting essential oil-containing PAMAM G4.0 showed higher antifungal activity in comparison with PAMAM G4.0 and marjoram volatile oil prepared using the same concentration. The enhanced antimicrobial activity was due to the restricted evaporation of the essential oil in the encapsulating systems, which suggests that the encapsulation of marjoram oil in PAMAM G4.0s can be useful in combating late blight in tomatoes and the use of marjoram oil is very promising in the field of plant disease control as a fungicidal alternative to chemical pesticides in agriculture.

In summary, the Special Issue *Environmentally Friendly Polymers and Polymer Composites* compiles the most recent research works in biopolymers, their blends and composites, and the use of natural additives, such as vegetable oils and other renewable and waste derived liquids, with a marked environmental efficiency devoted to developing novel sustainable materials. The research studies gathered in the present Special Issue can certainly help to reveal the real potential of these materials in different applications. It is also highly expected that they will contribute and trigger the transfer of the current knowledge to industry, particularly in the food packaging sector, which is currently requesting the use of renewable and biodegradable materials to reduce the dependence of polymers on fossil fuels and mitigate the effect of plastic wastes on the environment. However, further studies on biopolymers and the optimization of their processes will be still needed to better control the development of the resulting environmentally friendly materials.

Author Contributions: All the guest editors wrote and reviewed this editorial letter. All authors have read and agreed to the published version of the manuscript.

Funding: This research work was funded by the Spanish Ministry of Science and Innovation (MICI) project number MAT2017-84909-C2-2-R.

Acknowledgments: S.T.-G. acknowledges MICI for his Ramón y Cajal contract (RYC2019-027784-I). The guest editors thank all the authors for submitting their work to this Special Issue and for its successful completion. We also acknowledge all the reviewers participating in the peer-review process of the submitted manuscripts for enhancing their quality and impact. We are also grateful to Dexin Wang and the editorial assistants of Materials who made the entire Special Issue creation a smooth and efficient process.

Conflicts of Interest: The authors declare no conflict of interest.

References

1. Mellinas, C.; Ramos, M.; Jiménez, A.; Garrigós, M.C. Recent Trends in the Use of Pectin from Agro-Waste Residues as a Natural-Based Biopolymer for Food Packaging Applications. *Materials* **2020**, *13*, 673. [CrossRef] [PubMed]
2. Taokaew, S.; Ofuchi, M.; Kobayashi, T. Size Distribution and Characteristics of Chitin Microgels Prepared via Emulsified Reverse-Micelles. *Materials* **2019**, *12*, 1160. [CrossRef] [PubMed]
3. Ahlawat, J.; Deemer, E.M.; Narayan, M. Chitosan Nanoparticles Rescue Rotenone-Mediated Cell Death. *Materials* **2019**, *12*, 1176. [CrossRef] [PubMed]
4. Potivara, K.; Phisalaphong, M. Development and Characterization of Bacterial Cellulose Reinforced with Natural Rubber. *Materials* **2019**, *12*, 2323. [CrossRef] [PubMed]
5. Lascano, D.; Quiles-Carrillo, L.; Balart, R.; Boronat, T.; Montanes, N. Toughened Poly (Lactic Acid)—PLA Formulations by Binary Blends with Poly(Butylene Succinate-*co*-Adipate)—PBSA and Their Shape Memory Behaviour. *Materials* **2019**, *12*, 622. [CrossRef] [PubMed]

6. Quiles-Carrillo, L.; Montanes, N.; Pineiro, F.; Jorda-Vilaplana, A.; Torres-Giner, S. Ductility and Toughness Improvement of Injection-Molded Compostable Pieces of Polylactide by Melt Blending with Poly(ε-caprolactone) and Thermoplastic Starch. *Materials* **2018**, *11*, 2138. [CrossRef] [PubMed]
7. Melendez-Rodriguez, B.; Torres-Giner, S.; Aldureid, A.; Cabedo, L.; Lagaron, J.M. Reactive Melt Mixing of Poly(3-Hydroxybutyrate)/Rice Husk Flour Composites with Purified Biosustainably Produced Poly(3-Hydroxybutyrate-*co*-3-Hydroxyvalerate). *Materials* **2019**, *12*, 2152. [CrossRef] [PubMed]
8. Bolcu, D.; Stănescu, M.M. The Influence of Non-Uniformities on the Mechanical Behavior of Hemp-Reinforced Composite Materials with a Dammar Matrix. *Materials* **2019**, *12*, 1232. [CrossRef] [PubMed]
9. Khalili, P.; Liu, X.; Zhao, Z.; Blinzler, B. Fully Biodegradable Composites: Thermal, Flammability, Moisture Absorption and Mechanical Properties of Natural Fibre-Reinforced Composites with Nano-Hydroxyapatite. *Materials* **2019**, *12*, 1145. [CrossRef] [PubMed]
10. Larraza, I.; Ugarte, L.; Fayanas, A.; Gabilondo, N.; Arbelaiz, A.; Corcuera, M.A.; Eceiza, A. Influence of Process Parameters in Graphene Oxide Obtention on the Properties of Mechanically Strong Alginate Nanocomposites. *Materials* **2020**, *13*, 1081. [CrossRef] [PubMed]
11. Dominici, F.; Samper, M.D.; Carbonell-Verdu, A.; Luzi, F.; López-Martínez, J.; Torre, L.; Puglia, D. Improved Toughness in Lignin/Natural Fiber Composites Plasticized with Epoxidized and Maleinized Linseed Oils. *Materials* **2020**, *13*, 600. [CrossRef] [PubMed]
12. Gonzalez, L.; Agüero, A.; Quiles-Carrillo, L.; Lascano, D.; Montanes, N. Optimization of the Loading of an Environmentally Friendly Compatibilizer Derived from Linseed Oil in Poly(Lactic Acid)/Diatomaceous Earth Composites. *Materials* **2019**, *12*, 1627. [CrossRef] [PubMed]
13. Liminana, P.; Quiles-Carrillo, L.; Boronat, T.; Balart, R.; Montanes, N. The Effect of Varying Almond Shell Flour (ASF) Loading in Composites with Poly(Butylene Succinate (PBS) Matrix Compatibilized with Maleinized Linseed Oil (MLO). *Materials* **2018**, *11*, 2179. [CrossRef] [PubMed]
14. Liminana, P.; Garcia-Sanoguera, D.; Quiles-Carrillo, L.; Balart, R.; Montanes, N. Optimization of Maleinized Linseed Oil Loading as a Biobased Compatibilizer in Poly(Butylene Succinate) Composites with Almond Shell Flour. *Materials* **2019**, *12*, 685. [CrossRef] [PubMed]
15. Araújo, M.N.; Visconte, L.L.Y.; Barreto, D.W.; Escócio, V.A.; Silva, A.L.N.d.; Sousa, A.M.F.d.; Pacheco, E.B.A.V. The Use of Cashew Nut Shell Liquid (CNSL) in PP/HIPS Blends: Morphological, Thermal, Mechanical and Rheological Properties. *Materials* **2019**, *12*, 1904. [CrossRef] [PubMed]
16. Thanh, V.M.; Bui, L.M.; Bach, L.G.; Nguyen, N.T.; Thi, H.L.; Hoang Thi, T.T. Origanum majorana L. Essential Oil-Associated Polymeric Nano Dendrimer for Antifungal Activity against Phytophthora infestans. *Materials* **2019**, *12*, 1446. [CrossRef] [PubMed]

materials

MDPI

Review

Recent Trends in the Use of Pectin from Agro-Waste Residues as a Natural-Based Biopolymer for Food Packaging Applications

Cristina Mellinas, Marina Ramos, Alfonso Jiménez and María Carmen Garrigós *

Department of Analytical Chemistry, Nutrition & Food Sciences, University of Alicante, ES-03690 Alicante, San Vicente del Raspeig, Spain; cristina.mellinas@ua.es (C.M.); marina.ramos@ua.es (M.R.); alfjimenez@ua.es (A.J.)

* Correspondence: mc.garrigos@ua.es

Received: 31 December 2019; Accepted: 28 January 2020; Published: 3 February 2020

Abstract: Regardless of the considerable progress in properties and versatility of synthetic polymers, their low biodegradability and lack of environmentally-friendly character remains a critical issue. Pectin is a natural-based polysaccharide contained in the cell walls of many plants allowing their growth and cell extension. This biopolymer can be extracted from plants and isolated as a bioplastic material with different applications, including food packaging. This review aims to present the latest research results regarding pectin, including the structure, different types, natural sources and potential use in several sectors, particularly in food packaging materials. Many researchers are currently working on a multitude of food and beverage industry applications related to pectin as well as combinations with other biopolymers to improve some key properties, such as antioxidant/antimicrobial performance and flexibility to obtain films. All these advances are covered in this review.

Keywords: pectin; food packaging; active compounds; agro-waste residues; circular economy

1. Introduction

Biopolymers are gaining their market share in the plastics industry by their intrinsic biodegradable character combined with interesting properties for specific applications. Biopolymers can be obtained/ extracted from natural sources, biosynthesized by living organisms or chemically synthesized from biological materials [1]. In addition, their natural-based origin, i.e., from renewable sources represents a great advantage over plastic commodities since their use decrease dependence from petroleum while preserving and even improving important material properties. There has been an increasing interest for the use of biopolymers in packaging, medicine, agriculture, and other sectors. Different types of carbohydrates, such as starch and cellulose, as well as other polysaccharides, such as alginates and pectin, as well as their combinations with animal-protein-based biopolymers, such as silk, wood, gelatin, collagen, chitosan/chitin, gums, plant-based proteins and lipids offer the possibility of rendering interesting applications for these advanced sectors. All these biopolymers offer interesting advantages in their use, such as their renewable origin, biocompatibility, barrier properties to moisture and/or gases, non-toxicity, non-polluting characteristics, mechanical integrity and relative low cost.

The increase in the use of biopolymers has caused that their global market is expected to reach around 10 billion US Dollars by 2021, growing by almost 17% over the forecast period 2017–2021. Western Europe covers the largest market segment, accounting for 41.5% of the global market while other regions are rapidly increasing their market share [2].

In addition, another important possibility offered by the use of biopolymers is their potential to be synthesized from the non-edible parts of plants or animals, avoiding the risk of depleting food from local communities, most of them in under-developed regions. Figure 1 shows the main types of biopolymers that can be obtained from biomass waste as well as some examples of their sources. They can be divided into three large groups: proteins, lipids and polysaccharides. The protein-based biopolymers can be obtained from both animal and vegetable wastes. For example, slaughterhouse wastes are a good source of proteins from animal origin, like gelatin. These wastes comprise the inedible tissues/parts of the animals slaughtered for the production of meat [3]. Among the proteins from plant origin, soy protein isolate is a good option to develop new materials due to its composition and excellent processing ability by gelling, emulsifying ability and water and oil holding capacity [4].

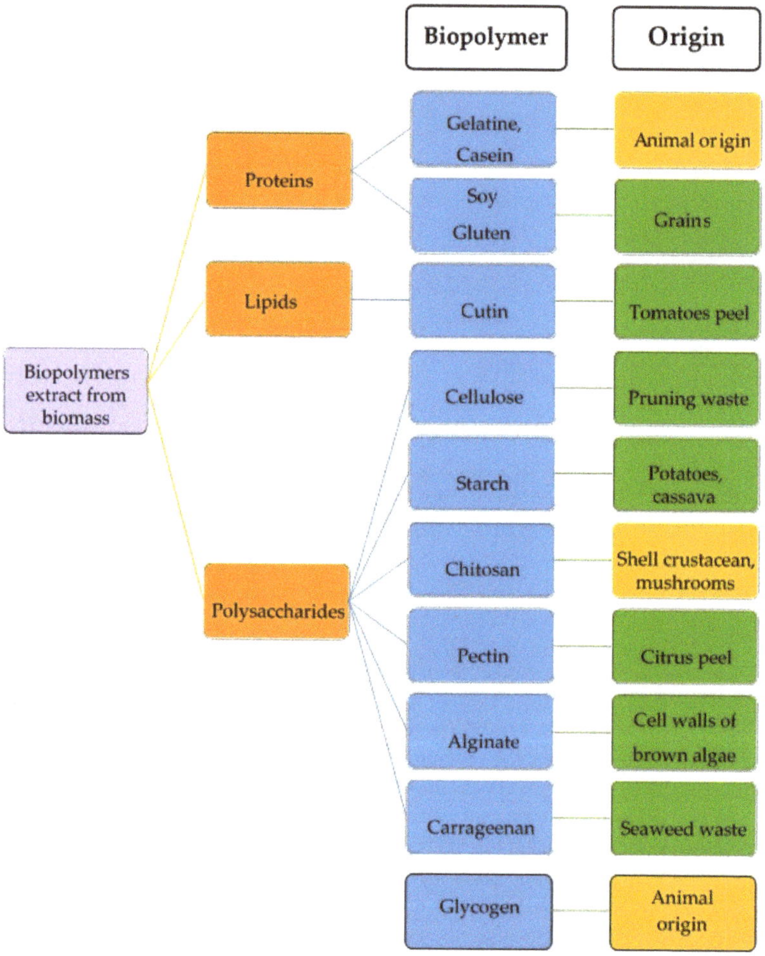

Figure 1. Different types of biopolymers obtained from animal and vegetable wastes.

Lipid-based polymers have been used in the last few years in food packaging or 3D printing materials. Their extraction from the natural sources is a necessary step to render isolated fatty acids to be further used in esterification reactions. For example, cutin extracted from tomato by-products [5] and lipids extracted from algae [6] have shown great potential to obtain specific fractions for the production of films with high barrier to water due to the repulsion caused by their high hydrophobic behaviour.

Polysaccharide-based polymers are the last group in Figure 1. These biopolymers are characterized by their biodegradable, biocompostable, sustainable and non-toxic characteristics. Additionally, polysaccharides are more thermally stable than other biopolymers, like lipids and proteins, since they are not irreversibly denatured via heating. However, their main disadvantages are the high sensitivity to moisture and low mechanical resistance [7]. In order to limit these problems, two different approaches have been proposed: the incorporation of different reinforcing additives to the polysaccharides matrices [8,9] and the combination with different polymers to obtain blends [10,11] or multilayer films [12,13]. The improvement in polysaccharides properties have permitted the extension of their use by the food industry in the last few years [14,15].

Pectin is one of the major structural polysaccharides present in many higher plant cells allowing primary cell wall extension and plant growth. It could be extracted and applied as an anionic biopolymer, soluble in water. A large number of recent articles have highlighted the advantages of using pectin over conventional polymers. Therefore, pectin is increasingly important for a multitude of food packaging applications, such as a thickening and gelling agent, colloidal stabilizer, texturizer, and emulsifier [16–18], a coating on fresh and cut fruits or vegetables [19–21] and as micro and nano-encapsulating agent for the controlled release of active principles with different functionalities [22]. Rodsamran et al. [16] reported that bioactive pectin films can retard soybean oil oxidation during 30 days of storage. Furthermore, Sucheta et al. [19] found that a pectin-corn flour-based coating significantly reduced the weight loss and decay per cent of tomatoes, delaying respiration with retention of biochemical quality of tomatoes. Additionally, polymeric blends of hydrocolloids obtained from chia seeds and apple pectin where developed with the aim to obtain antioxidant polymer blend films using the 2,2-diphenyl-1-picrylhydrazyl (DPPH) assay to estimate their antioxidant activity [10].

Pectin has been also extracted from waste biomass by using innovative methods, contributing to waste management in agriculture and food processing industries. Different pectin sources can be used, such as by-products of juice manufacturing as well as orange, mango, banana, lime and pomegranate peels and seeds. Therefore, this review aims to present and discuss the potential of pectin as a bio-based material in food packaging applications by its efficient extraction from waste biomass, while addressing a solution to the important environmental problems caused by the disposal of residues and by-products in the food sector.

2. Pectin

2.1. Pectin Structure

Pectin is a complex heteropolysaccharide and a major multifunctional component of the cell wall in many terrestrial plants. It is usually found in association with other compounds like cellulose, lignin or polyphenols present in the cell wall of plants [22]. Pectin is mainly composed of galacturonic acid units (Figure 2). The carboxyl groups of uronic acid residues can be present in different forms in the polymer structure, either free or as a salt form with sodium, calcium or other small counter-ions. In some cases, they can be also present as naturally-esterified groups, particularly with methanol, depending on the pectin source and/or the extraction method. Due to the presence of free carboxyl groups, pectin solutions exhibit acidic pH values. Galacturonic acid comprises approximately 70% of the pectin composition, depending on the plant species, and all the pectic polysaccharides contain galacturonic acid linked at the O-1 and the O-4 positions [23]. Pectin has a linear anionic backbone which regions showing no side chains known as "smooth regions" and regions with non-ionic side chains known as "hairy regions" [24].

Different pectin structural domains may be distinguished (Figure 2), influencing their properties depending on pectin proportions [23].

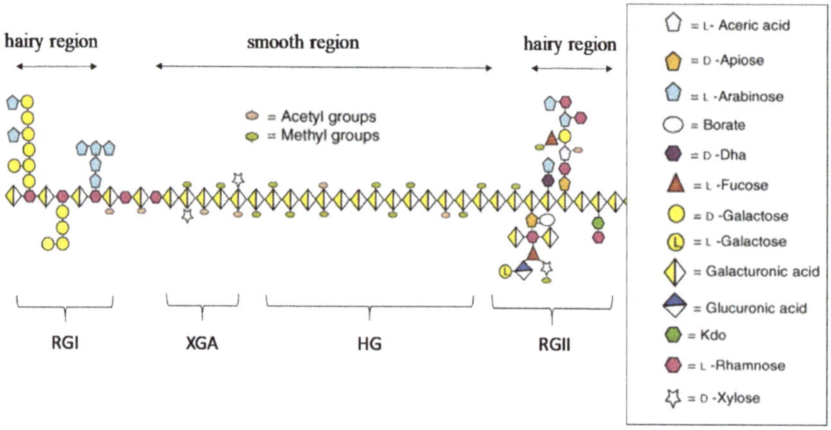

Figure 2. Schematic pectin structure, adapted from [23].

Homogalacturonan (HG): HG is the major domain of pectin in cell walls of plants, representing approximately 65% of the total pectin content. It is formed by galacturonic acid residues, linked by α (1→4) bonds, and their carboxyl groups are partially methyl esterified at position 6. Additionally, this domain may be acetylated at position 2 or 3 depending on the origin of the pectin. These domains are the main constituents of the above-mentioned "smooth regions" [24].

Rhamnogalacturonan I (RG-I) contains about 20%–35% of the total pectin and shows a more complex structure than HG. It contains repeated units of disaccharides consisting of L-rhamnose and galacturonic acid that can be also acetylated in the positions 2 or 3. It could have up to 100 units of (1,2)-α-L-Rha-(1,4)-α-D-GalA. In addition, large amount of L-rhamnose structures are substituted at O-4 by different neutral sugars such as L-galactose and L-arabinose [23].

Rhamnogalacturonan II (RG-II) contains about 10% of the total pectin and it is the structurally more complex component. Despite it is a relatively minor component in the pectin chain, RG-II plays a central role in the structure of plant cell walls. Small structure modifications of RG-II lead to reductions in the dimers formation and they can cause severe growth defects. So, dimerization of RG-II in the cell wall may be crucial for the normal growth and development of plants. This domain is composed of an HG backbone of (at least eight) 1,4-linked α-d-GalA residues decorated with side branches consisting of different types of sugars (rhamnose, fucose, xylose, galactose, apiose or aceric acid) in over 20 different types of linkages [23,25,26].

It is generally believed that the pectic polysaccharides are covalently bonded with high crosslinking densities since harsh chemical treatments or digestion by pectin-degrading enzymes are required to isolate HG, RG-I, and RG-II from each other. In addition, it has been reported that other components, such as xylogacturonan (XGA) and apiogalacturonan (AP), could replace the galacturonic acid units in some parts of the pectin chain [23]. The complexity of the pectin structures increases since it can be changed during the plant storage, extraction and processing, resulting in modifications of pectin functionalities and hindering its structural elucidation. It has been reported that the variations in chain lengths in each of the different domains are not the same, because HG and RGII have a highly homogeneous structure while RGI exhibits a wide heterogeneity in its composition [26].

2.2. Type of Pectins

The degree of esterification (DE) is an important parameter for the definition of the pectin applications and it is defined as the percentage of carboxyl groups esterified present in the structure of pectin. DE is often used to classify the different types of pectins (Figure 3). Depending on the DE different emulsifying, texturizing and gelling properties are observed. In general, when the DE increases, the water solubility decreases due to the hydrophobic nature of esters with long hydrocarbon chains. In contrast, when the DE increases the gelation rate also improves resulting in rapid gelation pectins [27]. Furthermore, the amount and composition of neutral sugars and the overall molecular weight of pectins have a great influence in their rheological properties [28].

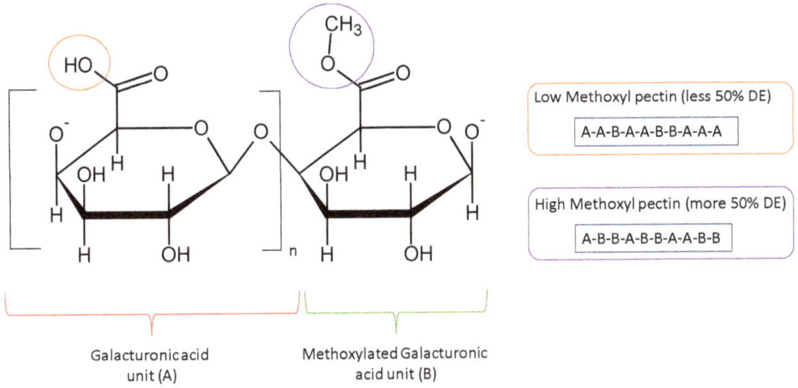

Figure 3. Structure of low and high methoxyl pectins.

High methoxyl pectin (HMP) shows DE higher than 50% and it is mainly used in the food industry by its thickening and gelling properties. It has been reported that HMP requires high amount of sugars for gelation and it is very sensitive to acidity [29]. HMP forms gels at low pH values and high concentrations of soluble solids due to the presence of hydrogen bonding and hydrophobic interactions between the pectin chains. Neutral sugars like sucrose play different roles in gelation through regulation of the hydrophobic interaction or directly binding to the polymer chains of HMP. Gels are formed when HG portions are cross-linked to form three dimensional crystalline networks in which water and other solutes are trapped [30]. The mechanism of formation of HMP gels is complex and it has been the subject of many investigations in the last decades. The glass transition theory has been proposed to explain the formation of HMP gels. Due to the high viscosity of molecules an arrest of the system kinetics occurs, giving as a result the formation of the gel due to the increased concentration of co-solutes and decrease in the water content. The transition from sol to gel behaviour is due to the combined effect of HMP and sucrose at pH 3, and occurs when excluded volume effects and attractive interactions are capable to give rise to an incipient three-dimensional network [31]. In addition, the effect of monovalent cations has been evaluated under different alkaline conditions (NaOH and KOH) at different pectin concentrations. It was suggested that HMP gel is formed through different mechanisms, such as de-esterification, self-aggregation, and entanglement under alkaline conditions. Na^+ or K^+ bind to dissociated carboxyl groups in HMP due to electronic attraction and this behaviour enables HMP molecules to move closer to each other, thereby improving gel network formation [32].

The emulsifying properties of HMP have been investigated by Jiang et al. in binary water-ethanol systems. They suggested that ethanol reduces the electrostatic repulsion and promotes pectin aggregation [33]. When HMP was mixed with the water-ethanol mixtures, the helix structure was broken, and electrostatic repulsion decreased. The compact and hydrophobic conformation enables pectin to adsorb better on the oil-water interface. The obtained emulsion showed good stability when using 21% of ethanol in the mixture.

Low methoxyl pectin (LMP) shows DE lower than 50% and it is generally formed by the de-esterification of HMP. Different agents can be used for the preparation of LMP from HMP, such as alkalis like sodium hydroxide or ammonia, enzymes (pectin methyl esterase) and concentrated acids [34].

LMP is widely used by the food industry to form low sugar-content jams as it does not require large amounts of sugar for gelation. It shows less sensitivity towards acidity and requires Ca^{2+} ions to form gels. The gelation mechanism in LMP is mediated by the formation of calcium bonds between two carboxyl groups from two chains in close contact [28]. Recently, Han et al. studied the effect of the calcium concentration, pH, soluble solids, and pectin concentration on the gel strength of LMP gels and they proposed different mechanisms of formation of pectin gels based on their rheological properties [35]. They observed that pH values close to the isoelectric point (pH = 3.50) and high calcium concentrations enhanced the storage modulus and gel strength by formation of calcium bridges at dissociated carboxyl groups. In addition, the sucrose content improved the gel strength because the neutral sugars provide hydroxyl groups to stabilize the gel and contribute to the formation of hydrogen bonds to immobilize free water. On the other hand, the formation of LMP gels under alkaline conditions was tested by Yang et al. [36]. They suggested that LMP can form relatively stable gels in the pH range of 3.5–9.5 using NaOH as pH regulator. In addition, they evaluated the effect of calcium concentration in the thermal and structural properties of the LMP gels and they concluded that the presence of calcium ions not only reduced the thermal stability but also the crystalline degree of LMP.

3. Sources of Pectin

Due to the high potential of pectin-based polymers, the extraction of pectin from biomass waste has been widely studied. Table 1 summarizes several published works based on the extraction of pectin from agro-waste sources.

Table 1. Different raw materials and extraction methods to obtain natural pectin.

Raw Material	Extraction Method	Conditions				Ref.
		T (°C)	Time (min)	LSR (mL/g)	Other Variables	
Eggplant peel waste	HAE	90	30	40	-	[37]
Orange peel waste	HAE	80	60	17.1	pH: 1.5	[38]
Orange peel waste	HAE	75	300	20	pH: 2.5	[39]
Orange peel waste	HAE	95	120	6	pH: 1.6	[40]
Pomelo peel waste	HAE	90	120	30	pH: 2	[41]
Pomegranate peel waste	HAE	70	120	10	-	[42]
Apple peel waste	HAE	85	120	25	-	[43]
Cashew apple pulp	HAE	100	120	5.15	-	[44]
Chamomile Waste	HAE	90	60	20	pH: 1.2	[45]
Durian rind waste	HAE	85	60	9	pH: 2.5	[46]
Hibiscus (*sabdariffa* L.)	HAE	100	30	20	pH: 2.5	[47]
Banana peel waste	HAE	86	360	50	pH: 2	[48]
Banana peel waste	HAE	90	30	20	pH: 1.5, 6	[49]
Mango peel waste	HAE	90	120	20	-	[50]
Jackfruit peel waste	HAE	138	9	17	-	[51]
Jackfruit peel waste	HAE	90	60	10	-	[52]
Jackfruit peel waste	HAE	90	60	20	-	[53]
Passion fruit rind	HAE	98	90	50	-	[54]
Tomato husk waste	HAE	100	15–25	30	-	[28]
Orange peel waste	HC	14.6–96	270	2.86	-	[55]
Artichoke (*Cynara scolymus* L.)	EAE	50	2880	15.4	pH: 5, Enzyme: 10.1 Ug^{-1}	[56]
Sisal Waste	EAE	50	1200	15	Enzyme: 88 Ug^{-1}, pH:5	[57]
Tobacco waste	MAE	-	4	20	550 W, pH: 1.8	[58]
Cocoa Pod Husk waste	SWE	121	30	27.5	103.4 bar	[59]
Custard apple peel waste	UAE	63	18	21	pH: 3	[60]
Mango peel waste	UAE	85	10	7.6	497.4 W/cm^2, pH: 2	[61]
Sisal Waste	UAE	-	60	15	450W, pH: 4	[57]
Passion fruit rind	UAE	-	10	20	135W	[62]
Jackfruit peel waste	UAE-MAE	86	29	48	-	[63]

HAE: Hydrothermal-assisted extraction; UAE: Ultrasound-assisted extraction; HC: Hydrodynamic cavitation; MAE: Microwave-assisted extraction; SWE: Subcritical water extraction; EAE: Enzyme-assisted extraction.

The peels of citrus fruits have been reported as the main source to obtain pectin at the industrial scale due to their good properties and high extraction yield. Hydrothermal extraction is the most usual method to obtain pectin from orange peels and it involves high temperatures (75–95 °C) and extraction times (60–300 min). Additionally, in all cases, the hydrothermal extraction of pectin takes place under acidic conditions using water as solvent. Pectin is very soluble in water and the acid medium decrease the presence of other compounds like polyphenols increasing extraction yields and helping to maintain the quality of the extracted pectin [38–41]. Other methods have been tested to reduce extraction times in citrus by-products. For example, microwave-assisted extraction (MAE) has been used in lime [64] and pomelo peels [65] reducing the extraction times to five and two minutes, respectively. However, high microwave powers (700–1100 W) were required to achieve these results. The hydrodynamic cavitation method was also used to obtain pectin derived from orange peel waste. Although a large decrease in the amount of solvent (2.86 mL/g of dry waste) was observed, long extraction times were also needed (270 min) [55].

The use of other sources to obtain pectin-based polymers in good grade and quality has been proposed in the last few years, such as eggplant peel [37], chamomile waste [45], cocoa pod husk [59,66], banana peel [49], mango peel [50,61,67] or tomato husk [28]. Tropical fruits have been also studied in the last years to obtain HMP. For example, passion fruit rind [54,62], durian rind [46] or jackfruit peels [51,52,63] have been proposed as interesting sources of pectins. Hydrothermal extraction is also the most used method in these types of wastes. Ultrasound-assisted extraction (UAE) has been also tested in passion fruit rind using 450 W and a water to dry sample ratio of 20 mL/g for 10 minutes. Results showed that the obtained pectin was mainly formed by homogalacturonans. Furthermore, their high degree of methylation indicated that the passion fruit pectin could be applied in gel forming products [62].

The use of innovative and sustainable extraction techniques is heading towards the study of hybrid techniques with the objective of combining their advantages, such as in the case of MAE and UAE. Pectin has been obtained from sisal waste by the combination of enzymatic and ultrasonic processes as an efficient strategy for the production of high-quality pectins since the enzymatic treatment disrupt the links between cellulose and xyloglucans in the cell wall of sisal and then the ultrasonic treatment produces mechanical destruction of the sisal structure to improve the release of pectin [57].

Finally, the introduction of new extraction techniques can be a great initial investment for companies since they offer the possibilities to get specific extractions of high added value purified compounds, although the costs of microwave or ultrasonic based equipment are higher than those of conventional extraction equipment, but in the long term, these devices are more profitable since the energy consumption, extraction time and the amount of expensive reagents used during pectin extraction are reduced [68,69].

4. Pectin-Based Materials for Food Packaging Applications

Pectin is a versatile compound that can be used to develop different materials in many food applications such as thickening and gelling agent, colloidal stabilizer, texturizer and emulsifier. These important applications are not limited to food processing, but also to packaging, coatings on fresh and cut fruits or vegetables and as microencapsulating agents (Table 2). Pectin is soluble in pure water and insoluble in organic solvents. Moreover, when dry pectin is mixed with water it tends to hydrate very rapidly, forming clumps. This behaviour is due to the formation of dry spheres of pectin contained in a highly hydrated outer coating. In order to eliminate these clumps, a vigorous and long agitation time is required [70]. In general terms, diluted pectin solutions present a Newtonian behaviour, but at high concentrations they show non-Newtonian behaviour, corresponding to pseudo-plastic characteristics. It was observed that the decrease in solubility and increase in viscosity contribute to increase the gelation capacity, i.e., the pectin concentration has a positive effect in gelation capacity and viscosity but a negative effect in solubility. Although it was stated that pectin properties are mainly dependent on structure, particularly DE [22,71], film forming, gelling and emulsifying properties should be also considered.

Table 2. Different types of pectin-based materials used in food packaging applications.

Type	Polymer Matrix	Additive	Application	Ref.
Film	LMP-bitter vetch protein	Transglutaminase	Drug delivery system	[72]
Film	LMP	Ascorbic acid	AO system	[73]
Film	HMP	Clove EO	AM system	[74]
Film	HMP	Marjoram EO	AO system	[75]
Film	HMP-Gluconaman	Tea extract	AO/AM system	[76]
Film	Pectin-Pullulan	AgNPs	AM system	[77]
Film	HM-Apple pectin	Chia seed hydrocolloid	AO system	[10]
Film	Chitosan-Starch-Pectin	Mint and rosemary oils Nisin	AO/AM system	[78]
Film	Fish gelatine-HMP	Hydroxytyrosol, dihydroxyphenylglycol	Preservation of beef meat	[79]
Film	HMP	Red cabbage extract	pH indicator	[80]
Film	Chitosan-HMP	Anthocyanin	pH indicator	[81]
Nanocomposite	Pectin	AgNPs, laponite	Coating polypropylene to improve barrier/AM properties	[82]
Nanocomposite	Pectin	Ag/AgCl-ZnONPs	AM system	[83]
Nanofiber	HMP	AgNPs	Reinforcement, AM	[84]
Nanofiber	LMP Polyethylene oxide	-	Reinforcement	[85]
Aerogel	Amidated pectin	TiO_2, NPs	AM under dark and UV illumination conditions	[86]
Hydrogel	LMP-Chitosan	Garlic and holy basil EOs	Incorporate to cellulose bag to improve AM properties	[87]
Oleogel	HMP	Camelia oil Tp-Palmitate	Drug delivery system	[88]
Emulsion	HMP	Clove EO	Bream fillets coating	[89]
Microemulsion	Chitosan-HMP	Cinnamaldehyde	AM system	[90]
Nanoemulsion	Food-grade pectin	Curcumin and garlic EOs	Coating chicken fillets	[91]
Nanoemulsion	HMP	Oregano, thyme, lemongrass, mandarin EOs	AM system	[92]
Nanoemulsion	HMP	Lemongrass EO	Addition in Cassava starch film to improve biodegradation properties	[93]
Multilayer emulsion	HMP-Chitosan	Astaxanthin	Release of hydrophobic carotenoids	[94]

EO: essential oil; AM: antimicrobial; AO: antioxidant.

4.1. Pectin-Based Films

Casting is the most used technique to obtain pectin-based films [10,73–75,79]. Pectin solutions (around 2–3 wt%) are mixed with the appropriate amount of plasticizer, commonly glycerol [95]. Then, the film forming solution is dried under controlled conditions of temperature and humidity forming a thin film. The incorporation of active agents, such as antimicrobial and/or antioxidant compounds, is performed after the incorporation of the plasticizer to obtain good compatibility between all components during the film processing [96–98]. Recently, Gouveira et al. [99] have reported the successful production of pectin-based films by using thermo-compression moulding of raw pectin with a natural deep eutectic solvent. The visual aspect of the obtained films was acceptable, since they were yellowish, visually homogenous, semi-transparent and without apparent pores, also showing high tensile strength and water resistance.

Pectin offers good compatibility with other biopolymers, such as proteins [73], lipids [100], other natural polysaccharides [101] or even synthetic biopolymers [82]. All these combinations represent alternatives when considering the final application of the obtained films. Both types of pectin (HMP and LMP) can form thin films under specific conditions. For example, LMP has been used as an appropriate matrix in new antioxidant systems with ascorbic acid as active additive by using casting as the processing method [72]. LMP was heated to 90 °C and then, glycerol and ascorbic acid were incorporated to the solution. Finally calcium chloride was added as the crosslinking agent to permit the formation of consistent and homogeneous pectin-based active films. In contrast, the addition of calcium ions is not necessary to develop films based on HMP, but low pH values and high sugar concentrations are needed to produce thin films. Nisar et al. [74] produced HMP films with antimicrobial properties incorporating clove essential oil by the casting method. Film forming solutions (3% *w*/*v*) were prepared by rehydrating pectin in sterile deionized water for 12 h at 20 °C. Glycerol was used as plasticizer at 30 wt % with magnetic stirring at 70 °C while pH was adjusted to 4.5. The clove essential oil with an emulsifier to improve the oil dispersion in the film aqueous solution were incorporated into the film forming solutions at different concentrations. A great integration of the clove essential oil into the polymer matrix was observed with a positive significant influence on the physico-chemical and functional properties, in particular barrier, mechanical, antioxidant and antimicrobial.

Marjoram, mint and rosemary essential oils are some of the active additives incorporated into the pectin matrix to get functionalities to these biopolymer materials. Almasi et al. [75] evaluated the effect on physico-chemical properties of marjoram essential oil in pectin films for food packaging applications. In order to prevent the degradation of the highly volatile essential oil and to control its release into food, it was incorporated using nanoemulsions and Pickering emulsions. Both types of emulsions combined the use of the essential oil and a low molecular weight surfactant with whey protein isolate or inulin as nanocarriers. Results obtained through X-ray diffraction (XRD), Fourier transformed infrared spectroscopy (FTIR) and field emission scanning electronic microscopy (FESEM) confirmed the high compatibility between pectin and both emulsions. The encapsulation of the essential oil through Pickering emulsions provided significantly slower releasing rates through the films when compared to nanoemulsions. For these reasons, authors concluded that the active pectin films containing Pickering emulsions showed the best potential to be used in active food packaging due to the slow release of the essential oil increasing food shelf-life. On the other hand, the synergic effect between mint and rosemary essential oils and nisin was investigated by Akhter et al. [78] using chitosan, starch and pectin blends. These authors concluded that the inclusion of rosemary essential oil and nisin improved the water barrier properties, tensile strength and thermal stability of the active biocomposites. Additionally, the combination of these compounds in a pectin matrix showed high antimicrobial action against some pathogenic strains (*Bacillus subtilis*, *Escherichia coli* and *Listeria monocytogenes*).

Furthermore, extracts derived from plants have been proposed to improve the functional properties of pectin. For example, tea extracts were incorporated into a HMP/Glucomannan blend. The influence of the addition of tea extracts at different concentrations (from 1% to 5% wt % on a dry basis) on the structural and physical properties of the blend, as well as on the antioxidant and antimicrobial activities were evaluated [76]. The authors found that concentrations of tea extract lower than 2 wt % improved all these properties but the effect was negative at high concentrations since some aggregation in the biopolymer macromolecules was observed. Red cabbage extract has been proposed for the development of a smart film based on HMP for meat and fish products. [80]. Red cabbage extract is rich in anthocyanins, showing the ability to change the colour of the biopolymer matrix at different pH values. It is known that the degradation of animal proteins produces an increase in pH due to the liberation of nitrogen compounds that can be monitored using a colorimetric sensor based on pectin and the red cabbage extract, offering innovative applications of pectin to the food industry [80]. These results showed the significant colour change in edible films when they are exposed to the headspace of meat and fish products at 21 °C and 4 °C, respectively.

The physical and functional properties of pectin-based films can be also modified by the combination of commercial pectin with corn flour and beetroot powder to minimize post-harvest decay, reducing ripening and improving sensorial properties of tomatoes [19]. In this study, results showed that pectin-based films protect from losses of polyphenols improving the antioxidant activity of these materials. In addition, other properties are modified due the presence of the edible coating. Pectin can modify the atmosphere around the fruit and/or vegetables, altering oxygen levels inside the fruit, retarding production of ethylene and, thus, limiting their physiological decay. In this work, pectin-based films showed low hydrophobicity to get optimum gas and water vapour permeability; reducing the ripening induced quality degradation in terms of texture and loss of bioactive compounds during storage.

Finally, the incorporation of nanoparticles to improve the physical and functional properties of pectin-based films has been evaluated. Biocomposites formed by a biopolymer matrix with metal or metal oxide nanoparticles are gaining importance in active food packaging since they could play a double role. On one hand, nanoparticles can act as nanofillers to enhance the mechanical and barrier properties of the biopolymer matrix and, on the other hand, they can interact directly with food due to their potential antimicrobial/antioxidant activity [102]. The effect of silver nanoparticles (AgNPs) has been tested in pectin, pullulan (a polysaccharide produced by fermentation by the *Aureobasidium pullulans* fungus), and pectin/pullulan blends [77]. Silver nanoparticles improved the mechanical properties of pullulan/AgNPs and pullulan/AgNPs/pectin composites while also showing high antimicrobial activity against foodborne pathogens, especially *Salmonella Typhimurium*, *Escherichia coli* and *Listeria monocytogenes*. AgNPs have been also proposed to develop nanocomposites based on pectin to be used as coatings for other polymer matrices with the aim to improve their barrier and mechanical properties as well as providing antimicrobial/antioxidant properties. Nanocomposites based on pectin with AgNPs and laponite have been evaluated to get a significant reduction in the oxygen transmission rate and water vapour transmission rate respect to neat polypropylene films taken as control [82]. The application of these new films showed antimicrobial activity against Gram-negative and Gram-positive bacteria, *Escherichia coli* and *Staphylococcus aureus*, respectively.

Other types of nanoparticles have been tested in pectin as the polymer matrix. Titanium oxide nanoparticles (TiO_2NPs) were incorporated at low concentrations (0–2 wt %) into biodegradable starch–pectin (3:1) films to improve their mechanical and barrier properties as well as their potential as antioxidant systems for food packaging applications [103]. In addition, visible and UV radiation was completely absorbed or scattered in these films by the addition of TiO_2NPs to get starch-pectin films with potential as UV screening packaging materials. On the other hand, the addition of halloysite nanotubes (HNT) offers great advantages to develop advanced food packaging materials. The effect of HNTs with salicylic acid [104], rosemary [97] and peppermint [105] essential oils in pectin films has been reported. HNTs showed high compatibility with pectin films improving their mechanical, thermal and moisture barrier properties [97,104]. The antimicrobial performance of these films was also improved due to the increase in the release rate of active compounds [97]. In fact, the antimicrobial activity of pectin-based films against Gram-negative *Escherichia coli* ATCC 25922, *Salmonella Typhimurium* ATCC 14028, *Pseudomonas aeruginosa* ATCC 10145, and Gram-positive *Staphylococcus aureus* ATCC 29213 was studied by the disk diffusion method, suggesting the effective antimicrobial properties of these functionalized films [104].

4.2. Emulsions and Gels

Pectins are widely used in the food industry as emulsifier and gelling agents. The ability of pectins to form gels under specific conditions has been used to obtain aerogels [86,106], hydrogels [87,107] or oleogels [88,108]. In particular, hydrogels are the most popular gel compositions used in food packaging, since they are able to absorb large amounts of water or other biological fluids inside their structure. For example, Torpol et al. studied the encapsulation of two different antimicrobial compounds: garlic and holy basil essential oils in chitosan-pectin hydrogel beads [87]. The entrapment of essential oils in the matrix structure was successful and it showed antimicrobial capacity against *Bacillus cereus*, *Clostridium perfringens*, *Escherichia coli*, *Pseudomonas fluorescens*, *Listeria monocytogenes* and *Staphylococcus aureus*, but not against *Lactobacillus plantarum* and *Salmonella Typhimurium*. Hydrogel coatings have been also proposed to reduce the deterioration of fresh fruit, meat or fish, as they can provide a semi-permeable protection to gases and water vapour and some other environmental factors that could damage food. By promoting food perspiration, these films also help to reduce enzymatic browning and water loss. Furthermore, this protection may also be enhanced by the addition of other ingredients, such as minerals, antioxidants, nutrients, vitamins or probiotics [109]. On the other hand, when the extraction of solvents at their supercritical state is produced in hydrogels or alcogels, the resultant material is called aerogels. Due to their unique properties, such as high porosity, high specific surface area, low relative density and thermal conductivity, these pectin-based biopolymers represent an innovative approach as advanced materials for food packaging since they can be used as internal layers, oxygen scavengers or drug delivery systems [110,111]. Recently, pectin-based aerogels have been developed for the storage of temperature-sensitive food. In this regard, TiO_2 nanoparticles were incorporated into the pectin matrix to improve the mechanical, thermal and antimicrobial properties of pectin when compared to control films [86].

As it has been mentioned above, pectin can be used as nanoemulsions, which are kinetically stable, but thermodynamically unstable, systems whose production requires emulsifiers to stabilize the dispersed phase [93]. Different types of essential oils have been encapsulated using nanoemulsions to delay and control the release processes. Several authors have studied the use of nanoemulsions in pectin matrices with different essential oils extracted from curcumin [91], lemongrass [92,93] and oregano [92]. Although essential oils are especially interesting in food packaging applications due to their antioxidant and/or antimicrobial properties, Mendes et al. [93] incorporated pectin nanoemulsions with the lemongrass essential oil into cassava starch film to improve the biodegradation rate of these formulations. Results obtained for the film with nanoemulsions showed a suitable degradation in vegetal compost, ensuring their complete biodegradation in a short time increasing their potential application in the food industry.

5. Conclusions

Researchers and scientists have achieved great success in the development of new systems based on pectins, as a natural bio-based biopolymer that can be obtained from agro-waste products, contributing to the implementation of the circular economy concept by improving waste management. This review article has considered the latest results obtained by researchers on the extraction, functionalization and potential applications in the food industry (including packaging), such as the production of films, emulsions and gels. However, due to the variability in the pectin structure the final application of pectin matrices is very diverse but very promising in many fields related to food packaging, particularly when active formulations are searched. Further studies on pectin matrices and optimization of polymer processes will be needed to better control the resulting pectin-based products.

Author Contributions: Conceptualization: C.M., M.R., A.J. and M.C.G.; methodology: C.M., M.R., A.J. and M.C.G.; formal analysis, discussion and supervision: C.M., M.R., A.J. and M.C.G. All authors have read and agreed to the published version of the manuscript.

Funding: This research received no external funding.

Acknowledgments: Authors would like to thank the Spanish Ministry of Science, Innovation and Universities for their support through the project referenced MAT2017-84909-C2-1-R.

Conflicts of Interest: The authors declare no conflict of interest.

References

1. Jha, A.; Kumar, A. Biobased technologies for the efficient extraction of biopolymers from waste biomass. *Bioprocess Biosyst. Eng.* **2019**, *42*, 1893–1901. [CrossRef] [PubMed]
2. Martău, G.A.; Mihai, M.; Vodnar, D.C. The Use of Chitosan, Alginate, and Pectin in the Biomedical and Food Sector—Biocompatibility, Bioadhesiveness, and Biodegradability. *Polymers* **2019**, *11*, 1837. [CrossRef] [PubMed]
3. Adhikari, B.B.; Chae, M.; Bressler, D.C. Utilization of slaughterhouse waste in value-added applications: Recent advances in the development of wood adhesives. *Polymers* **2018**, *10*, 176. [CrossRef]
4. Nishinari, K.; Fang, Y.; Guo, S.; Phillips, G.O. Soy proteins: A review on composition, aggregation and emulsification. *Food Hydrocoll.* **2014**, *39*, 301–318. [CrossRef]

5. Benítez, J.J.; Castillo, P.M.; del Río, J.C.; León-Camacho, M.; Domínguez, E.; Heredia, A.; Guzmán-Puyol, S.; Athanassiou, A.; Heredia-Guerrero, J.A. Valorization of Tomato Processing by-Products: Fatty Acid Extraction and Production of Bio-Based Materials. *Materials* **2018**, *11*, 2211. [CrossRef]
6. Tran, D.-T.; Lee, H.R.; Jung, S.; Park, M.S.; Yang, J.-W. Lipid-extracted algal biomass based biocomposites fabrication with poly(vinyl alcohol). *Algal Res.* **2018**, *31*, 525–533. [CrossRef]
7. Damm, T.; Commandeur, U.; Fischer, R.; Usadel, B.; Klose, H. Improving the utilization of lignocellulosic biomass by polysaccharide modification. *Process Biochem.* **2016**, *51*, 288–296. [CrossRef]
8. Valdés, A.; Mellinas, A.C.; Ramos, M.; Garrigós, M.C.; Jiménez, A. Natural additives and agricultural wastes in biopolymer formulations for food packaging. *Front. Chem.* **2014**, *2*. [CrossRef]
9. Shankar, S.; Tanomrod, N.; Rawdkuen, S.; Rhim, J.-W. Preparation of pectin/silver nanoparticles composite films with UV-light barrier and properties. *Int. J. Biol. Macromol.* **2016**, *92*, 842–849. [CrossRef] [PubMed]
10. da Silva, I.S.V.; de Sousa, R.M.F.; de Oliveira, A.; de Oliveira, W.J.; Motta, L.A.C.; Pasquini, D.; Otaguro, H.; da Silva, I.S.V.; de Sousa, R.M.F.; de Oliveira, A.; et al. Polymeric blends of hydrocolloid from chia seeds/apple pectin with potential antioxidant for food packaging applications. *Carbohydr. Polym.* **2018**, *202*, 203–210. [CrossRef]
11. Correa, J.P.; Molina, V.; Sanchez, M.; Kainz, C.; Eisenberg, P.; Massani, M.B. Improving ham shelf life with a polyhydroxybutyrate/polycaprolactone biodegradable film activated with nisin. *Food Packag. Shelf Life* **2017**, *11*, 31–39. [CrossRef]
12. Xia, C.; Wang, W.; Wang, L.; Liu, H.; Xiao, J. Multilayer zein/gelatin films with tunable water barrier property and prolonged antioxidant activity. *Food Packag. Shelf Life* **2019**, *19*, 76–85. [CrossRef]
13. Wang, H.; Gong, X.; Miao, Y.; Guo, X.; Liu, C.; Fan, Y.-Y.; Zhang, J.; Niu, B.; Li, W. Preparation and characterization of multilayer films composed of chitosan, sodium alginate and carboxymethyl chitosan-ZnO nanoparticles. *Food Chem.* **2019**, *283*, 397–403. [CrossRef] [PubMed]
14. Mellinas, C.; Valdés, A.; Ramos, M.; Burgos, N.; Del Carmen Garrigós, M.; Jiménez, A. Active edible films: Current state and future trends. *J. Appl. Polym. Sci.* **2016**, *133*. [CrossRef]
15. Abdul Khalil, H.P.S.; Chong, E.W.N.; Owolabi, F.A.T.; Asniza, M.; Tye, Y.Y.; Rizal, S.; Nurul Fazita, M.R.; Mohamad Haafiz, M.K.; Nurmiati, Z.; Paridah, M.T. Enhancement of basic properties of polysaccharide-based composites with organic and inorganic fillers: A review. *J. Appl. Polym. Sci.* **2019**, *136*. [CrossRef]
16. Rodsamran, P.; Sothornvit, R. Lime peel pectin integrated with coconut water and lime peel extract as a new bioactive film sachet to retard soybean oil oxidation. *Food Hydrocoll.* **2019**, *97*. [CrossRef]
17. Sun, X.; Cameron, R.G.; Bai, J. Effect of spray-drying temperature on physicochemical, antioxidant and antimicrobial properties of pectin/sodium alginate microencapsulated carvacrol. *Food Hydrocoll.* **2020**, *100*. [CrossRef]
18. de Oliveira Alves Sena, E.; Oliveira da Silva, P.S.; de Aragão Batista, M.C.; Alonzo Sargent, S.; Ganassali de Oliveira Junior, L.F.; Almeida Castro Pagani, A.; Gutierrez Carnelossi, M.A. Calcium application via hydrocooling and edible coating for the conservation and quality of cashew apples. *Sci. Hortic* **2019**, *256*. [CrossRef]
19. Sucheta; Chaturvedi, K.; Sharma, N.; Yadav, S.K. Composite edible coatings from commercial pectin, corn flour and beetroot powder minimize post-harvest decay, reduces ripening and improves sensory liking of tomatoes. *Int. J. Biol. Macromol.* **2019**, *133*, 284–293. [CrossRef]
20. Pizato, S.; Chevalier, R.C.; Dos Santos, M.F.; Da Costa, T.S.; Arévalo Pinedo, R.; Cortez Vega, W.R. Evaluation of the shelf-life extension of fresh-cut pineapple (Smooth cayenne) by application of different edible coatings. *Br. Food J.* **2019**, *121*, 1592–1604. [CrossRef]
21. Jiang, Y.; Li, F.; Li, D.; Sun-Waterhouse, D.; Huang, Q. Zein/Pectin Nanoparticle-Stabilized Sesame Oil Pickering Emulsions: Sustainable Bioactive Carriers and Healthy Alternatives to Sesame Paste. *Food Bioprocess Technol.* **2019**, *12*, 1982–1992. [CrossRef]
22. Noreen, A.; Nazli, Z.-H.; Akram, J.; Rasul, I.; Mansha, A.; Yaqoob, N.; Iqbal, R.; Tabasum, S.; Zuber, M.; Zia, K.M. Pectins functionalized biomaterials; a new viable approach for biomedical applications: A review. *Int. J. Biol. Macromol.* **2017**, *101*, 254–272. [CrossRef] [PubMed]
23. Mohnen, D. Pectin structure and biosynthesis. *Curr. Opin. Plant Biol.* **2008**, *11*, 266–277. [CrossRef] [PubMed]
24. Naqash, F.; Masoodi, F.A.; Rather, S.A.; Wani, S.M.; Gani, A. Emerging concepts in the nutraceutical and functional properties of pectin—A Review. *Carbohydr. Polym.* **2017**, *168*, 227–239. [CrossRef]

25. O'Neill, M.A.; Ishii, T.; Albersheim, P.; Darvill, A.G. RHAMNOGALACTURONAN II: Structure and Function of a Borate Cross-Linked Cell Wall Pectic Polysaccharide. *Annu. Rev. Plant Biol.* **2004**, *55*, 109–139. [CrossRef]
26. Voragen, A.G.J.; Coenen, G.J.; Verhoef, R.P.; Schols, H.A. Pectin, a versatile polysaccharide present in plant cell walls. *Struct. Chem.* **2009**, *20*, 263–275. [CrossRef]
27. Sañudo Barajas, J.A.; Ayón, M.; Velez, R.; Verdugo-Perales, M.; Lagarda, J.; Allende, R. *Pectins: From the Gelling Properties to the Biological Activity*; Nova Publishers: Hauppauge, NY, USA, 2014; pp. 203–224.
28. Morales-Contreras, B.E.; Rosas-Flores, W.; Contreras-Esquivel, J.C.; Wicker, L.; Morales-Castro, J. Pectin from Husk Tomato (Physalis ixocarpa Brot.): Rheological behavior at different extraction conditions. *Carbohydr. Polym.* **2018**, *179*, 282–289. [CrossRef]
29. Giacomazza, D.; Bulone, D.; San Biagio, P.L.; Marino, R.; Lapasin, R. The role of sucrose concentration in self-assembly kinetics of high methoxyl pectin. *Int. J. Biol. Macromol.* **2018**, *112*, 1183–1190. [CrossRef]
30. do Nascimento, G.E.; Simas-Tosin, F.F.; Iacomini, M.; Gorin, P.A.J.; Cordeiro, L.M.C. Rheological behavior of high methoxyl pectin from the pulp of tamarillo fruit (Solanum betaceum). *Carbohydr. Polym.* **2016**, *139*, 125–130. [CrossRef]
31. Giacomazza, D.; Bulone, D.; San Biagio, P.L.; Lapasin, R. The complex mechanism of HM pectin self-assembly: A rheological investigation. *Carbohydr. Polym.* **2016**, *146*, 181–186. [CrossRef]
32. Wang, H.; Wan, L.; Chen, D.; Guo, X.; Liu, F.; Pan, S. Unexpected gelation behavior of citrus pectin induced by monovalent cations under alkaline conditions. *Carbohydr. Polym.* **2019**, *212*, 51–58. [CrossRef] [PubMed]
33. Jiang, W.; Qi, J.-R.; Huang, Y.; Zhang, Y.; Yang, X.-Q. Emulsifying properties of high methoxyl pectins in binary systems of water-ethanol. *Carbohydr. Polym.* **2020**, *229*, 115420. [CrossRef] [PubMed]
34. Fishman, M.L.; Chau, H.K.; Qi, P.X.; Hotchkiss, A.T.; Garcia, R.A.; Cooke, P.H. Characterization of the global structure of low methoxyl pectin in solution. *Food Hydrocoll.* **2015**, *46*, 153–159. [CrossRef]
35. Han, W.; Meng, Y.; Hu, C.; Dong, G.; Qu, Y.; Deng, H.; Guo, Y. Mathematical model of Ca^{2+} concentration, pH, pectin concentration and soluble solids (sucrose) on the gelation of low methoxyl pectin. *Food Hydrocoll.* **2017**, *66*, 37–48. [CrossRef]
36. Yang, X.; Nisar, T.; Liang, D.; Hou, Y.; Sun, L.; Guo, Y. Low methoxyl pectin gelation under alkaline conditions and its rheological properties: Using NaOH as a pH regulator. *Food Hydrocoll.* **2018**, *79*, 560–571. [CrossRef]
37. Kazemi, M.; Khodaiyan, F.; Hosseini, S.S.; Najari, Z. An integrated valorization of industrial waste of eggplant: Simultaneous recovery of pectin, phenolics and sequential production of pullulan. *Waste Manag.* **2019**, *100*, 101–111. [CrossRef]
38. Senit, J.J.; Velasco, D.; Gomez Manrique, A.; Sanchez-Barba, M.; Toledo, J.M.; Santos, V.E.; Garcia-Ochoa, F.; Yustos, P.; Ladero, M. Orange peel waste upstream integrated processing to terpenes, phenolics, pectin and monosaccharides: Optimization approaches. *Ind. Crops Prod.* **2019**, *134*, 370–381. [CrossRef]
39. Hilali, S.; Fabiano-Tixier, A.S.; Ruiz, K.; Hejjaj, A.; Ait Nouh, F.; Idlimam, A.; Bily, A.; Mandi, L.; Chemat, F. Green Extraction of Essential Oils, Polyphenols, and Pectins from Orange Peel Employing Solar Energy: Toward a Zero-Waste Biorefinery. *ACS Sustain. Chem. Eng.* **2019**, *7*, 11815–11822. [CrossRef]
40. Tovar, A.K.; Godínez, L.A.; Espejel, F.; Ramírez-Zamora, R.-M.; Robles, I. Optimization of the integral valorization process for orange peel waste using a design of experiments approach: Production of high-quality pectin and activated carbon. *Waste Manag.* **2019**, *85*, 202–213. [CrossRef]
41. Roy, M.C.; Alam, M.; Saeid, A.; Das, B.C.; Mia, M.B.; Rahman, M.A.; Eun, J.B.; Ahmed, M. Extraction and characterization of pectin from pomelo peel and its impact on nutritional properties of carrot jam during storage. *J. Food Process. Preserv.* **2018**, *42*, 1–9. [CrossRef]
42. Shakhmatov, E.G.; Makarova, E.N.; Belyy, V.A. Structural studies of biologically active pectin-containing polysaccharides of pomegranate Punica granatum. *Int. J. Biol. Macromol.* **2019**, *122*, 29–36. [CrossRef] [PubMed]
43. Cho, E.-H.; Jung, H.-T.; Lee, B.-H.; Kim, H.-S.; Rhee, J.-K.; Yoo, S.-H. Green process development for apple-peel pectin production by organic acid extraction. *Carbohydr. Polym.* **2019**, *204*, 97–103. [CrossRef] [PubMed]
44. Tamiello-Rosa, C.S.; Cantu-Jungles, T.M.; Iacomini, M.; Cordeiro, L.M.C. Pectins from cashew apple fruit (Anacardium occidentale): Extraction and chemical characterization. *Carbohydr. Res.* **2019**, *483*, 107752. [CrossRef] [PubMed]
45. Slavov, A.; Yantcheva, N.; Vasileva, I. Chamomile Wastes (Matricaria chamomilla): New Source of Polysaccharides. *Waste Biomass Valor.* **2018**, *10*, 1–12. [CrossRef]

46. Hasem, N.H.; Mohamad Fuzi, S.F.Z.; Kormin, F.; Abu Bakar, M.F.; Sabran, S.F. Extraction and partial characterization of durian rind pectin. *IOP Conf. Ser. Earth Environ. Sci.* **2019**, *269*, 012019. [CrossRef]
47. Esparza-Merino, R.M.; Macías-Rodríguez, M.E.; Cabrera-Díaz, E.; Valencia-Botín, A.J.; Estrada-Girón, Y. Utilization of by-products of Hibiscus sabdariffa L. as alternative sources for the extraction of high-quality pectin. *Food Sci. Biotechnol.* **2019**, *28*, 1003–1011. [CrossRef]
48. Marenda, F.R.B.; Colodel, C.; Canteri, M.H.G.; de Olivera Müller, C.M.; Amante, E.R.; de Oliveira Petkowicz, C.L.; de Mello Castanho Amboni, R.D. Investigation of cell wall polysaccharides from flour made with waste peel from unripe banana (Musa sapientum) biomass. *J. Sci. Food Agric.* **2019**, *99*, 4363–4372. [CrossRef]
49. Maneerat, N.; Tangsuphoom, N.; Nitithamyong, A. Effect of extraction condition on properties of pectin from banana peels and its function as fat replacer in salad cream. *J. Food Sci. Technol.* **2017**, *54*, 386–397. [CrossRef]
50. Banerjee, J.; Singh, R.; Vijayaraghavan, R.; MacFarlane, D.; Patti, A.F.; Arora, A. A hydrocolloid based biorefinery approach to the valorisation of mango peel waste. *Food Hydrocoll.* **2018**, *77*, 142–151. [CrossRef]
51. Li, W.J.; Fan, Z.G.; Wu, Y.Y.; Jiang, Z.G.; Shi, R.C. Eco-friendly extraction and physicochemical properties of pectin from jackfruit peel waste with subcritical water. *J. Sci. Food Agric.* **2019**, *99*. [CrossRef]
52. Sundarraj, A.A.; Thottiam Vasudevan, R.; Sriramulu, G. Optimized extraction and characterization of pectin from jackfruit (Artocarpus integer) wastes using response surface methodology. *Int. J. Biol. Macromol.* **2018**, *106*, 698–703. [CrossRef] [PubMed]
53. Fazio, A.; La Torre, C.; Dalena, F.; Plastina, P. Screening of glucan and pectin contents in broad bean (Vicia faba L.) pods during maturation. *Eur. Food Res. Technol.* **2019**, *246*, 333–347. [CrossRef]
54. Inayati, I.; Puspita, R.I.; Fajrin, V.L. Extraction of pectin from passion fruit rind (Passiflora edulis var. flavicarpa Degener) for edible coating. *AIP Conf. Proc.* **2018**, *1931*. [CrossRef]
55. Meneguzzo, F.; Brunetti, C.; Fidalgo, A.; Ciriminna, R.; Delisi, R.; Albanese, L.; Zabini, F.; Gori, A.; dos Santos Nascimento, L.B.; De Carlo, A.; et al. Real-Scale Integral Valorization of Waste Orange Peel via Hydrodynamic Cavitation. *Processes* **2019**, *7*, 581. [CrossRef]
56. Sabater, C.; Corzo, N.; Olano, A.; Montilla, A. Enzymatic extraction of pectin from artichoke (Cynara scolymus L.) by-products using Celluclast®1.5L. *Carbohydr. Polym.* **2018**, *190*, 43–49. [CrossRef] [PubMed]
57. Yang, Y.; Wang, Z.; Hu, D.; Xiao, K.; Wu, J.-Y. Efficient extraction of pectin from sisal waste by combined enzymatic and ultrasonic process. *Food Hydrocoll.* **2018**, *79*, 189–196. [CrossRef]
58. Zhang, M.; Zeng, G.; Pan, Y.; Qi, N. Difference research of pectins extracted from tobacco waste by heat reflux extraction and microwave-assisted extraction. *Biocatal. Agric. Biotechnol.* **2018**, *15*, 359–363. [CrossRef]
59. Muñoz-Almagro, N.; Valadez-Carmona, L.; Mendiola, J.A.; Ibáñez, E.; Villamiel, M. Structural characterisation of pectin obtained from cacao pod husk. Comparison of conventional and subcritical water extraction. *Carbohydr. Polym.* **2019**, *217*, 69–78. [CrossRef]
60. Shivamathi, C.S.; Moorthy, I.G.; Kumar, R.V.; Soosai, M.R.; Maran, J.P.; Kumar, R.S.; Varalakshmi, P. Optimization of ultrasound assisted extraction of pectin from custard apple peel: Potential and new source. *Carbohydr. Polym.* **2019**, *225*, 115240. [CrossRef]
61. Guandalini, B.B.V.; Rodrigues, N.P.; Marczak, L.D.F. Sequential extraction of phenolics and pectin from mango peel assisted by ultrasound. *Food Res. Int.* **2019**, *119*, 455–461. [CrossRef]
62. de Souza, C.G.; Rodrigues, T.H.S.; e Silva, L.M.A.; Ribeiro, P.R.V.; de Brito, E.S. Sequential extraction of flavonoids and pectin from yellow passion fruit rind using pressurized solvent or ultrasound. *J. Sci. Food Agric.* **2018**, *98*, 1362–1368. [CrossRef] [PubMed]
63. Xu, S.-Y.; Liu, J.-P.; Huang, X.; Du, L.-P.; Shi, F.-L.; Dong, R.; Huang, X.-T.; Zheng, K.; Liu, Y.; Cheong, K.-L. Ultrasonic-microwave assisted extraction, characterization and biological activity of pectin from jackfruit peel. *LWT* **2018**, *90*, 577–582. [CrossRef]
64. Rodsamran, P.; Sothornvit, R. Microwave heating extraction of pectin from lime peel: Characterization and properties compared with the conventional heating method. *Food Chem.* **2019**, *278*, 364–372. [CrossRef] [PubMed]
65. Wandee, Y.; Uttapap, D.; Mischnick, P. Yield and structural composition of pomelo peel pectins extracted under acidic and alkaline conditions. *Food Hydrocoll.* **2019**, *87*, 237–244. [CrossRef]
66. Lu, F.; Rodriguez-Garcia, J.; Van Damme, I.; Westwood, N.J.; Shaw, L.; Robinson, J.S.; Warren, G.; Chatzifragkou, A.; McQueen Mason, S.; Gomez, L.; et al. Valorisation strategies for cocoa pod husk and its fractions. *Curr. Opin. Green Sustain. Chem.* **2018**, *14*, 80–88. [CrossRef]

67. Maran, J.P.; Swathi, K.; Jeevitha, P.; Jayalakshmi, J.; Ashvini, G. Microwave-assisted extraction of pectic polysaccharide from waste mango peel. *Carbohydr. Polym.* **2015**, *123*, 67–71. [CrossRef]
68. Wang, W.; Ma, X.; Jiang, P.; Hu, L.; Zhi, Z.; Chen, J.; Ding, T.; Ye, X.; Liu, D. Characterization of pectin from grapefruit peel: A comparison of ultrasound-assisted and conventional heating extractions. *Food Hydrocoll.* **2016**, *61*, 730–739. [CrossRef]
69. Albuquerque, B.R.; Prieto, M.A.; Vazquez, J.A.; Barreiro, M.F.; Barros, L.; Ferreira, I.C.F.R. Recovery of bioactive compounds from Arbutus unedo L. fruits: Comparative optimization study of maceration/microwave/ ultrasound extraction techniques. *Food Res. Int.* **2018**, *109*, 455–471. [CrossRef]
70. Raj, S. A Review on Pectin: Chemistry due to General Properties of Pectin and its Pharmaceutical Uses. *Sci. Rep.* **2012**, *1*, 550. [CrossRef]
71. Alistair, M.;.Stephen, G.O.P. *Food Polysaccharides and Their Applications*; CRC Press: Boca Raton, FL, USA, 2006.
72. Porta, R.; Di Pierro, P.; Sabbah, M.; Regalado-Gonzales, C.; Mariniello, L.; Kadivar, M.; Arabestani, A. Blend films of pectin and bitter vetch (Vicia ervilia) proteins: Properties and effect of transglutaminase. *Innov. Food Sci. Emerg. Technol.* **2016**, *36*, 245–251. [CrossRef]
73. Chiarappa, G.; De'Nobili, M.D.; Rojas, A.M.; Abrami, M.; Lapasin, R.; Grassi, G.; Ferreira, J.A.; Gudiño, E.; de Oliveira, P.; Grassi, M. Mathematical modeling of L-(+)-ascorbic acid delivery from pectin films (packaging) to agar hydrogels (food). *J. Food Eng.* **2018**, *234*, 73–81. [CrossRef]
74. Nisar, T.; Wang, Z.-C.; Yang, X.; Tian, Y.; Iqbal, M.; Guo, Y. Characterization of citrus pectin films integrated with clove bud essential oil: Physical, thermal, barrier, antioxidant and antibacterial properties. *Int. J. Biol. Macromol.* **2018**, *106*, 670–680. [CrossRef] [PubMed]
75. Almasi, H.; Azizi, S.; Amjadi, S. Development and characterization of pectin films activated by nanoemulsion and Pickering emulsion stabilized marjoram (*Origanum majorana* L.) essential oil. *Food Hydrocoll.* **2020**, *99*, 105338. [CrossRef]
76. Lei, Y.; Wu, H.; Jiao, C.; Jiang, Y.; Liu, R.; Xiao, D.; Lu, J.; Zhang, Z.; Shen, G.; Li, S. Investigation of the structural and physical properties, antioxidant and antimicrobial activity of pectin-konjac glucomannan composite edible films incorporated with tea polyphenol. *Food Hydrocoll.* **2019**, *94*, 128–135. [CrossRef]
77. Lee, J.H.; Jeong, D.; Kanmani, P. Study on physical and mechanical properties of the biopolymer/silver based active nanocomposite films with antimicrobial activity. *Carbohydr. Polym.* **2019**, *224*, 115159. [CrossRef] [PubMed]
78. Akhter, R.; Masoodi, F.A.; Wani, T.A.; Rather, S.A. Functional characterization of biopolymer based composite film: Incorporation of natural essential oils and antimicrobial agents. *Int. J. Biol. Macromol.* **2019**, *137*, 1245–1255. [CrossRef] [PubMed]
79. Bermúdez-Oria, A.; Rodríguez-Gutiérrez, G.; Rubio-Senent, F.; Fernández-Prior, Á.; Fernández-Bolaños, J. Effect of edible pectin-fish gelatin films containing the olive antioxidants hydroxytyrosol and 3,4-dihydroxyphenylglycol on beef meat during refrigerated storage. *Meat Sci.* **2019**, *148*, 213–218. [CrossRef]
80. Dudnyk, I.; Janeček, E.-R.; Vaucher-Joset, J.; Stellacci, F. Edible sensors for meat and seafood freshness. *Sens. Actuators B Chem.* **2018**, *259*, 1108–1112. [CrossRef]
81. Maciel, V.B.V.; Yoshida, C.M.P.; Franco, T.T. Chitosan/pectin polyelectrolyte complex as a pH indicator. *Carbohydr. Polym.* **2015**, *132*, 537–545. [CrossRef]
82. Vishnuvarthanan, M.; Rajeswari, N. Food packaging: Pectin—laponite—Ag nanoparticle bionanocomposite coated on polypropylene shows low O2 transmission, low Ag migration and high antimicrobial activity. *Environ. Chem. Lett.* **2019**, *17*, 439–445. [CrossRef]
83. Yu, N.; Peng, H.; Qiu, L.; Wang, R.; Jiang, C.; Cai, T.; Sun, Y.; Li, Y.; Xiong, H. New pectin-induced green fabrication of Ag@AgCl/ZnO nanocomposites for visible-light triggered antibacterial activity. *Int. J. Biol. Macromol.* **2019**, *141*, 207–217. [CrossRef] [PubMed]
84. Li, K.; Cui, S.; Hu, J.; Zhou, Y.; Liu, Y. Crosslinked pectin nanofibers with well-dispersed Ag nanoparticles: Preparation and characterization. *Carbohydr. Polym.* **2018**, *199*, 68–74. [CrossRef] [PubMed]
85. McCune, D.; Guo, X.; Shi, T.; Stealey, S.; Antrobus, R.; Kaltchev, M.; Chen, J.; Kumpaty, S.; Hua, X.; Ren, W.; et al. Electrospinning pectin-based nanofibers: A parametric and cross-linker study. *Appl. Nanosci.* **2018**, *8*, 33–40. [CrossRef]

86. Nešić, A.; Gordić, M.; Davidović, S.; Radovanović, Ž.; Nedeljković, J.; Smirnova, I.; Gurikov, P. Pectin-based nanocomposite aerogels for potential insulated food packaging application. *Carbohydr. Polym.* **2018**, *195*, 128–135. [CrossRef] [PubMed]
87. Torpol, K.; Sriwattana, S.; Sangsuwan, J.; Wiriyacharee, P.; Prinyawiwatkul, W. Optimising chitosan–pectin hydrogel beads containing combined garlic and holy basil essential oils and their application as antimicrobial inhibitor. *Int. J. Food Sci. Technol.* **2019**, *54*, 2064–2074. [CrossRef]
88. Luo, S.-Z.; Hu, X.-F.; Jia, Y.-J.; Pan, L.-H.; Zheng, Z.; Zhao, Y.-Y.; Mu, D.-D.; Zhong, X.-Y.; Jiang, S.-T. Camellia oil-based oleogels structuring with tea polyphenol-palmitate particles and citrus pectin by emulsion-templated method: Preparation, characterization and potential application. *Food Hydrocoll.* **2019**, *95*, 76–87. [CrossRef]
89. Nisar, T.; Yang, X.; Alim, A.; Iqbal, M.; Wang, Z.-C.; Guo, Y. Physicochemical responses and microbiological changes of bream (Megalobrama ambycephala) to pectin based coatings enriched with clove essential oil during refrigeration. *Int. J. Biol. Macromol.* **2019**, *124*, 1156–1166. [CrossRef]
90. Gong, C.; Lee, M.C.; Godec, M.; Zhang, Z.; Abbaspourrad, A. Ultrasonic encapsulation of cinnamon flavor to impart heat stability for baking applications. *Food Hydrocoll.* **2020**, *99*, 105316. [CrossRef]
91. Abdou, E.S.; Galhoum, G.F.; Mohamed, E.N. Curcumin loaded nanoemulsions/pectin coatings for refrigerated chicken fillets. *Food Hydrocoll.* **2018**, *83*, 445–453. [CrossRef]
92. Guerra-Rosas, M.I.; Morales-Castro, J.; Cubero-Márquez, M.A.; Salvia-Trujillo, L.; Martín-Belloso, O. Antimicrobial activity of nanoemulsions containing essential oils and high methoxyl pectin during long-term storage. *Food Control* **2017**, *77*, 131–138. [CrossRef]
93. Mendes, J.F.; Norcino, L.B.; Martins, H.H.A.; Manrich, A.; Otoni, C.G.; Carvalho, E.E.N.; Piccoli, R.H.; Oliveira, J.E.; Pinheiro, A.C.M.; Mattoso, L.H.C. Correlating emulsion characteristics with the properties of active starch films loaded with lemongrass essential oil. *Food Hydrocoll.* **2020**, *100*, 105428. [CrossRef]
94. Liu, C.; Tan, Y.; Xu, Y.; McCleiments, D.J.; Wang, D. Formation, characterization, and application of chitosan/pectin-stabilized multilayer emulsions as astaxanthin delivery systems. *Int. J. Biol. Macromol.* **2019**, *140*, 985–997. [CrossRef] [PubMed]
95. Sganzerla, W.G.; Paes, B.B.; Azevedo, M.S.; Ferrareze, J.P.; da Rosa, C.G.; Nunes, M.R.; Veeck, A.P.L. Bioactive and biodegradable film packaging incorporated with acca sellowiana extracts: physicochemical and antioxidant characterization. *Chem. Eng. Trans.* **2019**, *75*, 445–450.
96. Spatafora Salazar, A.S.; Sáenz Cavazos, P.A.; Mújica Paz, H.; Valdez Fragoso, A. External factors and nanoparticles effect on water vapor permeability of pectin-based films. *J. Food Eng.* **2019**, *245*, 73–79. [CrossRef]
97. Gorrasi, G. Dispersion of halloysite loaded with natural antimicrobials into pectins: Characterization and controlled release analysis. *Carbohydr. Polym.* **2015**, *127*, 47–53. [CrossRef] [PubMed]
98. Bernhardt, D.C.; Pérez, C.D.; Fissore, E.N.; De'Nobili, M.D.; Rojas, A.M. Pectin-based composite film: Effect of corn husk fiber concentration on their properties. *Carbohydr. Polym.* **2017**, *164*, 13–22. [CrossRef] [PubMed]
99. Gouveia, T.I.A.; Biernacki, K.; Castro, M.C.R.; Gonçalves, M.P.; Souza, H.K.S. A new approach to develop biodegradable films based on thermoplastic pectin. *Food Hydrocoll.* **2019**, *97*, 105175. [CrossRef]
100. Manrich, A.; Moreira, F.K.V.; Otoni, C.G.; Lorevice, M.V.; Martins, M.A.; Mattoso, L.H.C. Hydrophobic edible films made up of tomato cutin and pectin. *Carbohydr. Polym.* **2017**, *164*, 83–91. [CrossRef]
101. Gao, H.-X.; He, Z.; Sun, Q.; He, Q.; Zeng, W.-C. A functional polysaccharide film forming by pectin, chitosan, and tea polyphenols. *Carbohydr. Polym.* **2019**, *215*, 1–7. [CrossRef]
102. Ramos, M.; Fortunati, E.; Peltzer, M.; Kenny, J.M.; Garrigós, M.C. Characterization and disintegrability under composting conditions of PLA-based nanocomposite films with thymol and silver nanoparticles. *Polym. Degrad. Stab.* **2016**, *132*, 2–10. [CrossRef]
103. Dash, K.K.; Ali, N.A.; Das, D.; Mohanta, D. Thorough evaluation of sweet potato starch and lemon-waste pectin based-edible films with nano-titania inclusions for food packaging applications. *Int. J. Biol. Macromol.* **2019**, *139*, 449–458. [CrossRef]
104. Makaremi, M.; Pasbakhsh, P.; Cavallaro, G.; Lazzara, G.; Aw, Y.K.; Lee, S.M.; Milioto, S. Effect of Morphology and Size of Halloysite Nanotubes on Functional Pectin Bionanocomposites for Food Packaging Applications. *ACS Appl. Mater. Interfaces* **2017**, *9*, 17476–17488. [CrossRef] [PubMed]
105. Biddeci, G.; Cavallaro, G.; Di Blasi, F.; Lazzara, G.; Massaro, M.; Milioto, S.; Parisi, F.; Riela, S.; Spinelli, G. Halloysite nanotubes loaded with peppermint essential oil as filler for functional biopolymer film. *Carbohydr. Polym.* **2016**, *152*, 548–557. [CrossRef]

106. Groult, S.; Budtova, T. Thermal conductivity/structure correlations in thermal super-insulating pectin aerogels. *Carbohydr. Polym.* **2018**, *196*, 73–81. [CrossRef] [PubMed]
107. Mehrali, M.; Thakur, A.; Kadumudi, F.B.; Pierchala, M.K.; Cordova, J.A.V.; Shahbazi, M.-A.; Mehrali, M.; Pennisi, C.P.; Orive, G.; Gaharwar, A.K.; et al. Pectin Methacrylate (PEMA) and Gelatin-Based Hydrogels for Cell Delivery: Converting Waste Materials into Biomaterials. *ACS Appl. Mater. Interfaces* **2019**, *11*, 12283–12297. [CrossRef]
108. Wijaya, W.; Sun, Q.-Q.; Vermeir, L.; Dewettinck, K.; Patel, A.R.; Van der Meeren, P. pH and protein to polysaccharide ratio control the structural properties and viscoelastic network of HIPE-templated biopolymeric oleogels. *Food Struct.* **2019**, *21*, 100112. [CrossRef]
109. Sanchís, E.; Ghidelli, C.; Sheth, C.C.; Mateos, M.; Palou, L.; Pérez-Gago, M.B. Integration of antimicrobial pectin-based edible coating and active modified atmosphere packaging to preserve the quality and microbial safety of fresh-cut persimmon (*Diospyros kaki* Thunb. cv. Rojo Brillante). *J. Sci. Food Agric.* **2017**, *97*, 252–260. [CrossRef]
110. Ahmadzadeh, S.; Nasirpour, A.; Keramat, J.; Desobry, S. Powerful Solution to Mitigate the Temperature Variation Effect: Development of Novel Superinsulating Materials. *Food Packag. Preserv.* **2018**, 137–176.
111. de Oliveira, J.P.; Bruni, G.P.; el Halal, S.L.M.; Bertoldi, F.C.; Dias, A.R.G.; Zavareze, E.d.R. Cellulose nanocrystals from rice and oat husks and their application in aerogels for food packaging. *Int. J. Biol. Macromol.* **2019**, *124*, 175–184. [CrossRef]

materials

MDPI

Article

Size Distribution and Characteristics of Chitin Microgels Prepared via Emulsified Reverse-Micelles

Siriporn Taokaew *, Mitsumasa Ofuchi and Takaomi Kobayashi

Department of Materials Science and Technology, School of Engineering, Nagaoka University of Technology, 1603-1, Kamitomioka, Nagaoka, Niigata 940-2188, Japan; technomare3156@gmail.com (M.O.); takaomi@vos.nagaokaut.ac.jp (T.K.)

* Correspondence: t.siriporn@mst.nagaokaut.ac.jp; Tel.: +81-258-47-9383

Received: 11 March 2019; Accepted: 8 April 2019; Published: 10 April 2019

Abstract: Chitin was extracted from local snow crab shell waste and used as a raw material in the fabrication of porous spherical microgels. The chitin microgels were obtained using a batch process of emulsification and, afterward, gelation. The effects of chitin concentrations, oil and water phase ratios (O:W), surfactants, and gelation on the size distribution and morphology of the microgels were investigated. The extracted chitin possessed α-chitin with a degree of acetylation of ~60% and crystallinity of 70%, as confirmed by Fourier Transform Infrared Spectroscopy (FTIR) and X-Ray Powder Diffraction (XRD). In the reverse-micellar emulsification, different chitin concentrations in NaOH solution were used as aqueous phases, and n-hexane media containing Span 80-based surfactants were used as dispersion phases. Various HCl solutions were used as gelling agents. Microgels with sizes ranging from ~5–200 μm were obtained relying on these studied parameters. Under the condition of 3% *w*/*w* chitin solution using O:W of 15:1 at 5% *w*/*w* of Span 80 (hydrophilic-lipophilic balance; HLB of 4.3), the gelation in the emulsified reverse micelles was better controlled and capable of forming spherical microgel particles with a size of 7.1 ± 0.3 μm, when 800 μL of 1 M HCl was added. The prepared chitin microgel exhibited macro-pore morphology and swelling behavior sensitive to the acidic pH.

Keywords: crab shell; chitin; spherical microgels; reverse micelle; gelation

1. Introduction

Chitin (*ß*-(1,4)-*N*-acetyl-D-glucosamine) is a natural polysaccharide found in shells of marine animals such as shrimp, lobster, and crabs. As a marine fishery product, about 30 thousand tons of crabs were annually caught in Japan [1]. This means that large amounts of crab shells were disposed in landfills without being recycled. It is known that crab shells are a good biomass source of chitin [2]. Due to biocompatibility and non-allergenicity, chitin has been widely used in pharmaceutics and bio-medical drugs. In addition, chitin extracted from crab shells has other characteristics such as antibacterial properties and protein affinity that are useful for wound dressing and controlled-drug release applications [3,4]. As a drug carrier, small spheres of gel forming chitin have been recognized as having high drug loading capacity, efficient drug control at the target site, sustained drug release, and high stability compared to micelles and lipid-based carriers [5,6]. Moreover, microgels can be applied as adsorbents, chemical/biological sensors, enzyme immobilization, and gene delivery vehicles [5,7].

The applications of chitin and its derivatives as microgels/microparticles were reported as biological filling [8] and drug delivery agents [6]. Chitin microparticles could regulate the depletion of cholesterol by cellular macrophage activation [9]. A fragmented physical hydrogel suspension of chitin derivatives was indicated to support reepithelization of spinal tissue and vasculature with minimal fibrous glial scaring [8]. Moreover, the fragmented chitin microgels loaded with an anti-metabolite drug for delivery in psoriasis treatment exhibited higher skin permeable efficacy than those of the control drug solution and the conventional drug gel. The drug-loaded fragmented chitin microgels also exhibited greater swelling and drug release at acidic pH than in neutral and alkaline conditions [6]. However, the time window of the use of microgels with an unspecific-shape was difficult to determine with high precision because the shape and the stimuli responsiveness influence the biodistribution, the circulation dynamics, the drug release, and the intracellular uptake of the microgels [8,10]. Hence, the fabrication of microgels that have precise geometries and stimuli-responsiveness has been significant in particle transportation and therapeutic agent delivery.

It has been reported that several microgel preparation methods including solid-phase organic synthesis [11], microfluidics [12], and emulsification [13,14] were feasible. However, solid-phase organic synthesis and microfluidics have numerous problems associated with the use of cross-linked insoluble polymers, the fluctuation of reaction rates, and the longer time-consumption in such processes [15,16]. In contrast, a reverse-micellar emulsification technique simplifies the process, making it an effective tool to synthesize small particles with controllable size and shape. An emulsion-based method is also energy-efficient, non-destructive, and attractive for large-scale production [13,14]. As compared to the other approaches, the reverse-micellar emulsification can enhance uniformity and dispersity of the polymeric particles, can be operated at low temperature, and provide a stable dispersion for a water in oil emulsion system [17]. Therefore, such a method is used to prompt self-assembly of surfactant in organic media, whereby the oil region having a nonpolar nature faces the outside surface of the micelle, and the polar region forms the core for polymeric microgels [18]. In such a structure, the tiny aqueous droplets with varied sizes are encapsulated, and the different-sized microgels are produced within the reverse micelle after gelation [13]. Accordingly, chitin microgels with the same size prepared by the reverse-micellar emulsification method can be described.

The aim of the present study was to prepare chitin microgel by using a reverse micelle system at various compositions of water in oil (W/O) emulsions. The chitin used in this study was extracted from shell waste of red snow crabs, which was collected from the local area in Niigata prefecture, Japan. The extracted chitin was then characterized and compared to commercial chitin and chitosan. The synthesis of the microgel was performed using a batch process of W/O emulsion. The effects of chitin concentrations (water phase) of 1–3% *w*/*w* and oil:water phase ratios of 3–15:1 were studied. It was known that hydrophilic-lipophilic balance (HLB) values of surfactant ranging from 3.5 to 6 were more suitable for a W/O emulsion system [19]. For Span 80, a nonionic-based surfactant, the HLB value of 4.3 was adjusted to 5 and 6 in this study. Concentrations of Span 80 (3–7% *w*/*w*) containing n-hexane (oil phase), and gelation using HCl were also investigated in terms of their size distribution and morphology.

2. Materials and Methods

2.1. Materials

Dried, cleaned snow crab shells, *Chionoecetes opilio*, were obtained from Teradomari port, Teradomari, Niigata prefecture, Japan. Chemicals were purchased from Wako Pure Chemical Industries, Ltd, Osaka, Japan. Distilled and ion-exchanged water was used in all the experiments.

2.2. Extraction of Chitin

The coarse flakes of crab shells (30 g) were hydrolyzed using 900 mL of 1.0 M HCl under stirring at room temperature (20 ± 5 °C) for 24 h. The reaction was stopped by adding water and filtered through a mesh sieve to remove small contaminants. Protein residuals were removed by heating the hydrolyzed chitin at 90 °C in 900 mL of 1.0 M NaOH under stirring for 5 h. Pigments in chitin were removed by stirring in 900 mL of ethanol for 5 h at 60 °C. The extracted chitin was dried in vacuum oven at 60 °C for 24 h, and ground in a blender.

2.3. Preparation of Microgels

The chitin powder was dissolved in 20% w/v NaOH at −20 °C under periodic stirring to obtain 1, 2, and 3 % w/w of chitin aqueous solution. Before emulsion formation, the oil layer solution consisted of n-hexane and surfactants were used as a dispersion phase and prepared in a 50 mL amber vial. According to the critical micelle concentration (CMC) of Span 80 in n-hexane, see Figure 1, below the CMC in the presence of surfactant monomer, there was no peak throughout the spectrum, see Figure 1a, but the peaks at 270 nm appeared at above the CMC. The CMC was approximately 0.25% w/w, determined by the change of the trend in Figure 1b, which indicates the initial formation of micelles. Above the CMC value (about 10 times), a number of surfactant molecules were able to gather and form stable micelles in the bulk liquid [20]. Therefore, 3, 5, and 7 % w/w of Span 80 were adopted for this study. A Span 80 (Sorbitan monooleate)-based surfactant was mixed with sodium cholate (HLB 18) to obtain HLB values of 4.3, 5, and 6. The chitin solution was dropped into the dispersion phase with vigorous stirring at 1500 rpm at room temperature (20 ± 5 °C) for 45 min. The W/O emulsion was then heated to 65 °C. Aqueous HCl solution in the range of 0.01–0.1 M concentration was used as counter-ions. In the gelation process of chitin microgel, 400–1200 μL of aqueous HCl solution was periodically dropped into the emulsion under stirring at 150 rpm. The parameters tested in the preparation of microgels are shown in Table 1. The chitin microgels were coagulated in the liquid medium and precipitated. The microgel was purified to remove the surfactant and residual n-hexane using dialysis (Molecular weight cut-off of 12 kDa, 0.5 nm, AS ONE corporation, Osaka Japan) in 1L of distilled deionized water for 72 h.

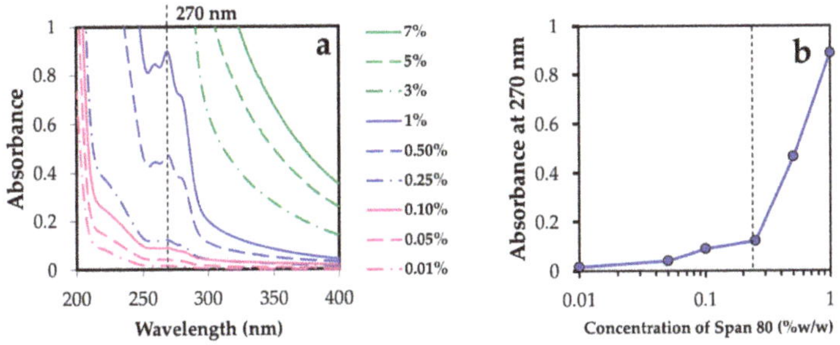

Figure 1. UV-Visible absorbance versus concentration profile of Span 80 in n-hexane at 20 ± 5 °C (**a**), and absorbance at 270 nm of Span 80 at various concentrations (**b**). Span 80 was dissolved in n-hexane at a concentration of 0.01–7% w/w. After thorough mixing, the solution was transferred to a 1.0 cm quartz cell and the spectrum was recorded at wavelengths of 200–400 nm using UV-visible near-infrared spectrophotometer (Jasco V570, Jasco Corporation, Tokyo, Japan). Blank n-hexane was used as a reference. The vertical dashed line in (**b**) marks the critical micelle concentration.

2.4. Characterization

2.4.1. X-Ray Fluorescence Spectroscopy (XRF)

An elemental study of the extracted chitin was performed using an X-ray fluorescence spectrometer (Rigaku ZSX Primus II, Tokyo, Japan) using ZSX software. This spectrometer contains a 50 keV and 50 mA X-ray tube, providing the detection of diverse elements of the Periodic Table. Prior to characterization, the sample pellets were prepared by using pressed powder method under a pressure of 500 kgf/cm^2.

2.4.2. X-Ray Powder Diffraction (XRD)

X-ray diffractograms were obtained using an X-ray diffractometer (Rigaku Smart Lab 3 kW, Tokyo, Japan) under operation conditions of 40 kV and 30 mA with Cu Kα radiation. The relative intensity was recorded in steps of 0.1° and at a speed of 3.0 °/min. The crystallinity index (*CrI*) was determined by integrated X-ray powder diffraction software (Rigaku PDXL2, Rigaku Corporation, Tokyo, Japan). The quantitative analysis was performed based on the Rietveld refinement and an ab-initio crystal structure determination using crystal structure information of α-chitin provided by the software. The degree of acetylation (*DA*) of chitin [21] was calculated by:

$$DA\ (\%) = 100 - \frac{(103.97 - CrI)}{0.7529} \quad (1)$$

Table 1. Parameters in preparation of chitin microgels.

Parameters	Concentration of Chitin Solution (% *w*/*w*)	O:W Volume Ratio	HLB of Surfactant	Concentration of Span 80 (% *w*/*w*)	Concentration of HCl (M)	Volume of HCl (μL)
Experiment 1	**1** **2** **3**	15:1	4.3	5	1.0	800
Experiment 2	3	**3:1** **7:1** **15:1**	4.3	5	1.0	800
Experiment 3	3	15:1	**4.3** **5** **6**	5	1.0	800
Experiment 4	3	15:1	4.3	**3** **5** **7**	1.0	800
Experiment 5	3	15:1	4.3	5	**0.05** **0.1** **1.0**	800
Experiment 6	3	15:1	4.3	5	1.0	**400** **800** **1200**

2.4.3. Fourier Transform Infrared Spectroscopy (FTIR)

FTIR spectra were obtained using a FTIR spectrometer (Jasco 4100, Jasco Corporation, Tokyo, Japan). The sample pellets were prepared using the KBr method. The absorption bands were scanned between 4000–400 cm^{-1}. The degree of acetylation (DA) was calculated by:

$$DA\ (\%) = \frac{1}{1.33}\left(\frac{A_{1655}}{A_{3450}}\right) \times 100 \quad (2)$$

where A_{1625} and A_{3450} are values of absorbance measured at 1625 and 3450 cm^{-1}, respectively [22].

2.4.4. Differential Scanning Calorimetry (DSC)

Thermograms were carried out using differential scanning calorimetry (DSC) (Rigaku, Thermo Plus EVO DSC823, Tokyo, Japan) under an air atmosphere. Dried samples (3–5 mg) were placed in hermetically sealed Al pans and immediately loaded in the DSC chamber. A sealed empty pan was used as a reference. Samples were scanned at the heating rate of 5 °C/min through the temperature range of 50–400 °C.

2.4.5. Dynamic Light Scattering (DLS)

The size distributions of the microgel samples subjected to the tested parameters and swelling test at different pH values of 2, 4, 7, and 10 were analyzed by dynamic light scattering (DLS Shimadzu SALD-7000, Tokyo, Japan). pH values in the swelling test were adjusted by using 0.1 M HCl and 0.1 M NaOH.

2.4.6. Optical Microscopy, Scanning Electron Microscopy (SEM), and Transmission Electron Microscopy (TEM)

The optical microscopic morphologies of the reverse micelles in the emulsion and the microgels were visualized using an optical microscope (Olympus CKX41 Inverted Phase Contrast Microscope, Tokyo, Japan) at the magnification of 20×. Morphologies of the freeze-dried microgels were studied by scanning electron microscopy (Desktop SEM Hitachi TM3030 Plus, Tokyo, Japan) and transmission electron microscopy (TEM Hitachi HT7700, Tokyo, Japan). The microgels were freeze-dried by immersing in liquid N_2 for 1 h before immediately loading the frozen samples into a chamber of a freeze dryer (Eyela Freeze Dryer FDU-1200, Tokyo, Japan). The freeze-drying process was operated at a condenser temperature of −40 °C under high vacuum. For SEM, the freeze-dried microgels were coated with gold using a gold sputter (Quick cool coater SC-701MC, Tokyo, Japan) under a high-vacuum condition. The surface morphology of the coated microgels was then observed at a voltage of 15 kV using a back-scatter detector (BSE) mode at 2000×. For TEM, the freeze-dried microgels were stained with Osmium tetroxide for 20 s before observing the sample morphology at 100 kV and 4000×.

2.4.7. Zeta-Potential

Zeta-potential of the microgels was determined using a zeta-potential analyzer equipped with auto-titrator, stirrer, and inbuilt peristatic pump (Otsuka ELSZ, Tokyo, Japan). The zeta-potential was recorded at the pH values ranging from 2 to 10 adjusted using 0.1 M HCl and 0.1 M NaOH. All measurements were carried out at room temperature (20 ± 5 °C).

2.4.8. Brunauer-Emmett-Teller (BET)

N_2 adsorption-desorption isotherms of freeze-dried chitin microgels were carried out using a surface area and porosity analyzer (Micromeritics TriStar II, Norcross, GA, USA) at 77 K using Brunauer-Emmett-Teller (BET) and Barrett-Joyner-Halenda (BJH) analyses. Before analysis, samples were degassed at 30 °C on a vacuum line for 24 h.

3. Results and Discussion

3.1. Properties of the Extracted Chitin

Chitin extraction, in this work, included acid-base hydrolysis and decoloration processes. The crab shell waste was extracted for chitin having yield of 25 ± 8% dry weight.

From XRF analysis, see Table 2, the extracted chitin retained high contents of C and O of the organic compound. The other sea contaminants in the extracted chitin were mainly removed from the crab shells and the quality was similar to the commercial chitin. While the heavy metals in the extracted chitin were not detected as compared to chitin from red shrimp shell [23].

Table 2. Elemental composition of crab shell, extracted chitin, commercial chitin, and commercial chitosan analyzed by X-ray fluorescence spectroscopy (XRF).

Elements	Crab Shell (mass%)	Extracted Chitin (mass%)	Commercial Chitin (mass%)	Commercial Chitosan (mass%)
C	29.2	52.6	52.2	50.8
O	46.3	47.2	47.6	49.1
Na	0.968	trace	trace	trace
Mg	1.19	trace	trace	0.0135
P	2.93	0.0293	0.0022	0.0074
S	0.344	0.0158	0.0123	0.0044
Cl	0.971	0.0167	0.112	trace
Ca	17.2	0.0782	0.0072	0.0457
Fe	0.016	0.0104	0.0053	0.0084

Figure 2 shows XRD patterns of the extracted chitin obtained in the 2θ range of 5–40°. The diffraction peaks of the extracted chitin, see Figure 2a, and the commercial chitin, see Figure 2b, at 9.4°, 12.8°, 19.4°, 20.8°, 23.5°, and 26.4° were observed with indices of (020), (101), (110), (120), (130), and (013). These parameters define the crystallographic planes of α-chitin. This indicated that chitin has high molecular packing with inter- or intramolecular hydrogen bonds, imparting a high degree of crystallinity [23–25]. The intensities of the (020) and (110) planes decreased and moved to higher angles with a reduction in the degree of acetylation (DA) [21]. In this work, characteristic peaks of chitosan indexed as (020) and (110) appear at 10.4 and 20.3°, respectively, see Figure 2d. The extracted chitin exhibited a crystallinity value of 70.3%, as shown in Table 3. The DA of the extracted chitin obtained by XRD and confirmed by FTIR techniques were 55.3% and 60.9%, respectively, see Table 3. This meant that the extracted chitin was partially deacetylated.

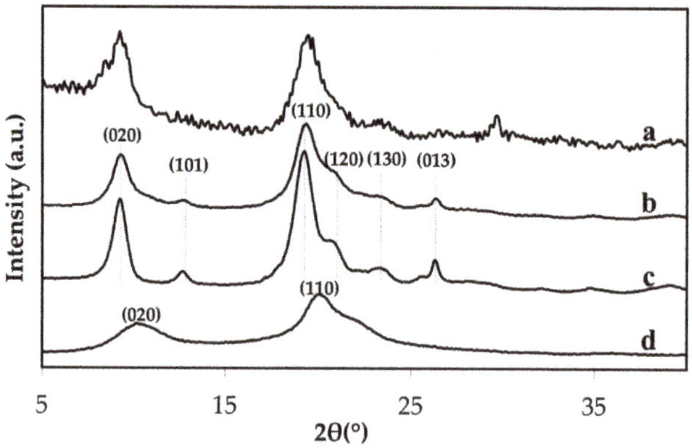

Figure 2. X-ray diffractograms of crab shell (**a**), extracted chitin (**b**), commercial chitin (**c**), and commercial chitosan (**d**).

Table 3. Crystallinity index (%CrI), degree of acetylation (%DA), and peak temperature of extracted chitin, commercial chitin, and commercial chitosan.

Samples	%CrI	%DA from XRD	%DA from FTIR	Peak Temperature (°C)
Extracted chitin	70.3	55.3	60.9	330
Commercial chitin	74.7	61.1	62.6	340
Commercial chitosan	44.7	21.2	45.7	295

As seen in Figure 3, the FTIR spectrum of the extracted chitin had a broad peak at about 3450 cm^{-1} assigned to OH stretching. Amide I, II, and III appeared at the observed absorption bands around 1652, 1557, and 1310 cm^{-1}, respectively. It was observed that the amide I band of the extracted chitin is split into two 1652 and 1623 cm^{-1}. The existence of these interchain bonds of carbonyl groups of amide I and II are responsible for the high chemical stability of the α-chitin structure [23,26]. DSC thermograms of the crab shell, extracted chitin, commercial chitin, and chitosan were compared, see Figure 4. The wide and weak endothermic peak of the extracted chitin in Figure 4b was noticed at about 50–90 °C and ascribed to the loss of bound water. The exothermic peak of the crab shell and extracted chitin was observed at 330 °C due to the crystalline α-chitin structure. This indicated that the extraction process of chitin retained the α-structure of the resulting product. The extracted chitin had a higher temperature at which the exothermic peak appeared than the chitosan, see Figure 4d. The exothermic peak observed for chitosan at 295 °C is the characteristic peak of amine (GlcN) unit decomposition [27].

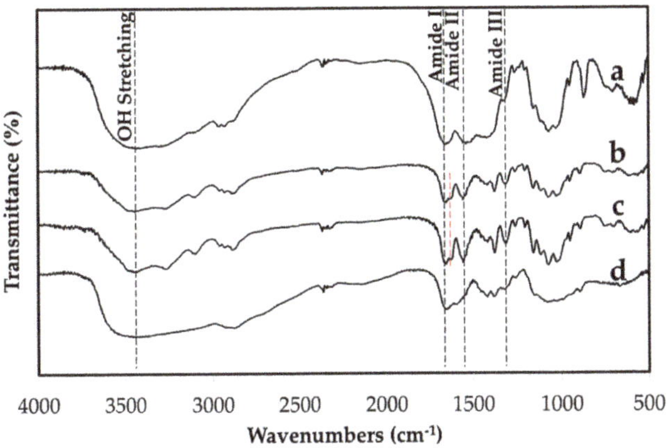

Figure 3. Fourier transform infrared (FTIR) spectra of crab shell (**a**), extracted chitin (**b**), commercial chitin (**c**), and commercial chitosan (**d**).

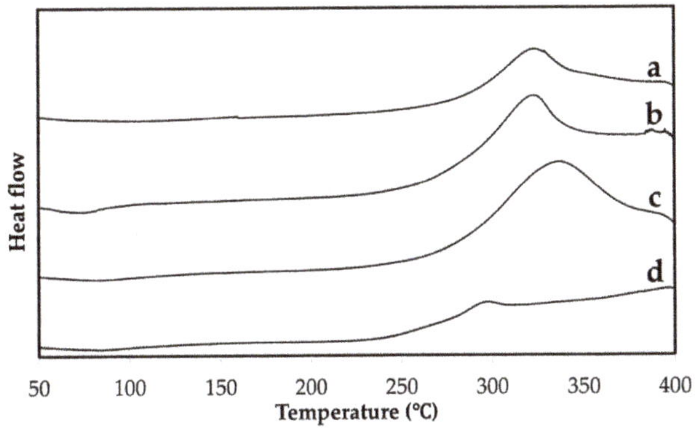

Figure 4. Differential scanning calorimetry (DSC) thermogram of crab shell (**a**), extracted chitin (**b**), commercial chitin (**c**), and commercial chitosan (**d**).

3.2. Reverse Micelle Emulsification for the Fabrication of Chitin Microgels

3.2.1. Effect of Water, Oil Phase, and Surfactant

In the reverse micelle emulsification, chitin in the alkali solution was prepared at 1, 2, and 3 % *w*/*w*, and then added dropwise into the oil phase. The microgels produced from 1 and 2 % *w*/*w* of the chitin solutions became small in size, but rather aggregated, see Figure 5a,b.

Increasing the chitin concentration to 3% provided more dispersed microgels with an average size of 7.1 ± 0.3 μm, see Figure 5c,g. Nevertheless, microgels produced from 3% chitin appeared as a weak gel with a less uniform size, as seen in Figure 5d,e. These differences are due to the low ratios of oil and water phases (O:W) at 3:1 and 7:1. At O:W of 15:1, the microgel appeared to be more dispersed, see Figure 5f. From the dynamic light scattering analysis, chitin microgels prepared from 1–3% *w*/*w* of chitin and O:W of 15:1 yielded a narrower size distribution (5–10 μm), see Figure 5g. However, there were wider size distributions (10–100 μm) of microgels when O:W of 3:1 and 7:1 were used (Figure 5h). Due to the low O:W of 3:1 and 7:1, the reverse micelles of chitin could not properly disperse in the oil phase during agitation. This might be due to the bigger microgels yielded after gelation, meaning that, low volume of the dispersion phase caused a high incidence of micelle breaking collisions during agitation.

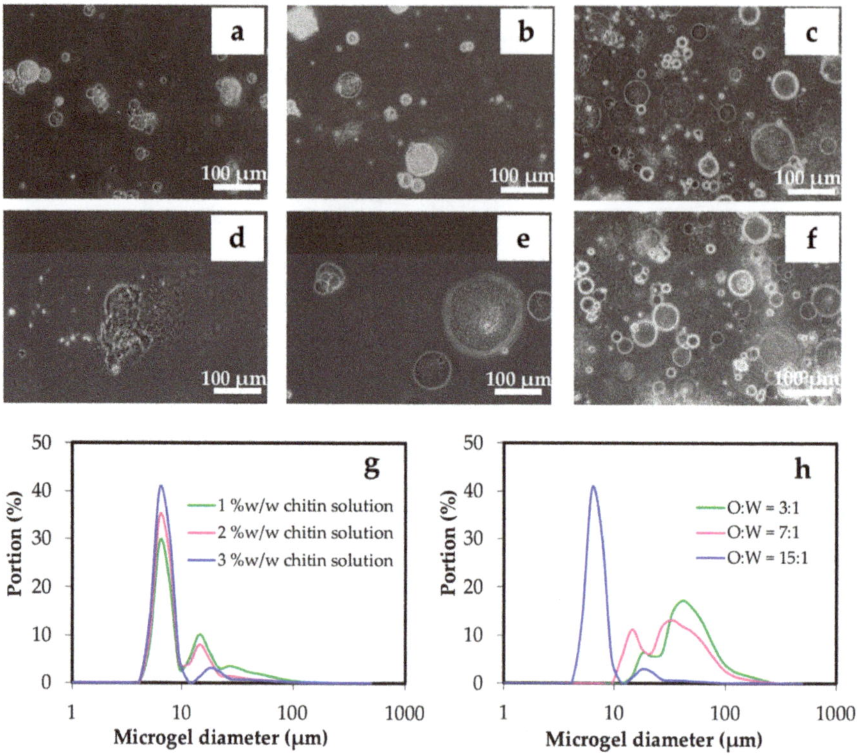

Figure 5. Optical microscopic images of microgels prepared from chitin solution at the concentrations of 1% *w*/*w* (**a**), 2% *w*/*w* (**b**), and 3% *w*/*w* (**c**), and O:W of 3:1 (**d**), 7:1 (**e**), and 15:1 (**f**) by controlling the concentration of Span 80 (HLB 4.3) at 5% *w*/*w* in oil phase. Gelation was carried out using 800 µL of 1.0 M HCl. The representative size distributions of microgels prepared by different chitin concentrations and O:W ratios are shown in (**g**) and (**h**), respectively.

It can be clearly observed in Figure 6a,g that HLB 4.3 was suitable for preparing chitin microgels in this study as compared to mixed surfactants having HLB 5 and 6 due to the balance of the size and strength of hydrophilic and lipophilic moieties of surfactant molecules. The bigger microgels with wider size distribution were the result of using mixed surfactants at HLB of 5 and 6, see Figure 6b,c,g. This was due to the higher hydrophilic portion in the surfactant that allowed the chitin aqueous solution to form larger and stable cores inside the reverse micelles.

Span 80 concentrations (HLB 4.3) of 3, 5, and 7 % *w*/*w* were varied in the preparation of chitin microgels. In this range of surfactant concentrations, the morphology of the resulting microgels observed through the optical microscope were similar in size, see Figure 6d–f. The size distributions were also comparable, when the Span 80 concentrations were in the range of 3–7% *w*/*w*, see Figure 6h. However, as seen in Figure 6, the microgel prepared under the condition of 3% *w*/*w* surfactant likely exhibited aggregation, in which slightly a larger portion of ~20 µm microgel was observed.

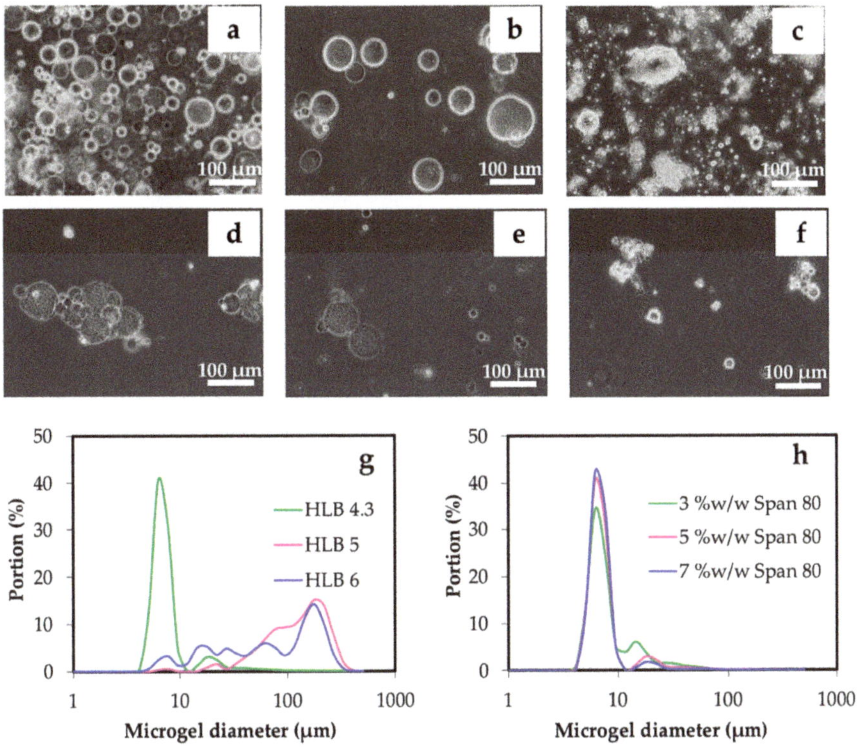

Figure 6. Optical microscopic images of microgels prepared from hydrophilic-lipophilic balance (HLB) values of surfactant at 4.3 (**a**), 5 (**b**), and 6 (**c**), and Span 80 at the concentrations of 3 % *w*/*w* (**d**), 5 % *w*/*w* (**e**), and 7 % *w*/*w* (**f**) by controlling the concentration of the chitin solution at 3 % *w*/*w* and an O:W ratio at 15:1. Gelation was carried out using 800 μL of 1.0 M HCl. The representative size distributions of microgels prepared by different HLB values and Span 80 concentrations are shown in (**g**) and (**h**), respectively.

3.2.2. Effect of Gelation

The gelation was implemented while the alkali chitin solution inside the emulsified reverse micelles was surrounded by an oil phase. When HCl solution was used as the gelling agent, the alkali conditions of the chitin solution have a neutralizing acid-base reaction. From Figure 7a–c,g, the concentration of HCl greatly affected the size distribution. When the HCl concentration was changed to 0.05, 0.1, and 1 M, a size variation (~5, 20, and 40 μm) was observed as the diluted HCl concentration was 0.05 M. It was seen that the size was less variant with increased HCl concentrations. As seen in Figure 7g, concentrations of 0.1 M and 1.0 M provided 50 and 5 μm diameter microgel spheres, respectively. This was related to the effect of the water phase on the microgel preparation, in which the initial concentration of chitin solution and external water affected the size distribution of the microgels. It was possibly owing to the involvement of a higher amount of water in the emulsion system; Span 80 was able to hold/absorb water into the core of reverse micelles [14]. The effect of a volume of 1.0 M HCl was further studied, see Figure 7d–f,h. It was found that only 400 μL was sufficient to yield the narrow size distribution of ~5 μm microgels. However, increasing the volume to 800–1200 μL slightly increased the portion of ~5 μm microgels.

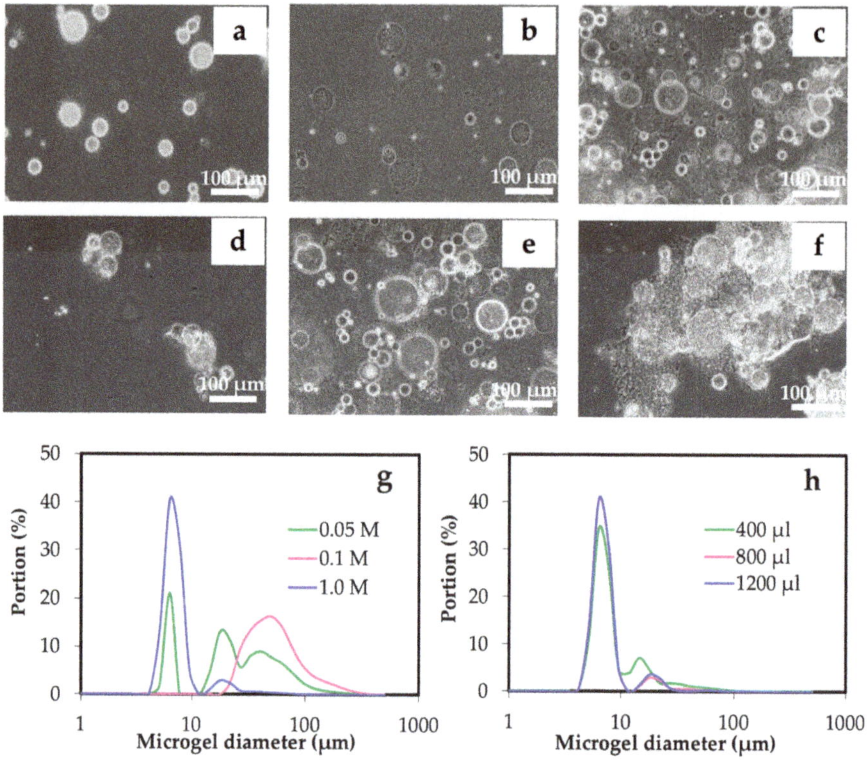

Figure 7. Optical microscopic images of microgels prepared from gelation using HCl at the concentrations of 0.05 M (**a**), 0.1 M (**b**), and 1.0 M (**c**), and 400 µL (**d**), 800 µL (**e**), and 1200 µL (**f**) of 1.0 M HCl by controlling the concentration of chitin solution of 3% *w*/*w* at an O:W ratio of 15:1, and concentration of Span 80 (HLB 4.3) at 5% *w*/*w* in the oil phase. The representative size distributions of microgels prepared by different concentrations and volume of HCl are shown in (**g**) and (**h**), respectively.

3.3. Properties of Chitin Microgels Prepared by the Reverse Micellar Method

The appearance of chitin solution containing reverse micelles is shown in Figure 8a, corresponding to microgels prepared with 3% *w*/*w* chitin solution, O:W of 15:1, 5% *w*/*w* Span 80 (HLB 4.3), and gelation by 800 µL of 1 M HCl. As compared to the size of reverse micelles, no remarkable change in size was observed after gelation of chitin microgels. Figure 8b showed no microgel breakage induced by collision during stirring. Electron microscopic images of the freeze-dried samples revealed a spherical chitin microgel with macropores on the surface, see Figure 8c, and an internal porous structure, see Figure 8d. The formation mechanism of macropores on microspheres prepared by an emulsion system when using Span 80 for the Poly(styrene-divinyl benzene) system has been reported [14]. Similarly, macropores of the resultant chitin microgels are strongly related to the absorption of water from the external aqueous phase into the reverse micelles. Since Span 80 with a HLB of 4.3 was less hydrophobic, it has a stronger ability to absorb water.

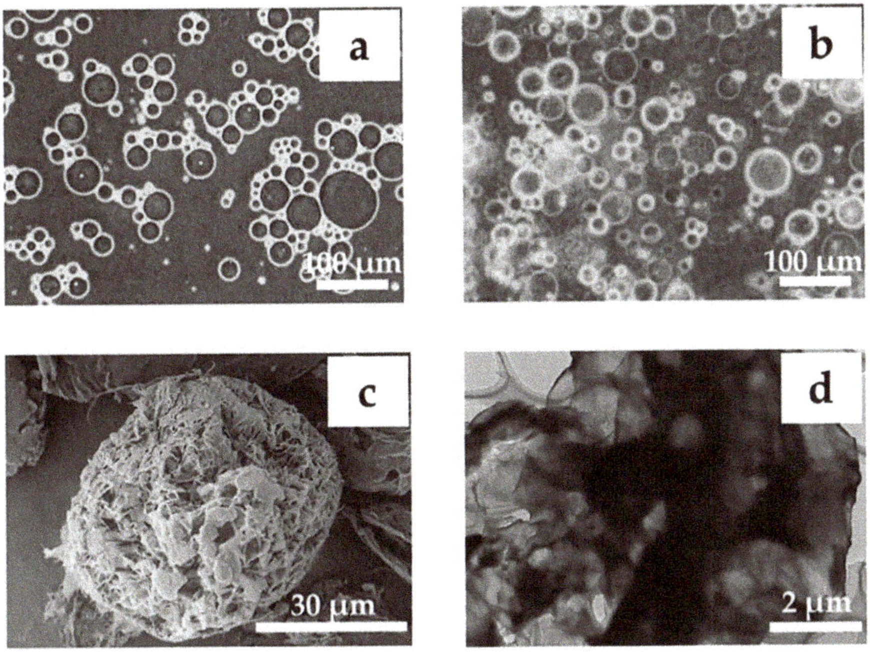

Figure 8. Optical microscopic images of reverse micelles in emulsion (**a**) and microgels (**b**) prepared under 3 % *w*/*w* chitin solution, O:W = 15:1, 5% *w*/*w* Span 80 (HLB 4.3), and gelation by 800 µL of 1.0 M HCl. Morphologies of the freeze-dried microgels are shown at magnifications of 2000× by scanning electron microscopy (SEM) (**c**) and 4000× by transmission electron microscopy (TEM) (**d**).

The porosity data of the chitin microgels was characterized by BET. The BET isotherms of the microgels are shown in Figure 9. The chitin microgels showed a type II isotherm for a microporous material according to the IUPAC classification [28]. The surface area and pore volume were 22.6 m^2/g and 0.03 cm^3/g, respectively.

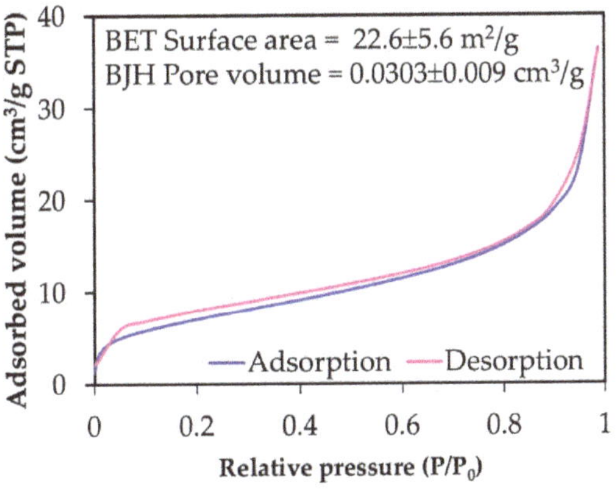

Figure 9. Brunauer-Emmett-Teller (BET) isotherm of freeze-dried microgels prepared under conditions of 3 % *w*/*w* chitin solution, O:W = 15:1, 5% *w*/*w* Span 80 (HLB 4.3), and gelation by 800 µl of 1.0 M HCl.

Under the optimal conditions of microgel preparation, the charge of partially deacetylated chitin microgels was investigated by measuring the zeta potential according to various pHs. Figure 10 shows the zeta potential of the chitin microgels prepared at 3% *w*/*w* chitin solution, O:W of 15:1, 5% *w*/*w* Span 80 (HLB 4.3), and gelation by 800 μL of 1.0 M HCl. The positive zeta potential at pH below 6 clearly indicated partially deacetylated chitin with different %DA due to the different number of amine groups to be protonated, leading to the positive charges. The isoelectric point (IP), that is the null zeta potential, increases to a higher pH value with lower %DA. Accordingly, IP values of chitosan (DA < 50%) and chitin (DA > 50%) were detected at pH values of 8.2–8.8 and 7.3–7.6, respectively [24,29]. In the present work, the IP value of the chitin microgels was observed at about pH 7.6, see Figure 10. This confirms that the chitin microgels were not further deacetylated during the preparation process of the chitin microgels.

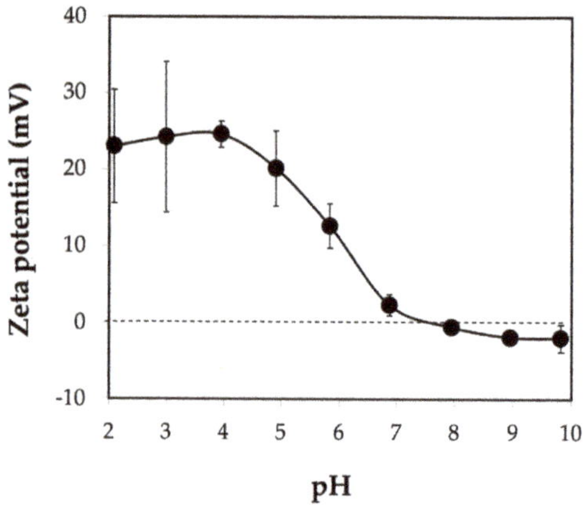

Figure 10. Zeta potential of chitin microgels prepared under conditions of 3 % *w*/*w* chitin solution, O:W = 15:1, 5% *w*/*w* Span 80 (HLB 4.3), and gelation by using 800 μL of 1.0 M HCl.

Since the microgels presented pH dependence, it showed different swelling of the microgels tested over the entire range of pH from 10 to 2. From Figure 11a, the chitin microgels swelled at low pH (pH < IP). Microgels that were approximately 6 μm in size increased to ~60 μm. However, the gradual decrease was observed between pH 2 and pH 4 due to the similar zeta potential of approximately +25 mV, as seen in Figure 10. This was possibly the impact of the degree of acetylation, which controls the characteristics and activities of chitin [30]. In the extracted chitin with a moderate degree of acetylation (~60%), a number of amino groups in the chitin polymer chains were protonated while exposed to a specific pH. The protonation leads to the repulsion of polymer chains and allows more water to enter into the microgel network; consequently, swelling occurs [31]. After adjusting the pH backward, from 2 to 10, a reversible swelling-shrinking behavior was noticed. The chitin microgels began to shrink to a smaller size as the pH increased (pH > IP), see Figure 11b. This was because the deprotonation made the electrostatic interactions in the microgel network reconstruct [31].

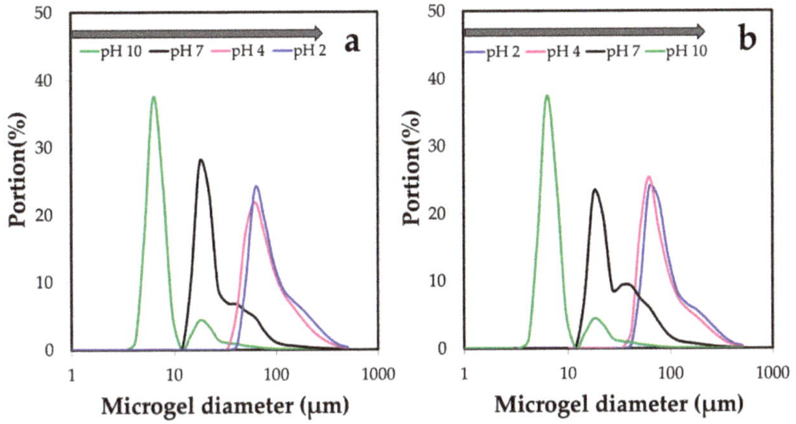

Figure 11. Reversible swelling-shrinking behavior of chitin microgels prepared under conditions of 3 % *w*/*w* chitin solution, O:W = 15:1, 5% *w*/*w* Span 80 (HLB 4.3), and gelation by using 800 μL of 1.0 M HCl. Responsiveness to pH was tested from pH 10 towards pH 2 (**a**) and pH 2 towards pH 10 (**b**).

4. Conclusions

Chitin extracted from crab shell waste was used for microgel fabrication. Simple gelation inside the emulsified reverse-micelles with low energy consumption was applied to prepare the chitin microgels. The spherical size distribution and the morphology of the microgels were greatly affected by the volume of the dispersion phase, hydrophilic-lipophilic balance of the used Span 80 surfactant, and concentration of the gelation agent. As a result, the chitin microgel with narrow size distribution (average size of 7.1 ± 0.3 μm) and porous spherical morphology was achieved under the condition of 3 % *w*/*w* chitin solution, O:W of 15:1, 5% *w*/*w* Span 80 (HLB 4.3), and gelation by 800 μL of 1.0 M HCl. Moreover, the prepared chitin microgels exhibited pH-dependent swelling-shrinking behavior over a wide range of pH values of between 2–10.

Author Contributions: Conceptualization, S.T.; methodology, M.O.; validation, S.T. and T.K.; formal analysis, M.O.; investigation, M.O. and S.T.; resources, S.T. and T.K.; writing—original draft preparation, S.T.; writing—review and editing, S.T.; supervision, S.T. and T.K.; project administration, S.T.

Funding: This research was funded by Nagaoka University of Technology.

Acknowledgments: This work was supported by Nagaoka University of Technology. The authors thank Analysis and Instrumentation Center of Nagaoka University of Technology for the technical assistance.

Conflicts of Interest: The authors declare no conflict of interest.

References

1. Quantity of Catches of Marine Fisheries by Species. Available online: http://www.stat.go.jp/english/data/nenkan/66nenkan/1431-08.html (accessed on 27 February 2019).
2. Yan, N.; Chen, X. Sustainability: Don't waste seafood waste. *Nature* **2015**, *524*, 155–157. [CrossRef]
3. Younes, I.; Rinaudo, M. Chitin and chitosan preparation from marine sources. Structure, properties and applications. *Mar. Drugs* **2015**, *13*, 1133–1174. [CrossRef] [PubMed]
4. Rinaudo, M. Chitin and chitosan: Properties and applications. *Prog. Polym. Sci.* **2006**, *31*, 603–632. [CrossRef]
5. Thorne, J.B.; Vine, G.J.; Snowden, M.J. Microgel applications and commercial considerations. *Colloid Polym. Sci.* **2011**, *289*, 625–646. [CrossRef]
6. Panonnummal, R.; Sabitha, M. Anti-psoriatic and toxicity evaluation of methotrexate loaded chitin nanogel in imiquimod induced mice model. *Int. J. Biol. Macromol.* **2018**, *110*, 245–258. [CrossRef]

7. Zhang, X.; Lu, S.; Gao, C.; Chen, C.; Zhang, X.; Liu, M. Highly stable and degradable multifunctional microgel for self-regulated insulin delivery under physiological conditions. *Nanoscale* **2013**, *5*, 6498–6506. [CrossRef] [PubMed]
8. Chedly, J.; Soares, S.; Montembault, A.; von Boxberg, Y.; Veron-Ravaille, M.; Mouffle, C.; Benassy, M.-N.; Taxi, J.; David, L.; Nothias, F. Physical chitosan microhydrogels as scaffolds for spinal cord injury restoration and axon regeneration. *Biomaterials* **2017**, *138*, 91–107. [CrossRef]
9. Nishiyama, A.; Shinohara, T.; Pantuso, T.; Tsuji, S.; Yamashita, M.; Shinohara, S.; Myrvik, Q.N.; Henriksen, R.A.; Shibata, Y. Depletion of cellular cholesterol enhances macrophage MAPK activation by chitin microparticles but not by heat-killed Mycobacterium bovis BCG. *Am. J. Physiol. Cell Physiol.* **2008**, *295*, 341–349. [CrossRef]
10. Caldorera-Moore, M.; Guimard, N.; Shi, L.; Roy, K. Designer nanoparticles: Incorporating size, shape, and triggered release into nanoscale drug carriers. *Expert Opin. Drug Deliv.* **2010**, *7*, 479–495. [CrossRef]
11. Vaino, A.R.; Janda, K.D. Solid-phase organic synthesis: A critical understanding of the resin. *J. Comb. Chem.* **2000**, *2*, 579–596. [CrossRef]
12. Seiffert, S. Microgel capsules tailored by droplet-based microfluidics. *ChemPhysChem* **2013**, *14*, 295–304. [CrossRef] [PubMed]
13. Lim, L.Y.; Wan, L.S.C.; Thai, P.Y. Chitosan microspheres prepared by emulsification and ionotropic gelation. *Drug Dev. Ind. Pharm.* **1997**, *23*, 981–985. [CrossRef]
14. Zhou, W.-Q.; Gu, T.-Y.; Su, Z.-G.; Ma, G.-H. Synthesis of macroporous poly(styrene-divinyl benzene) microspheres by surfactant reverse micelles swelling method. *Polymer* **2007**, *48*, 1981–1988. [CrossRef]
15. Spanka, C.; Clapham, B.; Janda, K.D. Preparation of new microgel polymers and their application as supports in organic synthesis. *J. Org. Chem.* **2002**, *67*, 3045–3050. [CrossRef]
16. Guerzoni, L.P.B.; Bohl, J.; Jans, A.; Rose, J.C.; Koehler, J.; Kuehne, A.J.C.; de Laporte, L. Microfluidic fabrication of polyethylene glycol microgel capsules with tailored properties for the delivery of biomolecules. *Biomater. Sci.* **2017**, *5*, 1549–1557. [CrossRef]
17. Uskokovic, V.; Drofenik, M. Sysnthesis of materials within reverse micelles. *Surf. Rev. Lett.* **2005**, *12*, 239–277. [CrossRef]
18. Fulton, J.L.; Smith, R.D. Reverse micelle and microemulsion phases in supercritcal fluids. *J. Phys. Chem.* **1988**, *92*, 2903–2907. [CrossRef]
19. Zheng, Y.; Zheng, M.; Ma, Z.; Xin, B.; Guo, R.; Xu, X. *Sugar Fatty Acid Esters*, 1st ed.; Elsevier: Saint Louis, IL, USA, 2015; pp. 215–243.
20. Rhein, L. *Surfactant Action on Skin and Hair: Cleansing and Skin Reactivity Mechanisms*; Elsevier Science: Amsterdam, The Netherlands, 2007; pp. 305–369.
21. Zhang, Y.; Xue, C.; Xue, Y.; Gao, R.; Zhang, X. Determination of the degree of deacetylation of chitin and chitosan by X-ray powder diffraction. *Carbohydr. Res.* **2005**, *340*, 1914–1917. [CrossRef]
22. Kasaai, M.R. A review of several reported procedures to determine the degree of N-acetylation for chitin and chitosan using infrared spectroscopy. *Carbohydr. Polym.* **2008**, *71*, 497–508. [CrossRef]
23. Cárdenas, G.; Cabrera, G.; Taboada, E.; Miranda, S.P. Chitin characterization by SEM, FTIR, XRD, and 13C cross polarization/mass angle spinning NMR. *J. Appl. Polym. Sci.* **2004**, *93*, 1876–1885. [CrossRef]
24. Pereira, A.G.B.; Muniz, E.C.; Hsieh, Y.-L. ^{1}H NMR and ^{1}H–^{13}C HSQC surface characterization of chitosan-chitin sheath-core nanowhiskers. *Carbohydr. Polym.* **2015**, *123*, 46–52. [CrossRef]
25. Wu, Y.; Sasaki, T.; Irie, S.; Sakurai, K. A novel biomass-ionic liquid platform for the utilization of native chitin. *Polymer* **2008**, *49*, 2321–2327. [CrossRef]
26. Jang, M.-K.; Kong, B.-G.; Jeong, Y.-I.; Lee, C.H.; Nah, J.-W. Physicochemical characterization of α-chitin, β-chitin, and γ-chitin separated from natural resources. *J. Polym. Sci. A* **2004**, *42*, 3423–3432. [CrossRef]
27. Guinesi, L.S.; Cavalheiro, É.T.G. The use of DSC curves to determine the acetylation degree of chitin/chitosan samples. *Thermochim. Acta* **2006**, *444*, 128–133. [CrossRef]
28. Alothman, A.Z. A Review: Fundamental aspects of silicate mesoporous materials. *Materials* **2012**, *5*, 2874–2902. [CrossRef]
29. Pereira, A.G.B.; Muniz, E.C.; Hsieh, Y.-L. Chitosan-sheath and chitin-core nanowhiskers. *Carbohydr. Polym.* **2014**, *107*, 158–166. [CrossRef]

30. Alabaraoye, E.; Achilonu, M.; Hester, R. Biopolymer (Chitin) from various marine seashell wastes: Isolation and Characterization. *J. Polym. Environ.* **2017**, *26*, 1–12. [CrossRef]
31. Kumirska, J.; Weinhold, M.X.; Thöming, J.; Stepnowski, P. Biomedical activity of chitin/chitosan based materials—influence of physicochemical properties apart from molecular weight and degree of n-acetylation. *Polymers* **2011**, *3*, 1875–1901. [CrossRef]

materials

MDPI

Article

Chitosan Nanoparticles Rescue Rotenone-Mediated Cell Death

Jyoti Ahlawat [1], Eva M. Deemer [2] and Mahesh Narayan [1,*]

1 Department of Chemistry & Biochemistry, The University of Texas at El Paso, El Paso, TX 79968, USA; jahlawat@miners.utep.edu
2 Material Science & Engineering department, The University of Texas at El Paso, El Paso, TX 79968, USA; emdeemer@utep.edu
* Correspondence: mnarayan@utep.edu; Tel.: +1-(915)-747-6614; Fax: +1-(915)-7478383

Received: 13 March 2019; Accepted: 4 April 2019; Published: 11 April 2019

Abstract: The aim of the present investigation was to study the anti-oxidant effect of chitosan nanoparticles on a human SH-SY5Y neuroblastoma cell line using a rotenone model to generate reactive oxygen species. Chitosan nanoparticles were synthesized using an ionotropic gelation method. The obtained nanoparticles were characterized using various analytical techniques such as Dynamic Light Scattering, Scanning Electron Microscopy, Transmission Electron Microscopy, Fourier Transmission Infrared spectroscopy and Atomic Force Microscopy. Incubation of SH-SY5Y cells with 50 μM rotenone resulted in 35–50% cell death within 24 h of incubation time. Annexin V/Propidium iodide dual staining verified that the majority of neuronal cell death occurred via the apoptotic pathway. The incubation of cells with chitosan nanoparticles reduced rotenone-initiated cytotoxicity and apoptotic cell death. Given that rotenone insult to cells causes oxidative stress, our results suggest that Chitosan nanoparticles have antioxidant and anti-apoptotic properties. Chitosan can not only serve as a novel therapeutic drug in the near future but also as a carrier for combo-therapy.

Keywords: chitosan (CS); anti-oxidant; anti-apoptotic activity; rotenone; Parkinson's disease (PD)

1. Introduction

Parkinson's disease (PD) is a multifocal progressive neurodegenerative disorder clinically defined by the presence of akinesia, postural instability, muscular rigidity, and tremor [1]. It is the second most common neurodegenerative disease and is prevalent in 0.1–0.3% population with an increased frequency observed in patients ≥65 years [2]. Interestingly, PD patients often display non-motor signs and symptoms such as sleep disturbances, mood deflection, anosmia, gastrointestinal dysfunction (e.g., 80% of patients suffer from constipation), sexual-urinary dysfunction, thermoregulation changes, neuropsychiatric problems, cardiovascular disturbances, and pain [1,3]. PD is characterized by the selective loss of dopaminergic neurons in the substantia nigra pars compacta and arises due to the deposition of insoluble polymers of α-synuclein in the neurons, forming spherical intracytoplasmic inclusions known as Lewy bodies [1]. These lamellated cytoplasmic bodies eventually result in neurodegeneration and the death of dopaminergic neurons [1]. This neuronal death is associated with disruption in cellular hemostasis, resulting in disruption of the nuclear membrane integrity, signaling α-synuclein aggregation which later propagates to other neurons by direct or indirect means [4]. Furthermore, studies have shown that α-synuclein aggregation impairs axonal transport and exerts a detrimental effect on the health of neurons due to the activation of neighboring inflammatory microglial cells [3].

Parkinson's disease can be either inherited or sporadic in nature. Although, familial PD accounts for around 10% of the cases, sporadic PD has been found to account for the remaining ones. Moreover, the etiology of PD is not completely understood but sporadic PD is believed to originate from interaction of individual genetic susceptibility and environmental exposure [5,6]. Probably, there is not one single factor that is solely responsible for causing the disease. Rather, there exists several factors acting simultaneously [7]. Previous studies have suggested that pesticides such as rotenone are involved in the increased risk of Parkinson's disease [5]. In addition to being a pesticide, it is a potent, highly specific inhibitor of complex I of the mitochondrial electron transport chain (ETC) [6]. Unlike N-methyl-4-phenyl-1,2,3,6-tetrahydropyridine (MPTP), which causes defects in complex I of ETC in catecholaminergic neurons, rotenone causes complex I inhibition, uniformly, across the brain [8]. Its hydrophobic structure allows easy penetration through the blood–brain barrier and cell membrane. The selective toxicity of this lipophilic compound is relevant because of its wide usage as a herbicide in gardens and as a delousing agent for animals and humans [7]. Furthermore, studies have shown degradation of selective nigrostriatal dopaminergic neurons upon rotenone infusion, reproducing pathological features of clinical Parkinson's disease [6]. Moreover, a study by Niyanyu et al. showed that rotenone can induce mitochondrial reactive oxygen species (ROS) production which is closely related to rotenone-induced apoptosis [9].

Since no viable treatment exists for PD, there is an urgent and unmet need for the development of novel therapeutic agents to either cease or reverse the symptoms or the progression of this progressive age-related disorder [10]. Synthetic compounds are associated with various side-effects [11]. Therefore, there is a need to find some natural neuroprotective agent that has the ability to scavenge ROS and hence defer the progression of Parkinson's disease [10]. Chitosan is a cationic polysaccharide, composed of a linear chain of D-glucosamine and N-acetyl-D-glucosamine linked via a β (1,4) bond, obtained from an alkaline N-deacetylation of chitin [12]. This marine shrimp-derived carbohydrate possesses well-documented antioxidant properties with minimal or no side-effects observed. Moreover, they also exhibit neuroprotective, anti-hemorrhagic, anti-tumor, anti-diabetic, anti-viral, and antibacterial effects. Furthermore, it has mucoadhesive properties allowing easy penetration of this carbohydrate through the well-organized epithelia [13]. In a study, in 2001, Gilgun-Sherki et al. showed that elevated ROS production and an imbalance between pro-oxidant and antioxidant activity (e.g., superoxide dismutase, catalase, and glutathione peroxidase enzyme) leads to neuronal death and hence a diseased condition [14]. In a different study, Guo et al. in 2006 reported that fucoidan (sulfated polysaccharide) could reverse changes such as superoxide dismutase activity and alleviate the reactive oxygen species level in PC 12 cells when exposed to hydrogen peroxide [15]. Later, Gao et al. (2012) showed the antioxidant effect of fucoidan on hydrogen-peroxide-treated PC12 cells and the pathway associated with it [16]. In a different study, Xie et al. (2014) reported antioxidants could alleviate the reactive oxygen species level [17]. Further, in 2016, Wang et al. reported fucoidan pretreatment could rescue the cells from oxidative stress, protein carbonyl lipid peroxidation, and mitochondrial dysfunction [18,19]. Related to this, Liu et al. (2016) showed the effect of sulfonated chitosan on the differentiation of neuronal cells and exhibited immunomodulatory effects [20]. Recently, Magnigandan et al. (2018) reported the anti-oxidant and ROS scavenging activity of low molecular weight sulphonated chitosan, where they found that rotenone insult resulted in antioxidant depletion and lipid oxidation causing cellular damage, oxidative stress, mitochondrial dysfunction, and hence, a diseased state which were reversed by the action of low molecular weight sulfated chitosan [11].

However, there are no reports on the in vitro neuroprotective effects of bare chitosan nanoparticles. Therefore, the goal of this study was to exploit the antioxidant and anti-apoptotic activity of the prepared chitosan nanoparticles for evaluation in vitro, against a human SH-SY5Y neuroblastoma cell line. Hence, rotenone was used as the causative agent for inducing PD in the SH-SY5Y cell line and then, the therapeutic neuroprotective efficacy of the synthesized chitosan nanoparticles was evaluated. Therefore, we propose that chitosan nanoparticles might be a potential candidate for the prevention of PD.

2. Materials and Methods

2.1. Materials

Chitosan (>75% deacetylated), and Sodium tripolyphosphate (Na-TPP) were purchased from Sigma-Aldrich (Saint Louis, MO, USA). Whereas, Dimethyl sulfoxide and acetic acid were ordered from Fisher Chemical (Hampton, NH, USA).

2.2. Preparation of Chitosan Nanoparticles Using Ionotropic Gelation Method

Chitosan nanoparticles were synthesized using the Calvo et al. 1997 method [21]. In this, 0.175% (w/v) chitosan powder was dissolved in 1% (v/v) acetic acid and kept on a magnetic stirrer for overnight stirring at a temperature between 25–28 °C. Later, the pH of the solution was adjusted to 5.2 using 1M NaOH followed by addition of 0.1% (w/v) sodium tripolyphosphate in a dropwise fashion. The chitosan solution was then stirred at 1000 rpm for 10 min. The solution was centrifuged at 20,000 rpm (Sorvall RC-5B refrigerated centrifuge, Fisher Scientific, Hampton, NH, USA) for 90 min at a temperature of 4 °C to pelletize the chitosan nanoparticles. After centrifugation, the supernatant was discarded and the pellet was washed with deionized water three times. Ultrasonication was performed using a probe sonicator (Branson Sonifier 450, Emerson Electric Company, St. Louis, MO, USA) in an ice bath for 10 min at an amplitude of 30%. The obtained suspension was then immediately freeze dried using lyophilizer (Labconco, Kansas City, MO, USA). After freeze-drying, the samples were stored at 4 °C to carry out further analysis.

2.3. Characterization of Nanoparticles

2.3.1. Determination of Average Size, Polydispersity Index, and Zeta Potential Using Dynamic Light Scattering

The average size, polydispersity index, and surface charge of the nanoparticles was determined using DLS (Dynamic light scattering) at 25 °C. A total of 1 mL CS NP solution was diluted ten times in deionized water and the sample was analyzed using a Malvern Zetasizer ZS90 (Malvern Panalytical Ltd., Malvern, UK).

2.3.2. Scanning Electron Microscopy (SEM)

The surface morphology of the prepared chitosan nanoparticles was analyzed using the S-3400N Type II scanning electron microscope, Pleasanton, USA (Hitachi High-Technologies Corporation, Tokyo, Japan). The liquid samples on a glass slide were dried overnight in a sterilized fume hood and later, the glass slide was mounted on a stainless stub using double-sided carbon tape. The sample was then coated with gold using a gold sputterer for 30 s to make the sample conductive. The images were recorded at an accelerating voltage of 2 kV and probe current below 20 μA.

2.3.3. Transmission Electron Microscopy (TEM)

The morphology of the nanoparticles was studied using the H-7650 transmission electron microscope, manufactured by Hitachi High-Technologies Corporation in Pleasanton, CA, USA. A single drop of the CS Nanoparticle dispersion was placed on carbon 400 mesh copper grid (Ted Pella, Redding, CA, USA). The copper grids were then dried overnight in a sterile fume hood and the images were taken using the Quartz PCI version 8 software (Quartz Imaging Company, Vancouver, BC, Canada) in TEM mode (200 kV).

2.3.4. Atomic Force Microscopy (AFM)

The surface roughness and morphology of the chitosan nanoparticles was further investigated using AFM (NT-MDT NTEGRA) in non-contact mode using Ted Pella TAP 150AL-G tip (Redding, CA,

USA) with a radius of <10 nm. After capturing the images, the data analysis was performed using NOVA software (NOVA Company, ChongQing, China).

2.3.5. Fourier Transform Infrared Spectroscopy (FTIR)

The IR spectra of the Chitosan nanoparticle sample was obtained using a Nicolet, Thermo Scientific FTIR instrument (Waltham, MA USA). The sample was ground to fine powder along with a KBr pellet. The scanning range was from 500–4000 cm^{-1}. The data were analyzed using OMNIC software (Fisher Scientific, Hampton, NH, USA).

2.4. *Cellular Behavior of Human SH-SY5Y Neuroblastoma Cell Lines on Treatment with Samples*

2.4.1. Cell Culture

Annexin V- FITC apoptosis kit (Beckman Coutler, Brea, CA, USA), Hoechst 33342 fluorescent satin (Invitrogen, Carlsbad, CA, USA), propidium iodide (PI) (Invitrogen, Eugene, OR, USA), Fetal Bovine Serum (FBS) (Atlanta Biologicals, Atlanta, GA, USA), and a human neuroblastoma cell line SH-SY5Y (ATCC, Manassas, VA, USA) were purchased. SH-SY5Y cells were cultured in DMEM and Hans's F12 media mixture (1:1) comprising of 10% FBS (v/v) supplemented with 1% (v/v) penicillin-streptomycin and maintained at 37 °C in an incubator with 5% CO2 atmosphere. Cells were sub-cultured every 48 h and Trypsin-EDTA 0.25% (1×) was used to detach cells from the culture surface when needed.

2.4.2. Differential Nuclear Staining Cytotoxicity Assay

The cytotoxicity of different concentrations of chitosan nanoparticles, chitosan powder, and rotenone were evaluated. Cells were first cultured on 96-well plates and incubated for 24 h to allow attachment to the culture surface. Later, cells were treated with a different concentration of rotenone and chitosan nanoparticles to determine the possible cytotoxic effect of the added treatments. Subsequently, untreated cells were taken as a negative control and hydrogen-peroxide-treated cells were taken as a positive control. Moreover, to determine the cytotoxic effect of rotenone on the cells, the cells were treated with chitosan nanoparticles (different concentrations) 6 h prior to rotenone exposure. Subsequently, cells were further incubated for 24 h. Later, 1 μg/mL mixture of PI/Hoechst 33342 was added to each well in the 96-well plates 1 h prior to the imaging process [22]. The images were captured using a Bioimager system (BD Biosciences Rockville, Montgomery, MD, USA). Five images were taken per well using a 10× objective lens. Subsequently, BD AttoVision v1.6.2 software (BD Biosciences Rockville, Montgomery, MD, USA) was used to determine the percentage cell death per well.

2.4.3. Flow Cytometric Assay

The SH-SY5Y cell lines were seeded at a density of 20,000 cells per well in a 24-well plate and incubated for 24 h to allow attachment of the cells to the culture surface. Cells were then incubated with various concentrations of chitosan nanoparticles prior to rotenone exposure and subsequently, cells were incubated for an additional 24 h. Cells from each well were then collected, washed, and processed [23]. Briefly, cells were concurrently stained by suspending them in a solution containing annexin V-FITC (PI) in 100 μL of binding buffer (Beckman Coulter, Brea, CA, USA). After incubation for a time interval of 15 min on ice, in a sterilized Lab Safety Cabinet II in dark, 400 μL of ice-cold binding buffer was added to the cells. The resulting suspension was then homogenized gently and subsequently, analyzed using a Cytomics FC 500 Beckman Coulter Flow cytometer. For each sample, approximately 10,000 events were captured and data analysis was performed using Beckman Coulter CXP software.

2.4.4. Mitochondrial Membrane Potential (ΔΨm)

The loss of mitochondrial membrane potential is a hallmark for cellular apoptosis. Briefly, SH-SY5Y cells were seeded onto a 96-well plate for 12 h. Subsequently, cells were treated with chitosan nanoparticles 6h prior to rotenone treatment. Later, after 24 h of incubation, the cells were incubated with rhodamine 123 dye at 37 °C for 30 min. Finally, the mitochondrial membrane potential was evaluated quantitatively using a Bioimager system (BD Biosciences Rockville, Montgomery, MD, USA) and the fluorescence intensity was measured at 485/530 nm.

2.5. Statistical Significance

The experimental data were expressed as the mean ± standard deviation of one or more individual experiments wherever applicable. The analysis of experimental data was performed with the students t-test using Graph Pad Prism 6.0 (San Diego, CA, USA) and statistically significant values were indicated as $p < 0.05$.

3. Results and Discussion

3.1. Dynamic Light Scattering

Particle size and surface charge are two important factors determining the size, stability, and effective delivery of the drug to the target site [24]. Figure 1A depicts the average particle size distribution of the synthesized CS Np using an ionotropic gelation method. As can be observed, the bare nanoparticles have an average size of 197.8 ± 49.18 nm with a PDI 0.244.

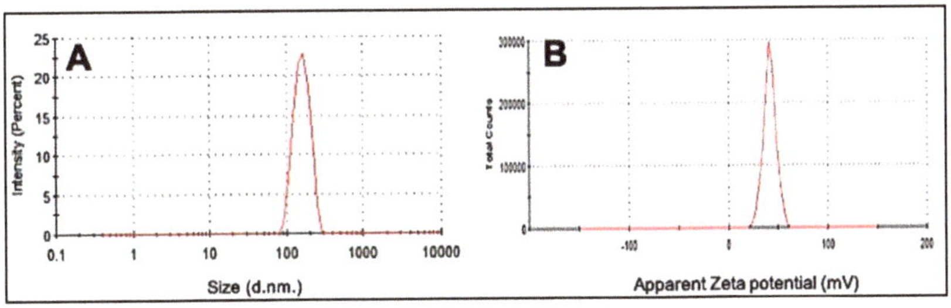

Figure 1. Dynamic light scattering (DLS) images of (**A**) average particle size distribution of CS Nanoparticle and (**B**) Zeta potential of synthesized CS Nanoparticle.

The zeta potential is a measure of the stability and surface charge on the particles. A high zeta potential is indicative of high electric surface charge allowing strong repulsion between the surrounding particles and hence preventing aggregation [24]. Figure 1B depicts the average zeta potential value of CS Np which was observed to be +36.0 ± 4.68. A previous study on chitosan nanoparticles showed that the blank nanoparticles in the ratio 5:1 CS/TPP had a particle size in the range of 300 to 390 nm and a zeta potential of + 44 ± 5.2, which strongly supports our findings [25].

3.2. Morphological Characterization

Figure 2 displays the scanning electron micrograph images of chitosan powder, freshly prepared nanoparticles, and lyophilized chitosan nanoparticles sample. It was observed that the nanoparticles displayed spherically compact structures with an average diameter of 220 ± 40 nm, see Figure 2E, which almost coincides with our dynamic light scattering study. Similar results were reported in another study where the CS/TPP in a 5:1 ratio displayed an average size of 200 ± 24 nm [26,27]. The freshly prepared CS Np appeared as clusters. This can be attributed to the fusion of particles through hydrogen bonding [24]. Moreover, the stability of chitosan nanoparticles was tested in buffer solution (pH 7.4). The morphology of the nanoparticles was observed to be spherical with an average diameter of 200 nm, see Supplementary Figure S1.

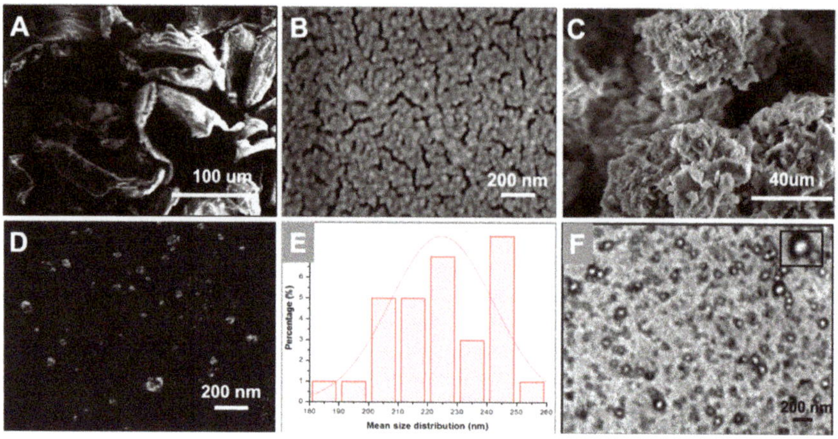

Figure 2. Scanning Electron Microscopy images of (**A**) chitosan powder; (**B**) freshly prepared CS Np and (**C**,**D**) lyophilized CS Np. Further, (**E**) depicts the average size distribution of the nanoparticles obtained from scanning electron microscopy. Whereas, (**F**) depicts the transmission electron microscopic image of chitosan nanoparticles.

Transmission electron microscopy is a technique which provides information on the morphology and size of the particles. Figure 3 depicts a TEM micrograph of small and spherical chitosan nanoparticles with a diameter of around 120 $\pm$ 30 nm. The size of the nanoparticles in the TEM image, see Figure 2F, are smaller than that represented by SEM and DLS. This can be attributed to the aggregation of nanoparticles due to their high surface area and energy which generates a larger entity [28]. However, dimensions of freshly prepared single particles can be observed clearly in the TEM image of the CS Np sample.

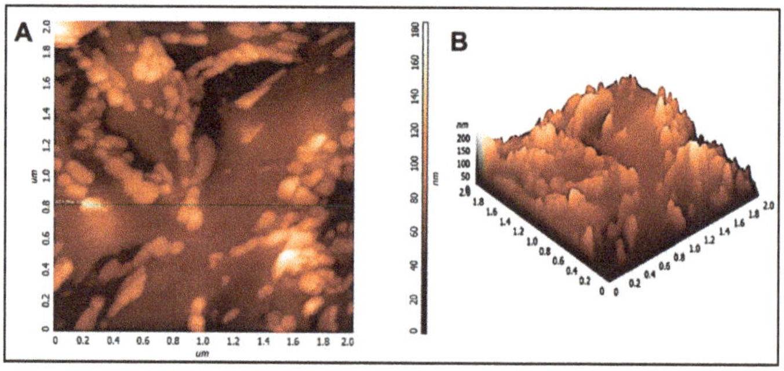

Figure 3. (**A**) 2D and (**B**) 3D atomic force micrograph of chitosan nanoparticles in a 2-micron $\times$ 2-micron area.

3.3. *AFM Analysis*

Scanning electron microscopy and transmission electron microscopy provides a two-dimensional projection or image of the nanoparticles. However, atomic force microscopy is a powerful technique which enables a three-dimensional surface profile of the nanoparticles to be viewed. In addition, it can also provide accurate heights of the nanoparticles. Figure 3 shows the atomic force micrograph of chitosan nanoparticles. As can be observed, the nanoparticles appeared to be spherical in shape. Furthermore, the size distribution histogram was performed for the nano-chitosan in a 2-micron $\times$ 2-micron area. Analysis shows that the average particle size is around 200 nm as the histogram peaks are at 0.2 microns, see Supplementary Figure S2, which coincides with our DLS, SEM, and TEM data.

Similar results were observed in previous reports [29–31]. In addition, a small peak appeared at 600 nm which can be attributed to the fusion of particles through hydrogen bonding [24].

3.4. Spectroscopic Characterization

Fourier-transform infrared spectroscopy (FTIR) is a powerful spectroscopic technique to determine the chemical composition and presence of a drug inside the nanoparticles. Figure 4 represents the FTIR spectra of chitosan powder and chitosan nanoparticles. In the FTIR spectrum of chitosan nanoparticles, a broad peak at 3436 cm^{-1} corresponds to the -NH and -OH stretching vibrations. The weak band at 2930 cm^{-1} corresponds to –CH stretching, whereas vibrational bands at 1640 cm^{-1}, 1560 cm^{-1}, and 1320 cm^{-1} may be attributed to the amide carbonyl stretch and -NH bend of the amine groups in the chitosan nanoparticles [32]. The band at 1099 cm^{-1} represents C–O bond stretching. Moreover, the vibrational band at 860 cm^{-1} corresponds to the CH_2OH group in the pyranose ring of chitosan. The only difference between the spectra of chitosan powder and chitosan nanoparticles occurred at 1155 cm^{-1}, which could be assigned to the linkage between phosphate groups of the sodium tripolyphosphate and ammonium ions of the chitosan group. Furthermore, our results agree with report of Gopalakrishnan et al. (2014) and Jafary et al. (2016) [24,32].

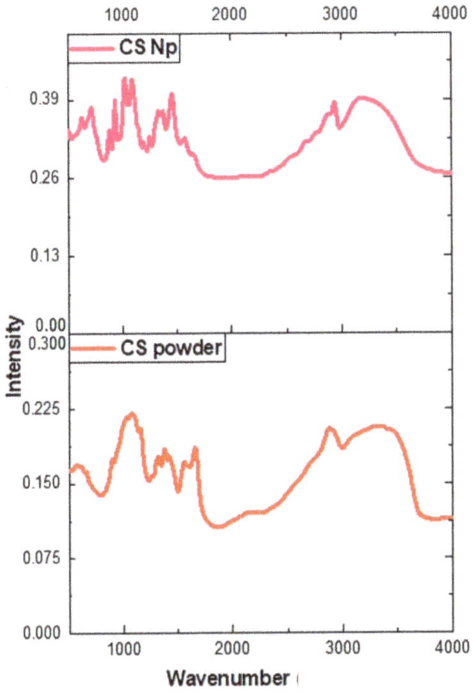

Figure 4. Fourier-transform infrared spectroscopy (FTIR) spectra of chitosan nanoparticles and chitosan powder.

3.5. Effect of Chitosan Nanoparticles on Rotenone-Induced Cell Death

The cytotoxicity of chitosan nanoparticles and its protective function against rotenone insult in SH-SY5Y cell line were checked using a high-throughput screening assay. Figure 5i shows cells exposed to an increasing concentration (1–20 μM) of chitosan nanoparticles exhibited cytotoxicity from 9% to 21%. A 10 μM concentration did not display much difference compared to the untreated and vehicle controls. However, on addition of 50 μM rotenone, around 35–50% of the cell death was reported in the SH-SY5Y cell line after 24 h of incubation time, see Figure 5ii. In contrast, the addition of a 10 μM chitosan nanoparticle solution prior to rotenone exposure resulted in 14–20% cell death compared to 35–50% cell death upon rotenone administration. These data show that treatment of cells with chitosan prior to rotenone exposure attenuated cell death by 25–30%. The morphology of cells after rotenone treatment and cells treated with chitosan prior to rotenone exposure further displays the protective aspect of chitosan nanoparticles against rotenone insult. Bright field microscopy images and Hoechst 33342-propidium iodide staining pictures, see Figure 5iii, further support the protective aspect of chitosan nanoparticles. The pretreatment of cells with a 10 μM chitosan nanoparticles solution for 6 h enabled the cells to retain their cellular morphology, as depicted by bright field microscopy images. In contrast, cells treated with 50 μM rotenone showed a morphology similar to that of cells exposed to 50 μM hydrogen peroxide, see Figure 5iv. Hydrogen peroxide was used as a positive control at a concentration of 50 μM.

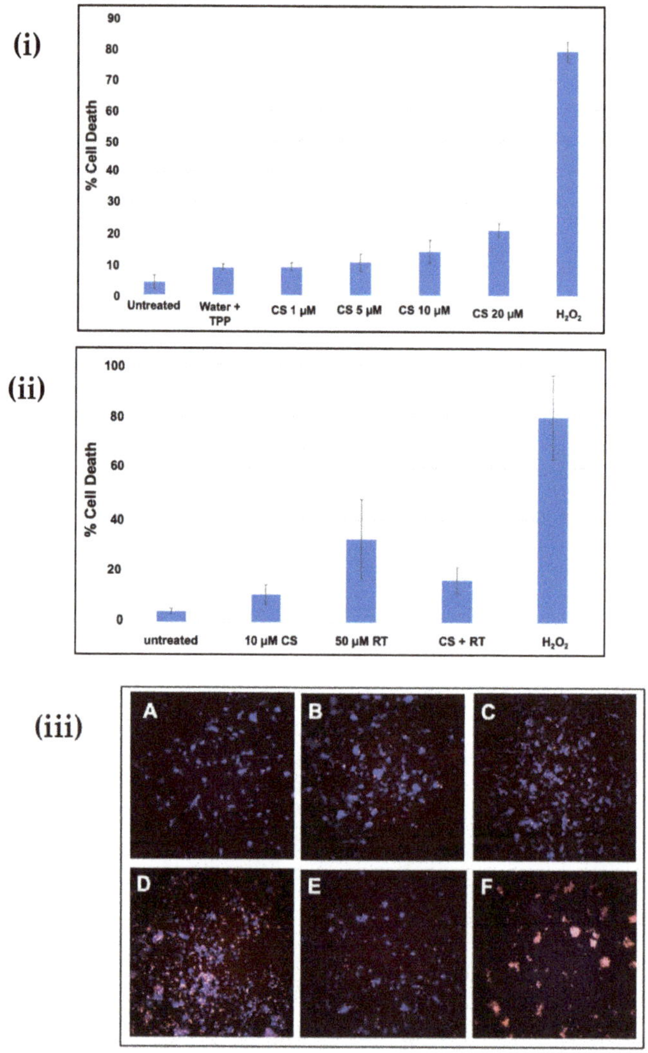

Figure 5. *Cont.*

(iv)

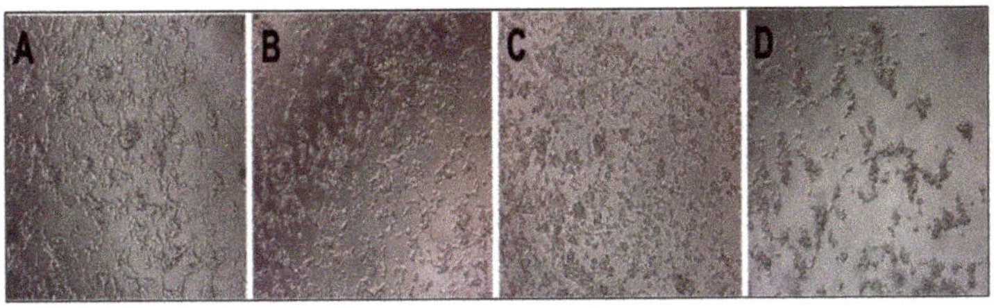

Figure 5. (**i**) Cytotoxicity of chitosan nanoparticles at different concentrations after 24 h of incubation; (**ii**) Protective effect of chitosan nanoparticles against rotenone insult in SH-SY5Y cells; (**iii**) Hoechst-propidium iodide staining images of (**a**) untreated cells, (**b**) water + Tripoly phosphate; (**c**) chitosan nanoparticles 10 µM concentration, (**d**) 50 µM rotenone, (**e**) CS Np + Rotenone and (**f**) hydrogen-peroxide-treated cells after 24 h of treatment; (**iv**) Bright field microscopy images of (**a**) untreated cells, (**b**) CS Np 10 µM, (**c**) CS Np + RT, & (**d**) 50 µM RT cells visualized using a compound microscope after 24 h of treatment. Each experimental point was assessed in quintuplicate.

3.6. Flow Cytometry Analysis

The mechanism by which cell-death occurs (apoptotic pathway or necrotic pathway) was examined. Figure 6i,ii shows cells pre-treated with 20 µM chitosan powder, 10 µM chitosan nanoparticles, and 20 µM chitosan nanoparticles. A concentration of 50 µM hydrogen peroxide was used as a positive control whereas untreated cells were used as a negative control. As can be observed, cells treated with 50 µM rotenone showed a substantial increase in early and late apoptosis (lower right and top right quadrants of the plot) 24 h after administration. A similar result was observed when cells were treated with 50 µM hydrogen peroxide. However, prior treatment of cells with 20 µM chitosan powder and 10 µM and 20 µM chitosan nanoparticles resulted in a notable rescue of the cells from rotenone insult (apoptotic-cell death). A notable difference was observed when 50 µM rotenone and CS Np 10 µM + rotenone treatments were compared. Thus, we can conclude that chitosan nanoparticles exhibit anti-apoptotic activity.

(i)

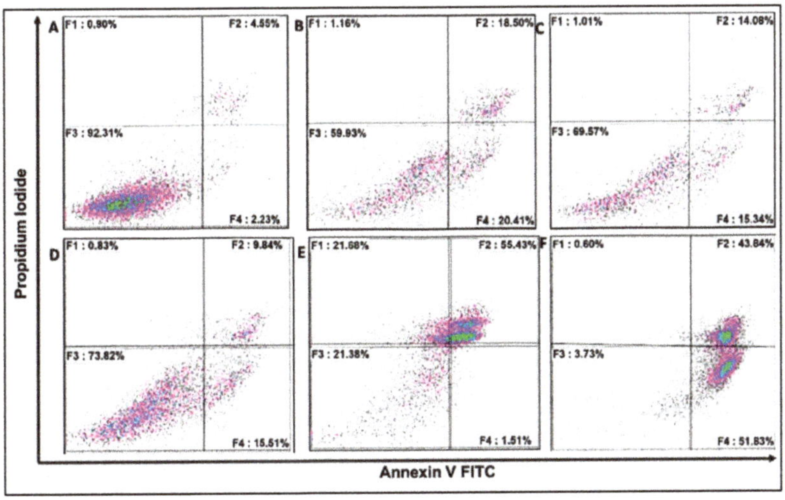

Figure 6. *Cont.*

(ii)

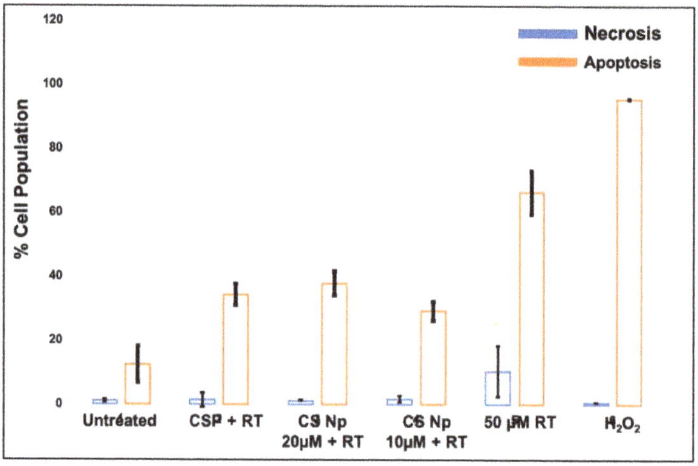

Figure 6. (**i**) Flow cytometry matrix plots used to measure apoptosis/necrotic distribution for (**A**) untreated, (**B**) 20 μM Chitosan powder + RT, (**C**) CS Np 20 μM, (**D**) CS Np+ RT, (**E**) 50 μM RT, and (F) hydrogen peroxide after 24 h of treatment; (**ii**) quantification of apoptotic/necrotic assay under the previously mentioned conditions. Each experimental point was assessed in triplicate.

3.7. Chitosan Np Prevents Rotenone-Induced Mitochondrial Dysfunction

The untreated cells, see Figure 7(iA), and cells treated with chitosan nanoparticles, see Figure 7(iB), exhibited green fluorescence, indicating that a large fraction of mitochondria inside the cell were in an energized state. However, cells upon treatment with rotenone, see Figure 7(iD), displayed decreased mitochondrial energy transduction as observed by the disappearance of green fluorescence. Further, cells treated with chitosan nanoparticles prior to rotenone treatment, see Figure 7(iC), exhibited green fluorescence which can be attributed to the inhibition of the collapse of the membrane potential via rescue of mitochondrial membrane depolarization by chitosan nanoparticles. Thus, these results suggest that chitosan nanoparticles may inhibit rotenone-induced cellular apoptosis through a mitochondria-involved pathway, as revealed by the considerable increase in green fluorescence intensity.

(i)

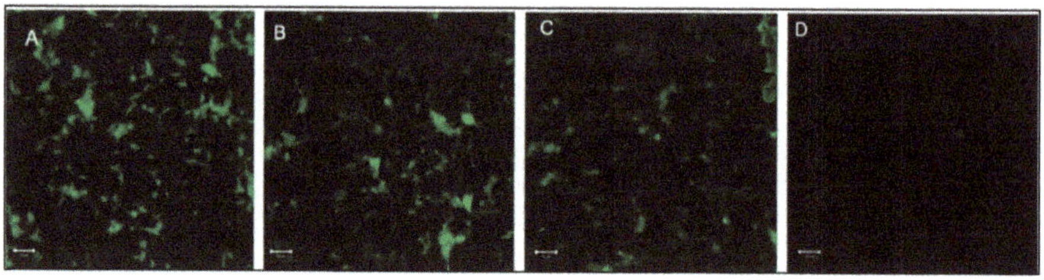

Figure 7. *Cont.*

(ii)

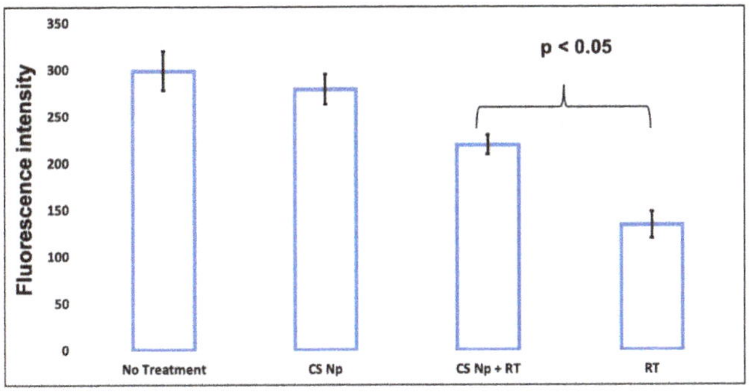

Figure 7. The effect of CS Np on rotenone insult determined using rhodamine 123 dye. (**i**) Cells were incubated with rhodamine 123 and the mitochondrial membrane potential was monitored using a fluorescent microscope for (**A**) untreated, (**B**) CS Np 20 μM, (**C**) 20 μM Chitosan powder + RT, and (**D**) 50 μM RT; (**ii**) quantification of Mitochondrial membrane potential assay under previously described conditions. Each experimental point was assessed in triplicate. The scale bar in the image is 50 μm.

4. Proposed mechanism

We henceforth propose a mechanistic model for the action of chitosan nanoparticles (Figure 8). Firstly, chitosan nanoparticles are internalized by the cells through endocytosis via an endosome–lysosome pathway. Later, these nanoparticles are released into the cytoplasm from the lysosome due to the low pH environment and enzymatic activity. These nanoparticles then scavenge the reactive oxygen species generated by the inhibition of complex I of the ETC of mitochondria by the rotenone insult. Thus, the apoptotic stimuli exerted on mitochondria is reduced. Subsequently, cytochrome c cannot be released through the permeability transition pore. Further, caspase 3 cannot be activated by caspase 9, resulting in inhibition of Poly ADP ribose polymerase (PARP) cleavage and hence, cell survival [22,23]. Therefore, chitosan nanoparticles protect the cell from rotenone insult due to its anti-oxidant and anti-apoptotic activity.

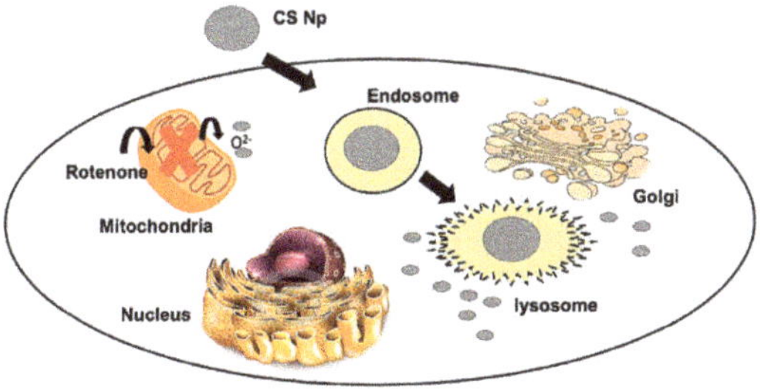

Figure 8. Mechanistic model of the action of chitosan nanoparticles.

5. Conclusions

The literature is flooded with various number of reports suggesting that oxidative stress and consequent cellular apoptosis are important factors in rotenone-induced cell death. Therefore, pre-treatment of cells with antioxidants such as low molecular weight sulphonated chitosan, fucoidan, quercetin, flavonoids, etc., likely reduces the adverse effects arising from rotenone insult. Although pre-treatment of cells with the above mentioned anti-oxidants were observed to alleviate reactive oxygen species production, the antioxidant effects of chitosan nanoparticulate system against rotenone insult had not been studied previously. In this regard, our study demonstrates the synthesis of chitosan nanoparticles by an ionic gelation method using TPP as the cross-linking agent. The optimized ratio of CS/TPP was 5:1, which produced spherical nanoparticles with an average size of 200 nm as observed by SEM, TEM, DLS, and AFM. Further, the cytotoxicity of the chitosan nanoparticles and its anti-oxidant and anti-apoptotic effects against rotenone-induced cell death were determined using a differential nuclear staining cytotoxicity assay and flow cytometric analysis. Therefore, Chitosan Nanoparticles could inhibit the rotenone-induced mitochondria involved apoptosis pathway, as revealed by the considerable increase in the green fluorescence intensity in the MMP assay. These findings suggest that chitosan nanoparticles might be a useful and promising neuroprotective agent for the prevention of Parkinson's disease.

Supplementary Materials: The following are available online at http://www.mdpi.com/1996-1944/12/7/1176/s1, Figure S1: SEM image of Chitosan nanoparticles in buffer solution, Figure S2: Size distribution analysis from AFM measurement.

Author Contributions: Conceptualization, J.A.; Methodology, J.A.; Software, E.D.; Validation, J.A., and M.N.; Formal Analysis, M.N.; Investigation, M.N.; Resources, M.N.; Data Curation, J.A. and E.D.; Writing-Original Draft Preparation, J.A. and M.N.; Writing-Review & Editing, J.A. and M.N.; Visualization, E.D.; Supervision, M.N.; Project Administration, M.N.; Funding Acquisition, M.N.

Funding: This research was funded by NIH grant number 1SC3 GM111200 01A1.

Acknowledgments: M.N. and J.A. are thankful to staff of cytometry, screening and imaging core facility of the Border Biomedical Research Center at the University of Texas at El Paso (UTEP). J.A. would also like to thank Ms. Lois Mendez for her help while performing experiments relating to cell studies.

Conflicts of Interest: Authors declare that there is no conflict of interest.

References

1. Parashar, A.; Udayabanu, M. Gut microbiota: Implications in Parkinson's disease. *Parkinsonism Relat. Disord.* **2017**, *38*, 1–7. [CrossRef] [PubMed]
2. Fil, A.; Cano-de-la-Cuerda, R.; Muñoz-Hellín, E.; Vela, L.; Ramiro-González, M.; Fernández-de-las-Peñas, C. Pain in Parkinson disease: A review of the literature. *Parkinsonism Relat. Disord.* **2013**, *19*, 285–294. [CrossRef] [PubMed]
3. Comi, C.; Magistrelli, L.; Oggioni, G.D.; Carecchio, M.; Fleetwood, T.; Cantello, R.; Antonini, A. Peripheral nervous system involvement in Parkinson's disease: Evidence and controversies. *Parkinsonism Relat. Disord.* **2014**, *20*, 1329–1334. [CrossRef]
4. Forno, L.S. The neuropathology of Parkinson's disease. In *Progress in Parkinson Research*; Springer: Boston, MA, USA, 1988; pp. 11–21.
5. Sherer, T.B.; Betarbet, R.; Testa, C.M.; Seo, B.B.; Richardson, J.R.; Kim, J.H.; Greenamyre, J.T. Mechanism of toxicity in rotenone models of Parkinson's disease. *J. Neurosci.* **2003**, *23*, 10756–10764. [CrossRef] [PubMed]
6. Cannon, J.R.; Tapias, V.; Na, H.M.; Honick, A.S.; Drolet, R.E.; Greenamyre, J.T. A highly reproducible rotenone model of Parkinson's disease. *Neurobiol. Dis.* **2009**, *34*, 279–290. [CrossRef] [PubMed]
7. Alam, M.; Schmidt, W.J. Rotenone destroys dopaminergic neurons and induces parkinsonian symptoms in rats. *Behav. Brain Res.* **2002**, *136*, 317–324. [CrossRef]
8. Sherer, T.B.; Kim, J.H.; Betarbet, R.; Greenamyre, J.T. Subcutaneous rotenone exposure causes highly selective dopaminergic degeneration and α-synuclein aggregation. *Exp. Neurol.* **2003**, *179*, 9–16. [CrossRef]
9. Li, N.; Ragheb, K.; Lawler, G.; Sturgis, J.; Rajwa, B.; Melendez, J.A.; Robinson, J.P. Mitochondrial complex I inhibitor rotenone induces apoptosis through enhancing mitochondrial reactive oxygen species production. *J. Biol. Chem.* **2003**, *278*, 8516–8525. [CrossRef]
10. Ahlawat, J.; Henriquez, G.; Narayan, M. Enhancing the Delivery of Chemotherapeutics: Role of Biodegradable Polymeric Nanoparticles. *Molecules* **2018**, *23*, 2157. [CrossRef]
11. Manigandan, V.; Nataraj, J.; Karthik, R.; Manivasagam, T.; Saravanan, R.; Thenmozhi, A.J.; Guillemin, G.J. Low Molecular Weight Sulfated Chitosan: Neuroprotective Effect on Rotenone-Induced In Vitro Parkinson's Disease. *Neurotoxicity Res.* **2019**, *35*, 505–515. [CrossRef]
12. Yen, M.T.; Yang, J.H.; Mau, J.L. Antioxidant properties of chitosan from crab shells. *Carbohydr. Polym.* **2008**, *74*, 840–844. [CrossRef]
13. Alonso, D.; Castro, A.; Martinez, A. Marine compounds for the therapeutic treatment of neurological disorders. *Expert Opin. Ther. Pat.* **2005**, *15*, 1377–1386. [CrossRef]

14. Gilgun-Sherki, Y.; Melamed, E.; Offen, D. Oxidative stress induced- neurodegenerative diseases: The need for antioxidants that penetrate the blood brain barrier. *Neuropharmacology* **2001**, *40*, 959–975. [CrossRef]
15. Guo, S.S.; Cui, X.L.; Rausch, W.D. Ganoderma Lucidum polysaccharides protect against MPP+ and rotenone-induced apoptosis in primary dopaminergic cell cultures through inhibiting oxidative stress. *Am. J. Neurodegener. Dis.* **2016**, *5*, 131. [PubMed]
16. Gao, Y.; Dong, C.; Yin, J.; Shen, J.; Tian, J.; Li, C. Neuroprotective effect of fucoidan on H_2O_2-induced apoptosis in PC_{12} cells via activation of PI_3K/Akt pathway. *Cell. Mol. Neurobiol.* **2012**, *32*, 523–529. [CrossRef] [PubMed]
17. Xie, Z.; Ding, S.; Shen, Y. Silibinin activates AMP-activated protein kinase to protect neuronal cells from oxygen and glucose depriva- tion-re-oxygenation. *Biochem. Biophys. Res. Commun.* **2014**, *454*, 313–319. [CrossRef]
18. Wang, J.; Liu, H.; Jin, W.; Zhang, H.; Zhang, Q. Structure–activity relationship of sulfated hetero/galactofucan polysaccharides on do- paminergic neuron. *Int. J. Biol. Macromol.* **2016**, *82*, 878–883. [CrossRef]
19. Wang, T.; Zhu, M.; He, Z.Z. Low-molecular-weight fucoidan attenuates mitochondrial dysfunction and improves neurological out- come after traumatic brain injury in aged mice: Involvement of Sirt3. *Cell. Mol. Neurobiol.* **2016**, *36*, 1257–1268. [CrossRef]
20. Liu, S.; Zhou, J.; Zhang, X.; Liu, Y.; Chen, J.; Hu, B.; Song, J.; Zhang, Y. Strategies to optimize adult stem cell therapy for tissue regeneration. *Int. J. Mol. Sci.* **2016**, *17*, 982. [CrossRef]
21. Calvo, P.; Remunan-Lopez, C.; Vila-Jato, J.L.; Alonso, M.J. Novel hydrophilic chitosan-polyethylene oxide nanoparticles as protein carriers. *J. Appl. Polym. Sci.* **1997**, *63*, 125–132. [CrossRef]
22. Kabiraj, P.; Marin, J.E.; Varela-Ramirez, A.; Zubia, E.; Narayan, M. Ellagic acid mitigates SNO-PDI induced aggregation of Parkinsonian biomarkers. *ACS Chem. Neurosci.* **2014**, *5*, 1209–1220. [CrossRef]
23. Kabiraj, P.; Marin, J.E.; Varela-Ramirez, A.; Narayan, M. An 11-mer Amyloid Beta Peptide Fragment Provokes Chemical Mutations and Parkinsonian Biomarker Aggregation in Dopaminergic Cells: A Novel Road Map for "Transfected" Parkinson's. *ACS Chem. Neurosci.* **2016**, *7*, 1519–1530. [CrossRef]
24. Gopalakrishnan, L.; Ramana, L.N.; Sethuraman, S.; Krishnan, U.M. Ellagic acid encapsulated chitosan nanoparticles as anti-hemorrhagic agent. *Carbohydr. Polym.* **2014**, *111*, 215–221. [CrossRef]
25. Grenha, A.; Seijo, B.; Remunán-López, C. Microencapsulated chitosan nanoparticles for lung protein delivery. *Eur. J. Pharm. Sci.* **2005**, *25*, 427–437. [CrossRef]
26. Arulmozhi, V.; Pandian, K.; Mirunalini, S. Ellagic acid encapsulated chitosan nanoparticles for drug delivery system in human oral cancer cell line (KB). *Colloids Surf. B Biointerfaces* **2013**, *110*, 313–320. [CrossRef]
27. Rampino, A.; Borgogna, M.; Blasi, P.; Bellich, B.; Cesàro, A. Chitosan nanoparticles: Preparation, size evolution and stability. *Int. J. Pharm.* **2013**, *455*, 219–228. [CrossRef]
28. Divya, K.; Jisha, M.S. Chitosan nanoparticles preparation and applications. *Environ. Chem. Lett.* **2018**, *16*, 101–112. [CrossRef]
29. Agudelo, D.; Kreplak, L.; Tajmir-Riahi, H.A. tRNA conjugation with chitosan nanoparticles: An AFM imaging study. *Int. J. Biol. Macromol.* **2016**, *85*, 150–156. [CrossRef]
30. Ghadi, A.; Mahjoub, S.; Tabandeh, F.; Talebnia, F. Synthesis and optimization of chitosan nanoparticles: Potential applications in nanomedicine and biomedical engineering. *Caspian J. Intern. Med.* **2014**, *5*, 156.
31. Yuan, Y.; Tan, J.; Wang, Y.; Qian, C.; Zhang, M. Chitosan nanoparticles as non-viral gene delivery vehicles based on atomic force microscopy study. *Acta Biochim. Biophys. Sin.* **2009**, *41*, 515–526. [CrossRef]
32. Jafary, F.; Panjehpour, M.; Varshosaz, J.; Yaghmaei, P. Stability improvement of immobilized alkaline phosphatase using chitosan nanoparticles. *Braz. J. Chem. Eng.* **2016**, *33*, 243–250. [CrossRef]

materials

MDPI

Article

Development and Characterization of Bacterial Cellulose Reinforced with Natural Rubber

Kornkamol Potivara and Muenduen Phisalaphong *

Department of Chemical Engineering, Faculty of Engineering, Chulalongkorn University, Phayathai Road, Pathumwan, Bangkok 10330, Thailand

* Correspondence: muenduen.p@chula.ac.th; Tel.: +66-2-218-6875

Received: 30 May 2019; Accepted: 17 July 2019; Published: 21 July 2019

Abstract: Films of bacterial cellulose (BC) reinforced by natural rubber (NR) with remarkably high mechanical strength were developed by combining the prominent mechanical properties of multilayer BC nanofibrous structural networks and the high elastic hydrocarbon polymer of NR. BC pellicle was immersed in a diluted NR latex (NRL) suspension in the presence of ethanol aqueous solution. Effects of NRL concentrations (0.5%–10% dry rubber content, DRC) and immersion temperatures (30–70 °C) on the film characteristics were studied. It was revealed that the combination of nanocellulose fibrous networks and NR polymer provided a synergistic effect on the mechanical properties of NR–BC films. In comparison with BC films, the tensile strength and elongation at break of the NR–BC films were considerably improved ~4-fold. The NR–BC films also exhibited improved water resistance over that of BC films and possessed a high resistance to non-polar solvents such as toluene. NR–BC films were biodegradable and could be degraded completely within 5–6 weeks in soil.

Keywords: bacterial cellulose; natural rubber; reinforcing; biodegradable polymers

1. Introduction

Pollution deriving from plastic materials is becoming one of the most prominent environmental concerns of recent years. Accumulation of plastic products in the environment has harmful effects on wildlife and their environment. Plastics dispersed in ocean ecosystems have become a major pollutant that has led to the direct deaths of marine animals. Therefore, developing renewable and biodegradable materials to replace conventional plastic materials is an increasingly important research area to reduce the level of plastic waste.

Thailand is the largest producer and exporter of natural rubber (NR) globally. Global natural rubber production in 2015 was 12.3 million tons, 92% of which was produced in the Asia-Pacific region. Thailand produced around 4.5 million tons and exported about 3.7 million tons in 2015 [1]. According to a recent release from the Association of Natural Rubber Producing Countries, global production of natural rubber was up in 2017 to ~13.3 million metric tons, but global consumption of NR dropped to ~12.9 million tons [2]. Therefore, research and development to expand commercial utilization of NR is required. NR latex (NRL) is a concentrated colloidal suspension produced by rubber trees. NRL is mainly composed of polyisoprene (poly(2-methyl-1,3-butadiene)). NR is a natural polymer of isoprene and is biodegradable. The most important property of NR is elasticity, which is the ability to return to its original shape and size. However, there are properties of NR that need to be improved, for example, hardness, Young's modulus, and abrasion resistance [3]. Various natural fibers have been used as a reinforcement material in NR matrices such as sisal/oil palm hybrid fibers [4], pineapple fibers [5], coconut fibers [6], bamboo fibers [7], and grass fibers [8].

Bacterial cellulose (BC) is nanocellulose produced by bacteria, principally of genera Acetobacter, such as *Acetobacter xylinum*, in the form of interconnected networks of cellulose nanofibers. BC possesses unique properties such as high purity (excluding hemicellulose and lignin), high crystallinity, excellent mechanical strength (very high modulus and tensile strength), excellent biodegradability, high water uptake capacity (up to 100 cc/g), and excellent biological affinity [9]. The unique morphological alignment with Nano nonwoven structure of BC resulted in large surface area as compared with plant cellulose fibers or electrospun cellulose nanofibers [10]. With numerous advantageous characteristics, BC has found use across multiple industries. BC is adopted by the paper industry as an emulsion-stabilizing compound. Furthermore, BC can be applied in the medical area as artificial skin for patients with burns and ulcers [11], as artificial blood vessels [12], for drug delivery, and for tissue engineering and wound healing [13].

Although BC possesses numerous useful properties, one disadvantage is the low-breaking elongation. Conversely, NR is well-known for possessing excellent elastic properties. Reinforcements of NR achieved using graphene, carbon nanotubes, or nanocellulose fibers, such as bacterial cellulose, have been previously reported [14–17]. From our previous study, reinforcement of NR with BC was performed via a latex aqueous micro dispersion process and the films of BC–NR by incorporation of BC fibers into NR matrices demonstrated good water affinity and increased mechanical properties when compared with pure NR matrices [17]. However, the composite film of NR incorporated into BC matrices has not been reported so far. Herein, to obtain films with excellent mechanical properties of BC, characterized by high tensile strength, and NR, characterized by high elasticity, NR–BC films possessing a high degree of mechanical strength were developed by immersing BC pellicles into a diluted NRL suspension. Operational parameters such as temperature and NR concentration were studied to optimize the immersion process. NR–BC films were characterized for their chemical and mechanical properties. To the best of our knowledge, this is the first report of such NR–BC films.

2. Materials and Methods

The *Acetobacter xylinum* (AGR60) was isolated from nata de coco. The stock culture was kindly supplied by Pramote Tammarat, the Institute of Food Research and Product Development, Kasetsart University, Bangkok, Thailand. Natural rubber latex (NRL) with 60% dry rubber content (DRC) was purchased from the Rubber Research Institute of Thailand (RRIT, Bangkok, Thailand). Sucrose and ammonium sulfate were purchased from Ajax Finechem Pty Ltd (New South Wales, Australia). Acetic acid was purchased from Mallinckrodt Chemicals (Paris, KY, USA). Absolute ethanol was purchased from QRec (Chonburi, Thailand).

2.1. Film Preparation

For BC biosynthesis, the medium for the inoculum was coconut-water supplemented with 5.0% (w/v) sucrose, 0.5% (w/v) ammonium sulfate, and 1.0% (w/v) acetic acid. The medium was sterilized at 110 °C for 5 min. Precultures were prepared by a transfer of 50 mL stock culture to 1000 mL in 1500 mL bottle and incubated statically at 30 °C for 7 days. After the surface pellicle was removed, a 5% (v/v) preculture broth was added to sterile medium and statically incubated at 30 °C for five days in a Petri-dish. All sample BC pellicles were purified by washing with deionized water (DI) for 30 min and, then treated with 1% NaOH (w/v) at room temperature to remove bacterial cells for 24 h, followed by a rinse with water until the pH became 7.0. The BC pellicles were soaked in DI water and stored at 4 °C until use.

The procedure for the preparation of BC reinforced with NR (NR–BC) films was developed as follows. NRL was diluted with DI water to form NRL suspension with concentrations of 0.5%–10% dry rubber content (DRC) (expressed as weight per volume). In order to reduce the viscosity of NRL suspension, 6 mL of 50% (v/v) aqueous ethanol solution was slowly added into 300 mL NRL suspension. BC pellicle was then immersed in NRL suspension with concentrations of 0.5%–10% DRC for 48 h as the immersion temperature varied from 30–70 °C. Then, it was washed with DI water, air-dried at 30 °C for 48 h, and stored in plastic film at room temperature. BC was defined as the unmodified BC and NR–BC was defined as the modified BC by immersing in an NRL suspension. The xNR–BCy film was defined as the NR–BC film modified by immersing in an NRL suspension at x% DRC and at an immersion temperature of y °C. For example, 0.5NR–BC50 was the NR–BC film modified by immersing in an NRL suspension at 0.5% DRC and at a temperature of 50 °C.

2.2. Characterization

2.2.1. Field Emission Scanning Electron Microscopy (FESEM)

The examination of the surface morphology was performed by field emission scanning electron microscopy (FESEM). Scanning electron micrographs were taken with JEOL JSM-7610F microscope (Tokyo, Japan). The films were frozen in liquid nitrogen, immediately snapped, and vacuum-dried. Then, the films were sputtered with gold and photographed. The coated specimens were kept in dry place before the analysis. The FE-SEM was obtained at 10 kV, which was considered to be a suitable condition for these samples. The average thickness of the dried BC films and NR–BC films was measured using the ImageJ program.

2.2.2. Laser Particle Size Distribution (PSD)

The particle sizes of rubber in NRL and NRL added with ethanol were investigated by laser diffraction technique. Particle size distribution curves were taken by Mastersizer 3000 (Malvern, UK). The operating size classes were recorded in the range of 0.01–3000 μm at a stirrer speed of 2000 rpm.

2.2.3. Fourier Transform Infrared Spectroscopy (FTIR)

The chemical structures of the films were analyzed and recorded by FTIR with a Nicolet SX-170 FTIR spectrometer (Thermo Fisher Scientific, Waltham, MA, USA) in the region of 4000–500 cm^{-1} at a resolution of 4 cm^{-1}.

2.2.4. Water Absorption Capacity (WAC)

Water absorption capacity (WAC) was determined by immersing the pre-weight of dry films of 20 mm × 20 mm in DI water at room temperature (30 °C) until equilibrium. Then, the films were removed from water and blotted out with Kim wipes. The weights of the hydrate films were then measured, and the procedure was repeated until there was no further weight change. WAC was calculated by the following Equation (1):

$$\text{WAC\%} = \left[\frac{W_h - W_d}{W_d}\right] \times 100, \tag{1}$$

where W_h and W_d denote the weights of hydrate and dry films, respectively.

2.2.5. Toluene Uptake (TU)

Specimens of BC and NR–BC films of 20 mm × 20 mm were weighed and immersed in toluene at room temperature. After that, the specimens were weighed. The procedure was repeated until there was no further weight change. The toluene uptake (TU) was calculated by the following Equation (2):

$$\text{TU\%} = \left[\frac{W_t - W_0}{W_o}\right] \times 100, \tag{2}$$

where W_0 and W_t denote the weights of films before and after the immersion in toluene, respectively.

2.2.6. X-Ray Diffraction (XRD)

The examination of the crystal structures of the films was performed by X-ray diffractometer (model D8 Discover, Bruker AXS, Karlsruhe, Germany). The films were cut into strip-shaped specimens of 4 cm in width and 5 cm in length. The operating voltage and current were 40 kV and 40 mA, respectively. Samples were scanned from 5–40° 2θ using CuKa radiation.

2.2.7. Differential Scanning Calorimetry (DSC)

DSC analysis was used to measure the thermal properties of the films, such as glass transition temperature (Tg) and crystalline melting temperature (Tm). A sample of about 4 mg was sealed in aluminum pan for DSC analysis under nitrogen gas. In addition, the curing behavior of the films was determined using a NETZSCH DSC 204 F1 Phoenix (Selb, Germany). The scanning range was −100 to 200 °C with a heating rate of 5 °C/min.

2.2.8. Thermal Gravimetric Analysis (TGA)

The thermal weight changes of BC, NR, and the films were determined using TGA (Q50 V6.7 Build 203, Universal V4.5A TA Instruments, New Castle, DE, USA) in a nitrogen atmosphere. The scanning range was 30 °C to 700 °C with a heating rate of 10 °C/min. The initial weight of each sample was around 10 mg and percentage weight loss versus decomposition temperature by TGA analysis was determined.

2.2.9. Mechanical Properties Testing

The tensile strength of the films was measured by Instron Testing Machine (ASD8-82A.TSX, NY, USA). The test conditions followed ASTM D882. The determination of elongation at break, tensile strength, and Young's modulus was performed using films in strip-shaped specimens of 10 mm in width and 10 cm in length. The mechanical properties of each sample were the average values determined from five specimens.

2.2.10. Biodegradation in Soil

Biodegradation of BC and NR–BC films in soil for six weeks was evaluated. The samples were cut into square pieces of 3 cm × 3 cm and were buried in 10 cm soil depth under uncontrolled temperature (24–35 °C). Samples were taken out every week and washed with DI water. Then, the samples were dried at 50 °C for 24 h and weighed. The specific biodegradation rates based on the mass loss of films were determined by the following Equation (3):

$$\text{Biodegradation\%} = \left[\frac{W_1 - W_2}{W_1}\right] \times 100, \tag{3}$$

where W_1 and W_2 denote the initial dry weight of the samples (g) and the residue dry weight of films after biodegradation in soil, respectively.

3. Results

3.1. NR Particle Size Distribution

The particle size distribution (PSD) of NR was analyzed by a laser diffraction particle size technique. The rubber particle size from NRL suspension varied from 0.01–2 μm, as shown in Figure 1. A bimodal PSD having two peaks at 0.08 μm and 0.7 μm was observed, with corresponding volume densities of about 2.6% and 9.6%, respectively. A solution of 50% (v/v) ethanol was slowly added into NRL suspension at 2.0% (v/v) to promote the penetration of NR into the BC nanofibrous network structure. No significant change in particle size distribution of NR in NRL suspension with the addition of the ethanol solution was observed.

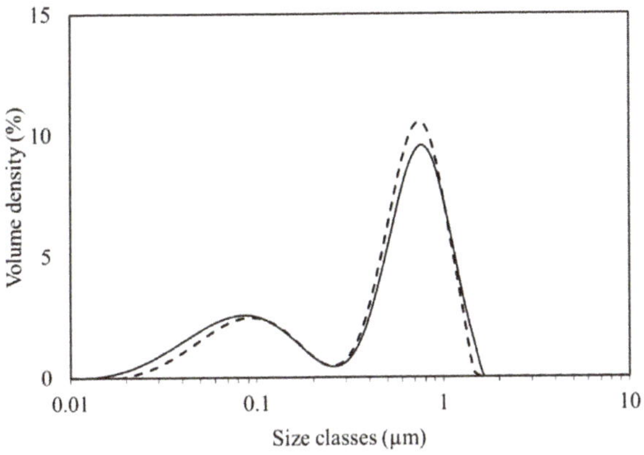

Figure 1. Particle size distribution of natural rubber (NR) in NR latex (NRL) suspension: in the absence of ethanol aqueous solution (solid line) and with the addition of ethanol solution (dash line).

3.2. *Effects of NRL Concentration and Temperature on Integration of NR into BC*

Surface morphology of a never-dried BC film is shown in Figure 2a. Surface and cross section of dried BC films are shown in Figure 2b,c, respectively. According to the observation from the FESEM images, the surface of never-dried BC films comprises microporous structures of fibrous networks with nanocellulose fiber diameters of ~50–100 nm. Never-dried BC films possess pore sizes ranging between 0.1–2 μm located between the fibers. A similar pore structure of unmodified BC hydrogel was previously reported [18]. After air drying at 30 °C, a dense nanocellulose layer structure was obtained, as shown in Figure 2b. Water loss during drying could result in shrinkage and a compact structure. The cross section of dried BC films in Figure 2c shows that the structures are composed of multilayers of thin sheets, which is the characteristic feature of BC film previously reported [13]. During the immersion, NR diffused into and gradually filled the never-dried BC pores, and by doing so, might coat the surface. FESEM images of the cross section of the dried NR–BC films by immersion in NRL suspensions of 0.5%, 2.5%, 5%, and 10% DRC at 50 °C are shown in Figure 3. The NR–BC films comprise dense nanocellulose layers incorporated with NR. The amount of integration of NR into BC was in association with the increase in weight (Figure 4) and thickness of the dried films (Table 1). The accumulated NR in the composite was estimated from the change of dried weight of the films before and after the imersion in NRL suspension, as shown in Figure 4. The average weight of the dried BC film was 0.0091 ± 0.0001 g, whereas the diameter and thickness were 14.01 ± 0.13 cm and 12.05 ± 0.26 μm, respectively. The diffusion and integration of NR into BC increased as a function of NRL concentration up to around 2.5%–5% DRC. The high amount of NR accumulated in the composite films was obtained by the immersion in NRL suspension at 2.5% DRC (at 50 °C), 5% DRC (at 50 °C), 2.5% DRC (at 60 °C), and 5% DRC (at 60 °C), where the estimated ratios of NR/BC in the composite films were 53.9%, 42.9%, 73.6%, and 70.3%, respectively (the amounts of NR in the composite films were 35.0%, 30.0%, 42.4%, and 41.3%, respectively). However, further increasing the NR concentration in the suspension to 7.5% and 10% DRC resulted in a significant decrease of NR adsorption. Higher weights and thicknesses of films were also observed as a function of increased immersion temperature from 30 °C to 50–60 °C. The rate of diffusion of NR into nanocellulose fiber networks should generally increase with temperature up to a certain point as a result of increased kinetic energy. Similar changes to film thickness and surface area have previously been reported in relation to the modification of BC by impregnation with aqueous alginate solutions at 1%–3% (w/v) at the immersion temperature of 30 °C to 50 °C [19].

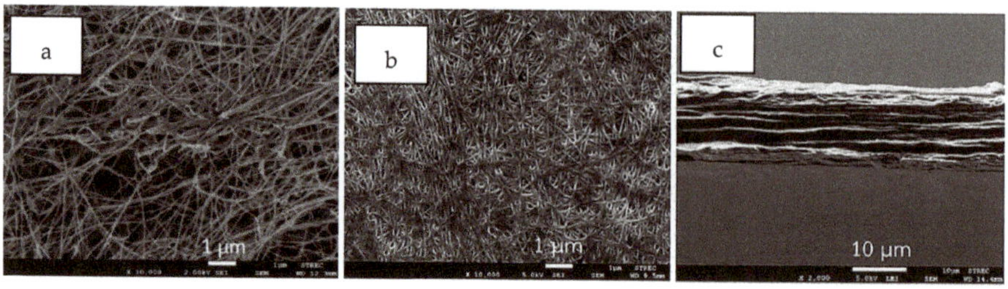

Figure 2. Field emission scanning electron microscopy (FESEM) images of surface morphologies of never dried films of bacterial cellulose (BC) (**a**) and dried BC films: surface (**b**) and cross section (**c**).

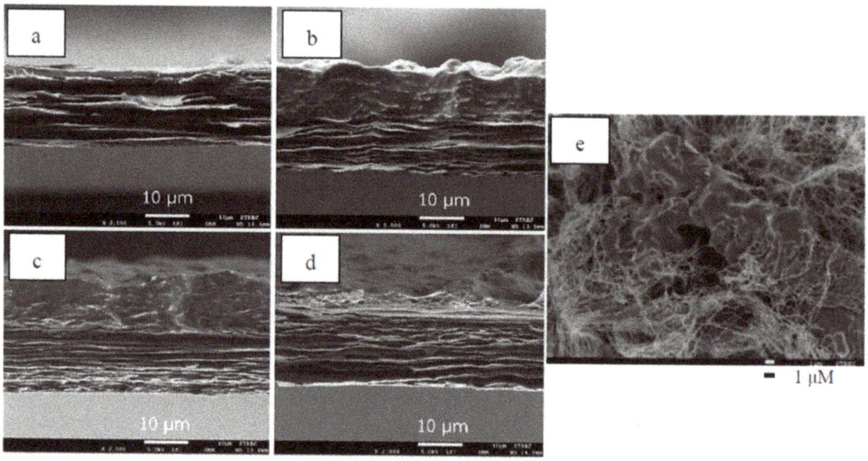

Figure 3. FESEM images of the cross section of dried films of 0.5NR–BC50 (**a**); 2.5NR–BC50 (**b**); 5NR–BC50 (**c**); 10NR–BC50 (**d**); and a closer look of the surface of dried NR–BC film (**e**).

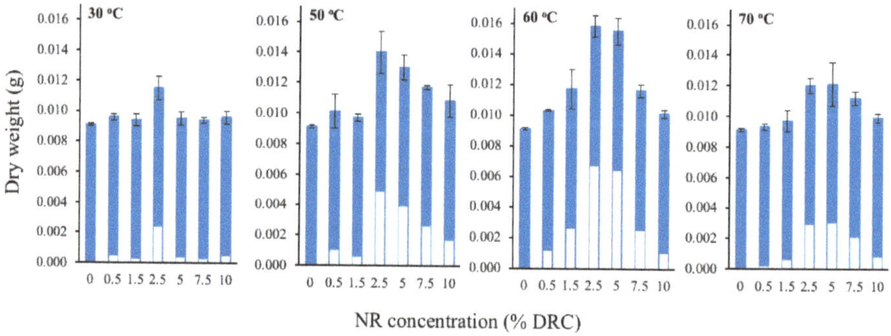

Figure 4. Dry weight of NR–BC films modified by immersing in an NRL suspension at various concentrations (0%–10% dry rubber content (DRC)) at immersion temperatures of 30 °C, 50 °C, 60 °C, and 70 °C. Values were expressed as mean ±SD (n = 3); blue bar = the dry weights of BC before the immersion and white bar = the estimated NR dry weights in the composites.

Table 1. Thickness of natural rubber (NR)–bacterial cellulose (BC) films.

Samples	Thickness (μm)	Samples	Thickness (μm)
0.5NR–BC30	12.74 ± 0.58	0.5NR–BC60	13.86 ± 0.13
2.5NR–BC30	19.08 ± 0.38	2.5NR–BC60	24.97 ± 0.90
5NR–BC30	15.85 ± 0.53	5NR–BC60	27.21 ± 0.39
10NR–BC30	14.66 ± 0.49	10NR–BC60	24.15 ± 1.62
0.5NR–BC50	14.84 ± 0.64	0.5NR–BC70	13.32 ± 0.58
2.5NR–BC50	20.61 ± 0.24	2.5NR–BC70	20.98 ± 0.69
5NR–BC50	21.99 ± 0.96	5NR–BC70	25.27 ± 0.29
10NR–BC50	18.54 ± 0.63	10NR–BC70	21.73 ± 0.35

Values are expressed as mean ±SD (n = 3).

3.3. Mechanical Properties

The mechanical properties of BC and NR–BC films were analyzed in terms of elongation at break, tensile strength, and Young's modulus, as shown in Figure 5. According to our previous studies, the tensile strength, elongation at break, and Young's modulus of unmodified BC films could be varied around 70–300 MPa, 0.5%–5%, and 5–17 GPa, respectively, depending on many factors, such as culture conditions, drying conditions, and bacteria strain [19–21]. In this study, the unmodified BC film possessed an elongation at break, tensile strength, and Young's modulus of 0.6%, 112.4 MPa, and 9.14 GPa, respectively; meanwhile, the corresponding values for the uncured NR film developed from pure NRL were around 100%–111%, 0.8–1.2 MPa, and 1.6–2.4 MPa, respectively [17,22]. The NR–BC films have considerably higher elongation at break compared with the unmodified BC films. The maximum values of elongation at break, at 3.0%–3.5%, were obtained when immersing BC in NRL at a concentration range of 2.5%–5% DRC between 50 and 60 °C. The incorporation of NR into the BC matrices resulted in the ability of the fibers to absorb more energy, and thus prevent the nanocellulose fibers from breaking. Consequently, when compared with the BC films, the films were demonstrated to elongate more prior to breaking. The NR–BC film tensile strengths are also significantly higher when compared with those of the BC-only film (Figure 5b. The optimal conditions to achieve maximum tensile strength of the NR-BC films are during the preparation of 2.5NR–BC50, in which the tensile strength increased to 392.4 MPa (a ~3.5-fold increase over that of the unmodified BC film). Additionally, the films treated with NRL suspension of 5% DRC and at the immersion temperature of 50–60 °C exhibited considerably higher tensile strengths when compared with the unmodified BC film. The tensile strengths of 2.5NR–BC50, 2.5NR–BC60, 5NR–BC50, and 5NR–BC60 showed a 3.5-, 1.7-, 2.3-, and 2.0-fold increase, respectively, over that of the unmodified BC film. The Young's modulus of pure BC and the films are shown in Figure 5c. The unmodified BC film exhibits a Young's modulus of 9.14 GPa, whereas the NR–BC films display higher Young's modulus values. Under the optimal preparation conditions, 2.5NR–BC50 displayed a Young's modulus value of 20.05 GPa or an ~2-fold increase over that of the BC film. As the maximum enhancement of the mechanical properties was obtained by immersing at 50 °C, the results of the following studies were performed on NR–BC films prepared by immersing in NRL suspensions at 50 °C.

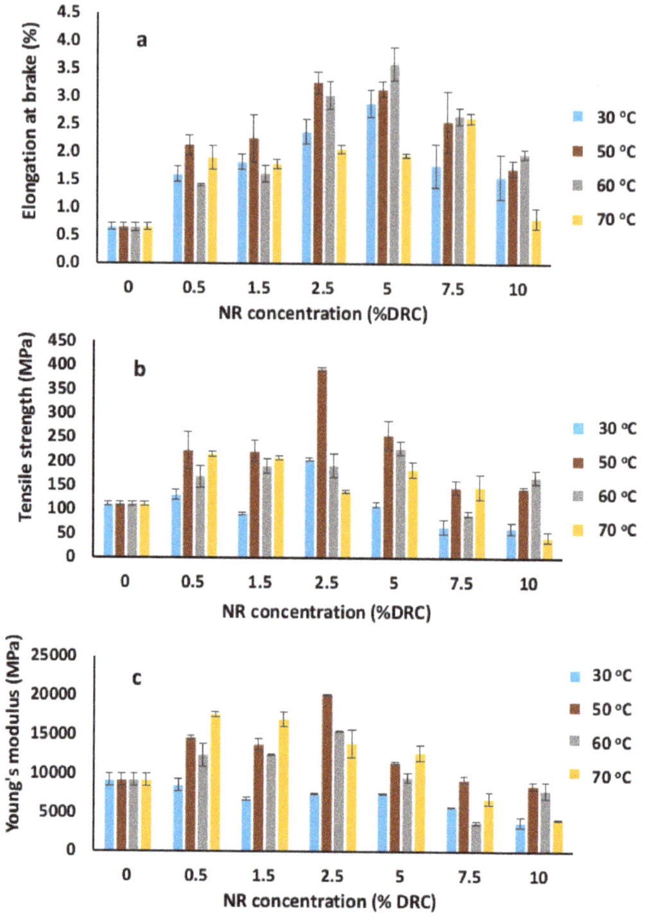

Figure 5. Elongation at break (**a**), tensile strength (**b**), and Young's modulus (**c**) of the dried BC and NR–BC films modified by the immersion in NRL of various concentrations (0%–10% DRC) at various immersion temperatures.

3.4. FTIR Analysis

As shown in Figure 6, the FTIR spectrum of pure BC shows characteristic peaks at 3347 cm^{-1} attributed to O–H stretching vibrations from hydroxyl groups, and at 2900–2800 cm^{-1} corresponding to C–H stretching, while a peak at 1650–1640 cm^{-1} corresponds to the vibration of the carbonyl group (C=O). A peak at 1440 cm^{-1} is attributed to CH_2 bending, and a peak located at 1165–1060 cm^{-1} is attributed to C–O stretching [17,23,24]. The FTIR spectrum of NR reveals several characteristic peaks. The peak at 2960 cm^{-1} is attributed to the vibration of C–H stretching, while the peak located at 2917 cm^{-1} corresponds to the symmetric stretching of methylene ($-CH_2$). The asymmetric stretching of the CH_3 group is observed at 2847 cm^{-1}. A peak located at 1637 cm^{-1} is assigned to the vibration of C=C, while the peak located at 1465 cm^{-1} is attributed to the symmetric bending of CH_2. The peak at 1375–1450 cm^{-1} is attributed to the vibration of CH_2 asymmetric bending and stretching [17,25]. The FT-IR spectra of the NR–BC films display characteristic peaks of both BC and NR. No occurrence of new peaks was observed.

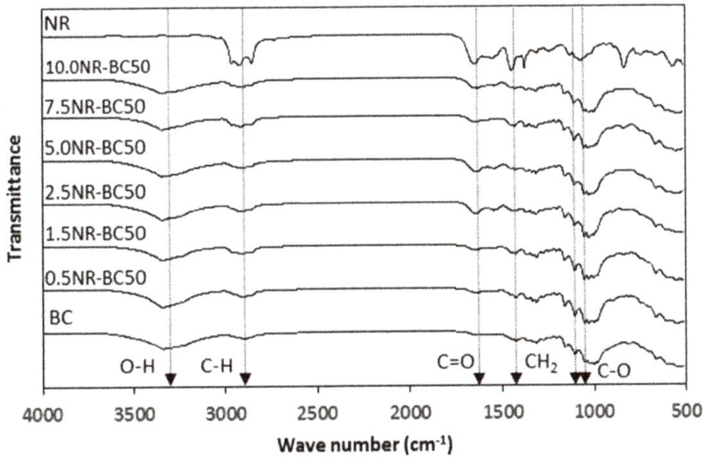

Figure 6. Fourier transform infrared spectroscopy (FTIR) spectra of BC, NR, and NR–BC films treated by the immersion in NRL of various concentrations at 50 °C.

3.5. X-Ray Diffraction (XRD) Analysis

The XRD patterns of BC and the NR–BC films are shown in Figure 7. The characteristic peaks of BC comprise three main peaks at two theta angles of 14.5°, 17.0°, and 22.8°, associated with the typical profile of cellulose [26]. The diffraction patterns of the NR–BC films exhibit a higher degree of order, as seen by the sharp peaks, whereas the diffraction pattern relating to the BC-only film displays relatively broad peaks, indicating a less crystalline material. The NR film displays a typical diffraction pattern of an amorphous polymer having a prominent broad hump located at a two-theta angle of 18°. The NR–BC films, prepared at various concentrations and immersion temperatures, display the same characteristics of the BC-only film diffraction pattern, indicating the presence of the BC crystalline structure. The peaks at two-theta ≈ 17° and 22.8° became slightly more intense with increasing NR, which might imply that the integration of NR into fibrous structure of BC might have some effect on the crystalline structure of BC. The amorphous broad hump associated with NR was not observed in the XRD patterns, as there was a small proportion of NR in the BC matrices.

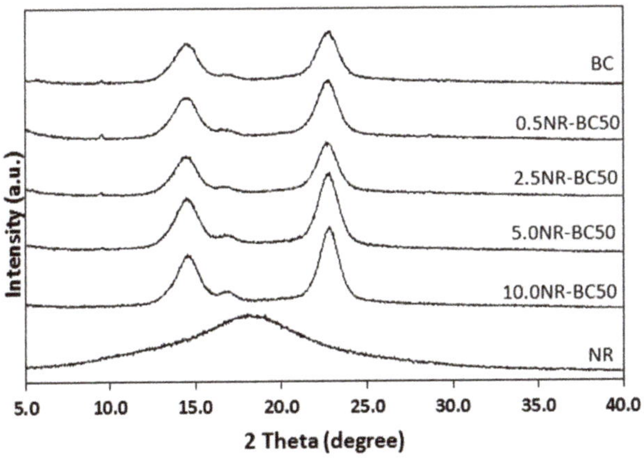

Figure 7. X-ray diffraction (XRD) patterns of BC, NR, and NR–BC films treated by the immersion in NRL of various concentrations at 50 °C.

3.6. *Differential Scanning Calorimetry (DSC)*

The DSC thermograms of dried BC, NR, and the NR–BC films are shown in Figure 8. The glass transition temperature (Tg) of pure BC was barely detected because the highly crystalline structure of nanocellulose exhibited flat heat flow curves, which are difficult to separate from the baseline. For the thermal degration, the BC film displays two exothermic broad peaks. The first peak around 217 °C is associated with the thermal degradation of proteinaceous matter in the BC film [19,26]. The second peak at ~304 and 340 °C is ascribed to the partial pyrolysis of cellulose [27]. The DSC thermogram of NR exhibited a Tg at −68.1 °C. The DSC thermograms of 5.0NR–BC50 and 10.0NR–BC50 exhibited a Tg at around −67 °C, slightly higher than the Tg of NR. The endothermic peaks exhibited by the NR–BC films between 30–150 °C are attributed to water loss or dehydration of the films [19,28]. NR loading into the BC matrices (2.5NR–BC50, 5.0NR–BC50, and 10.0NR–BC50) resulted in a slight change to the position of the exothermic peaks.

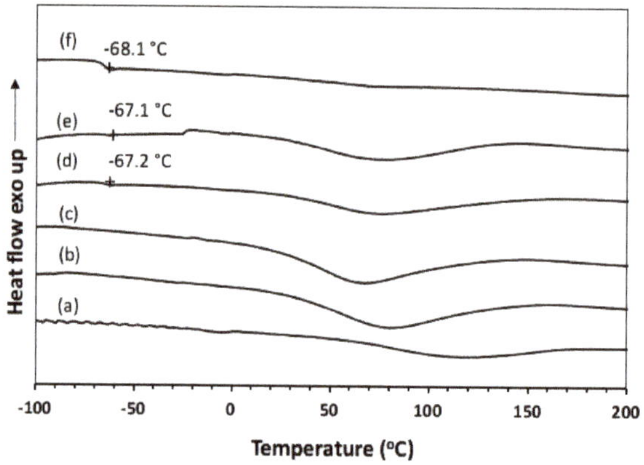

Figure 8. Differential scanning calorimetry (DSC) chromatograms of BC (**a**), NR (**f**), and NR–BC films of 0.5 NR–BC50 (**b**), 2.5 NR–BC50 (**c**), 5.0 NR–BC50 (**d**), and 10.0 NR–BC50 (**e**).

3.7. *Thermal Gravimetric Analysis (TGA) and Differential Thermal Analysis (DTA)*

The TGA and DTA curves of BC, NR, and the NR–BC films are shown in Figure 9a,b, respectively. The slight weight loss from 30–200 °C is associated with the vaporization of water. The percentage weight loss of pure BC across a temperature range of 210–240 °C is ~6 wt.%, indicating the decomposition of proteins [29]. The major pyrolysis of BC, resulting from the decomposition of cellulose [30], occurs at ~300–360 °C, and results in a residual char product (≈20 wt.%). Conversely, the pyrolysis temperature of NR occurs at ~340–440 °C, at which about half of the mass loss is observed for pyrolysis at 380 °C. The thermal decomposition of the NR–BC films is divided into three weight loss stages. At temperatures <200 °C, the weight loss is associated with water loss. The second and third weight loss stages occur across a temperature range of 300–400 °C. The maximum rate of weight loss of the films occur at ~320 °C and ~380 °C in accordance with the decomposition of BC and NR, respectively. Films of 5.0NR–BC50 and 10.0NR–BC50 presented slightly increased thermal stability when compared with the BC film and other NR–BC films.

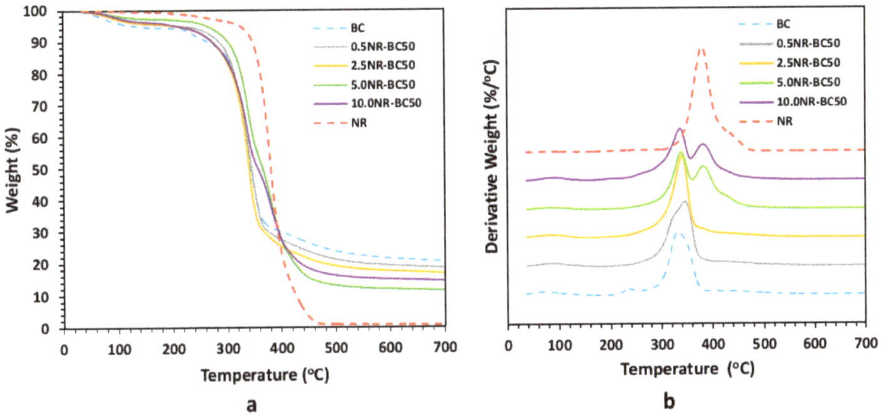

Figure 9. Thermal gravimetric analysis (TGA) (**a**) and differential thermal analysis (DTA) (**b**) curves of BC, NR, and NR–BC films.

3.8. Water Absorption Capacity (WAC)

Water absorption capacities of BC, NR, and NR–BC films treated by the immersion in NRL of various concentrations at 50 °C are shown in Table 2 and Figure 10. Overall, all films showed rapid adsorption of water in the first 20 min and then slow adsorption until concentrations reached equilibrium at around 1 h. The BC film has highly water absorption capacity at 610.5% because of the hydrophilic property of the hydroxyl group in its structure [31,32]. Because of the hydrophobic structure of NR, the water absorption capacity of the NR film was very low (10.9%). As compared with BC, NR–BC films had significantly higher resistance to water.

Table 2. Water absorption capacity (WAC%) and toluene uptake (TU%) of NR–BC50 films.

Solvent	Absorption (WAC%, TU%)							
	BC	0.5 NR–BC	1.5 NR–BC	2.5 NR–BC	5 NR–BC	7.5 NR–BC	10 NR–BC	NR
Water	610.5 ± 2.6	213.9 ± 6.4	208.2 ± 2.4	88.4 ± 3.3	88.6 ± 1.8	115.9 ± 3.3	142.5 ± 1.2	10.9 ± 0.3
Toluene	8.7 ± 1.0	33.2 ± 6.8	34.2 ± 12.6	70.9 ± 24.0	68.6 ± 13.5	30.5 ± 11.7	47.3 ± 3.4	2,652.8 ± 300.0

Values are expressed as mean ±SD (n = 3).

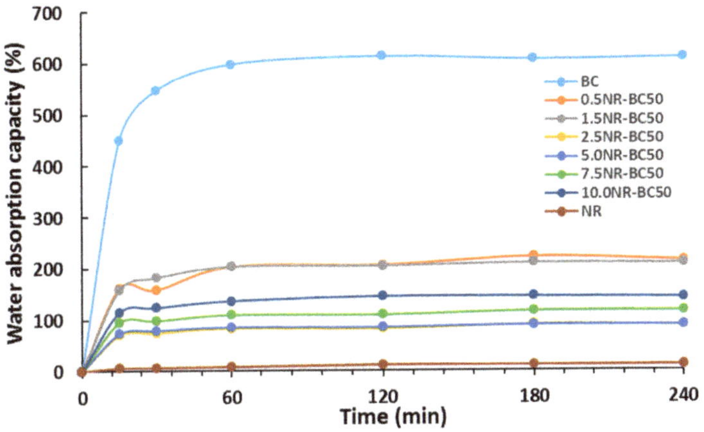

Figure 10. Water absorption capacity (WAC %) with time of NR–BC50 films.

3.9. Toluene Uptake (TU)

NR is a nonpolar material; therefore, it is soluble in non-polar solvents. Toluene is an aromatic solvent that is mostly used in many rubber industries. Thus, the effect of toluene uptake on NR-BC films was investigated (Table 2). At initial absorption, the toluene uptake of the NR film rapidly increased and reached the maximum value at 2642% in 1 h. After that, the degradation of the NR film by dissolving in toluene was observed [33]. Because of its hydrophilic nature, the BC film had high resistance to non-polar solvents and exhibited very low toluene uptake at around 6.4%. The toluene uptake of NR–BC films is higher than BC film, but so much less than NR (0.02–0.05 of NR film).

3.10. Biodegradation in Soil

Biodegradability is an essential property when considering environment issues, and is a critical property for the application of green packaging materials. BC structures comprise crystalline nanocellulose fibers and a minor amount of amorphous cellulose chains, which can be attacked by multiple microorganisms in the soil through enzymatic degradation [34,35]. The conformation of NR is relatively resistant to biodegradation through microorganisms when compared with many other natural polymers [36]. However, there are known microorganisms in soil such as bacteria and fungi that have the ability to degrade NR [36,37]. Natural latex rubber is biodegradable, as is claimed by numerous products and manufacturers. Natural rubber latex gloves can be disposed of by either landfill or incineration, which are not harmful to the environment. Recently, it was shown that the mixed culture isolated from soil samples collected from rubber contaminated ground in Songkhla province, Thailand had potential in degrading rubber, in which significant changes could be detected within 30 days [38]. In this study, the biodegradability of BC and the NR–BC films in soil is shown in Figures 11 and 12. The films underwent soil burial experiments and the average soil temperature was 35.1 ± 2.0 °C. The BC film exhibited a higher weight loss percentage when compared with the other films, and was completely decomposed within four weeks. The NR–BC films were completely decomposed within 4–6 weeks. Overall, higher NR loadings were observed to slow the decomposition rate of the films. Films of 2.5NR–BC50 and 5NR–BC50 demonstrated a higher resistance to microorganism degradation when compared with 0.5NR–BC50 and 10NR–BC50. The films of 2.5NR–BC50 and 5NR–BC50 were completely decomposed within six and five weeks, respectively, whereas 0.5NR–BC50 and 10NR–BC50 were completely decomposed within four weeks.

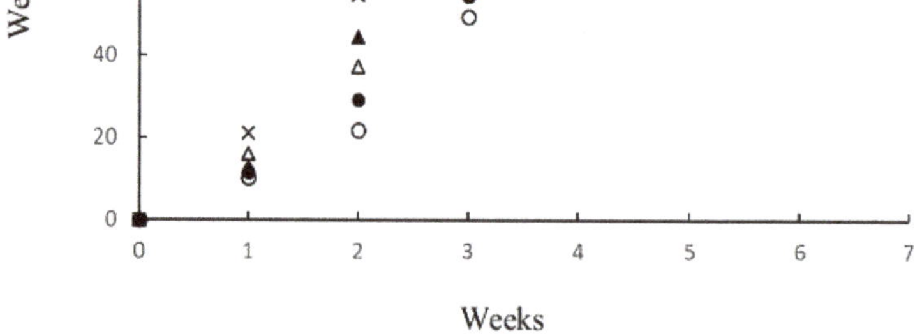

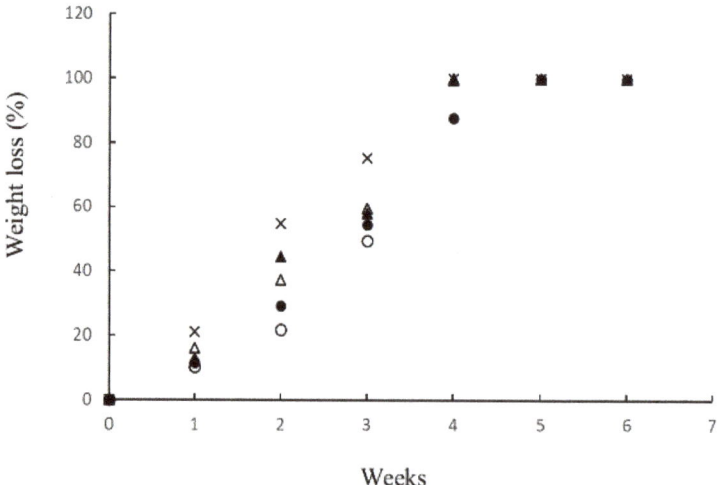

Figure 11. Biodegradability of films in soil after 0–6 weeks. (BC = ×; 0.5 NR–BC50 = white triangle; 2.5 NR–BC50 = white circle; 5 NR–BC50 = black circle; 10 NR–BC50 = black triangle).

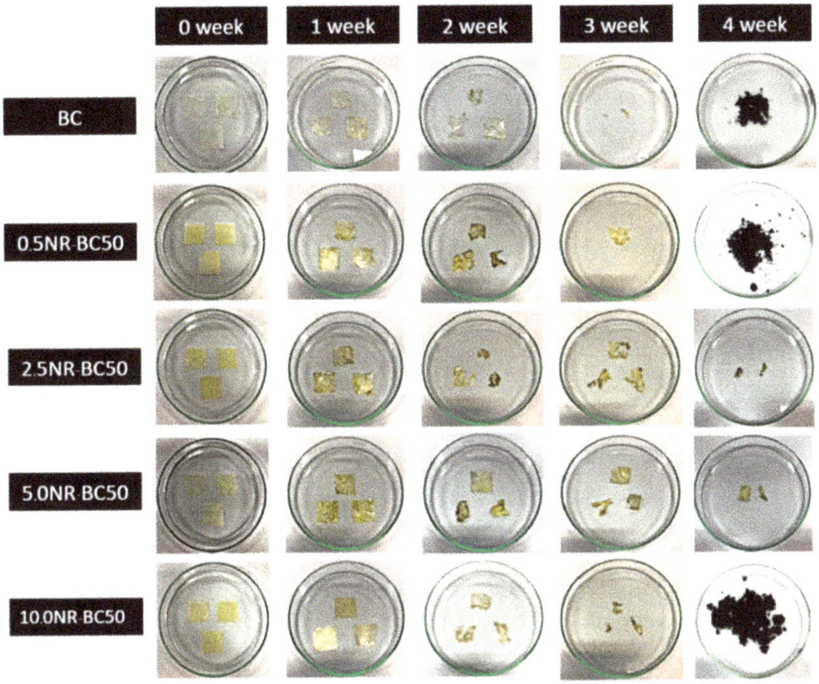

Figure 12. The images of degraded materials at the different weeks in the study of biodegradation in soil.

4. Discussion

BC pellicle was immersed in diluted NR latex (NRL) suspensions with the supplement of ethanol aqueous solution. The slow addition of a 50% (v/v) ethanol solution into the NRL suspension at 2.0% (v/v) did not show a significant effect on NR particle size. However, it could result in a decrease in viscosity of the NRL suspension as a result of the reduction of the gel content of rubbers from NRL [39]. It was found that the addition of ethanol in NRL at a specific fraction could promote the penetration of NR into the BC nanofibrous network structure. However, as ethanol is an organic polar solvent, the addition at too high a fraction into the NRL suspension resulted in the coagulation of NR molecules [40]. From our preliminary test, the addition of ethanol at concentration ≥70% v/v into NRL for 2.0% (v/v) or the addition of 50% v/v ethanol into NRL for ≥3.0% v/v caused coagulation of NR.

After the modification by immersing BC films in diluted NRL suspensions, it was found that the maximum dry weight of the NR–BC films was around 0.014–0.016 g, which was about 1.5–1.8 of that of the BC film. The maximum thickness of the NR–BC films was 20–27 μm, or about 1.7–2.3 of that of the BC film. The maximum amount of integration of NR into BC was obtained by immersing BC in NRL suspensions of 2.5%–5.0% DRC at an immersion temperature of 50–60 °C. However, further increasing the NRL concentration above 5.0% DRC led to the agglomeration of NR molecules, resulting in a lower diffusion of NR molecules into the BC film. At higher concentrations, the close vicinity of the NR molecules with respect to each other could result in increased interaction and agglomeration. As the BC pellicle pore size was relatively small, the diffusion of agglomerated NR into the BC fibrous network was hindered. Considering how diffusion is influenced by immersion temperature, at low NRL concentrations (< 2.5% DRC), no significant increase of NR–BC film thickness was observed when increasing the immersion temperature from 30 °C to 70 °C. At low NR concentrations, low adhesion of NR to the BC matrices was observed. When BC was immersed in NRL suspensions at medium to high concentrations (2.5%–10% DRC), the NR–BC film thickness significantly increased as a function of increased immersion temperature from 30 °C to 60 °C. Increased kinetic energy of the NR molecules at high temperature was considered to promote the diffusion of NR into the BC matrices. At high temperature, the agitated particles were subjected to stronger and more frequent collisions. However, NR–BC film thickness and weight decreased as a function of elevated immersion temperature from 60 °C to 70 °C, which implied less accumulation of NR in BC films. When subjected to a high immersion temperature of 70 °C, the NR molecules, as a result of a high collision rate, could form particle agglomeration. The coating of agglomerated NR on the BC surface and in the BC fibrous networks was thought to prevent the diffusion of small NR particles into the BC pores. The low degree of NR penetration resulted in a smaller NR–BC film thickness when the immersion temperature increased from 60 °C to 70 °C.

BC films usually possess a high mechanical strength (high modulus and tensile strength), but low elongation at break (or low fracture strain). Conversely, NR films show higher elongation at break, but possess a relatively lower mechanical strength as compared with BC films. In this study, it was shown that the mechanical properties of the BC films were considerably improved by the addition of NR into BC matrices. The important factors that affect the diffusion of NR into BC matrices are the concentration of NRL suspension and temperature. The concentration of NRL suspension at 2.5%–5.0% DRC and the immersion temperature at 50–60 °C are the optimal conditions for high diffusion of NR into BC matrices. The integration of NR into BC matrices resulted in improved mechanical properties of the films. When subjected to a high immersion temperature of 70 °C, the NR molecules, as a result of a high collision rate, could form particle agglomeration. The agglomerated NR particle could cover some parts of the BC surface and filled in the pores of BC matrices. The large particles could prevent the diffusion of small NR particles into the pores of the inner part. At high temperature, NR agglomeration could be generated to a greater extent, especially in the condition with higher concentration of NRL suspension. This agglomeration might cause a problem of poor distribution of NR inside the BC matrices. As a result, at an immersion temperature of 70 °C, NR–BC films exhibited the largest Young's modulus in the 0.5% DRC group, but exhibited the smallest Young's modulus in the 10% DRC group. However, under the optimal condition, NR integrated into the BC films and was well distributed in BC matrices. NR could bind the nanocellulose fibers together and, consequently, the mechanical properties of the NR–BC films, with respect to their modulus and strength, were enhanced compared with the BC film. NR possessed high structural regularity and typically crystallizes spontaneously when stretched [41]. The NR bonds in the nanocellulose fibrous network restricted the movement of the nanocellulose fibers and enhanced mechanical strength [42]. It was revealed that the presence of NR on the BC matrices in the NR–BC films induced superior mechanical properties such as high elongation at break and high tensile strength. Therefore, the combination of a nanocellulose fibrous network and NR could result in a synergistic effect on the mechanical properties.

The FTIR and XRD results showed that there was no chemical interaction between NR and BC; however, the integration of NR into fibrous structure of BC might improve crystalline structure of BC. At high NR diffusion into BC fibers, NR–BC films exhibit relatively high structural and thermal stability. It was shown that 5.0NR–BC50 and 10.0NR–BC50 presented relatively increased thermal stability when compared with that of the BC film and the other NR–BC films. On the other hand, it was shown that XRD peaks of 5.0NR–BC50 and 10.0NR–BC50 are also slightly sharper than the others. Therefore, the integration of NR into BC matrices at a certain content (an optimal concentration range) might have some positive effects on the crystalline structure of BC. The crystalline structure could affect thermal properties of the composites. The result of TGA residual mass (Figure 9) showed that the remaining mass by the higher order was BC > 0.5 NR–BC50 > 2.5 NR–BC 50 > 10NR–BC50 > 5.0 NR–BC 50 > NR. According to our previous study [17], the char yield of BC was higher than that of NR, and the char yields increased along with the ratio of BC in NR composites. In this study, the ratio of BC/NR by the higher order was 0.5NR–BC50 > 10 NR–BC50 > 5.0NR–BC50 > 2.5NR–BC50 (Figure 4). Compared with the other composite films, the remaining mass of the composite film of 0.5 NR–BC50 was the highest because the ratio of BC/NR of this composite film was higher than the others. For the same reason, the residual of 10.0NR–BC50 was greater than that of 5.0NR–BC50. However, it is noticed that the remaining mass of 2.5 NR–BC50, which has the highest ratio of NR (or the lowest ratio of BC) is quite high when compared with the others. As a result, it was suggested that not only the composition, but also the structure of the composites could also have an effect on the thermal properties.

BC has a hydrophilic structure and NR has a hydrophobic structure; therefore, the values of WAC% decreased with the ratio of NR in the NR–BC composites. The films of 2.5NR–BC50 and 5NR–BC50 had the highest water resistance (the lowest WAC%). Because of NR binding into BC fibers, only small amounts of water could diffuse or be adsorbed into NR–BC films. Consequently, the water resistance of NR–BC films increased with NR concentration in the films. The toluene uptake (TU%) of NR–BC50 films was in range of 30%–70% and the increase in toluene uptake was related to the concentration of NR in the films. However, the toluene uptake of NR–BC films was much lower than that of NR films and, after the immersion of NR–BC films in toluene for 4 h, no significant change in overall outlook of the NR–BC films was observed. The good resistance to toluene should be attributed to the hydrophilic nature and high stability of nanocellulose network structure of BC in nonpolar solvents.

The evaluation of biodegradation capability of NR–BC50 films in comparison with BC films was conducted in soil environment for six weeks. BC fiber is biodegradable by various bacteria and fungi in soil. The microorganism first attacks the nanocellulose amorphous region and, thereafter, decomposes all the crystalline regions. On the other hand, it has been previously reported for a slow process of biodegradation of NR and related compounds by some microorganisms in soil [43]. In this study, during the biodegradation test, the NR–BC films transitioned to a loose structure. The films were lumpy and highly stretched as a result of the decomposition of nanocellulose by the microorganisms, while the NR particles in the NR–BC films underwent a relatively slower rate of decomposition than the nanocellulose of BC. However, all NR–BC films were biodegradable and could be degraded completely in soil environment within 5–6 weeks.

5. Conclusions

Films of BC reinforced with NR, prepared by immersing BC into a diluted NRL suspension, demonstrated superior mechanical properties when compared with BC-only films. The combination of a nanocellulose fibrous network and NR polymer synergistically improves the film mechanical properties. Films of 2.5NR-BC50 demonstrated considerably enhanced tensile strength and elongation at break. The NR–BC films also exhibit high structural and thermal stability and are completely degraded in soil within 5–6 weeks.

Author Contributions: Conceptualization, M.P.; Methodology, K.P. and M.P.; Validation, M.P.; Formal Analysis, K.P. and M.P.; Investigation, K.P. and M.P.; Resources, M.P.; Writing—Original Draft Preparation, K.P. and M.P.; Writing—Review & Editing, M.P.; Supervision, M.P.; Project Administration, M.P.; Funding Acquisition, M.P.

Funding: This research was funded by Thailand Science Research and Innovation, grant number RGU6280004.

Conflicts of Interest: The authors declare no conflict of interest.

References

1. Industry Focus. Thailand: The World's Leader in Natural Rubber Production. Thailand Investment Review. Available online: https://www.boi.go.th/upload/content/TIR_AUGUST_PROOF_10_44271.pdf (accessed on 28 May 2019).
2. Natural Rubber Production Up, but Consumption Dips in 2017, European Rubber Journal Report (2018). Available online: http://www.rubbernews.com/article/20180118/NEWS/180119950/natural-rubber-production-up-but-consumption-dips-in-2017 (accessed on 28 May 2019).
3. Zhang, Y.; Liu, Q.; Zhang, Q.; Lu, Y. Gas barrier properties of natural rubber/kaolin composites prepared by melt blending. *Appl. Clay Sci.* **2010**, *50*, 255–259. [CrossRef]
4. Jacob, M.; Thomas, S.; Varughese, K.T. Mechanical properties of sisal/oil palm hybrid fiber reinforced natural rubber composites. *Compos. Sci. Technol.* **2004**, *64*, 955–965. [CrossRef]

5. George, J.; Bhagawan, S.S.; Prabhakaran, N.; Thomas, S. Short pineapple-leaf-fiber-reinforced low-density polyethylene composites. *J. Appl. Polym. Sci.* **1995**, *57*, 843–854. [CrossRef]
6. Arumugam, N.; Selvy, K.T.; Rao, K.V.; Rajalingam, P. Coconut-fiber-reinforced rubber composites. *J. Appl. Polym. Sci.* **1989**, *37*, 2645–2659. [CrossRef]
7. Ismail, H.; Edyham, M.; Wirjosentono, B. Bamboo fibre filled natural rubber composites: The effects of filler loading and bonding agent. *Polym. Test.* **2002**, *21*, 139–144. [CrossRef]
8. De, D.; De, D.; Adhikari, B. Curing characteristics and mechanical properties of alkali-treated grass-fiber-filled natural rubber composites and effects of bonding agent. *J. Appl. Polym. Sci.* **2006**, *101*, 3151–3160. [CrossRef]
9. Shoda, M.; Sugano, Y. Recent advances in bacterial cellulose production. *Biotechnol. Bioproc. E.* **2005**, *10*, 1–8. [CrossRef]
10. Jiji, S.; Thenmozhi, S.; Kadirvelu, K. Comparison on properties and efficiency of bacterial and electrospun cellulose nanofibers. *Fiber. Polym.* **2018**, *19*, 2498–2506. [CrossRef]
11. Ougiya, H.; Watanabe, K.; Morinaga, Y.; Yoshinaga, F. Emulsion-stabilizing effect of bacterial cellulose. *Biosci. Biotech. Bioch.* **1997**, *61*, 1541–1545. [CrossRef]
12. Klemm, D.; Schumann, D.; Udhardt, U.; Marsch, S. Bacterial synthesized cellulose-artificial blood vessels for microsurgery. *Prog. Polym. Sci.* **2001**, *26*, 1561–1603. [CrossRef]
13. Svensson, A.; Nicklasson, E.; Harrah, T.; Panilaitis, B.; Kaplan, D.L.; Brittberg, M.; Gatenholm, P. Bacterial cellulose as a potential scaffold for tissue engineering of cartilage. *Biomaterials* **2005**, *26*, 419–431. [CrossRef] [PubMed]
14. Wu, X.; Lin, T.F.; Tang, Z.H.; Guo, B.C.; Huang, G.S. Natural rubber/graphene oxide composites: effect of sheet size on mechanical properties and strain induced crystallization behavior. *Express Polym. Lett.* **2015**, *9*, 672–685. [CrossRef]
15. Fu, D.H.; Zhan, Y.H.; Yan, N.; Xia, H.S. A comparative investigation on strain induced crystallization for graphene and carbon nanotubes filled natural rubber composites. *Express Polym. Lett.* **2015**, *9*, 597–607. [CrossRef]
16. Trovatti, E.; Carvalho, A.J.F.; Ribeiro, S.J.L.; Gandini, A. Simple green approach to reinforce natural rubber with bacterial cellulose nanofibers. *Biomacromolecules* **2013**, *14*, 2667–2674. [CrossRef] [PubMed]
17. Phomrak, S.; Phisalaphong, M. Reinforcement of natural rubber with bacterial cellulose via a latex aqueous microdispersion process. *J. Nanomater.* **2017**, 4739793. [CrossRef]
18. Pircher, N.; Veigel, S.; Aigner, N.; Nedelec, J.M.; Rosenau, T.; Liebner, F. Reinforcement of bacterial cellulose aerogels with biocompatible polymers. *Carbohyd. Polym.* **2014**, *111*, 505–513. [CrossRef]
19. Suratago, T.; Taokaew, S.; Kanjanamosit, N.; Kanjanaprapakul, K.; Burapatana, V.; Phisalaphong, M. Development of bacterial cellulose/alginate nanocomposite membrane for separation of ethanol-water mixtures. *J. Ind. Eng. Chem.* **2015**, *32*, 305–312. [CrossRef]
20. Taokaew, S.; Seetabhawang, S.; Phisalaphong, M. Biosynthesis and characterization of nanocellulose-gelatin films. *Materials* **2013**, *6*, 782–794. [CrossRef] [PubMed]
21. Subtaweesin, C.; Woraharn, W.; Taokaew, S.; Chiaoprakobkij, N.; Sereemaspun, A.; Phisalaphong, M. Characteristics of curcumin-loaded bacterial cellulose films and anticancer properties against malignant melanoma skin cancer cells. *Appl. Sci.* **2018**, *8*, 1188. [CrossRef]
22. Panitchakarn, P.; Wikranvanich, J.; Phisalaphong, M. Synthesis and characterization of natural rubber/coal fly ash composites via latex aqueous microdispersion. *J. Ind. Eng. Chem.* **2019**, *21*, 134. [CrossRef]
23. Taokaew, S.; Nunkaew, N.; Siripong, P.; Phisalaphong, M. Characteristics and anticancer properties of bacterial cellulose films containing ethanolic extract of mangosteen peel. *J. Biomat. Sci. Polym. E.* **2014**, *25*, 907–922. [CrossRef] [PubMed]
24. Halib, N.; Amin, M.C.I.M.; Ahmad, I. Physicochemical properties and characterization of nata de coco from local food industries as a source of cellulose. *Sains Malays.* **2012**, *41*, 205–211.
25. Guidelli, E.J.; Ramos, A.P.; Zaniquelli, M.E.D.; Baffa, O. Green synthesis of colloidal silver nanoparticles using natural rubber latex extracted from *Hevea brasiliensis*. *Spectrochim. Acta A* **2011**, *82*, 140–145. [CrossRef] [PubMed]
26. Czaja, W.; Romanovicz, D.; Malcolm Brown, R. Structural investigations of microbial cellulose produced in stationary and agitated culture. *Cellulose* **2004**, *11*, 403–411. [CrossRef]

27. Barud, H.S.; Ribeiro, C.A.; Crespi, M.S.; Martines, M.A.U.; Dexpert-Ghys, J.; Marques, R.F.C.; Messaddeq, Y.; Ribeiro, S.J.L. Thermal characterization of bacterial cellulose-phosphate composite membranes. *J. Therm. Anal. Calorim.* **2007**, *87*, 815–818. [CrossRef]
28. Oliveira, R.L.; Vieira, J.G.; Barud, H.S.; Assunção, R.M.N.; Filho, G.R.; Ribeiro, S.J.L.; Messadeqq, Y. Synthesis and characterization of methylcellulose produced from bacterial cellulose under heterogeneous condition. *J. Brazil. Chem. Soc.* **2015**, *26*, 1861–1870. [CrossRef]
29. George, J.; Ramana, K.V.; Sabapathy, S.N.; Jagannath, J.H.; Bawa, A.S. Characterization of chemically treated bacterial (*Acetobacter xylinum*) biopolymer: Some thermo-mechanical properties. *Int. J. Biol. Macromol.* **2005**, *37*, 189–194. [CrossRef] [PubMed]
30. Barud, H.S.; de Araújo Júnior, A.M.; Santos, D.B.; de Assunção, R.M.N.; Meireles, C.S.; Cerqueira, D.A.; Filho, G.R.; Ribeiro, C.A.; Messaddeq, Y.; Ribeiro, S.J.L. Thermal behavior of cellulose acetate produced from homogeneous acetylation of bacterial cellulose. *Thermochim. Acta* **2008**, *471*, 61–69. [CrossRef]
31. Kanjanamosit, N.; Muangnapoh, C.; Phisalaphong, M. Biosynthesis and characterization of bacteria cellulose-alginate film. *J. Appl. Polym. Sci.* **2010**, *115*, 1581–1588. [CrossRef]
32. Lin, W.-C.; Lien, C.-C.; Yeh, H.-J.; Yu, C.-M.; Hsu, S.-h. Bacterial cellulose and bacterial cellulose-chitosan membranes for wound dressing applications. *Carbohyd. Polym.* **2013**, *94*, 603–611. [CrossRef]
33. Obasi, H.C.; Ogbobe, O.; Igwe, I.O. Diffusion characteristics of toluene into natural rubber/linear low density polyethylene blends. *Int. J. Polym. Sci.* **2009**, 140682. [CrossRef]
34. Pérez, J.; Munoz-Dorado, J.; de la Rubia, T.; Martinez, J. Biodegradation and biological treatments of cellulose, hemicellulose and lignin: An overview. *Int. Microbiol* **2002**, *5*, 53–63. [CrossRef]
35. Béguin, P.; Aubert, J.-P. The biological degradation of cellulose. *FEMS Microbiol. Rev.* **1994**, *13*, 25–58. [CrossRef] [PubMed]
36. Cherian, E.; Jayachandran, K. Microbial degradation of natural rubber latex by a novel species of Bacillus sp. SBS25 isolated from soil. *Int. J. Environ. Res.* **2009**, *3*, 599–604. [CrossRef]
37. Shah, A.A.; Hasan, F.; Shah, Z.; Kanwal, N.; Zeb, S. Biodegradation of natural and synthetic rubbers: A review. *Int. Biodeter. Biodegr.* **2013**, *83*, 145–157. [CrossRef]
38. Nawong, C.; Umsakul, K.; Sermwittayawong, N. Rubber gloves biodegradation by a consortium, mixed culture and pure culture isolated from soil samples. *Braz. J. Microbiol.* **2018**, *49*, 481–488. [CrossRef]
39. Chaikumpollert, O.; Yamamoto, Y.; Suchiva, K.; Kawahara, S. Protein-free natural rubber. *Colloid Polym. Sci.* **2012**, *290*, 331–338. [CrossRef]
40. Kawahara, S.; Kakubo, T.; Nishiyama, N.; Tanaka, Y.; Isono, Y.; Sakdapipanich, J.T. Crystallization behavior and strength of natural rubber: Skim rubber, deproteinized natural rubber, and pale crepe. *J. Appl. Polym. Sci.* **2000**, *78*, 1510–1516. [CrossRef]
41. Arayapranee, W. Rubber abrasion resistance. In *Abrasion Resistance of Materials*, 1st ed.; Adamiak, M., Ed.; In Tech: Rijeka, Croatia, 2012; Volume 8, pp. 147–166.
42. Ahmad, M.R.; Ahmad, N.A.; Suhaimi, S.A.; Bakar, N.A.A.; Ahmad, W.Y.W.; Salleh, J. Tensile and tearing strength of uncoated and natural rubber latex coated high strength woven fabrics. In Proceedings of the 2012 IEEE Symposium on Humanities, Science and Engineering Research, Kuala Lumpur, Malaysia, 24–27 June 2012; pp. 541–545.
43. Rose, K.; Steinbüchel, A. Biodegradation of natural rubber and related compounds: recent insights into a hardly understood catabolic capability of microorganisms. *Appl. Environ. Microb.* **2005**, *71*, 2803–2812. [CrossRef]

materials

MDPI

Article

Toughened Poly (Lactic Acid)—PLA Formulations by Binary Blends with Poly(Butylene Succinate-*co*-Adipate)—PBSA and Their Shape Memory Behaviour

Diego Lascano [1], Luis Quiles-Carrillo [2,*], Rafael Balart [2], Teodomiro Boronat [2] and Nestor Montanes [2]

[1] Escuela Politécnica Nacional, 17-01-2759 Quito, Ecuador; dielas@epsa.upv.es
[2] Technological Institute of Materials (ITM), Universitat Politècnica de València (UPV), Plaza Ferrándiz y Carbonell 1, 03801 Alcoy, Spain; rbalart@mcm.upv.es (R.B.); tboronat@dimm.upv.es (T.B.); nesmonmu@upvnet.upv.es (N.M.)
* Correspondence: luiquic1@epsa.upv.es; Tel.: +34-966-528-433

Received: 16 January 2019; Accepted: 16 February 2019; Published: 22 February 2019

Abstract: This study reports the effect of poly(butylene succinate-*co*-adipate) (PBSA) on the mechanical performance and shape memory behavior of poly(lactic acid) (PLA) specimens that were manufactured by injection molding and hot-press molding. The poor miscibility between PLA and PBSA was minimized by the addition of an epoxy styrene-acrylic oligomer (ESAO), which was commercially named Joncryl®. It was incorporated during the extrusion process. Tensile, impact strength, and hardness tests were carried out following international standards. PLA/PBSA blends with improved mechanical properties were obtained, which highlighted the sample that was compatibilized with ESAO, leading to a remarkable enhancement in elongation at break, but showing poor shape memory behaviour. Field Emission Scanning Electron Microscopy (FESEM) images showed how the ductile properties were improved, while PBSA loading increased, thus leading to minimizing the brittleness of neat PLA. The differential scanning calorimetry (DSC) analysis revealed the low miscibility between these two polymers and the improving effect of PBSA in PLA crystallization. The bending test carried out on the sheets of PLA/PBSA blends showed the direct influence that the PBSA has on the reduction of the shape memory that is intrinsically offered by neat PLA.

Keywords: poly (lactic acid) (PLA); poly(butylene succinate-co-adipate) (PBSA); binary blends; shape memory behaviour

1. Introduction

The use of polymers and plastics in our daily life is almost mandatory due to their huge range of properties. For this reason, the demand for these materials has remarkably increased in the last decades. Unlike their production, the treatment of these materials after their end-of-life has been neglected, resulting in the oversaturation of plastic wastes in the environment. Since most of these plastics are synthetic, petroleum-derived materials, they have a high resistance to microbial degradation, so their decomposition is complex and extremely slow [1]. The development of new environmentally friendly polymeric materials has become a leading force in this industry because of this. The environmentally friendly properties of a polymer could be related to their origin (bio-sourced) or to their end-of-life (biodegradable or disintegrable in controlled compost soil). Taking into account their origin, some of the polymers have been fully or partially obtained from renewable resources, such as poly(ethylene) (PE) and poly(propylene) (PP) from sugarcane, poly(amides) (PAs) from castor oil, poly(carbonate) (PC) from corn, and so on [2–4]. These biobased polymers are identical to their counterpart petroleum-derived polymers and despite that they are not biodegradable, they have a positive effect on carbon footprint [3]. Another interesting group of environmentally friendly polymers is that which includes some petroleum-derived poly(esters), but, due to the nature of the ester group, they can disintegrate in controlled compost conditions. This groups includes poly(ε-caprolactone) (PCL), poly(butylene succinate) (PBS), poly(butylene succinate-co-adipate) (PBSA), poly(glycolic acid) (PGA), among others [1,5–7]. Finally, the most interesting group of environmentally friendly polymers is that of biobased and biodegradable polymers. Polysaccharides (starch, cellulose, chitin, and so on), protein based polymers (gluten, soybean protein, casein, collagen, among others), and bacterial polymers, such as poly(3-hydroxybutyrate) (PHB) or poly(3-hydroxybutyrate-*co*-3-hydroxyvalerate) (PHBV) are included in this group. Although these polymers are very promising, their properties are still quite far from those of commodities and engineering plastics [8–11].

Nowadays, poly(lactic acid) (PLA) is one of the most studied aliphatic polyesters thanks to its good mechanical properties and it can be either obtained from petroleum or renewable resources. Bio-sourced PLA is produced by anaerobic fermentation of sugars that are derived starch-rich plants, such as corn, sugarcane, beet sugar, potato, and so on [12]. It can also be obtained through the direct condensation of lactic acid and by ring opening polymerization of cyclic lactide dimer (ROP) [6,13]. The excellent tensile strength that PLA presents is the reason why it is used as an alternative to conventional plastics, such as high and low-density poly(ethylenes) (HDPE/LDPE), poly(styrene) (PS), and poly(ethylene terephthalate) (PET) [6]. It can be manufactured in a wide range of processing methods, such as conventional extrusion, injection molding, blow molding, film forming, three-dimensional (3D)-printed, and so on [14–17]. All of these features make PLA exceptionally useful in the packaging industry, food containers [12,18], and in biomedicine for controlled drug delivery and tissue engineering [19]. Although PLA presents good balanced properties and remarkable environmental benefits, its use has been limited due to its cost [13] and, in addition, it is a quite brittle material. With the aim of improving the ductility and toughness of PLA [20], many research approaches have been used. One is plasticization with conventional plasticizers, such as poly(ethylene glycol) (PEG), poly(propylene glycol) (PPG), lactic acid oligomers (OLAs), modified vegetable oils (MVOs), esters from citric acid or adipic acid, and so on [21–23]. All of these plasticizers contribute to improving ductile properties due to a remarkable decrease in the glass transition temperature, T_g, but the mechanical resistant properties are highly reduced. Another approach to overcome (or minimize) the extremely high brittleness of PLA is by blending with other flexible polymers. Regarding blends, compatibility/miscibility issues must be taken into account [24]. Bhatia et al. [25] have reported the properties of binary blends of PLA with PBS. The addition of 30 wt % PBS to PLA resulted in a clear increase in toughness, but over 50 wt % PBS addition, the clarity of the materials is reduced. As PLA and PBS show restricted miscibility, Harada et al. [26] used an isocyanate as a reactive processing aid to increase the impact strength of PLA/PBS binary blends. PCL is another flexible aliphatic poly(ester) that can provide increased toughness to PLA, as reported by Matta et al. [6]. Poly(butylene adipate-*co*-terephthalate) (PBAT) has great flexibility and it maintains excellent biodegradability properties. As can be shown by Khatsee et al. [27], binary PLA/PBAT blends can be obtained by electrospinning for controlled antibiotic release. Poly(propylene carbonate) (PPC) has been successfully used in PLA blends with good shape memory polymers [28].

Taking advantage of the PLA structure and materials that are based on PLA, a new research field has emerged, which is shape memory. Because of its particular structure, this field is promising for PLA. PLA has crystalline domains that define the permanent shape and switching segments that fix the temporary shape [29]. This particular structure allows for PLA to switch from a temporary shape to its permanent shape. The switching segments are activated by an external stimulus that can be either physical or chemical, but temperature is the most common stimulus leading to a thermal-responsive memory shape by selecting the appropriate temperature cycle regarding the T_g [30–32]. Shape memory behaviour has been used in electronics [29], and one of most recent applications include pieces that are made by 3D printing with PLA matrix filaments [33]. As these 3D-printed parts can change their shape by heating/cooling above/below the T_g, this technology is known as four-dimensional (4D)-printing. Taking advantage of the PLA biocompatibility with the human body, recent applications of PLA are focused on tissue engineering by making scaffold systems of physical blends, with TPU for support material in cartilage or bone repair [32,34]. The aim of this research is to enhance PLA toughness, through physical blending with poly(butylene succinate-*co*-adipate) (PBSA) to obtain flexible shape memory polymers with improved toughness. Despite that PBSA is based on fossil resources, it can undergo disintegration in controlled compost [35,36]. PBSA could offer high flexibility and great impact strength properties to PLA [37]. As these two poly(esters) show restricted miscibility, an epoxy styrene-acrylic oligomer (ESAO) will be used.

2. Materials and Methods

2.1. Materials.

PLA was an IngeoTM Biopolymer 6201D that was obtained from NatureWorks (Minnetonka, MN, USA). It is a thermoplastic resin derived from annually renewable resources. Its glass transition temperature, T_g, is comprised between 55–60 °C and the melt peak temperature is located in the 155–170 °C range. Its density is 1.25 g cm^{-3} and it possesses a melt flow index, MFI of 15–30 g/10 min. With regard to the flexible polymer for binary blends, an aliphatic poly(ester) copolymer, PBSA, was used. A commercial Bionolle grade 3002 MD was obtained from Showa Denko Europe GmbH (Munich, Germany). It shows good processability for extrusion and blow molding. The density reported is for this PBSA grade is 1.23 g cm^{-3}. Regarding its thermal properties, its T_g is close to −45 °C and the melt peak temperature is 94 °C.

A multi-functional epoxy-based styrene-acrylic oligomer (ESAO) was used as a compatibilizer. A commercial ESAO grade JONCRYL® ADR-4300 that was distributed by Basf (Barcelona, Spain) was used. This compatibilizer has a glass temperature T_g of 56 °C and an epoxy equivalent weight of 445 g mol^{-1}. Figure 1 shows the schematic representation of the structures of both PLA and PBS and the ESAO generic structure.

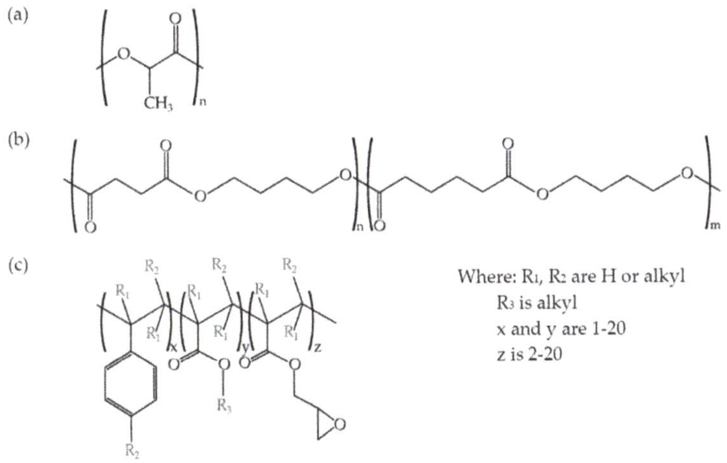

Figure 1. Schematic representation of (**a**) poly(lactic acid) (PLA), (**b**) poly(butylene succinate-*co*-adipate) (PBSA), and (**c**) epoxy styrene-acrylic oligomer (ESAO) (Joncryl® ADR-4300).

2.2. *Manufacturing of PLA/PBSA Binary Blends*

Prior to processing, as both of the poly(esters) are highly sensitive to hydrolysis, PLA and PBSA were dried at 50 °C for 48 h. Table 1 shows the labelling and the compositions of the developed blends. According to the literature [38], PLA blends with 20 wt % of PBS have a good balance between mechanical properties and shape memory behaviour. Thus, the addition of a compatibilizer to the $PLA_{80}PBSA_{20}$ mixture was decided to determine whether it causes any effect in its properties. To obtain homogeneous mixtures, all of the materials were subjected to a mechanical pre-mixing for 3 min in a zipper bag. These mixtures were extruded using a co-rotating twin-screw extruder ZSK-18 MEGAlab from Coperion (Stuttgart, Germany) that was equipped with a screw diameter of 18 mm with a length to diameter ratio, L/D of 48. The dosage of each component was controlled by a side twin-screw feeder ZS-B 18 from K-Tron (Pitman, NJ, USA). The screw speed was set to 180 rpm using a temperature profile of 145–155–160–180–185–190–190 °C from the feeding to the die. The feed rate was set to 2 kg h^{-1}. Once the different blends were extruded, they were cooled down to 15 °C in a water bath and subsequently pelletized using an air knife unit.

The pelletized compounds were shaped into standard samples for characterization by injection molding in a Meteor 270/75 from Mateu & Solé (Barcelona, Spain). The temperature profile was set at 175–180–185–190 °C for PLA-based blends, while PBSA was programmed with a lower temperature profile of 105 °C for the different heating barrels. A clamping force of 75 tons was applied. The cooling time was set to 10 s.

Table 1. Composition and labelling of PLA/PBSA binary blends.

Code	PLA (wt %)	PBSA (wt %)	Joncryl® ADR (phr *)
PLA	100	-	-
$PLA_{90}PBSA_{10}$	90	10	-
$cPLA_{80}PBSA_{20}$	80	20	0.5
$PLA_{80}PBSA_{20}$	80	20	-
$PLA_{70}PBSA_{30}$	70	30	-
PBSA	-	100	-

* phr denotes the weight parts of Joncryl® per hundred parts by weight of PLA/PBSA blend.

2.3. *Mechanical Properties*

The tensile test was carried out following the recommendation of the ISO 527 standard while using a mechanical universal testing machine ELIB 50 from S.A.E. Ibertest (Madrid, Spain). It was equipped with a 5 kN load cell and the selected crosshead speed of 10 mm min^{-1} was used. The impact strength test was performed on a Charpy pendulum from Metrotec (San Sebastián, Spain), following the recommendations of ISO 179, using a 6 J pendulum on unnotched samples and a 1 J pendulum on notched samples ("V" type, 2 mm depth and a radio of 0.5 mm). Hardness measurements were obtained in a durometer 676-D from J. Bot Instruments (Barcelona, Spain) using Shore D scale following ISO 868.

All of the tests were carried out at room temperature with at least five samples and the corresponding properties were averaged.

2.4. *Thermal Characterization*

The main thermal transitions were analyzed by differential scanning calorimetry (DSC) in an 821 DSC calorimeter from Mettler-Toledo, Inc. (Schwerzenbach, Switzerland). The samples weight was between 5–10 mg. The samples were placed into standard sealed aluminum pans (40 μL) and then subjected to a thermal cycle consisting of three steps: initial heating from 30 °C to 200 °C, followed by a cooling process to −60 °C and after that, a second heating process up to 350 °C. The heating/cooling rate was set to 10 °C min^{-1}. These tests were carried out under a dry atmosphere with a constant nitrogen flow of 30 mL min^{-1}. The glass temperature (T_g), cold crystallization temperature peak (T_{cc}), the melting temperature (T_m), and the melt enthalpy (ΔH_m) were obtained from second heating step in this analysis.

The thermal degradation (weight loss) and thermal stability were followed by thermogravimetric analysis (TGA) in a TGA/SDTA 851 thermobalance from Mettler-Toledo (Schwerzenbach, Switzerland). The average weight of the samples was between 5–10 mg. These samples were placed on standard alumina crucibles with a total volume capacity of 70 μL and subsequently exposed to a heating program from 30 °C up to 700 °C at a constant heating rate of 20 °C min^{-1} in an air atmosphere.

2.5. Morphology Characterization by Field Emission Scanning Electron Microscopy (FESEM)

To analyze the morphology of fractured surfaces from impact tests, field emission scanning electron microscopy (FESEM) was used. A ZEISS ULTRA 55 FESEM microscope from Oxford Instruments (Abingdon, UK) was used working at an accelerating voltage of 2 kV. Prior to the test, the samples' surfaces were coated by an ultrathin gold-palladium layer in a high vacuum sputter coater EM MED20 from Leica Microsystem (Milton Keynes, UK).

2.6. Shape Memory Behaviour Characterization

The shape memory behaviour of the materials was evaluated using a conventional bending test on sheets with dimensions of 65 × 10 × 1 mm^3. These sheets were obtained by hot-press molding at 140 °C in a hot press Hoytom M.N.1417 (Bilbao, Spain) from Robima S.A. The switch transition temperature that leads the programming and the recovery cycle was set at 70 °C (PLA glass transition). The cooling temperature was set to 22 °C and the stabilization time was 15 min. The temporary shape was fixed to bending angles of 120° and 90°.

3. Results

3.1. Mechanical Properties and Morphology of Binary PLA/PBSA Blends

The tensile behaviour of binary PLA/PBSA blends is shown in Table 2. The tensile modulus, E_T of neat PLA is relatively high, about 1165 MPa. Regarding its tensile strength (σ_T), PLA offers a relatively high value of 64.0 MPa, as compared to other commodities. While intrinsic mechanical resistant properties of neat PLA are high, its elongation (ε_b) at break is only of 9.23%, which is responsible for high brittleness and the fragility of this material. Blending PLA with a flexible PBSA polymer has noticeable effects on overall mechanical performance. So that, the addition of 10 wt % PBSA to PLA leads to an expected decrease on mechanical resistant properties, such as E_T and σ_T, down to values of 1012 MPa and 52.6 MPa, respectively. As the wt % PBSA increases, both tensile modulus and strength decrease, which means that the brittle behaviour is diminished. At room temperature, PBSA is above its corresponding T_g value −41 °C [36], which means a flexible behaviour, as observed in Table 2. Above its T_g, the PBSA chains can move freely, as reported by Ojijo et al. [39], thus leading to high ε_b value of 432.7%. This improved chain mobility can exert a positive effect on PLA toughness.

Table 2. Summary of mechanical properties of binary PLA/PBSA blends obtained from tensile tests.

Code	Tensile Strength, σ_T (MPa)	Elastic Modulus, E_T (MPa)	Elongation at Break, ε_b (%)
PLA	64.0 ± 1.2	1165 ± 44	9.2 ± 1.5
$PLA_{90}PBSA_{10}$	52.6 ± 0.8	1012 ± 21	12.2 ± 3.8
$cPLA_{80}PBSA_{20}$	41.2 ± 2.0	754 ± 47	121.2 ± 18.7
$PLA_{80}PBSA_{20}$	42.0 ± 3.6	849 ± 75	29.7 ± 6.3
$PLA_{70}PBSA_{30}$	37.7 ± 3.0	625 ± 72	56.5 ± 10.3
PBSA	18.3 ± 1.6	159 ± 61	432.7 ± 57.4

As expected, with PBSA addition, the elongation at break tends to increase. Only the addition of 10 wt % PBSA to PLA gives an increased ε_b value of 12.2% and this is still more evident for PLA/PBSA blends containing 20 wt % and 30 wt % PBSA with ε_b values of 29.7% and 56.5%, respectively, thus leading to a clear increase in mechanical ductile properties.

Although good mechanical properties can be obtained by blending PLA with PBSA, the poor miscibility between these two polymers does not allow good interface interactions, as reported by Nofar et al. [40]. For this reason, a compatibilizer agent, namely Joncryl®, has been added to the binary blend of 80 wt % PLA and 20 wt % PBSA. This compatibilizer agent has been extensively used as chain extender in poly(esters) due to the reaction of epoxy groups with hydroxyl terminal groups in poly(esters) [41,42]. This particular behaviour can be positive in binary blends of poly(esters), as this compatibilizer can react either with the hydroxyl terminal groups of PLA and PBSA, thus leading to a compatibilization effect. The relatively low compatibilizer loading (0.5 phr) allows for this, since higher loadings could lead to branching, gel formation, and, even, some crosslinking [43,44]. As observed in Table 2, the uncompatibilized blend containing 20 wt % PBSA gives an ε_b value of 29.7%, while the compatibilized blend with the same composition offers an ε_b increase up to values of 121.2%, thus giving clear evidences of chain extension/compatibilization, as reported by Eslami et al. [45]. A low percentage of chain extender in blends leads to obtaining an improvement in the interface with both materials, improving some properties. In this particular case, this blend takes advantages of PBSA, improving the elongation at break. The increase in the ductile behaviour observed in PLA/PBSA blends and, specifically in the compatibilized PLA/PBSA blend, suggest an increase in toughness.

Nevertheless, it is worthy to note that toughness is not uniquely linked to ductile properties (i.e. elongation at break), but also to mechanical resistant properties (tensile strength). In this work, toughness has been quantitatively measured through the determination of the impact-absorbed energy in impact test (Charpy). Table 3 shows the impact strength values as a function of the composition of the developed materials. In a very first attempt, a 6 J pendulum was used, it did not provide enough energy to break some specimens, in order to have all the measurements in the same conditions, a "V" notch was done in all samples and then tested with a 1 J pendulum. As one can see, PLA is a brittle material with very low energy absorption (2.48 kJ m^{-2}) when compared to PBSA (26.02 kJ m^{-2}). As expected, the impact strength increases with the PBSA loading on binary blends up to values of 5.75 kJ m^{-2} for the uncompatibilized PLA/PBSA blend containing 30 wt % PBSA. It is worthy to note the good impact strength that was achieved with the blend with 20 wt % PBSA (3.28 kJ m^{-2}) and the clear positive effect of the compatibilizing effect provided by Joncryl®, since the same blend is able to reach an impact strength of about 4.33 kJ m^{-2}. Figure 2 presents the stress-strain curves of the developed blends. As it was expected, the stress tends to decrease when PBSA is added. On the other hand, the tensile strain or percentage elongation tends to increase, with the case of the compatibilized blend (cPLA$_{80}$PBSA$_{20}$) being more noticeable, which shows strain values of about 120%. This results in an increase in the area below the curve, together with the increasing tendency of the impact strength, makes an improvement on toughness when PBSA is added. With regard to hardness, the tendency is similar to other mechanical resistant properties. A clear decreasing tendency in the Shore D hardness values can be observed when PBSA is added. Nevertheless, the standard deviation does not allow for identifying a clear effect of the compatibilizer on Shore D values.

Table 3. Impact absorbed energy (Charpy test) and Shore D hardness of binary PLA/PBSA blends.

Code	Impact Strength (kJ m^{-2}) ("V" notched)	Impact Strength (kJ m^{-2}) (unnotched)	Shore D Hardness
PLA	2.48 ± 0.22	28.10 ± 2.40	78.80 ± 0.84
PLA$_{90}$PBSA$_{10}$	2.54 ± 0.34	23.03 ± 2.80	74.00 ± 2.74
cPLA$_{80}$PBSA$_{20}$	4.33 ± 0.02	28.90 ± 0.85	75.00 ± 1.00
PLA$_{80}$PBSA$_{20}$	3.28 ± 0.28	27.52 ± 2.13	73.00 ± 1.41
PLA$_{70}$PBSA$_{30}$	5.75 ± 0.60	N/B	72.20 ± 1.60
PBSA	26.02 ± 0.60	N/B	57.00 ± 0.71

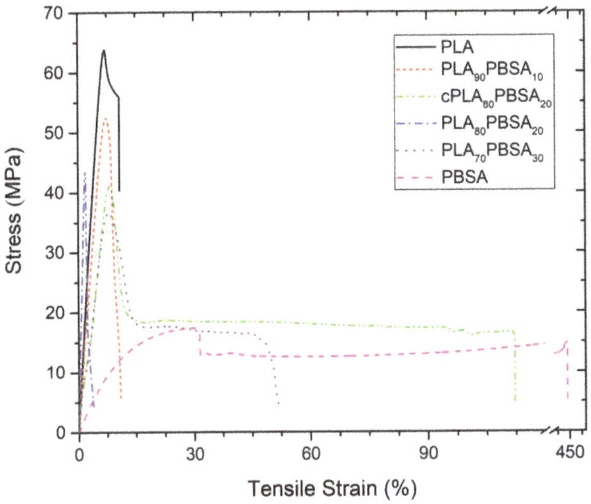

Figure 2. Strain–stress curves corresponding to binary PLA/PBSA blends.

This particular behaviour is directly related to the morphology of the developed materials (Figure 3). Figure 3a shows the FESEM images of the fractured surfaces corresponding to unblended PLA. As it can be seen, the surface is uniform and smooth typical of a brittle fracture, with different crack growths [41]. This morphology is in total agreement with previous mechanical results. A remarkable change in the fracture surface morphology can be observed in Figure 3e, which corresponds to the binary PLA/PBSA blend with 30 wt % PBSA. The surface is not as smooth and it shows increased roughness due to increased plastic deformation. Similar morphology can be observed for uncompatibilized PLA/PBSA blends with different PBSA loading. The compatibilized blend with 20 wt % PBSA shows different fracture morphology due to its higher elongation at break, which allows more deformation before fracture (Figure 3c). It can be seen in Figure 3c that the surface presents a greater tear on it, which suggests increased PLA-PBSA interaction, because the compatibilizer works as a bridge between these two polymers, as reported Eslami et al. [45]. Obviously, PBSA shows a typical ductile fracture (Figure 3f), with a wavy surface topography that is representative for plastic deformation (even in impact conditions).

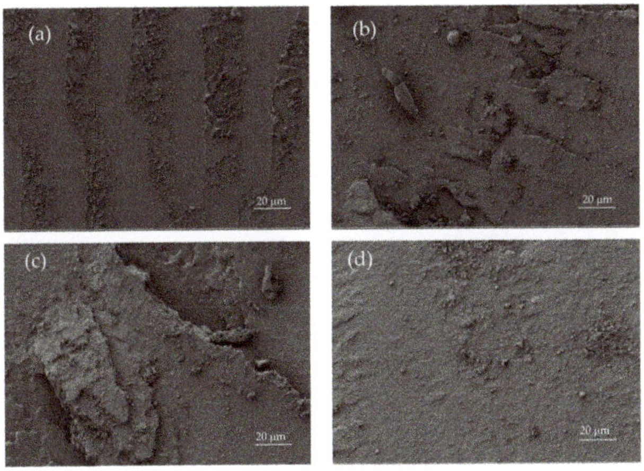

Figure 3. *Cont.*

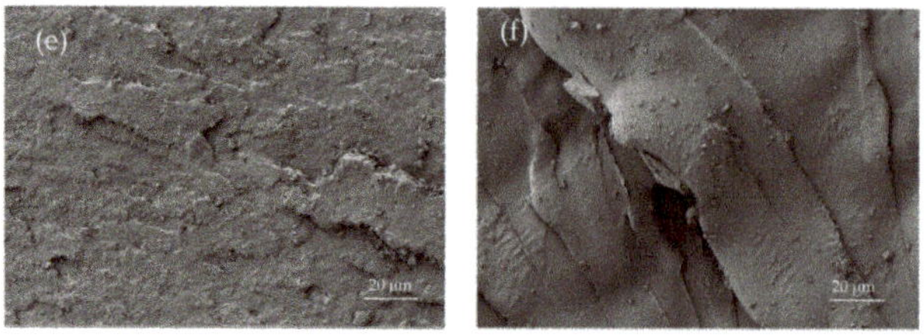

Figure 3. Field emission scanning electron microscopy (FESEM) images at ×500 corresponding to fractured surfaces from impact tests of (**a**) PLA, (**b**) $PLA_{90}PBSA_{10}$, (**c**) $cPLA_{80}PBSA_{20}$, (**d**) $PLA_{80}PBSA_{20}$, (**e**) $PLA_{70}PBSA_{30}$, and (**f**) PBSA.

3.2. Thermal Behaviour of Binary PLA/PBSA Blends

Figure 4 shows the DSC thermograms corresponding to the second heating cycles of the developed PLA/PBSA blends, while Table 4 gathers the main parameters that were obtained by DSC. The first thermal transition that can be observed in Figure 4 is the glass transition temperature, T_g of the PLA-rich phase. This is located between 60–70 °C. As the temperature rises above the T_g, an exothermal peak appears between 95–110 °C, which corresponds to the cold crystallization process of PLA chains. This cold crystallization process is related to a rearrangement of the PLA chains to an ordered structure that is activated by temperature. At higher temperatures comprised in the 155–170 °C range, an endothermic peak is observed, which is attributable to the melting process of the crystalline domains in PLA [46]. Regarding neat PBSA, it shows a melting process comprised between 70–100 °C that overlaps the cold crystallization process of PLA. Neat PLA shows a T_g of 63.4 °C and this is slightly reduced to 61.1 °C in the blend with 10 wt % PBSA, which suggests slight miscibility between these two poly(esters). Higher PBSA contents of 30 wt % only promote a slight decrease in T_g down to values of 60.6 °C, which corroborates the restricted miscibility between PLA and PBSA. This slight change in the T_g by the addition of PBSA to PLA is a clear indication of restricted miscibility of these two biopolymers, as reported Lee et al. [5]. Another important finding that can be outlined from DSC thermograms is the cold crystallization process. Although it overlaps with the melting process of PBSA, one important change can be identified. In particular, the cold crystallization process is moved toward lower temperatures, which means that the energy barrier for PLA crystallization is reduced. This could be due to partial miscibility between PLA and PBSA, but the most important mechanism that is responsible for this is the melting of the PBSA-rich phase that contributes to increase chain motion, thus allowing PLA chains to arrange into a packed structure at lower temperatures, due to the lubricant effect of the melted PBSA-rich phase, as reported by Lee et al. [5] in previous researches. In particular, the cold crystallization peak temperature, T_{cc} is reduced below 100 °C, while the typical T_{cc} for neat PLA is close to 109 °C. As expected, the compatibilization effect that Joncryl® provides leads somewhat to a restriction in chain motion, thus leading to a slightly higher T_{cc} value, as compared to all other blends. This similar behaviour has been reported in PLA-based materials by L. Quiles-Carrillo et al. [47] in PLA/Almond shell flour composites that were compatibilized with ESAO. The T_{cc} of PLA/ASF composites that were obtained were even higher than neat PLA. This shift of the cold crystallization process indicates that crystallites of PLA can be formed at lower temperatures due to the effect of PBSA, as suggested by Ojijo et al. [39]. With regard to the melt peak temperature of PLA (T_{m_PLA}), the changes are negligible. Neat PLA shows a T_m of 170.9 °C and the melt peak temperature of the PLA-rich phase in the binary PLA/PBSA blends decreases to 168 °C. Frenz et al. [48] have reported an increase in the melt strength of PLA and other poly(esters) by the addition of chain extenders, such as Joncryl®, but no remarkable changes in the peak temperature can be observed.

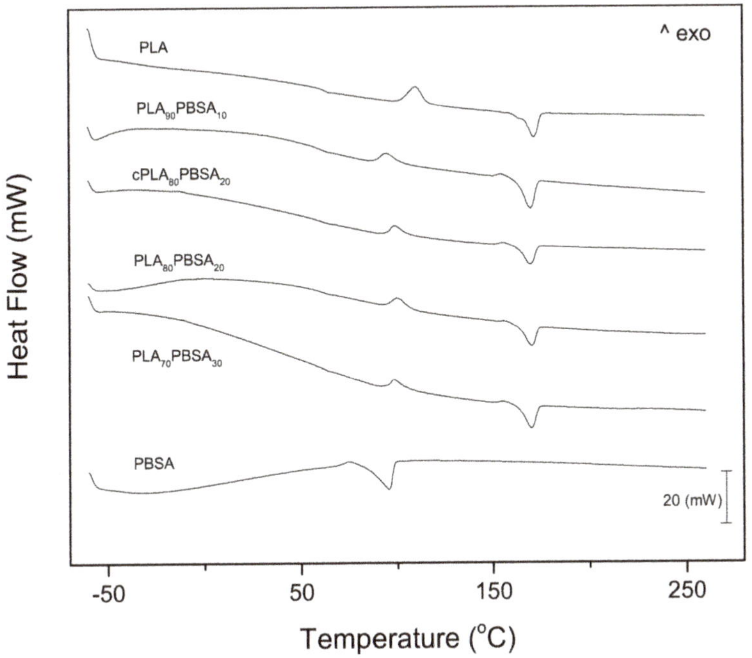

Figure 4. Differential scanning calorimetry (DSC) scans (second heating cycle) corresponding to binary PLA/PBSA blends.

Table 4. Main thermal parameters obtained by differential scanning calorimetry (DSC) and of binary PLA/PBSA blends.

Code	T_{g_PLA} (°C)	T_{cc_PLA} (°C)	T_{m_PLA} (°C)	T_{m_PBSA} (°C)
PLA	63.4 ± 0.6	109.9 ± 1.1	170.9 ± 3.3	-
$PLA_{90}PBSA_{10}$	61.1 ± 1.2	94.5 ± 1.7	168.9 ± 2.2	-
$cPLA_{80}PBSA_{20}$	61.1 ± 0.9	98.7 ± 1.4	168.8 ± 2.7	-
$PLA_{80}PBSA_{20}$	61.7 ± 0.6	100.3 ± 1.2	169.4 ± 2.6	-
$PLA_{70}PBSA_{30}$	60.6 ± 1.2	97.9 ± 1.0	168.7 ± 2.9	-
PBSA	-	-	-	95.2 ± 1.4

Regarding the thermal stability at high temperatures, Figure 5 gathers the comparative TGA degradation curves corresponding to the neat PLA, neat PBSA, and PLA/PBSA blends. The TGA thermograms indicate that neat PLA and PLA/PBSA blends degradation occurs in a single step process. Regarding neat PBSA, its degradation process takes place in two stages. The first one occurs at about 400 °C, with an associated weight loss of around 90%, and it is in accordance with Renoux et al. [38]. The second stage is about 500 °C and only a 10% weight loss occurs.

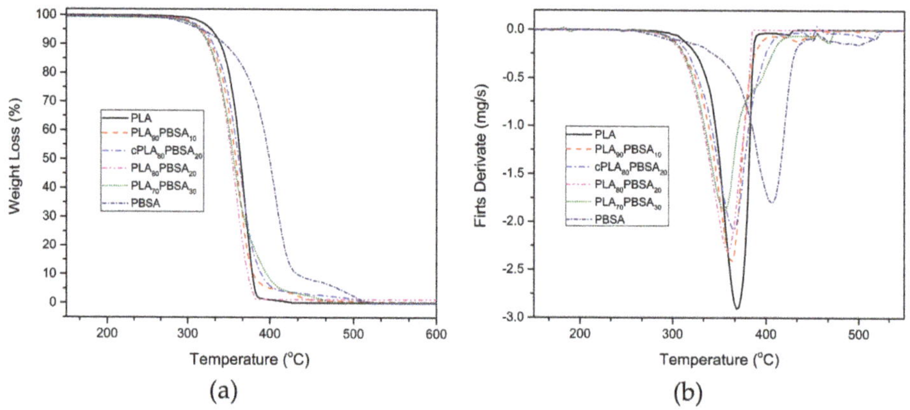

Figure 5. Thermogravimetric (TGA) degradation curves corresponding to binary PLA/PBSA blends, (**a**) thermogravimetry (TG) mass loss and (**b**) differential thermogravimetry (DTG) first derivative.

Table 5 shows the thermal parameters that are related to the different materials, specifically the characteristic temperatures at 5% weight loss ($T_{5\%}$) and the maximum degradation rate temperature (T_{max}). Even though a slight decrease in the thermal stability in $T_{5\%}$ can be detected by the presence of PBSA, in general, it does not cause a noticeable effect in the PLA matrix. In fact, the PLA/PBSA blends show typical PLA behaviour. At high temperatures, a slight increase in thermal stability can be observed, the influence of PBSA on the PLA/PBSA mixtures is evident, since they show a behavior close to the PBSA at a temperature range that is close to its degradation, with neat PBSA being the most thermally stable. The compatibilization effect that Joncryl® provides leads to gaining thermal stability and it can be observed in Figure 5. The uncompatilized blend containing 20 wt % PBSA degrades at earlier temperature when compared with its compatibilized counterpart, which corroborates what Frenz et al. [48] reported.

Table 5. Main thermal parameters obtained by thermogravimetric analysis (TGA) of binary PLA/PBSA blends.

Code	TGA		
	$T_{5\%}$ (°C)	T_{Max} (°C)	$Mass_{Residual}$ (%)
PLA	328.7 ± 5.25	368.1 ± 6.3	0.05 ± 0.01
$PLA_{90}PBSA_{10}$	317.0 ± 4.4	362.5 ± 4.3	0.02 ± 0.02
$cPLA_{80}PBSA_{20}$	319.3 ± 3.5	365.4 ± 5.2	0.12 ± 0.02
$PLA_{80}PBSA_{20}$	312.3 ± 5.2	359.5 ± 5.4	1.37 ± 0.01
$PLA_{70}PBSA_{30}$	314.7 ± 4.7	355.8 ± 5.7	0.20 ± 0.01
PBSA	317.0 ± 5.7	405.3 ± 7.8	0.01 ± 0.01

3.3. Shape Memory Behaviour of Binary PLA/PBSA Blends

Several authors have proposed different techniques to characterize the shape memory behaviour in biopolymers, such as tensile and DMA tests [49,50]. These are specifically used to determine the recovery rate after subjecting the sample to a specific thermal cycle. Despite this, the use of a conventional bending test is widely used because of its simplicity and the quality of the information that it can provide. The measurement of the recovery angle is a qualitative/quantitative method to visualize the shape memory behavior and it represents a way to understand the shape memory behaviour of the blends and the effect of the PBSA addition. Table 6 shows the recovery percentage of the neat PLA and PLA/PBSA blends after the programming and recovery cycle. As we expected, the neat PLA presents high values of shape memory reaching a recovery of 70% and 58% corresponding to the 90° and 120° flexural deformation, respectively. Notice that PLA shape memory effect works better for small deformations.

Table 6. Shape memory behaviour parameters corresponding to binary PLA/PBSA blends.

Code	Permanent Shape (º)	Temporal Shape (º)	Final Shape (º)	Recovery (%)	Temporal Shape (º)	Final Shape (º)	Recovery (%)
PLA			153	70		155	58
$PLA_{90}PBSA_{10}$			155	72		158	63
$cPLA_{80}PBSA_{20}$	180	90	108	20	120	141	35
$PLA_{80}PBSA_{20}$			144	60		146	43
$PLA_{70}PBSA_{30}$			109	21		140	33

In fact, with the addition of PBSA, the PLA shape memory capability tends to decrease. It is worthy to note the particular case of $PLA_{90}PBSA_{10}$ shown in Figure 6b,e, which presents the best recovery behaviour of all of the developed materials, even better than the neat PLA. Tcharkhtchi et al. [38] remarked that, to get an adequate shape memory effect, the PBS percentage must be of about 20%, which gives consistency to the obtained results.

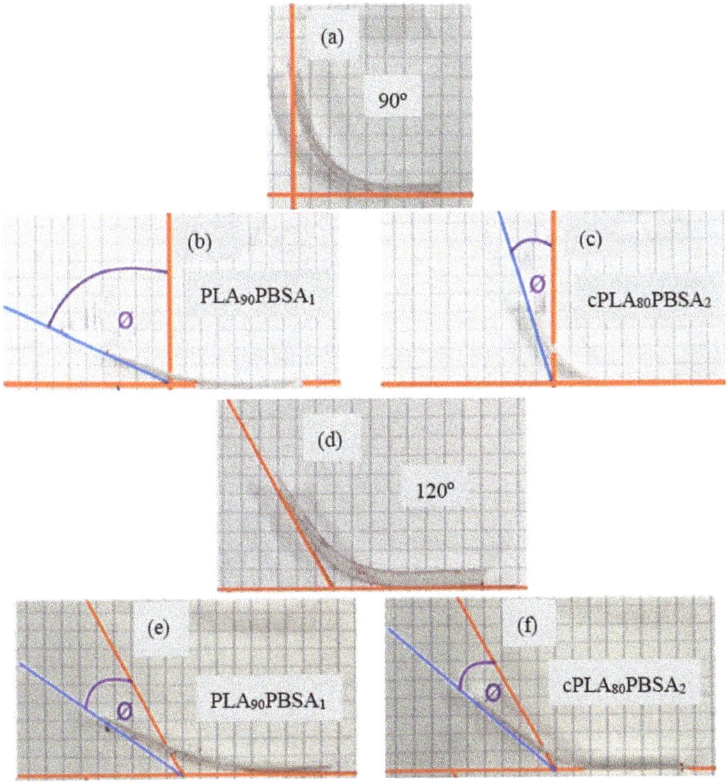

Figure 6. Images of the shape memory behaviour of binary PLA/PBSA blends in flexural conditions with a permanent shape after deformation and shape recovery of (**a**) temporary shape (90°), (**b**) shape recovery of PLA90PBSA10 (90°), (**c**) shape recovery of cPLA80PBSA20 (90°), (**d**) temporary shape (120°), (**e**) shape recovery of PLA90PBSA10 (120°), and (**f**) shape recovery of cPLA80PBSA20 (120°).

The clear negative effect that is provided by Joncryl® is noticeable, since almost all of the shape memory capability of the mixtures is lost, reaching values 20% and 35%, corresponding to the 90° and 120° deformation, respectively, as can be seen in Figure 6c,f.

4. Conclusions

Injection molding and the hot-press molding process made binary PLA/PBSA blends. The influence of PBSA has a noticeable effect on mechanical performance, decreasing the tensile modulus and strength, which means that the brittle behaviour of the blends has diminished, thus leading to an increase in mechanical ductile properties. The compatibilized blend shows elongation at break values of 121%, remarking the evidence of the compatibilization of these two polymers, adding up to the high values that were obtained in the impact test, suggesting the improvement on toughness.

The thermal analysis shows that the presence of PBSA induces a slightly decrease in PLA glass transition temperature (T_g), which corroborates the low miscibility between PLA and PBSA. The PBSA effect is noticeable in the PLA matrix, enhancing the PLA crystallization, and leading to a low cold crystallization temperature (T_{cc}). Although the presence of PBSA induced a decrease in the shape memory behavior on PLA, 10 wt % PBSA in a PLA/PBSA blends shows the best shape memory recovery, reaching values that were even better than PLA.

Author Contributions: The idea was proposed by N.M., D.L., and R.B.; Experimental stage was performed by D.L., L.Q.-C., N.M., and T.B.; The analysis of data was carried out by D.L., L.Q.-C., N.M; The stage of writing was performed by D.L. and L.Q.-C.; Examination and corrections, N.M., T.B., L.Q.-C and R.B.; Supervision throughout the project was carried out by R.B., N.M. and T.B.

Funding: This research was supported by the Spanish Ministry of Economy and Competitiveness (MINECO) program numbers MAT2017-84909-C2-2-R. L.Q.-C wants to thank GV for his FPI grant (ACIF/2016/182) and the MECD for his FPU grant (FPU15/03812).

Conflicts of Interest: The authors declare no conflict of interest.

References

1. Vroman, I.; Tighzert, L. Biodegradable polymers. *Materials* **2009**, *2*, 307–344. [CrossRef]
2. Axelsson, L.; Franzén, M.; Ostwald, M.; Berndes, G.; Lakshmi, G.; Ravindranath, N.H. Perspective: Jatropha cultivation in southern India: Assessing farmers' experiences. *Biofuels Bioprod. Biorefin.* **2012**, *6*, 246–256. [CrossRef]
3. Shen, L.; Haufe, J.; Patel, M.K. Product overview and market projection of emerging bio-based plastics PRO-BIP 2009. In *Report for European Polysaccharide Network of Excellence (EPNOE) and European Bioplastics*; Universitat Utrecht: Utrecht, The Netherlands, 2009; Volume 243.
4. Brodin, M.; Vallejos, M.; Opedal, M.T.; Area, M.C.; Chinga-Carrasco, G. Lignocellulosics as sustainable resources for production of bioplastics—A review. *J. Clean. Prod.* **2017**, *162*, 646–664. [CrossRef]
5. Lee, S.; Lee, J.W. Characterization and processing of Biodegradable polymer blends of poly (lactic acid) with poly (butylene succinate adipate). *Korea Aust. Rheol. J.* **2005**, *17*, 71–77.
6. Matta, A.K.; Rao, R.U.; Suman, K.N.S.; Rambabu, V. Preparation and Characterization of Biodegradable PLA/PCL Polymeric Blends. *MSPRO* **2014**, *6*, 1266–1270. [CrossRef]
7. Geueke, B. Dossier—Bioplastics as food contact materials. *Food Packag. Forum* **2014**, 1–8. [CrossRef]
8. Baran, A.; Vrábel, P.; Olčák, D.; Chodák, I. Solid state13C-NMR study of a plasticized PLA/PHB polymer blend. *J. Appl. Polym. Sci.* **2018**, *135*, 46296. [CrossRef]
9. Pokhrel, C.P.; Ohga, S. Submerged culture conditions for mycelial yield and polysaccharides production by Lyophyllum decastes. *Food Chem.* **2007**, *105*, 641–646. [CrossRef]
10. Bledzki, A.K.; Jaszkiewicz, A. Mechanical performance of biocomposites based on PLA and PHBV reinforced with natural fibres—A comparative study to PP. *Compos. Sci. Technol.* **2010**, *70*, 1687–1696. [CrossRef]
11. Duesterhoeft, D. Congress, Legislation, etc.: About Congress. *Blume Libr. LibGuide* **2016**. [CrossRef]
12. Sudesh, K.; Iwata, T. Sustainability of biobased and biodegradable plastics. *Clean Soil Air Water* **2008**, *36*, 433–442. [CrossRef]
13. Gruber, P.; O'Brien, M. Polylactides "NatureWorks® PLA". *Biopolym. Online* **2005**, 235–239. [CrossRef]

14. Saleem, M.; Tanveer, F.; Ahmad, A.; Gilani, S.A. Correlation between shoulder pain and functional disability among nurses. *Rawal Med. J.* **2018**, *43*, 483–485.
15. Koide, S.; Shi, J. Microbial and quality evaluation of green peppers stored in biodegradable film packaging. *Food Control* **2007**, *18*, 1121–1125. [CrossRef]
16. Jonoobi, M.; Harun, J.; Mathew, A.P.; Oksman, K. Mechanical properties of cellulose nanofiber (CNF) reinforced polylactic acid (PLA) prepared by twin screw extrusion. *Compos. Sci. Technol.* **2010**, *70*, 1742–1747. [CrossRef]
17. Harris, A.M.; Lee, E.C. Improving mechanical performance of injection molded PLA by controlling crystallinity. *J. Appl. Polym. Sci.* **2008**, *107*, 2246–2255. [CrossRef]
18. Serna, L.; Albán, F. Ácido Poliláctico (PLA): Propiedades y Aplicaciones. *Ingeniería y Competitividad* **2003**, *5*, 16–26. [CrossRef]
19. Hyon, S.-H.H. Biodegradable poly (lactic acid) microspheres for drug delivery systems. *Yonsei Med. J.* **2000**, *41*, 720–734. [CrossRef]
20. Jiang, L.; Wolcott, M.P.; Zhang, J. Study of biodegradable polylactide/poly(butylene adipate-co-terephthalate) blends. *Biomacromolecules* **2006**, *7*, 199–207. [CrossRef]
21. Piorkowska, E.; Kulinski, Z.; Galeski, A.; Masirek, R. Plasticization of semicrystalline poly(l-lactide) with poly(propylene glycol). *Polymer* **2006**, *47*, 7178–7188. [CrossRef]
22. Martin, O.; Averous, L. Plasticization and properties of biodegradable multiphase systems polymer. *Polymer* **2001**, *42*, 6209–6219. [CrossRef]
23. Burgos, N.; Martino, V.P.; Jiménez, A. Characterization and ageing study of poly(lactic acid) films plasticized with oligomeric lactic acid. *Polym. Degrad. Stab.* **2013**, *98*, 651–658. [CrossRef]
24. Li, Y.; Shimizu, H. Toughening of polylactide by melt blending with a biodegradable poly(ether)urethane elastomer. *Macromol. Biosci.* **2007**, *7*, 921–928. [CrossRef] [PubMed]
25. Bhatia, A.; Gupta, R.K.; Bhattacharya, S.N.; Choi, H.J. Compatibility of Biodegradable Poly (lactic acid) (PLA) and Poly (butylene succinate) (PBS) Blends for Packaging Application. *Korea Aust. Rheol. J.* **2007**, *19*, 125–131.
26. Harada, M.; Ohya, T.; Iida, K.; Hayashi, H.; Hirano, K.; Fukuda, H. Increased Impact Strength of Biodegradable Poly (lactic acid)/Poly (butylene succinate) Blend Composites by Using Isocyanate as a Reactive Processing Agent. *J. Appl. Polym. Sci.* **2007**, *106*, 1813–1820. [CrossRef]
27. Khatsee, S.; Daranarong, D.; Punyodom, W.; Worajittiphon, P. Electrospinning polymer blend of PLA and PBAT: Electrospinnability-solubility map and effect of polymer solution parameters toward application as antibiotic-carrier mats. *J. Appl. Polym. Sci.* **2018**, *135*, 46486. [CrossRef]
28. Qin, S.X.; Yu, C.X.; Chen, X.Y.; Zhou, H.P.; Zhao, L.F. Fully Biodegradable Poly(lactic acid)/Poly(propylene carbonate) Shape Memory Materials with Low Recovery Temperature Based on in situ Compatibilization by Dicumyl Peroxide. *Chin. J. Polym. Sci.* **2018**, *36*, 783–790. [CrossRef]
29. Fan, X.; Tan, B.H.; Li, Z.; Loh, X.J. Control of PLA Stereoisomers-Based Polyurethane Elastomers as Highly Efficient Shape Memory Materials. *ACS Sustain. Chem. Eng.* **2017**, *5*, 1217–1227. [CrossRef]
30. Mu, T.; Liu, L.; Lan, X.; Liu, Y.; Leng, J. Shape memory polymers for composites. *Compos. Sci. Technol.* **2018**, *160*, 169–198. [CrossRef]
31. Gil, E.S.; Hudson, S.M. Stimuli-reponsive polymers and their bioconjugates. *Prog. Polym. Sci.* **2004**, *29*, 1173–1222. [CrossRef]
32. Balk, M.; Behl, M.; Wischke, C.; Zotzmann, J.; Lendlein, A. Recent advances in degradable lactide-based shape-memory polymers. *Adv. Drug Deliv. Rev.* **2016**, *107*, 136–152. [CrossRef] [PubMed]
33. Liu, W.; Wu, N.; Pochiraju, K. Shape recovery characteristics of SiC/C/PLA composite filaments and 3D printed parts. *Compos. Part A Appl. Sci. Manuf.* **2018**, *108*, 1–11. [CrossRef]
34. Hong, H.; Wei, J.; Yuan, Y.; Chen, F.-P.; Wang, J.; Qu, X.; Liu, C.-S. A novel composite coupled hardness with flexibleness-polylactic acid toughen with thermoplastic polyurethane. *J. Appl. Polym. Sci.* **2011**, *121*, 855–861. [CrossRef]
35. Yasunobu, K.; Kazuhiko, F.; Yoshiharu, D.; Kunioka, M.; Kobayashi, G.; Shiotani, T.; Shima, Y.; Doi, Y.Y.; Eya, H.; Iwaki, N.; et al. *Enzymatic Degradation of Polymer Blends*; Doi, Y., Fukuda, K., Eds.; Elsevier: Amsterdam, The Netherlands, 1994; Volume 12, pp. 136–149.
36. Chen, Y.A.; Tsai, G.S.; Chen, E.C.; Wu, T.M. Thermal degradation behaviors and biodegradability of novel nanocomposites based on various poly[(butylene succinate)-co-adipate] and modified layered double hydroxides. *J. Taiwan Inst. Chem. Eng.* **2017**, *77*, 263–270. [CrossRef]

37. Pivsa-art, W.; Pivsa-art, S.; Fujii, K.; Nomura, K.; Ishimoto, K.; Aso, Y.; Yamane, H.; Ohara, H. Compression molding and melt-spinning of the blends of poly (lactic acid) and poly (butylene succinate- co -adipate). *J. Appl. Polym. Sci.* **2015**, *132*. [CrossRef]
38. Tcharkhtchi, A.; Elhirisia, S.A.; Ebrahimi, K.M.; Fitoussi, J.; Shirinbayan, M.; Farzaneh, S. Partial shape memory effect of polymers. *AIP Conf. Proc.* **2014**, *1599*, 278–281.
39. Ojijo, V.; Sinha Ray, S.; Sadiku, R. Role of specific interfacial area in controlling properties of immiscible blends of biodegradable polylactide and poly [(butylene succinate)-co-adipate]. *ACS Appl. Mater. Interfaces* **2012**, *4*, 6690–6701. [CrossRef] [PubMed]
40. Nofar, M.; Tabatabaei, A.; Sojoudiasli, H.; Park, C.B.; Carreau, P.J.; Heuzey, M.C.; Kamal, M.R. Mechanical and bead foaming behavior of PLA-PBAT and PLA-PBSA blends with different morphologies. *Eur. Polym. J.* **2017**, *90*, 231–244. [CrossRef]
41. Quiles-Carrillo, L.; Montanes, N.; Sammon, C.; Balart, R.; Torres-Giner, S. Compatibilization of highly sustainable polylactide/almond shell flour composites by reactive extrusion with maleinized linseed oil. *Ind. Crops Prod.* **2018**, *111*, 878–888. [CrossRef]
42. Zhang, Y.; Yuan, X.; Liu, Q.; Hrymak, A. The Effect of Polymeric Chain Extenders on Physical Properties of Thermoplastic Starch and Polylactic Acid Blends. *J. Polym. Environ.* **2012**, *20*, 315–325. [CrossRef]
43. Garcia-Campo, M.J.; Quiles-Carrillo, L.; Masia, J.; Reig-Pérez, M.J.; Montanes, N.; Balart, R. Environmentally friendly compatibilizers from soybean oil for ternary blends of poly(lactic acid)-PLA, poly(ε-caprolactone)-PCL and poly(3-hydroxybutyrate)-PHB. *Materials* **2017**, *10*, 1339. [CrossRef] [PubMed]
44. Villalobos, M.; Awojulu, A.; Greeley, T.; Turco, G.; Deeter, G. Oligomeric chain extenders for economic reprocessing and recycling of condensation plastics. *Energy* **2006**, *31*, 3227–3234. [CrossRef]
45. Eslami, H.; Kamal, M.R. Effect of a Chain Extender on the Rheological and Mechanical Properties of Biodegradable Poly (lactic acid)/Poly [(butylene succinate)-co-adipate] Blends. *J. Appl. Polym. Sci.* **2013**, *129*, 2418–2428. [CrossRef]
46. Quiles-Carrillo, L.; Duart, S.; Montanes, N.; Torres-Giner, S.; Balart, R. Enhancement of the mechanical and thermal properties of injection-molded polylactide parts by the addition of acrylated epoxidized soybean oil. *Mater. Des.* **2018**, *140*, 54–63. [CrossRef]
47. Quiles-Carrillo, L.; Montanes, N.; Garcia-Garcia, D.; Carbonell-Verdu, A.; Balart, R.; Torres-Giner, S. Effect of different compatibilizers on injection-molded green composite pieces based on polylactide filled with almond shell flour. *Compos. Part B Eng.* **2018**, *147*, 76–85. [CrossRef]
48. Frenz, V.; Scherzer, D.; Villalobos, M.; Awojulu, A.; Edison, M.; Van Der Meer, R. Multifunctional polymers as chain extenders and compatibilizers for polycondensates and biopolymers. *Tech. Pap. Reg. Tech. Conf. Soc. Plast. Eng.* **2008**, *3*, 1678–1682.
49. Dogan, S.K.; Boyacioglu, S.; Kodal, M.; Gokce, O.; Ozkoc, G. Thermally induced shape memory behavior, enzymatic degradation and biocompatibility of PLA/TPU blends: "Effects of compatibilization". *J. Mech. Behav. Biomed. Mater.* **2017**, *71*, 349–361. [CrossRef] [PubMed]
50. Peponi, L.; Navarro-Baena, I.; Sonseca, A.; Gimenez, E.; Marcos-Fernandez, A.; Kenny, J.M. Synthesis and characterization of PCL-PLLA polyurethane with shape memory behavior. *Eur. Polym. J.* **2013**, *49*, 893–903. [CrossRef]

materials

MDPI

Article

Ductility and Toughness Improvement of Injection-Molded Compostable Pieces of Polylactide by Melt Blending with Poly(ε-caprolactone) and Thermoplastic Starch

Luis Quiles-Carrillo [1], Nestor Montanes [1], Fede Pineiro [1], Amparo Jorda-Vilaplana [1] and Sergio Torres-Giner [1,2,*]

1 Technological Institute of Materials (ITM), Universitat Politècnica de València (UPV), Plaza Ferrándiz y Carbonell 1, 03801 Alcoy, Spain; luiquic1@epsa.upv.es (L.Q.-C.); nesmonmu@upvnet.upv.es (N.M.); fepival@epsa.upv.es (F.P.); amjorvi@upv.es (A.J.-V.)

2 Novel Materials and Nanotechnology Group, Institute of Agrochemistry and Food Technology (IATA), Spanish National Research Council (CSIC), Calle Catedrático Agustín Escardino Benlloch 7, 46980 Paterna, Spain

* Correspondence: storresginer@iata.csic.es; Tel.: +34-963-900-022

Received: 5 October 2018; Accepted: 26 October 2018; Published: 30 October 2018

Abstract: The present study describes the preparation and characterization of binary and ternary blends based on polylactide (PLA) with poly(ε-caprolactone) (PCL) and thermoplastic starch (TPS) to develop fully compostable plastics with improved ductility and toughness. To this end, PLA was first melt-mixed in a co-rotating twin-screw extruder with up to 40 wt % of different PCL and TPS combinations and then shaped into pieces by injection molding. The mechanical, thermal, and thermomechanical properties of the resultant binary and ternary blend pieces were analyzed and related to their composition. Although the biopolymer blends were immiscible, the addition of both PCL and TPS remarkably increased the flexibility and impact strength of PLA while it slightly reduced its mechanical strength. The most balanced mechanical performance was achieved for the ternary blend pieces that combined high PCL contents with low amounts of TPS, suggesting a main phase change from PLA/TPS (comparatively rigid) to PLA/PCL (comparatively flexible). The PLA-based blends presented an "island-and-sea" morphology in which the TPS phase contributed to the fine dispersion of PCL as micro-sized spherical domains that acted as a rubber-like phase with the capacity to improve toughness. In addition, the here-prepared ternary blend pieces presented slightly higher thermal stability and lower thermomechanical stiffness than the neat PLA pieces. Finally, all biopolymer pieces fully disintegrated in a controlled compost soil after 28 days. Therefore, the inherently low ductility and toughness of PLA can be successfully improved by melt blending with PCL and TPS, resulting in compostable plastic materials with a great potential in, for instance, rigid packaging applications.

Keywords: PLA; PCL; TPS; biopolymer blends; mechanical properties; compostable plastics

---

1. Introduction

The extensive use of petroleum-derived polymers is responsible for the increasing concern about the environmental impact of plastics due to both their origin and end-of-cycle, since most of them are not biodegradable. Worldwide polymer production was estimated to be 260 million metric tons per year in 2007 and it is considered that in 2020 each person will consume around 40 kg of plastic annually [1]. Bioplastics emerge as an alternative to conventional plastics, including both natural-sourced polymers and also petroleum-based polyesters that undergo biodegradation. Among biopolymers, polylactide

(PLA) is currently considered one of the most promising biopolyester at industrial scale due to its good balance between physicochemical properties, low price, and sustainability [2]. PLA is obtained from lactide derived from starch fermentation and it is fully biodegradable. The increasing use of PLA in the last years is noticeable with a current worldwide production of about 140,000 tons/year [3]. The main uses of PLA cover a wide variety of industrial sectors for instance automotive [4–6], biomedical applications [7,8], packaging [9,10] or, lately, the growing industry of 3D printing [11,12]. Despite this, PLA presents several intrinsic restrictions that are mainly related to its relatively high price, low heat resistance, and high fragility [13]. As a result, PLA cannot fulfill the technical requirements of some industries, limiting its expansion to commodity areas such as food packaging [14].

To overcome or, at least, minimize the low ductility and toughness of PLA, several approaches have been considered with excellent results. The first approach is copolymerization. For instance, the simultaneous polymerization of lactide acid (LA) with glycolic acid (GA) leads to the synthesis of poly(lactic acid-*co*-glycolic acid) (PLGA). In general terms, PLGA copolymers exhibit improved solubility as well as better ductile properties than both PLA and poly(glycolic acid) (PGA) homopolymers [15,16]. Nevertheless, copolymers are frequently expensive and their use is not yet generalized at industrial scale. A second strategy to increase PLA toughness is focused on the use of plasticizers. Some of the widely used plasticizers for PLA include poly(ethylene glycol) (PEG) [17], triethyl citrate (TEC) [18,19], and oligomers of lactic acid (OLAs) [20]. All these plasticizers contribute positively to increasing ductility by providing a relevant decrease in the glass transition temperature (T_g) of PLA but they can also reduce the heat resistance, tensile strength, and stiffness. In addition to these plasticizers, in recent years, new vegetable oil-derived plasticizers have been successfully developed for PLA formulations such as maleinized, acrylated, hydroxylated, and epoxidized vegetable oils [21–24]. Although their efficiency as primary plasticizers for PLA is lower than those indicated previously, the particular chemical structure of these multi-functionalized modified vegetable oils delivers chain extension, branching and, in some cases, cross-linking resulting in improved toughness without compromising the mechanical strength in a great extent [23]. The third route is related to the manufacturing of PLA-based blends. This represents a very cost-effective solution to reduce the intrinsic fragility of PLA materials without significantly decreasing their tensile strength. A wide variety of binary blends based on PLA has been extensively studied in the last years. For instance, it is worthy to note the interest in binary blends of PLA with polyhydroxyalkanoates (PHAs) [25,26], polyamides (PAs) [27,28], poly(butylene adipate-*co*-terephthalate) (PBAT) [29,30], thermoplastic starch (TPS) [31], poly(ε-caprolactone) (PCL), poly(butylene succinate) (PBS), and poly(butylene succinate-*co*-adipate) (PBSA) [32–34]. These previous studies are based on the fact that, to improve toughness, PLA is blended with flexible polymers that perform as a rubber-like phase inside a rigid polymer matrix as, for instance, polybutadiene rubbers (BRs) do in high-impact polystyrene (HIPS).

In addition to binary blends, a wide variety of ternary blends based on PLA have been proposed to tailor the desired properties, particularly in terms of improved toughness [35–37]. On the one hand, PCL is a well-known synthetic aliphatic biopolyester, characterized by a high crystallinity, relatively fast biodegradability, and high ductility. However, PCL shows a low melting temperature (T_m), of about 60 °C, which restricts its use in a wide range of applications [38]. PLA/PCL blends are attracting some industrial uses since flexible PCL domains can be finely dispersed into the rigid PLA matrix leading to improved toughness without compromising biodegradation [39]. In addition, the resultant blends are fully resorbable, finding interesting applications as biomedical devices. On the other hand, starch is a versatile and useful biopolymer. Starch has to be modified by means of plasticizers (e.g., glycerol and water) [40] and/or chemical reaction (e.g., esterification) [41] in order to be melt-processed, which then results in TPS. The role of plasticizers is to destructurize granular starch by breaking hydrogen bonds between the starch macromolecules, accompanying with a partial depolymerization of starch backbone. As a result, TPS leads to compostable plastic materials offering interesting opportunities in the packaging field due to its low cost and tailor-made mechanical behavior by selecting the appropriate plasticizers [42]. Blending of PLA with TPS is, therefore, a good way to balance the price and develop materials that has new performances.

The aim of this work was to prepare and characterize ternary blends of PLA with PCL and TPS to overcome the intrinsic brittleness of PLA. To this end, different PCL and TPS contents were blended by melt compounding with PLA to obtain PLA-based materials with tailor-made properties. The resultant PLA/PCL/TPS ternary blends were, thereafter, injection-molded into pieces and subjected to mechanical, morphological, thermal, and thermomechanical analysis while their potential compostability was also ascertained.

2. Materials and Methods

2.1. Materials

Commercial PLA Ingeo™ biopolymer 6201D was purchased from NatureWorks (Minnetonka, MN, USA). This PLA resin has a density of 1.24 $g \cdot cm^{-3}$, a met flow rate (MFR) of 15–30 $g \cdot 10\ min^{-1}$ (210 °C, 2.16 kg), a T_g value in the 55–60 °C range, and a T_m value in the 165–175 °C range. This MFR allows the manufacturing of PLA articles by both extrusion and injection molding. PCL was a Capa™ 6800 commercial grade supplied by Perstorp UK Ltd. (Warrington, UK) with a density of 1.15 $g \cdot cm^{-3}$, a T_g value of −60 °C, and a T_m value in the 58–62 °C range. The melt flow index (MFI) of PCL is 2–4 $g \cdot 10\ min^{-1}$ (160 °C, 2.16 kg). TPS Mater-Bi® NF 866 was obtained from Novamont SPA (Novara, Italy), which is derived from maize starch. Its MFI is 3.5 $g \cdot 10\ min^{-1}$ (150 °C, 2.16 kg). This TPS resin presents a density of 1.27 $g \cdot cm^{-3}$, a T_g value ranging from −35 °C to −40 °C, and a T_m value in the 110–120 °C range.

2.2. Manufacturing of Ternary PLA/PCL/TPS Blends

Prior to manufacturing, all the biopolymer pellets were dried at 45 °C for 48 h in a MDEO dehumidifier from Industrial Marsé (Barcelona, Spain). All blends contained 60 wt % PLA while PCL and TPS varied from 0 to 40 wt % to give a series of materials with different properties. The corresponding amounts of each biopolymer is summarized and coded in Table 1.

Initially, the biopolymer pellets were weighed and manually mixed in a zipper bag. Then, the different mixtures were melt-compounded in a co-rotating twin-screw extruder from Construcciones Mecánicas Dupra S.L. (Alicante, Spain) at a rotating speed of 30 rpm. The screws had a diameter of 25 mm with a length-to-diameter ratio (L/D) of 24. The temperature profile, from the feeding hopper to the extrusion die (circular), was set at 165 °C–170 °C–175 °C–180 °C. The extruded materials were pelletized in an air-knife unit.

The compounded pellets were finally processed by injection molding in a Meteor 270/75 injection machine from Mateu and Solé (Barcelona, Spain). The temperature profile during the injection molding process was: 160 °C (hopper)–165 °C–170 °C–180 °C (injection nozzle). A clamping force of 75 tons was applied while the cavity filling and cooling time were set at 1 and 10 s, respectively. Pieces with a mean thickness of 4 mm were produced.

Table 1. Composition and coding of the polylactide (PLA), poly(ε-caprolactone) (PCL), and thermoplastic starch (TPS) blends.

Sample	PLA (wt %)	PCL (wt %)	TPS (wt %)
PLA	100	0	0
$PLA_{60}PCL_{40}TPS_0$	60	40	0
$PLA_{60}PCL_{30}TPS_{10}$	60	30	10
$PLA_{60}PCL_{20}TPS_{20}$	60	20	20
$PLA_{60}PCL_{10}TPS_{30}$	60	10	30
$PLA_{60}PCL_0TPS_{40}$	60	0	40

2.3. Mechanical Characterization

Tensile and flexural tests were performed on the injection-molded pieces of PLA and its blends using a universal test machine ELIB 50 from S.A.E. Ibertest (Madrid, Spain). Tensile tests were carried out following the guidelines of ISO 527-1:2012 using a cross-head speed rate of 10 $mm \cdot min^{-1}$. Similarly, flexural tests were carried out according to ISO 178 and the speed rate was 5 $mm \cdot min^{-1}$. Both tests were carried out at 25 °C and with a load cell of 5 kN. At least six samples of each material were tested.

Shore D hardness of the biopolymer pieces were obtained in a Shore durometer 676-D from J. Bot Instruments (Barcelona, Spain), as recommended by ISO 868:2003. A type-D indenter with a load of 5 kg and an indentation time of 12–15 s was used to stabilize the measurement. The impact-absorbed energy, which is directly related to toughness, was estimated by using the Charpy impact test with a 1-J pendulum from Metrotec S.A. (San Sebastian, Spain). The average energy per unit cross-section area was obtained on V-notched samples with a radius of 0.25 mm, as recommended by ISO 179-1:2010. Both mechanical tests were carried out at room temperature, that is, 25 °C, and five different samples of each material were tested.

2.4. Morphological Characterization

The morphology of the fracture surfaces was studied on the broken samples after the impact tests by field emission scanning electron microscopy (FESEM) in a ZEISS ULTRA 55 microscope from Oxford Instruments (Abingdon, UK). Before placing the samples into the vacuum chamber, all surfaces were covered with a thin metallic layer of gold-palladium by sputtering in an EMITECH mod. SC7620 from Quorum Technologies, Ltd. (East Sussex, UK). The acceleration voltage for the FESEM study was 2 kV.

2.5. Solubility

The relative affinity of the biopolymers was estimated by measuring the solubility parameters (δ) according to the Small's method [43]. To consider the blend miscible, the δ values of the polymers should be of the same order. This parameter was determined according to Equation (1):

$$\delta = \frac{\rho \cdot \Sigma G}{M_n}, \quad (1)$$

where ρ is the density of the polymer, M_n is the molar mass of the repeating unit, and ΣG is the sum of the group contributions to the cohesive energy density.

2.6. Thermal Characterization

The thermal transitions of PLA and its blends were obtained by differential scanning calorimetry (DSC) in a Mettler-Toledo 821 calorimeter (Schwerzenbach, Switzerland). An average sample weight comprised in the 5–7 mg range was used for all DSC tests. The thermal program consisted of a first heating step from 25 °C to 190 °C, followed by a cooling step down to 25 °C, and a second heating step up to 300 °C. All heating rates were set at 10 $°C \cdot min^{-1}$. A constant nitrogen flow-rate of 66 $mL \cdot min^{-1}$ was used to achieve inert atmosphere. Aluminum pans with a total volume capacity of 40 μL were used.

Thermal stability was determined by thermogravimetric analysis (TGA) in a Mettler-Toledo TGA/SDTA 851 thermobalance (Schwerzenbach, Switzerland). Samples with an average size of 5–7 mg were placed into standard alumina crucibles with a total volume capacity of 70 μL and subjected to a heating program from 30 °C to 650 °C at a heating rate of 20 $°C \cdot min^{-1}$ in air atmosphere.

2.7. Thermomechanical Characterization

The effect of temperature on the mechanical properties was followed by dynamic mechanical thermal analysis (DMTA) in an oscillatory rheometer AR-G2 from TA Instruments (New Castle, DE,

USA). This rheometer is equipped with a special clamp system to work with solid samples in a combined torsion/shear mode. Injection-molded pieces with dimensions of 4 mm × 10 mm × 40 mm were subjected to a temperature sweep from −80 °C to 120 °C at a constant heating rate of 2 °C·min^{-1}. The selected frequency was 1 Hz and the maximum shear deformation was set at 0.1% (% γ).

The thermomechanical behavior of the ternary blends was also assessed by obtaining the Vicat softening temperature (VST) and the heat deflection temperature (HDT) in a Vicat/HDT station VHDT 20 from Metrotec S.A. (San Sebastián, Spain). VST was obtained following the procedure described in ISO 306, using the B50 heating method and applying a total force of 50 N at a heating rate of 50 °C·h^{-1}. Regarding HDT, ISO 75-1 recommendations were followed. To this end, samples sizing 4 mm × 10 mm × 80 mm were placed between two supports with a total span of 60 mm. After this, a load of 320 g was applied in the center using a heating rate of 120 °C·h^{-1}.

2.8. Disintegration Test

A disintegration test in controlled compost conditions was conducted following the guidelines of ISO 20200 at a temperature of 58 °C and a relative humidity (RH) of 55%. For this, squared thermo-compressed samples sizing 1 mm × 30 mm × 30 mm were placed in a carrier bag and buried in a controlled soil with the following composition (in dry weight): sawdust (40 wt %), rabbit-feed (30 wt %), ripe compost (10 wt %), corn starch (10 wt %), saccharose (5 wt %), corn seed oil (4 wt %), and urea (1 wt %). To follow the disintegration process, samples were periodically unburied, washed with distilled water, dried, and weighed in an analytic balance. In order to get a visual evolution of this process, pictures of the disintegration process were also collected. The weight loss due to disintegration in the controlled compost soil was calculated by means of Equation (2):

$$\text{Weight loss}(\%) = \left(\frac{W_0 - W_t}{W_0}\right)\cdot 100, \tag{2}$$

where W_t is the weight of the sample after a bury time t and W_0 is the initial dry weight of the sample. All tests were carried out in triplicate to ensure reliability.

2.9. Statistical Analysis

Ternary graphs were plotted using Origin Pro 2015 from OriginLab Corporation (Northampton, MA, USA) with the Ternary Contour function using the average and standard deviation values.

3. Results

3.1. Mechanical Properties

The injection-molded pieces of PLA and of the binary and ternary blends of PLA with PCL and TPS were tested in order to determine their mechanical properties. The tensile strength ($\sigma_{tensile}$) and elongation at break (ε_b) were obtained under tensile conditions, while the flexural modulus ($E_{flexural}$) and flexural strength ($\sigma_{flexural}$) were determined under flexural conditions. Figure 1 shows the resultant stress–strain curves of the injection-molded PLA-based pieces obtained during the tensile tests (Figure 1a) and flexural tests (Figure 1b).

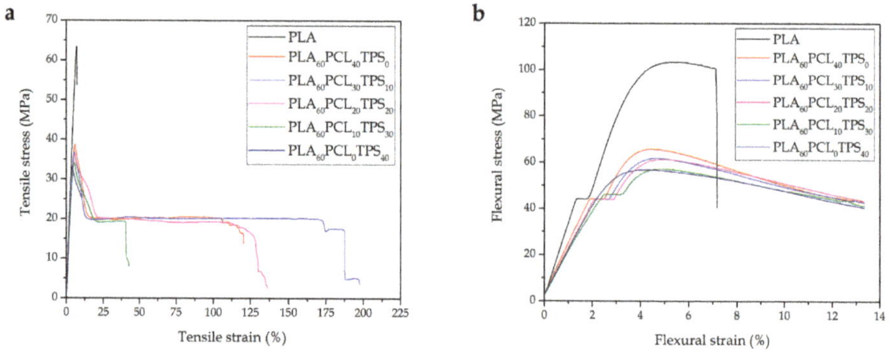

Figure 1. Stress–strain curves of the polylactide (PLA), poly(ε-caprolactone) (PCL), and thermoplastic starch (TPS) blend pieces obtained from: (**a**) tensile test; and (**b**) flexural test.

Figure 2 summarizes in ternary graphs the evolution of the tensile properties, that is, ε_b and $\sigma_{tensile}$, of the injection-molded PLA-based pieces with the addition of PCL and TPS. One can observe in Figure 2a that the neat PLA piece was very fragile, presenting a ε_b value of 4.9%. This value, together with a medium-to-high $\sigma_{tensile}$ value of 63.4 MPa, was responsible for its high brittleness. As one can see, the addition of both PCL and TPS provided a positive effect on the PLA's ductility, but this effect was much more pronounced with PCL due to its intrinsic higher flexibility compared to TPS. In particular, the $PLA_{60}PCL_{30}TPS_{10}$ and $PLA_{60}PCL_{20}TPS_{20}$ blend pieces showed a remarkable increase in ε_b with values of 196.7% and 134.3%, respectively, which were noticeably higher than that of the neat PLA piece. It is also worthy to note that these two ternary blend pieces presented higher ductility than the binary blend piece of PLA with PCL, that is, $PLA_{60}PCL_{40}TPS_0$, which suggests a synergistic effect of both PCL and TPS on the overall material's ductility. With regard to the mechanical strength of the PLA-based pieces, as shown in Figure 2b, one can observe that the pieces presented lower $\sigma_{tensile}$ values after the addition of PCL and TPS. In the case of the binary blend piece of PLA with PCL, that is, $PLA_{60}PCL_{40}TPS_0$, the value of $\sigma_{tensile}$ was reduced to 39.1 MPa, which is remarkable lower than that observed for the neat PLA piece. The binary blend piece of PLA with TPS, that is, $PLA_{60}PCL_0TPS_{40}$, resulted in even a lower $\sigma_{tensile}$ value, that is, 33.6 MPa. All intermediate compositions showed a proportional decrease depending on the PCL and TPS content. With regard to the blend pieces containing 30 wt % and 40 wt % TPS, that is, $PLA_{60}PCL_{10}TPS_{30}$ and $PLA_{60}PCL_0TPS_{40}$, respectively, the ductility was poor when compared to the ternary blend piece with the highest PCL content, that is, $PLA_{60}PCL_{30}TPS_{10}$. This suggests that both individual PCL and TPS biopolymers have a positive effect on the ductile properties of PLA but the best results were obtained for the ternary blends that combined a high PCL content with low amounts of TPS. The addition of 40 wt % TPS to PLA without PCL, that is, $PLA_{60}PCL_0TPS_{40}$, produced the piece with the poorest mechanical performance. Although this piece doubled the ductility of the neat PLA piece, that is, ε_b increased to 8.8%, $\sigma_{tensile}$ also decreased to a value of 33.6 MPa. As previously indicated, the binary blend piece made of PLA with 40 wt % PCL, that is, $PLA_{60}PCL_{40}TPS_0$, also provided non-optimum results showing a value of ε_b of 114.3%. However, interestingly, the ternary blend pieces containing 20–30 wt % PCL and 20–10 wt % TPS, that is, $PLA_{60}PCL_{20}TPS_{20}$ and $PLA_{60}PCL_{30}TPS_{10}$, offered the best ductile properties with remarkable high ε_b values.

The above-described observation suggests that a main phase change, from PLA/TPS (comparatively rigid) to PLA/PCL (comparatively flexible), occurred in the ternary blends when relative high contents of PCL and low contents of TPS are blended with PLA. In this sense, other authors have reported that the ductility of PLA/TPS blends can be drastically increased by the incorporation of high amounts of flexible polyesters. For instance, Zhen et al. [44] observed that the addition of PBS led to a mechanical strength decrease and ductility increase in TPS/PLA blends. Whereas the

$\sigma_{tensile}$ values decreased from 28.54 MPa to 14.60 MPa with the increase of PBS content from 0% to 50 wt %, the values of ε_b of the ternary blends also increased from 1.82% to 45.17%. However, the most significant mechanical changes were obtained for PBS contents above 20 wt %, which was ascribed to the main phase change from TPS/PLA (comparatively rigid) to TPS/PBS (comparatively flexible). Similar results were previously obtained by Ren et al. [45] for ternary TPS/PLA/PBAT blends, in which the main phase changed from TPS/PLA (comparatively rigid) to TPS/PBAT (comparatively flexible) when the PBAT reached contents between 20 and 30 wt %.

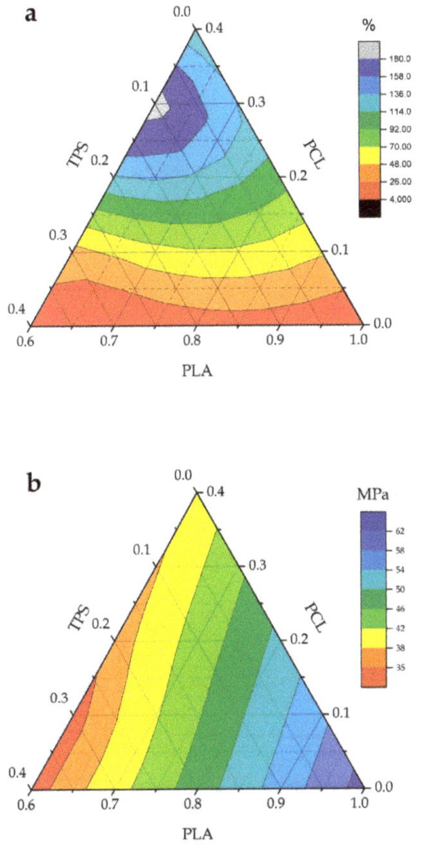

Figure 2. Ternary graphs showing the evolution of the mechanical properties of the polylactide (PLA), poly(ε-caprolactone) (PCL), and thermoplastic starch (TPS) blend pieces in terms of: (**a**) elongation at break (ε_b); and (**b**) tensile strength ($\sigma_{tensile}$).

Figure 3 shows a ternary graph with the evolution of flexural properties of the injection-molded PLA-based pieces, that is, $E_{flexural}$ and $\sigma_{flexural}$, when varying the composition of the blends. With regard to $E_{flexural}$, in Figure 3a it can be seen that a clear reduction was observed after the incorporation of PCL and/or TPS in comparison to the neat PLA piece. In fact, it was reduced from 3200 MPa, for the neat PLA piece, to 2100 MPa, for the binary blend piece of PLA with 40 wt % PCL, that is, $PLA_{60}PCL_{40}TPS_0$. The value of $E_{flexural}$ followed the same tendency as reported by Ferry et al. [46], decreasing as the TPS content increased in PLA/TPS blends. In particular, $E_{flexural}$ presented the lowest value, that is, 1780 MPa, for the binary blend piece of PLA with 40 wt % TPS, that is, $PLA_{60}PCL_0TPS_{40}$. As shown in Figure 3b, $\sigma_{flexural}$ decreased from 103 MPa, for the neat PLA piece, down to values of 65 MPa and 57 MPa for the binary blend pieces containing 40 wt % PCL, that is, $PLA_{60}PCL_{40}TPS_0$, and 40 wt % TPS, that is, $PLA_{60}PCL_0TPS_{40}$, respectively. Intermediate compositions of the ternary blends showed a proportional decrease in the $\sigma_{flexural}$ values as a function of their composition. Similar results were reported by García-Campo et al. [47] where intermediate compositions of the ternary PLA/PHB/PCL blends presented an intermediate mechanical behavior between the binary PLA/PHB and PLA/PCL blends.

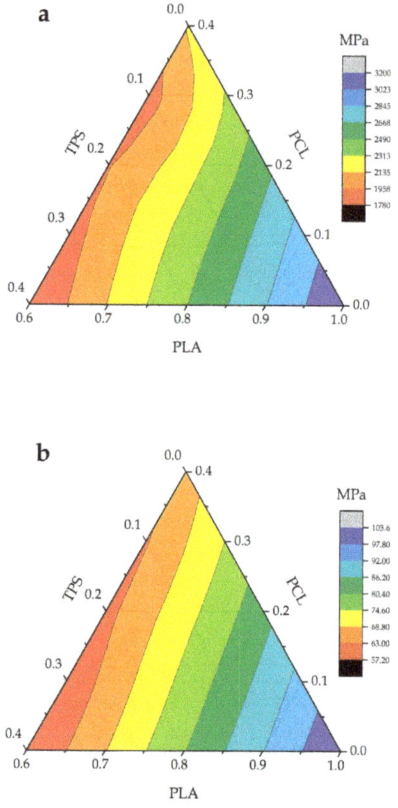

Figure 3. Ternary graphs showing the evolution of the mechanical properties of the polylactide (PLA), poly(ε-caprolactone) (PCL), and thermoplastic starch (TPS) blend pieces in terms of: (**a**) flexural modulus ($E_{flexural}$); and (**b**) flexural strength ($\sigma_{flexural}$).

As stated above, one of the main drawbacks of PLA is its low toughness. Table 2 summarizes the main results obtained from the Charpy impact test as well as the Shore D hardness measurements. As it can be observed, the typical energy absorption of the V-notched neat PLA piece was very low, of about 2.14 $kJ \cdot m^{-2}$. With regard to the binary blend piece with 40 wt % PCL, that is, $PLA_{60}PCL_{40}TPS_0$, it resulted in an impact energy per unit cross-section of 6.52 $kJ \cdot m^{-2}$, which represents an increase of more than three times compared to the neat PLA piece. Similar findings were reported for instance by Chen et al. [48], showing a remarkable improvement in the PLA toughness by the addition of PCL. The ternary blend pieces with 10–30 wt % PCL also showed relatively high values of impact strength, thus, supporting the good effect of PCL on the overall PLA toughness. It is also important to remark that the binary blend piece of PLA with 40 wt % TPS, that is, $PLA_{60}PCL_0TPS_{40}$, provided increased toughness with an impact strength value of 5.46 $kJ \cdot m^{-2}$. However, as observed above for other mechanical properties, the effect of PCL was more intense than that of TPS. In relation to the Shore D hardness, the hardness value of the neat PLA piece was 73.1. The Shore D hardness values decreased by approximately 10 units in all the developed blend pieces, thus, reaching a plateau at values of 63–64.

Table 2. Impact strength and Shore D hardness of the polylactide (PLA), poly(ε-caprolactone) (PCL), and thermoplastic starch (TPS) blend pieces.

Sample	Impact Strength (kJ·m^{-2})	Shore D Hardness
PLA	2.14 ± 0.28	73.1 ± 1.3
$PLA_{60}PCL_{40}TPS_{0}$	6.52 ± 0.62	63.0 ± 1.0
$PLA_{60}PCL_{30}TPS_{10}$	6.46 ± 0.39	63.6 ± 1.1
$PLA_{60}PCL_{20}TPS_{20}$	6.51 ± 0.27	63.7 ± 1.2
$PLA_{60}PCL_{10}TPS_{30}$	6.33 ± 0.24	64.3 ± 1.1
$PLA_{60}PCL_{0}TPS_{40}$	5.46 ± 0.88	64.6 ± 1.1

3.2. Morphology

Figure 4 shows the FESEM images corresponding to fracture surfaces of the different injection-molded PLA-based pieces obtained after the impact tests. Figure 4a, which corresponds to the neat PLA piece, shows the typical fracture surface of a brittle material with low roughness, that is, a smooth and relatively flat surface. Regarding the binary blend piece of PLA with 40 wt % PCL, that is, $PLA_{60}PCL_{40}TPS_{0}$, shown in Figure 4b, a clearly different fracture surface can be observed. In particular, the surface roughness was higher and the flat surface changed to an "island-and-sea" morphology that was based on finely dispersed PCL-rich domains, sizing 1–5 µm, into the PLA matrix. Although PLA and PCL are thermodynamically immiscible [49], this particular structure positively contributed to improving toughness as the enclosed microdroplets of PCL were able to absorb energy, acting as a rubber-like phase dispersed in a brittle matrix [50]. Plastic deformation provided by PCL can be also observed by the presence of some filaments along the PLA matrix. Addition of 10 wt % TPS in the ternary blend piece, that is, $PLA_{60}PCL_{30}TPS_{10}$, also produced a noticeable change in the morphology, which can be observed in Figure 4c. In particular, one can observe that the TPS-rich domains presented a higher size, in the 1–35 µm range. A similar morphology was previously reported by Sarazin et al. [31]. In Figure 4d–f one can observe that, as the TPS content increased, the TPS-rich domains increased both in number and size, which is an indication of their poor interfacial interaction with the PLA-based matrix [51]. With regard to the blend pieces with the highest TPS contents, that is, both $PLA_{60}PCL_{10}TPS_{30}$ and $PLA_{60}PCL_{0}TPS_{40}$, the domains changed from spherical to a ribbon-like morphology due to stretching of the TPS phase during fracture. This morphological changes were also observed by Carmona et al. [52] in TPS/PCL/PLA blends at high TPS contents, that is, 33.3 wt % TPS. Ferri et al. [49] have previously related the formation of TPS flakes to the crystalline plane growth or "crystalline lamellae" located at the amylopectin branches that fold up during fracture. In particular, the mechanically-induced flakes structures form parallel-plane blocks and clusters, resulting in granules separated by porous of amorphous areas in which amylose and plasticizers can be allocated. Since PLA is a hydrophobic biopolymer whilst TPS is highly hydrophilic, indeed one of the main drawbacks of TPS is related to its extremely high moisture sensitiveness, this results in the lack (or very low) affinity between the two biopolymers that frequently leads to a strong phase separation [53].

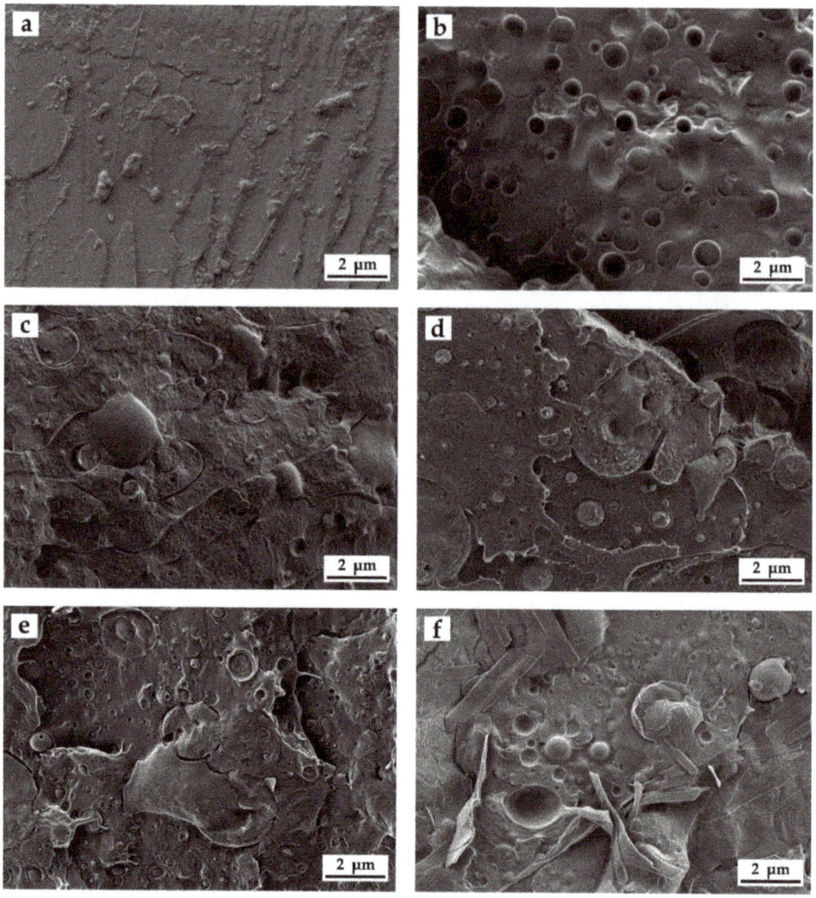

Figure 4. Field emission scanning electron microscopy (FESEM) images of the fracture surfaces of the polylactide (PLA), poly(ε-caprolactone) (PCL), and thermoplastic starch (TPS) blend pieces: (**a**) Neat PLA; (**b**) $PLA_{60}PCL_{40}TPS_0$; (**c**) $PLA_{60}PCL_{30}TPS_{10}$; (**d**) $PLA_{60}PCL_{20}TPS_{20}$; (**e**) $PLA_{60}PCL_{10}TPS_{30}$; and (**f**) $PLA_{60}PCL_0TPS_{40}$. Images were taken at 5000× and scale markers are of 2 μm.

To further study the compatibility of the developed blends and also to ascertain their resultant morphologies, the miscibility of the biopolymer formulations was evaluated using the Small's method. According to this, the closer the δ values, the higher the miscibility of the polymers in the blend. Table 3 shows the chemical structure and the resultant δ values of the here-studied biopolymers. One can observe that both PLA and PCL presented a relatively similar δ value while TPS presented a considerably lower value, which support the above-described mechanical and morphological results. This difference in the δ values can be mainly related to the higher density of oxygen atoms in the chemical structure of TPS, mainly hydroxyl groups (–OH), which are certainly responsible for its high hydrophilicity. However, it is also worthy to mention that the δ values obtained for TPS can also vary considerably due to the thermoplastic carbohydrate is obtained by mixing with large quantities of plasticizers. The here-reported δ values are in agreement with Samper et al. [54] who obtained values for PLA and TPS of 19.1–20.1 and 8.4, respectively. Similarly, Bordes et al. [55] reported a δ value of 17 $MPa^{1/2}$ for PCL.

Table 3. Values of the solubility parameters (δ) obtained for polylactide (PLA), poly(ε-caprolactone) (PCL), and thermoplastic starch (TPS).

Biopolymer	Chemical Structure	ΣG $(cal/cc)^{1/2}$ [56]	δ $(MPa^{1/2})$
PLA	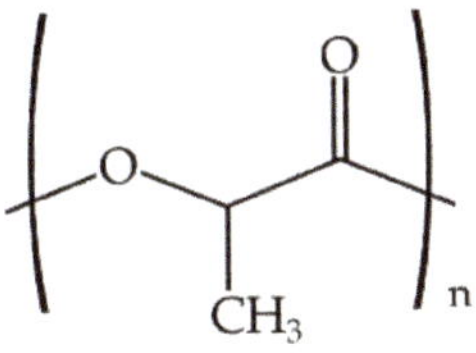	587	20.8
PCL	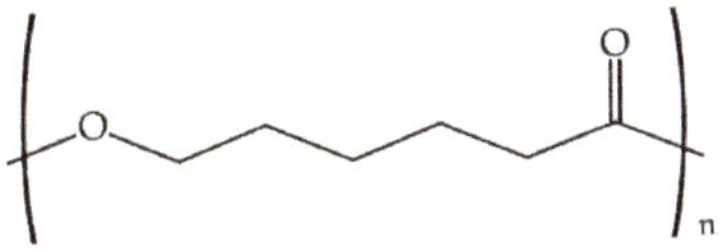	1010	19.4
TPS	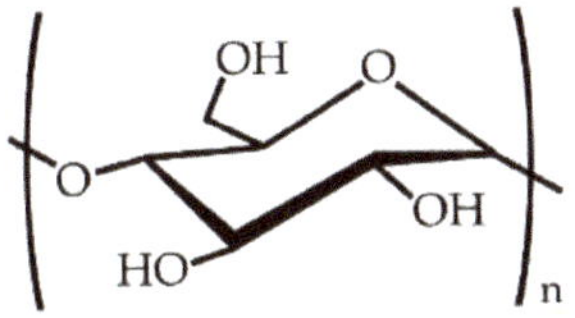	662	11.2

3.3. Thermal Properties

Figure 5 shows a comparison plot of the DSC curves obtained during the second heating cycle performed on the injection-molded PLA-based pieces. One can observe that the neat PLA piece showed a T_m value of 169.5 °C. In addition, PLA developed cold crystallization with a cold crystallization temperature (T_{CC}) located at approximately 103 °C and a value of T_g of around 63 °C. In the DSC curve for the binary blend piece of PLA with 40 wt % PCL, that is, $PLA_{60}PCL_{40}TPS_0$, it can be observed that the melt peak intensity for PLA was lower due to the diluting effect of PCL. An additional melting process with a peak located at ~59 °C appeared, which is attributable to the PCL's T_m. This melting process overlapped with the glass transition region of PLA so that it was not possible to separate both processes by conventional DSC. Similar results were also obtained by, for instance, Navarro-Baena et al. [57] for PLA/PCL blends using dynamic DSC measurements. In addition, the value of T_m for PLA did not remarkably change in the blends. As the PCL content in the ternary blends decreased, the corresponding peak intensity, that is, the melting enthalpy (ΔH_m), also decreased. In the case of the blend piece of PLA with 40 wt % TPS, that is, $PLA_{60}PCL_0TPS_{40}$, it also showed a slight shift of the cold crystallization region towards lower temperatures, which can be ascribed to a plasticizing effect of the PLA matrix by TPS. In this sense, it is worthy to note that TPS contains high amounts of plasticizers, such as glycerol, which can contribute to plasticizing PLA. The resultant plasticization is also evident by observing the PLA's T_g, which moved down to 61.2 °C. The glass transition regions of both PCL and TPS were not registered using the present thermal program since these peaks are located below room temperature, in particular from −50 °C to −65 °C for PCL [42] and from −75 °C to 10 °C for TPS [58,59].

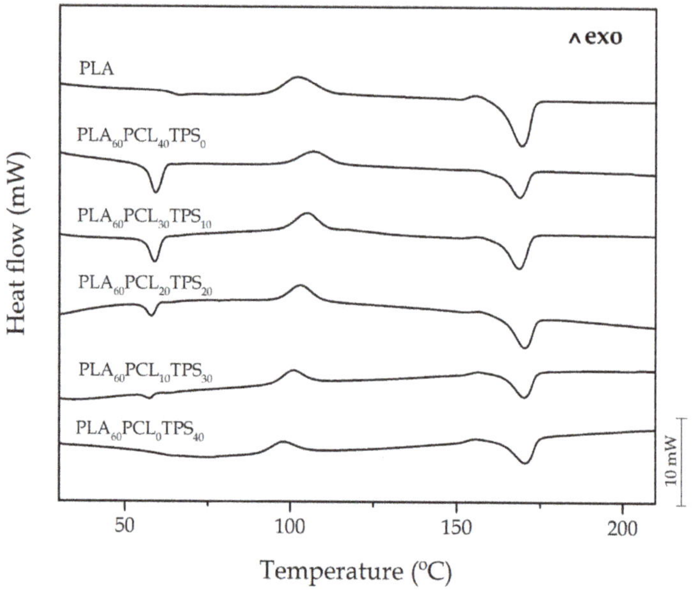

Figure 5. Comparative plot of the differential scanning calorimetry (DSC) curves of the polylactide (PLA), poly(ε-caprolactone) (PCL), and thermoplastic starch (TPS) blend pieces.

Figure 6 gathers the TGA thermograms (Figure 6a) and their corresponding first derivative thermogravimetric (DTG) curves (Figure 6b) of the injection-molded PLA-based pieces. Additionally, Table 4 summarizes the main thermal values obtained from these curves. It can be observed that the $PLA_{60}PCL_{30}TPS_{10}$ piece showed the lowest thermal stability, having the decomposition process in two stages. Its typical thermal degradation parameters, that is, the onset degradation temperature ($T_{5\%}$) and degradation temperature (T_{deg}), were 303.5 °C and 348 °C, respectively. Regarding the neat PLA piece, although it showed a high $T_{5\%}$ value, that is, 322 °C, its degradation occurred in a single step at a relatively low T_{deg} value, that is, 360 °C. In contrast, the $PLA_{60}PCL_{30}TPS_{10}$ piece and, in particular, the $PLA_{60}PCL_{10}TPS_{30}$ piece, improved the thermal stability by having lower mass losses at high temperatures while their T_{deg} values showed an increase of up to 15 °C with regard to the neat PLA. Therefore, the addition of both PCL and TPS led to an increase of the thermal stability of PLA at high temperatures. In addition, the binary and ternary pieces presented a thermal degradation process in two steps. The first mass loss corresponds to the PLA degradation while the second, at higher temperatures, can be attributed to the PCL and TPS decompositions. Additionally, the PLA degradation onset was delayed by the presence of both PCL and TPS. In this sense, Patrício et al. [60] reported that the addition of PCL can successfully enhance the thermal stability of PLA. In particular, it was observed an increase in the T_{deg} value from 325 °C, for the neat PLA, up to 334 °C, for binary blends of PLA with PCL at different ratios. Mofokeng et al. [61] however suggested the lack of miscibility between PLA and PCL, indicating completely independent degradation stages for each biopolymer phase in the blend.

With regard to the residual mass, it can be observed that TPS contributed to generating a higher amount of residue. Whereas the neat PLA piece resulted in a very low char content, of approximately 1.5%, this value increased up to 6.4% in the binary blend of PLA with 40 wt % TPS, that is, $PLA_{60}PCL_{0}TPS_{40}$. Thus, intermediate compositions led to intermediate char residues. This result can be related to additives incorporated into the biopolymer by the manufacturer.

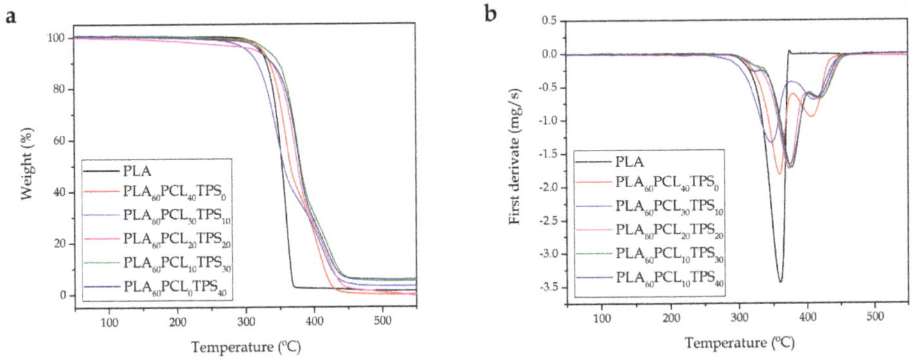

Figure 6. Comparative plot of the polylactide (PLA), poly(ε-caprolactone) (PCL), and thermoplastic starch (TPS) blend pieces in terms of: (**a**) Thermogravimetric analysis (TGA) curves; and (**b**) first derivative thermogravimetric (DTG) curves.

Table 4. Thermal degradation properties in terms of the onset degradation temperature ($T_{5\%}$), degradation temperature (T_{deg}), and residual mass at 650 °C of the polylactide (PLA), poly(ε-caprolactone) (PCL), and thermoplastic starch (TPS) blend pieces.

Sample	$T_{5\%}$ (°C)	T_{deg} (°C)	Residual Mass (%)
PLA	322.67 ± 1.36	359.74 ± 1.58	1.5 ± 0.3
$PLA_{60}PCL_{40}TPS_0$	325.03 ± 1.69	358.94 ± 2.14	0.4 ± 0.2
$PLA_{60}PCL_{30}TPS_{10}$	303.50 ± 1.74	347.99 ± 2.36	3.2 ± 0.4
$PLA_{60}PCL_{20}TPS_{20}$	315.33 ± 1.95	373.18 ± 1.74	1.2 ± 0.2
$PLA_{60}PCL_{10}TPS_{30}$	332.06 ± 1.41	373.21 ± 1.95	5.7 ± 0.5
$PLA_{60}PCL_0TPS_{40}$	320.34 ± 1.25	376.61 ± 1.78	6.4 ± 0.4

3.4. Thermomechanical Properties

DMTA allows estimating the effect of temperature on the mechanical performance. Additionally, it is a more sensitive technique to evaluate potential changes in T_g that, in turn, can be directly related to miscibility in polymer blends [62]. Figure 7 shows the evolution of the storage modulus (G′) and the dynamic damping factor (tan δ) as a function of temperature in the injection-molded PLA-based pieces. Figure 7a presents the G′ curves for the neat PLA piece and for the binary and ternary blend pieces of PLA with PCL and TPS. G′ is directly related to the stored elastic energy and, consequently, can be directly related to stiffness. Regarding the neat PLA piece, its G′ value was 1.69 GPa at −80 °C. One can also observe that the G′ value increased from 5.6 MPa, at 80 °C, to 76 MPa, above 90 °C. This stiffness increase is ascribed to the cold crystallization process of PLA due to the rearrangement of the biopolyester chains to give a more packed structure [63]. Addition of 40 wt % TPS led to lower G′ values. For instance, the $PLA_{60}PCL_0TPS_{40}$ piece presented a G′ value of 1.55 GPa at −80 °C and the same trend that in the case of PLA was observed at higher temperatures. The highest decrease in G′ was obtained for the binary blend piece of PLA with 40 wt % PCL, that is, $PLA_{60}PCL_{40}TPS_0$, with a value of 1.30 GPa at −80 °C. Therefore, the addition of both PCL and TPS represents an interesting strategy to obtain PLA-based toughened formulations. In relation to the intermediate compositions, for instance the $PLA_{60}PCL_{20}TPS_{20}$ piece, it is worthy to note the remarkable decrease in G′ observed at about −60 °C, which corresponds to the glass transition of PCL. Another important decrease in G′ was observed in the ternary blend pieces in the thermal region located from −20 °C to −30 °C, which is attributable to the glass transition of TPS.

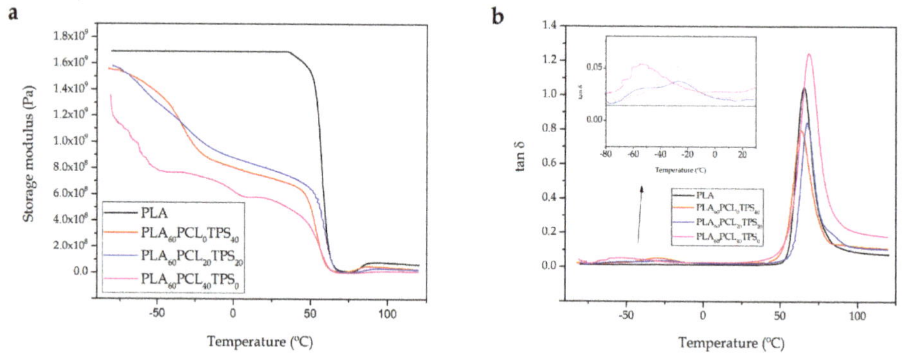

Figure 7. Comparative plot of the polylactide (PLA), poly(ε-caprolactone) (PCL), and thermoplastic starch (TPS) blend pieces in terms of: (**a**) Storage modulus (G′) versus temperature; and (**b**) dynamic damping factor (tan δ) versus temperature.

Figure 7b shows the evolution of tan δ that is, the ratio of G″ to G′, versus temperature. The alpha (α)-relaxation regions of each biopolymer can be clearly identified by the peaks of the tan δ plots, which are related to their T_gs and molecular motions [64]. One can observe that the α-peak of the PLA phase slightly changed in the pieces when it was melt blended with the other biopolymers. In particular, it increased from 65.2 °C, for the neat PLA piece, to 68.4 °C, for the binary blend piece containing 40 wt % PCL, that is, $PLA_{60}PCL_{40}TPS_0$, while it was reduced to 64.0 °C, for the binary blend piece containing 40 wt % TPS, that is, $PLA_{60}PCL_0TPS_{40}$. In this sense, Martin et al. [65] observed that the α-relaxation region of the PLA phase presented a gradual decrease with increasing the amounts of TPS. In particular, the T_g of PLA decreased from 67 °C, for neat PLA, to about 55 °C, for PLA blends containing 10 wt % TPS. Since it was observed that glycerol has a relatively low effect on the glass transition of PLA, the shift of the α-relaxation to lower temperatures suggested some interaction between TPS and PLA and, as a result, partial miscibility between the two biopolymers was inferred. However, since this reduction was moderate, a small degree of miscibility between the blend components was concluded. In relation PCL, Mittal et al. [66] showed that the α-relaxation region of PLA occurred at higher temperatures as the amount of PCL in the binary blends was increased. In particular, the α-peak of the neat PLA increased from approximately 55 °C to 61 °C. This effect was ascribed by the authors to a better intermixing of the phases in the presence of PCL. Additionally, one can also observe that the α-peak values for the PCL- and TPS-rich phases in the blend pieces were located at approximately −55 °C and −30 °C, respectively.

Furthermore, one can observe in Table 5 that the addition of both PCL and TPS yielded lower VST and HDT values than those observed for the neat PLA piece. These results are in agreement with the above-described mechanical and thermomechanical results due to both PCL and TPS provided increased ductility and, subsequently, the material's ability to deform was remarkably increased.

Table 5. Thermomechanical properties in terms of the Vicat softening temperature (VST) and heat deflection temperature (HDT) of the polylactide (PLA), poly(ε-caprolactone) (PCL), and thermoplastic starch (TPS) blend pieces.

Sample	VST (°C)	HDT (°C)
PLA	53.2 ± 0.5	47.9 ± 0.5
$PLA_{60}PCL_{40}TPS_0$	51.2 ± 0.6	43.2 ± 0.4
$PLA_{60}PCL_{30}TPS_{10}$	50.2 ± 0.5	46.4 ± 0.5
$PLA_{60}PCL_{20}TPS_{20}$	48.8 ± 0.3	46.6 ± 0.4
$PLA_{60}PCL_{10}TPS_{30}$	47.1 ± 0.5	46.2 ± 0.4
$PLA_{60}PCL_0TPS_{40}$	47.4 ± 0.4	46.8 ± 0.3

3.5. Disintegration in Controlled Compost Soil

Figure 8 shows the percentage of weight loss as a function of the elapsed time during disintegration in the controlled compost soil of the injection-molded PLA-based pieces. One can observe that all of the here-prepared biopolymer pieces presented a significant loss of mass after only one week while they were fully disintegrated at the end of the test, that is, after a period of 28 days. The sample with the highest degradation rate was the neat PLA piece. In fact, after 21 days in the controlled compost soil, this sample already lost 100% of its initial weight. The addition of both PCL and TPS slightly reduced the biodegradation rate of PLA and this effect was more marked for the binary blend pieces, that is, $PLA_{60}PCL_{40}TPS_0$ and $PLA_{60}PCL_0TPS_{40}$, than for the ternary blend pieces. For instance, after 21 days, whereas the ternary $PLA_{60}PCL_{20}TPS_{20}$ piece showed a mass loss of 89.9%, this value was only 57.3% for the binary $PLA_{60}PCL_0TPS_{40}$ piece. This suggests that the biodegradation rate of PCL and TPS was lower than that observed for PLA in the selected compost soil. Therefore, the use of ternary blends improved the compostability profile of the binary blends made of PLA with PCL or TPS since, as shown during the morphological analysis, the regions of the secondary phases in the ternary blend pieces were smaller. Previous research studies have reported, however, that the PCL and TPS degradation rates are faster than that of PLA [67,68]. These differences can be ascribed to the type of culture present in the medium during disintegration. For instance, Thakore et al. [69] described that the different compost soils from municipal yard waste sites, which generally contains various types of microorganisms, can strongly affect the biodegradation profile of compostable biopolymer articles in a different manner. In particular, it was observed that the TPS degradation pathway was mainly produced due to two enzymes secreted by the microbes. In particular, esterase cleaves the ester bond, releasing free phthalic acid and starch, while amylase acts on starch to produce reducing sugars.

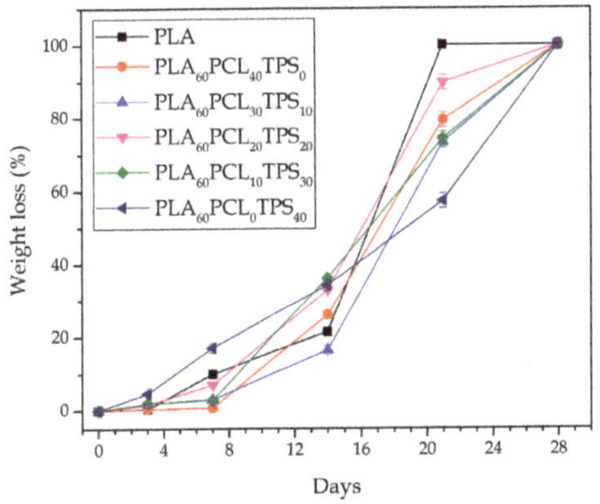

Figure 8. Evolution plot of the percentage of weight loss as a function of the elapsed time during disintegration in controlled compost soil of the polylactide (PLA), poly(ε-caprolactone) (PCL), and thermoplastic starch (TPS) blend pieces.

Figure 9 finally shows the visual aspect of the injection-molded PLA-based pieces during the disintegration test, giving some further information about their compostability profile. After analyzing the samples appearance, one can conclude that all the PLA-based pieces were either fully disintegrated or significantly fragmented after 21 days. Regarding neat PLA, one can observe that its piece become opaque after only 3 days of incubation in the controlled compost soil due to hydrolysis of the biopolyester [70]. Although a slight weight decrease was observed, no significant alterations from a physical point of view (e.g., color changes, presence of micro-cracks, etc.) were seen during the first week. Over the second week, however, the PLA-based pieces revealed significant evidences of biodegradation. At this incubation time, the PLA piece as well as the $PLA_{60}PCL_{20}TPS_{20}$ and $PLA_{60}PCL_{10}TPS_{30}$ ternary blend pieces were extensively biodegraded producing small fragments. Although the other blend pieces, that is, the $PLA_{60}PCL_{40}TPS_0$, $PLA_{60}PCL_{30}TPS_{10}$, and $PLA_{60}PCL_0TPS_{40}$, still remained into a single part, they visually presented a clear weight loss and develop a dark brown color. After 21 days, the neat PLA piece was fully biodegraded while the binary and ternary blend pieces were considerably disintegrated into small fragments, with the exemption of the binary blend piece of PLA with 40 wt % TPS, that is, $PLA_{60}PCL_0TPS_{40}$. Therefore, as explained above, the addition of PCL and TPS slightly slowed down the disintegration process of PLA. This delay was mostly visible in the PLA-based pieces with high contents of either PCL or TPS, thought in the case of the plasticized carbohydrate it was even more pronounced.

Figure 9. Visual aspect at selected disintegration times of the polylactide (PLA), poly(ε-caprolactone) (PCL), and thermoplastic starch (TPS) blend pieces.

4. Discussion

Binary and ternary blend pieces based on PLA with different PCL and TPS contents are herein presented as novel sustainable plastics with improved ductility and toughness. In the here-performed tensile and flexural tests, it was observed that the addition of PCL and TPS provided a positive effect on flexibility and impact strength but also a slight reduction in the mechanical strength properties. Although both biopolymers individually produced a positive effect on the ductile properties of PLA, the best results were obtained for the ternary blends that combined high PCL contents with low amounts of TPS. In particular, the ternary blend piece of PLA with 30 wt % PCL and 10 wt % TPS, that is, $PLA_{60}PCL_{30}TPS_{10}$, showed the highest flexibility with a ε_b value of 196.7%, approximately 40 times higher than that observed for the neat PLA piece. Similar findings were obtained in the impact tests, in which the ternary blends containing the highest PCL contents provided toughness increases of more than three times in comparison to the neat PLA piece. During the thermal analysis, DSC confirmed that the here-prepared binary and ternary blends are immiscible while TGA revealed that the ternary blend pieces present slightly higher thermal stability than the neat PLA piece and the binary blend pieces. Thermomechanical analysis, performed by means of DMTA, as well as VST and HDT measurements, also demonstrated that the blend pieces presented lower stiffness since both PCL and TPS effectively softened PLA. Finally, during the disintegration test in a controlled compost soil, it was observed that all PLA-based pieces presented a significant mass loss after only two weeks while the blend pieces disintegrated into small fragments after a period of 21 days. At the end of the test, that is, after 28 days, all pieces fully biodegraded. Although the addition of both PCL and TPS slightly reduced the PLA disintegration rate, this impairment was more marked for the binary blend pieces, that is, $PLA_{60}PCL_{40}TPS_0$ and $PLA_{60}PCL_0TPS_{40}$. Interestingly, the ternary blend pieces with intermediate contents of PCL and TPS presented a biodegradation rate close to that observed for the neat PLA piece.

5. Conclusions

The development of ternary blends based on PLA with relatively high contents of PCL and low contents TPS can be successfully applied for the development of compostable plastic articles with improved ductility and toughness. Potential uses of the here-described injection-molded pieces can be found in the rigid packaging industry, where for instance sustainable trays, bottles, and caps with high mechanical strength, but also sufficient ductility and impact strength, are currently required.

Author Contributions: Conceptualization was devised by S.T.-G.; methodology, validation, and formal analysis was carried out by L.Q.-C., N.M., F.P. and A.J.-V.; investigation, resources, data curation, and writing—original draft preparation was performed by L.Q.-C. and F.P.; writing—review and editing, L.Q.-C. and S.T.-G.; supervision, A.J.-V. and S.T.-G.; project administration, S.T.-G.

Funding: This research was supported by the Ministry of Science, Innovation, and Universities (MICIU) program numbers MAT2017-84909-C2-2-R and AGL2015-63855-C2-1-R, and by the EU H2020 project YPACK (reference number 773872).

Acknowledgments: L.Q.-C. wants to thank Generalitat Valenciana (GV) for his FPI grant (ACIF/2016/182) and the Spanish Ministry of Education, Culture, and Sports (MECD) for his FPU grant (FPU15/03812). S.T.-G. also acknowledges the MICIU for his Juan de la Cierva contract (IJCI-2016-29675).

Conflicts of Interest: The authors declare no conflict of interest.

References

1. Hopewell, J.; Dvorak, R.; Kosior, E. Plastics recycling: Challenges and opportunities. *Philos. Trans. R. Soc. B Biol. Sci.* **2009**, *364*, 2115–2126. [CrossRef] [PubMed]
2. Quiles-Carrillo, L.; Montanes, N.; Garcia-Garcia, D.; Carbonell-Verdu, A.; Balart, R.; Torres-Giner, S. Effect of different compatibilizers on injection-molded green composite pieces based on polylactide filled with almond shell flour. *Compos. Part B Eng.* **2018**, *147*, 76–85. [CrossRef]
3. Nampoothiri, K.M.; Nair, N.R.; John, R.P. An overview of the recent developments in polylactide (PLA) research. *Bioresour. Technol.* **2010**, *101*, 8493–8501. [CrossRef] [PubMed]
4. Kumar, N.; Das, D. Fibrous biocomposites from nettle (girardinia diversifolia) and poly(lactic acid) fibers for automotive dashboard panel application. *Compos. Part B Eng.* **2017**, *130*, 54–63. [CrossRef]
5. Bouzouita, A.; Notta-Cuvier, D.; Raquez, J.-M.; Lauro, F.; Dubois, P. Poly(Lactic Acid)-Based Materials for Automotive Applications. In *Industrial Applications of Poly(lactic acid)*; Springer: Berlin/Heidelberg, Germany, 2017.
6. Garces, J.M.; Moll, D.J.; Bicerano, J.; Fibiger, R.; McLeod, D.G. Polymeric nanocomposites for automotive applications. *Adv. Mater.* **2000**, *12*, 1835–1839. [CrossRef]

7. Lasprilla, A.J.; Martinez, G.A.; Lunelli, B.H.; Jardini, A.L.; Maciel Filho, R. Poly-lactic acid synthesis for application in biomedical devices—A review. *Biotechnol. Adv.* **2012**, *30*, 321–328. [CrossRef] [PubMed]
8. Torres-Giner, S.; Gimeno-Alcañiz, J.V.; Ocio, M.J.; Lagaron, J.M. Optimization of electrospun polylactide-based ultrathin fibers for osteoconductive bone scaffolds. *J. Appl. Polym. Sci.* **2011**, *122*, 914–925. [CrossRef]
9. Muller, J.; González-Martínez, C.; Chiralt, A. Combination of poly(lactic) acid and starch for biodegradable food packaging. *Materials* **2017**, *10*, 952. [CrossRef] [PubMed]
10. Kakroodi, A.R.; Kazemi, Y.; Nofar, M.; Park, C.B. Tailoring poly(lactic acid) for packaging applications via the production of fully bio-based in situ microfibrillar composite films. *Chem. Eng. J.* **2017**, *308*, 772–782. [CrossRef]
11. Kao, C.-T.; Lin, C.-C.; Chen, Y.-W.; Yeh, C.-H.; Fang, H.-Y.; Shie, M.-Y. Poly(dopamine) coating of 3d printed poly (lactic acid) scaffolds for bone tissue engineering. *Mater. Sci. Eng. C* **2015**, *56*, 165–173. [CrossRef] [PubMed]
12. Chen, Q.; Mangadlao, J.D.; Wallat, J.; De Leon, A.; Pokorski, J.K.; Advincula, R.C. 3D printing biocompatible polyurethane/poly(lactic acid)/graphene oxide nanocomposites: Anisotropic properties. *ACS Appl. Mater. Interfaces* **2017**, *9*, 4015–4023. [CrossRef] [PubMed]
13. Quiles-Carrillo, L.; Duart, S.; Montanes, N.; Torres-Giner, S.; Balart, R. Enhancement of the mechanical and thermal properties of injection-molded polylactide parts by the addition of acrylated epoxidized soybean oil. *Mater. Des.* **2018**, *140*, 54–63. [CrossRef]
14. Torres-Giner, S.; Gil, L.; Pascual-Ramírez, L.; Garde-Belza, J.A. Packaging: Food waste reduction. In *Encyclopedia of Polymer Applications*; Mishra, M., Ed.; CRC Press: Boca Raton, FL, USA, 2018.
15. Mooney, D.; Breuer, C.; McNamara, K.; Vacanti, J.; Langer, R. Fabricating tubular devices from polymers of lactic and glycolic acid for tissue engineering. *Tissue Eng.* **1995**, *1*, 107–118. [CrossRef] [PubMed]
16. Elsawy, M.A.; Kim, K.-H.; Park, J.-W.; Deep, A. Hydrolytic degradation of polylactic acid (PLA) and its composites. *Renew. Sustain. Energy Rev.* **2017**, *79*, 1346–1352. [CrossRef]
17. Pluta, M.; Piorkowska, E. Tough crystalline blends of polylactide with block copolymers of ethylene glycol and propylene glycol. *Polym. Test.* **2015**, *46*, 79–87. [CrossRef]
18. Maiza, M.; Benaniba, M.T.; Quintard, G.; Massardier-Nageotte, V. Biobased additive plasticizing polylactic acid (PLA). *Polímeros* **2015**, *25*, 581–590. [CrossRef]
19. Ljungberg, N.; Wesslen, B. The effects of plasticizers on the dynamic mechanical and thermal properties of poly (lactic acid). *J. Appl. Polym. Sci.* **2002**, *86*, 1227–1234. [CrossRef]
20. Darie-Niţă, R.N.; Vasile, C.; Irimia, A.; Lipşa, R.; Râpă, M. Evaluation of some eco-friendly plasticizers for PLA films processing. *J. Appl. Polym. Sci.* **2016**, *133*, 43223. [CrossRef]
21. Quiles-Carrillo, L.; Blanes-Martínez, M.; Montanes, N.; Fenollar, O.; Torres-Giner, S.; Balart, R. Reactive toughening of injection-molded polylactide pieces using maleinized hemp seed oil. *Eur. Polym. J.* **2018**, *98*, 402–410. [CrossRef]
22. Ferri, J.M.; Garcia-Garcia, D.; Montanes, N.; Fenollar, O.; Balart, R. The effect of maleinized linseed oil as biobased plasticizer in poly(lactic acid)-based formulations. *Polym. Int.* **2017**, *66*, 882–891. [CrossRef]
23. Carbonell-Verdu, A.; Garcia-Garcia, D.; Dominici, F.; Torre, L.; Sanchez-Nacher, L.; Balart, R. PLA films with improved flexibility properties by using maleinized cottonseed oil. *Eur. Polym. J.* **2017**, *91*, 248–259. [CrossRef]
24. Quiles-Carrillo, L.; Montanes, N.; Sammon, C.; Balart, R.; Torres-Giner, S. Compatibilization of highly sustainable polylactide/almond shell flour composites by reactive extrusion with maleinized linseed oil. *Ind. Crop. Prod.* **2018**, *111*, 878–888. [CrossRef]
25. Gerard, T.; Budtova, T. Morphology and molten-state rheology of polylactide and polyhydroxyalkanoate blends. *Eur. Polym. J.* **2012**, *48*, 1110–1117. [CrossRef]
26. Yu, L.; Dean, K.; Li, L. Polymer blends and composites from renewable resources. *Prog. Polym. Sci.* **2006**, *31*, 576–602. [CrossRef]
27. Gug, J.-I.; Tan, B.; Soule, J.; Downie, M.; Barrington, J.; Sobkowicz, M. Analysis of models predicting morphology transitions in reactive twin-screw extrusion of bio-based polyester/polyamide blends. *Int. Polym. Process.* **2017**, *32*, 363–377. [CrossRef]
28. Stoclet, G.; Seguela, R.; Lefebvre, J.-M. Morphology, thermal behavior and mechanical properties of binary blends of compatible biosourced polymers: Polylactide/polyamide11. *Polymer* **2011**, *52*, 1417–1425. [CrossRef]

29. Al-Itry, R.; Lamnawar, K.; Maazouz, A. Improvement of thermal stability, rheological and mechanical properties of PLA, PBAT and their blends by reactive extrusion with functionalized epoxy. *Polym. Degrad. Stab.* **2012**, *97*, 1898–1914. [CrossRef]
30. Wu, N.; Zhang, H. Mechanical properties and phase morphology of super-tough PLA/PBAT/EMA-GMA multicomponent blends. *Mater. Lett.* **2017**, *192*, 17–20. [CrossRef]
31. Sarazin, P.; Li, G.; Orts, W.J.; Favis, B.D. Binary and ternary blends of polylactide, polycaprolactone and thermoplastic starch. *Polymer* **2008**, *49*, 599–609. [CrossRef]
32. Valerio, O.; Misra, M.; Mohanty, A.K. Statistical design of sustainable thermoplastic blends of poly(glycerol succinate-co-maleate) (PGSMA), poly(lactic acid) (PLA) and poly(butylene succinate) (PBS). *Polym. Test.* **2018**, *65*, 420–428. [CrossRef]
33. Ostafinska, A.; Fortelný, I.; Hodan, J.; Krejčíková, S.; Nevoralová, M.; Kredatusová, J.; Kruliš, Z.; Kotek, J.; Šlouf, M. Strong synergistic effects in PLA/PCL blends: Impact of PLA matrix viscosity. *J. Mech. Behav. Biomed. Mater.* **2017**, *69*, 229–241. [CrossRef] [PubMed]
34. Wu, D.; Lin, D.; Zhang, J.; Zhou, W.; Zhang, M.; Zhang, Y.; Wang, D.; Lin, B. Selective localization of nanofillers: Effect on morphology and crystallization of PLA/PCL blends. *Macromol. Chem. Phys.* **2011**, *212*, 613–626. [CrossRef]
35. Liu, H.; Song, W.; Chen, F.; Guo, L.; Zhang, J. Interaction of microstructure and interfacial adhesion on impact performance of polylactide (PLA) ternary blends. *Macromolecules* **2011**, *44*, 1513–1522. [CrossRef]
36. Wokadala, O.C.; Ray, S.S.; Bandyopadhyay, J.; Wesley-Smith, J.; Emmambux, N.M. Morphology, thermal properties and crystallization kinetics of ternary blends of the polylactide and starch biopolymers and nanoclay: The role of nanoclay hydrophobicity. *Polymer* **2015**, *71*, 82–92. [CrossRef]
37. Zolali, A.M.; Favis, B.D. Partial to complete wetting transitions in immiscible ternary blends with PLA: The influence of interfacial confinement. *Soft Matter* **2017**, *13*, 2844–2856. [CrossRef] [PubMed]
38. Matzinos, P.; Tserki, V.; Kontoyiannis, A.; Panayiotou, C. Processing and characterization of starch/polycaprolactone products. *Polym. Degrad. Stab.* **2002**, *77*, 17–24. [CrossRef]
39. Maglio, G.; Malinconico, M.; Migliozzi, A.; Groeninckx, G. Immiscible poly(L-lactide)/poly(ε-caprolactone) blends: Influence of the addition of a poly(L-lactide)-poly(oxyethylene) block copolymer on thermal behavior and morphology. *Macromol. Chem. Phys.* **2004**, *205*, 946–950. [CrossRef]
40. Forssell, P.; Mikkilä, J.; Suortti, T.; Seppälä, J.; Poutanen, K. Plasticization of barley starch with glycerol and water. *J. Macromol. Sci. Part A* **1996**, *33*, 703–715. [CrossRef]
41. Raquez, J.M.; Nabar, Y.; Srinivasan, M.; Shin, B.Y.; Narayan, R.; Dubois, P. Maleated thermoplastic starch by reactive extrusion. *Carbohydr. Polym.* **2008**, *74*, 159–169. [CrossRef]
42. Averous, L.; Moro, L.; Dole, P.; Fringant, C. Properties of thermoplastic blends: Starch–polycaprolactone. *Polymer* **2000**, *41*, 4157–4167. [CrossRef]
43. Odelius, K.; Ohlson, M.; Höglund, A.; Albertsson, A.C. Polyesters with small structural variations improve the mechanical properties of polylactide. *J. Appl. Polym. Sci.* **2013**, *127*, 27–33. [CrossRef]
44. Zhen, Z.; Ying, S.; Hongye, F.; Jie, R.; Tianbin, R. Preparation, characterization and properties of binary and ternary blends with thermoplastic starch, poly(lactic acid) and poly(butylene succinate). *Polym. Renew. Resour.* **2011**, *2*, 49–62. [CrossRef]
45. Ren, J.; Fu, H.; Ren, T.; Yuan, W. Preparation, characterization and properties of binary and ternary blends with thermoplastic starch, poly(lactic acid) and poly(butylene adipate-co-terephthalate). *Carbohydr. Polym.* **2009**, *77*, 576–582. [CrossRef]
46. Ferri, J.; Garcia-Garcia, D.; Sánchez-Nacher, L.; Fenollar, O.; Balart, R. The effect of maleinized linseed oil (MLO) on mechanical performance of poly(lactic acid)-thermoplastic starch (PLA-TPS) blends. *Carbohydr. Polym.* **2016**, *147*, 60–68. [CrossRef] [PubMed]
47. García-Campo, M.; Boronat, T.; Quiles-Carrillo, L.; Balart, R.; Montanes, N. Manufacturing and characterization of toughened poly(lactic acid) (PLA) formulations by ternary blends with biopolyesters. *Polymers* **2018**, *10*, 3. [CrossRef]
48. Chen, C.-C.; Chueh, J.-Y.; Tseng, H.; Huang, H.-M.; Lee, S.-Y. Preparation and characterization of biodegradable PLA polymeric blends. *Biomaterials* **2003**, *24*, 1167–1173. [CrossRef]
49. Ferri, J.M.; Fenollar, O.; Jorda-Vilaplana, A.; García-Sanoguera, D.; Balart, R. Effect of miscibility on mechanical and thermal properties of poly(lactic acid)/polycaprolactone blends. *Polym. Int.* **2016**, *65*, 453–463. [CrossRef]

50. Tang, L.; Wang, L.; Chen, P.; Fu, J.; Xiao, P.; Ye, N.; Zhang, M. Toughness of ABS/PBT blends: The relationship between composition, morphology, and fracture behavior. *J. Appl. Polym. Sci.* **2018**, *135*, 46051. [CrossRef]
51. Muthuraj, R.; Misra, M.; Mohanty, A.K. Biodegradable compatibilized polymer blends for packaging applications: A literature review. *J. Appl. Polym. Sci.* **2018**, *135*, 45726. [CrossRef]
52. Carmona, V.B.; Corrêa, A.C.; Marconcini, J.M.; Mattoso, L.H.C. Properties of a biodegradable ternary blend of thermoplastic starch (TPS), poly(ε-caprolactone) (PCL) and poly(lactic acid) (PLA). *J. Polym. Environ.* **2015**, *23*, 83–89. [CrossRef]
53. Kim, H.-Y.; Park, S.S.; Lim, S.-T. Preparation, characterization and utilization of starch nanoparticles. *Colloid Surf. B Biointerfaces* **2015**, *126*, 607–620. [CrossRef] [PubMed]
54. Samper, M.; Marina Patricia, A.; Santiago, F.; Juan, L. Influence of biodegradable materials in the recycled polystyrene. *J. Appl. Polym. Sci.* **2014**, *131*, 41161.
55. Bordes, C.; Fréville, V.; Ruffin, E.; Marote, P.; Gauvrit, J.; Briançon, S.; Lantéri, P. Determination of poly(ε-caprolactone) solubility parameters: Application to solvent substitution in a microencapsulation process. *Int. J. Pharm.* **2010**, *383*, 236–243. [CrossRef] [PubMed]
56. Small, P. Some factors affecting the solubility of polymers. *J. Appl. Chem.* **1953**, *3*, 71–80. [CrossRef]
57. Navarro-Baena, I.; Sessini, V.; Dominici, F.; Torre, L.; Kenny, J.M.; Peponi, L. Design of biodegradable blends based on PLA and PCL: From morphological, thermal and mechanical studies to shape memory behavior. *Polym. Degrad. Stab.* **2016**, *132*, 97–108. [CrossRef]
58. Averous, L.; Boquillon, N. Biocomposites based on plasticized starch: Thermal and mechanical behaviours. *Carbohydr. Polym.* **2004**, *56*, 111–122. [CrossRef]
59. Zhang, Y.; Rempel, C.; Liu, Q. Thermoplastic starch processing and characteristics—A review. *Crit. Rev. Food Sci. Nutr.* **2014**, *54*, 1353–1370. [CrossRef] [PubMed]
60. Patrício, T.; Bártolo, P. Thermal stability of PCL/PLA blends produced by physical blending process. *Procedia Eng.* **2013**, *59*, 292–297. [CrossRef]
61. Mofokeng, J.; Luyt, A. Morphology and thermal degradation studies of melt-mixed poly(lactic acid) (PLA)/poly(ε-caprolactone) (PCL) biodegradable polymer blend nanocomposites with TiO_2 as filler. *Polym. Test.* **2015**, *45*, 93–100. [CrossRef]
62. Quiles-Carillo, L.; Montanes, N.; Lagaron, J.M.; Balart, R.; Torres-Giner, S. In situ compatibilization of biopolymer ternary blends by reactive extrusion with low-functionality epoxy-based styrene–acrylic oligomer. *J. Polym. Environ.* **2018**. [CrossRef]
63. Garcia-Campo, M.J.; Quiles-Carrillo, L.; Masia, J.; Reig-Pérez, M.J.; Montanes, N.; Balart, R. Environmentally friendly compatibilizers from soybean oil for ternary blends of poly(lactic acid)-PLA, poly(ε-caprolactone)-PCL and poly(3-hydroxybutyrate)-PHB. *Materials* **2017**, *10*, 1339. [CrossRef] [PubMed]
64. Torres-Giner, S.; Montanes, N.; Fenollar, O.; García-Sanoguera, D.; Balart, R. Development and optimization of renewable vinyl plastisol/wood flour composites exposed to ultraviolet radiation. *Mater. Des.* **2016**, *108*, 648–658. [CrossRef]
65. Martin, O.; Averous, L. Poly(lactic acid): Plasticization and properties of biodegradable multiphase systems. *Polymer* **2001**, *42*, 6209–6219. [CrossRef]
66. Mittal, V.; Akhtar, T.; Matsko, N. Mechanical, thermal, rheological and morphological properties of binary and ternary blends of PLA, TPS and PCL. *Macromol. Mater. Eng.* **2015**, *300*, 423–435. [CrossRef]
67. Di Franco, C.; Cyras, V.; Busalmen, J.; Ruseckaite, R.; Vázquez, A. Degradation of polycaprolactone/starch blends and composites with sisal fibre. *Polym. Degrad. Stab.* **2004**, *86*, 95–103. [CrossRef]
68. Iovino, R.; Zullo, R.; Rao, M.; Cassar, L.; Gianfreda, L. Biodegradation of poly(lactic acid)/starch/coir biocomposites under controlled composting conditions. *Polym. Degrad. Stab.* **2008**, *93*, 147–157. [CrossRef]
69. Thakore, I.; Desai, S.; Sarawade, B.; Devi, S. Studies on biodegradability, morphology and thermo-mechanical properties of LDPE/modified starch blends. *Eur. Polym. J.* **2001**, *37*, 151–160. [CrossRef]
70. Sikorska, W.; Musiol, M.; Nowak, B.; Pajak, J.; Labuzek, S.; Kowalczuk, M.; Adamus, G. Degradability of polylactide and its blend with poly[(R,S)-3-hydroxybutyrate] in industrial composting and compost extract. *Int. Biodeterior. Biodegrad.* **2015**, *101*, 32–41. [CrossRef]

materials

MDPI

Article

Reactive Melt Mixing of Poly(3-Hydroxybutyrate)/Rice Husk Flour Composites with Purified Biosustainably Produced Poly(3-Hydroxybutyrate-*co*-3-Hydroxyvalerate)

Beatriz Melendez-Rodriguez [1], Sergio Torres-Giner [1,*], Abdulaziz Aldureid [2], Luis Cabedo [2] and Jose M. Lagaron [1,*]

1 Novel Materials and Nanotechnology Group, Institute of Agrochemistry and Food Technology (IATA), Spanish Council for Scientific Research (CSIC), Calle Catedrático Agustín Escardino Benlloch 7, 46980 Paterna, Spain

2 Polymers and Advanced Materials Group (PIMA), Universitat Jaume I, 12071 Castellón, Spain

* Correspondence: storresginer@iata.csic.es (S.T.-G.); lagaron@iata.csic.es (J.M.L.)

Received: 31 May 2019; Accepted: 1 July 2019; Published: 4 July 2019

Abstract: Novel green composites based on commercial poly(3-hydroxybutyrate) (PHB) filled with 10 wt % rice husk flour (RHF) were melt-compounded in a mini-mixer unit using triglycidyl isocyanurate (TGIC) as compatibilizer and dicumyl peroxide (DCP) as initiator. Purified poly(3-hydroxybutyrate-*co*-3-hydroxyvalerate) (PHBV) produced by mixed bacterial cultures derived from fruit pulp waste was then incorporated into the green composite in contents in the 5–50 wt % range. Films for testing were obtained thereafter by thermo-compression and characterized. Results showed that the incorporation of up to 20 wt % of biowaste derived PHBV yielded green composite films with a high contact transparency, relatively low crystallinity, high thermal stability, improved mechanical ductility, and medium barrier performance to water vapor and aroma. This study puts forth the potential use of purified biosustainably produced PHBV as a cost-effective additive to develop more affordable and waste valorized food packaging articles.

Keywords: PHB; PHBV; rice husk; green composites; biosustainability; waste valorization

1. Introduction

The current concern to reduce the use of petroleum-derived materials has led to the search for natural and biodegradable polymers. Polyhydroxyalkanoates (PHAs) is a family of linear polyesters produced in nature by the action of bacteria during fermentation of sugar or lipids in famine conditions [1]. PHAs represent a good alternative to conventional polymers in the frame of the circular economy since they are fully bio-based and biodegradable [2]. Among the different commercially available PHAs, the most widely studied is poly(3-hydroxybutyrate) (PHB). This isotactic homopolyester presents a relatively high melting temperature (T_m) and good stiffness and strength due to its high crystallinity (>50%). As a result, PHB articles present similar performance or even greater than some commodities plastics such as polypropylene (PP) and barrier properties close to those of polyethylene terephthalate (PET) [3]. PHB undergoes rapid and complete disintegration within a maximum period of 6 months through the action of enzymes and/or chemical deterioration associated with living microorganisms. Moreover, PHB is biodegradable not only in composting conditions but also in other environments such as marine water [4].

However, the use of PHB for packaging applications is limited due to its excessive brittleness and narrow processing temperature window [5]. To overcome these limitations, its copolymer with 3-hydroxyvalerate (HV), that is, poly(3-hydroxybutyrate-*co*-3-hydroxyvalerate) (PHBV), can result advantageous since it shows higher ductility as well as reduced crystallinity and lower T_m [6]. PHBV articles have been proposed to be applied in the areas of food and cosmetic packaging such as shampoo bottles, plastic beverage bottles, milk cartons, cosmetic containers, among others, due to its renewability, biodegradability, and high water vapor barrier [7,8]. However, obtaining PHAs habitually requires large investments due to both the high cost of the carbon source and the lack of efficient cultivation techniques [9]. Indeed, the current production cost of PHAs is estimated to be up to 15-fold higher than conventional polyolefins [10]. Therefore, great efforts in their industrial production are currently focused on reducing the manufacturing cost to make it more competitive [11]. In this sense, biowaste derived PHAs are both economically and environmentally attractive, in particular, those that use food waste as the raw material source. For instance, fermented cheese whey (CW), which is mostly not fully valorized at present, can be used as the feeding solution for PHA production [12]. Nowadays, most significant research efforts are targeted to optimize the extraction methods, especially in mixed cultures, and also to reduce the amounts of chemicals used to make the process environmentally friendly [13].

Despite of the extraordinary suitability of PHAs as candidates for sustainable food packaging applications, they are still not cost-effective due to the fermentation and downstream processes during bioreactor production [14]. In this context, a possible strategy to reduce price is the use of agro-food waste derived fillers, which also allows a more sustainable packaging concept since they valorize residues obtained from the agricultural and food industries, and thus reduce the overall impact of the industrial production cycle [15]. The combination of a bio-based and biodegradable polymer with natural fillers is habitually termed "green composite", which means that the whole material is obtained from renewable resources and it is also biodegradable [16]. Over the past few years, the use of natural fillers to develop polymer composites has significantly increased because of their significant processing advantages, biodegradability, low cost, non-abrasive, low relative density, high specific strength, and renewable nature [17]. Moreover, these natural fillers represent an environmentally friendly solution since they decrease polluting emissions and energy requirements for processing as well as enhance energy recovery and end-of-life biodegradability [18–20]. In this context, different green composites have been obtained using lignocellulosic fillers derived from food, agricultural, industrial, and marine resources such as rice husk [21,22], almond husk [23–25], walnut shell [26], peanut shell [27], coconut fibers [28], orange peel [29], recycled cotton [30], *Posidonia oceanica* seaweed [31], etc.

Rice (*Oryza sativa* L.) is an important crop cultivated mostly in China, India, and Indonesia [32]. The annual world rice production is approximately 600 million tons, of which 20% is currently wasted as rice husk [33]. Most of this by-product is used as a bedding material for animals, burned, or landfilled, causing several environmental and health problems. Rice husk is a relatively hard material since it is typically composed of 20 wt % ash, 38 wt % cellulose, 22 wt % lignin, 18 wt % pentose, and 2 wt % of other organic components [34]. Therefore, rice husk has been used to reinforce several thermoplastics such as high-density polyethylene (HDPE) [35,36], PP [37], PP and HDPE [38], polylactide (PLA) [39], and also recently PHB [40–43]. However, the inherently poor interfacial adhesion between the lignocellulosic fillers and polymers generally yields a composite with low dispersion and a high content of particle aggregates [44]. This effect is related to the low chemical affinity between lignocellulosic fillers with most polymer and biopolymer matrices, which compromises the strength and also increases moisture absorption of the green composites [45]. To improve the interfacial adhesion between both composite components, compatibilizers or coupling agents are generally added or the filler surfaces are chemically pretreated [46]. Moreover, in the case of reactive compatibilizers, chemical bonds between the fillers and polymer matrix are formed and the overall performance of the composite can be remarkably improved [47]. For instance, the maleic anhydride (MA) grafting of PHBV matrix prior to extrusion has successfully increased the hydrophilicity of the biopolyester matrix making it more compatible with lignocellulosic fillers [48].

In this context, triglycidyl isocyanurate (TGIC) and dicumyl peroxide (DCP) can be effectively combined to compatibilize polymer composites. On the one hand, TGIC is a three-functional epoxy compound that plays a hinge-like role between lignocellulosic fillers and polyester matrices. The hydroxyl (–OH) groups of cellulose present on the fillers' surface and the ones from the end groups of the biopolyester molecular chains, namely hydroxyl or carboxyl, can readily react with the epoxy groups of TGIC during melt blending [49]. Also, TGIC has been reported to provide a chain-extension effect on the processability of PET, increasing its molecular weight (M_W) and potentially avoiding chain scission by hydrolysis [50]. On the other hand, DCP has been used as a free-radical grafting initiator in different polymer systems. In this sense, peroxides can form covalent carbon–carbon cross-links between the biopolymer chain segments, promoting the compatibilization of immiscible components in binary polymer blends [51] and also in polymer composites [52]. In the latter case, the addition of DCP to the composite mixture during melt mixing can give rise to both cross-linking of the polymer chains and grafting of natural fillers onto the polymer chains. Interestingly, due to the presence of three reactive –OH groups on each cellulose unit, the grafting of the cellulosic fillers onto the polymer chains dominates over the cross-linking of polymers because of the higher free radical reactivity of the –OH groups of cellulose [53]. Based on this phenomenon, different studies have for instance reported that DCP improved the mechanical properties of low-density polyethylene (LDPE)/wood fiber composites via peroxide-initiated cross-linking process [54–57].

The objective of this study is to develop highly sustainable materials with enhanced performance based on commercial PHB and rice husk fillers containing different amounts of purified PHBV that was produced by mixed bacterial cultures derived from wastes of the food industry. The green composites were prepared by melt compounding in a laboratory melt-mixer and shaped into films by thermo-compression. The resultant green composite films were characterized in terms of their morphology and optical characteristics as well as thermal, mechanical, and barrier properties in order to ascertain their potential in food packaging applications.

2. Materials and Methods

2.1. Materials

Commercial PHB homopolyester was supplied as P226F in the form of pellets by Biomer (Krailling, Germany). According to the manufacturer, this biopolymer resin presents a density of 1.25 g/cm^3 and a melt flow index (MFI) of 10 g/10 min (5 kg, 180 °C). Biowaste derived PHBV copolyester was produced at pilot-plant scale at Universidade NOVA (Lisbon, Portugal) using mixed microbial cultures fed with fermented fruit pulps supplied by SumolCompal S.A. (Carnaxide, Portugal) as an industrial residue of the juice industry. The molar fraction of HV in the copolymer was ~20 mol %. The PHBV was purified with chloroform (Sigma-Aldrich S.A., Madrid, Spain) to produce a solid powder. Further details about the biopolymer and its purification route can be found elsewhere [58].

Rice husk was kindly provided by Herba Ingredients (Valencia, Spain). It was delivered in the form of flakes as a by-product of the rice industry. D-limonene, with 98% purity, TGIC (reference 379506), with a M_W of 297.26 g/mol, and DCP (reference 329541), with a M_W of 270.37 g/mol and 98% assay, were all purchased from Sigma-Aldrich S.A. (Madrid, Spain).

2.2. Preparation of RHF

The procedure to obtain rice husk flour (RHF) consisted on a mechanical grinding following to sieving to ensure a low particle size. For this, the native rice husk was ground in a mechanical knife mill (Thermomix TM21, Vorwerk, Madrid, Spain) and then sieved in a 140-μm mesh (TED-0300, Filtra Vibración S.L., Badalona, Spain). The resultant powder was dried at 100 °C in oven (T3060, Heraeus Instruments, Hanau, Germany) for 24 h. Figure 1 shows the as-received flakes of rice husk (Figure 1a) and the resultant RHF (Figure 1b).

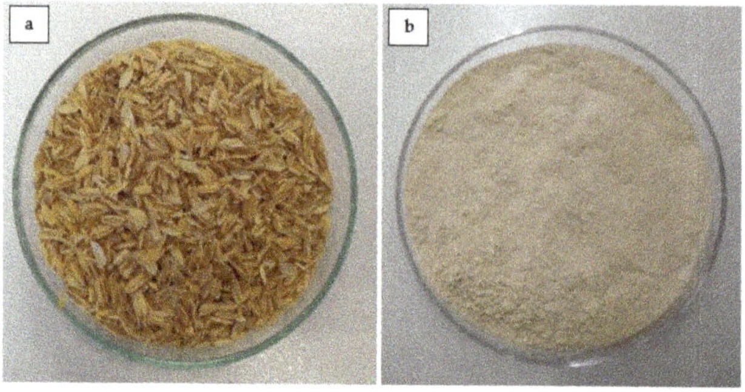

Figure 1. (**a**) As-received flakes of rice husk; (**b**) rice husk flour (RHF).

2.3. Melt Mixing

Prior to processing, both PHA resins were dried at 60 °C for 24 h in an oven (Digitheat, JP selecta S.A., Barcelona, Spain) to remove any residual moisture. Then, different amounts of purified PHBV, from 5 wt % to 50 wt %, were manually pre-mixed with commercial PHB in a zipper bag. A fixed content of RHF was added at 10 wt % to the mixture whereas the reactive compatibilizers, that is, TGIC and DCP, were incorporated at 1 part per hundred resin (phr) and 0.25 phr of PHB/PHBV/RHF composite, respectively. A PHB/RHF composite without PHBV and a PHB/PHBV blend without RHF were also prepared as control materials. Table 1 summarizes the different formulations prepared.

Table 1. Set of formulations prepared according to the weight content (wt %) of poly(3-hydroxybutyrate) (PHB), poly(3-hydroxybutyrate-*co*-3-hydroxyvalerate) (PHBV), and rice husk flour (RHF) in which triglycidyl isocyanurate (TGIC) and dicumyl peroxide (DCP) were added as parts per hundred resin (phr) of PHB/PHBV/RHF composite.

Sample	PHB (wt %)	PHBV (wt %)	RHF (wt %)	TGIC (phr)	DCP (phr)
PHB/RHF	90	0	10	1	0.25
PHB/PHBV	90	10	0	1	0.25
PHB/PHBV5/RHF	85	5	10	1	0.25
PHB/PHBV10/RHF	80	10	10	1	0.25
PHB/PHBV20/RHF	70	20	10	1	0.25
PHB/PHBV30/RHF	60	30	10	1	0.25
PHB/PHBV50/RHF	40	50	10	1	0.25

To prepare the samples, a total amount of 12 g of material was melt-compounded in a 16 cm^3 Brabender Plastograph Original E mini-mixer from Brabender GmbH & Co. KG (Duisburg, Germany). First the biopolymers and, then, the RHF powder were fed to the internal mixing chamber at a rotating speed of 60 rpm for 1 min. After this, TGIC and DCP were added and the whole composition was melt-mixed at 100 rpm for another 3 min. The processing temperature was set at 180 °C. Once the mixing process was completed, each batch was withdrawn from the mini-mixer and cooled in air to room temperature. The resultant doughs were stored in dissectors containing silica gel at 0% relative humidity (RH) and 25 °C for at least 48 h for conditioning.

The different doughs were, thereafter, thermo-compressed into films using a hot-plate hydraulic press (Carver 4122, Wabash, IN, USA). The samples were first placed in the plates at 180 °C for 1 min, without pressure, to remove any residual moisture and then hot-pressed at 4–5 bars for 3 min. Flat films with a total thickness of ~100 μm were obtained and stored in a desiccator at 25 °C and 0% RH for 15 days before characterization.

2.4. Characterization

2.4.1. Morphology

The morphologies of the RHF particles and the film cross-sections were observed by scanning electron microscopy (SEM) using an S-4800 device from Hitachi (Tokyo, Japan). For the cross-section observations, the films were cryo-fractured by immersion in liquid nitrogen. The samples were previously fixed to beveled holders using conductive double-sided adhesive tape and sputtered with a mixture of gold-palladium under vacuum. An accelerating voltage of 10 kV and a working distance of 15 mm were selected during SEM analysis. The estimation of the dimensions was performed by means of the ImageJ software v 1.41 (NIH, Bethesda, MD, USA) using a minimum of 20 SEM micrographs.

The particle size distribution was determined by dynamic light scattering (DLS) using a laser diffraction analyzer Mastersizer 2000 from Malvern Panalytical Ltd. (Malvern, UK). According to the manufacturer, the error for the equipment of median diameter (D_{50}) is 3%. Measurements were taken under stirring to avoid settling of large particles.

2.4.2. Transparency

The light transmission of the films was determined in specimens of 50 × 30 mm^2 by quantifying the absorption of light at wavelengths between 200 nm and 700 nm in an ultraviolet–visible (UV–vis) spectrophotometer VIS3000 from Dinko Instruments (Barcelona, Spain). The transparency (T) and opacity (O) were calculated using Equation (1) [59] and Equation (2) [60], respectively:

$$T = \frac{A_{600}}{L} \quad (1)$$

$$O = A_{500} \cdot L \quad (2)$$

where A_{500} and A_{600} are the absorbance values at 500 nm and 600 nm, respectively, and L is the film thickness (mm).

2.4.3. Color Measurements

The color of the films was determined using a chroma meter CR-400 (Konica Minolta, Tokyo, Japan). The color difference (ΔE^*) was calculated using the following Equation (3) [59], as defined by the Commission Internationale de l'Eclairage (CIE):

$$\Delta E^* = [(\Delta L^*)^2 + (\Delta a^*)^2 + (\Delta b^*)^2]^{0.5} \quad (3)$$

where ΔL^*, Δa^*, and Δb^* correspond to the differences in terms of lightness from black to white, from green to red, and from blue to yellow, respectively, between the film samples and the control film of PHB/PHBV.

2.4.4. Thermal Analysis

Thermal transitions were studied by differential scanning calorimetry (DSC) on a DSC-7 analyzer equipped with the cooling accessory Intracooler 2 from PerkinElmer, Inc. (Waltham, MA, USA). A two-step program under nitrogen atmosphere and with a flow rate of 20 mL/min was applied: first heating from −30 °C to 190 °C followed by cooling to −30 °C. The heating and cooling rates were both set at 10 °C/min and the typical sample weight was ~3 mg. An empty aluminum pan was used as reference whereas calibration was performed using an indium sample. The values of T_m and enthalpy of melting (ΔH_m) were obtained from the heating scan, while the crystallization temperature from the melt (T_c) and enthalpy of crystallization (ΔH_c) were determined from the cooling scan. All DSC measurements were performed in triplicate.

Thermogravimetric analysis (TGA) was performed in a TG-STDA model TGA/STDA851e/LF/1600 thermobalance from Mettler-Toledo, LLC (Columbus, OH, USA). The samples, with a weight of ~15 mg, were heated from 50 °C to 800 °C at a heating rate of 10 °C/min under a flow rate of 50 mL/min of nitrogen (N_2). All TGA measurements were also done in triplicate.

2.4.5. Mechanical Tests

Tensile tests were performed on stamped dumbbell-shaped film samples sizing 115 × 16 mm^2 using an Instron 4400 universal testing machine, equipped with a 1-kN load cell, from Instron (Norwood, MA, USA) according to the ASTM standard method D638. The tests were done using a cross-head speed of 10 mm/min. Samples were conditioned for 24 h prior to analysis and the tests were performed at room conditions, that is, 40% RH and 25 °C. A minimum of six specimens were tested for each sample.

2.4.6. Permeability Tests

The gravimetric method ASTM E96-95 was used to determinate the water vapor permeability (WVP) of the films. To this end, Payne permeability cups (diameter of 3.5 cm) from Elcometer Sprl (Hermallesous-Argenteau, Belgium) were filled with 5 mL of distilled water. The films were not in direct contact with water but exposed to 100% RH on one side and secured with silicon rings. They were placed within a desiccator and sealed with dried silica gel at 0% RH and 25 °C. The control samples consisted of cups with aluminum films to estimate solvent loss through the sealing. The cups were weighted periodically using an analytical balance (±0.0001 g). WVP was calculated from the regression analysis of weight loss data vs. time, whereas the weight loss was calculated as the total loss minus the loss through the sealing. The permeability was obtained by multiplying the permeance by the film thickness. All WVP measurements were performed in triplicate.

The limonene permeability (LP), similar as described above for WVP, was measured placing 5 mL of D-limonene inside the Payne permeability cups. The cups containing the films were placed at controlled room conditions of 40% RH and 25 °C. The limonene vapor permeation rate (LPRT) values were estimated from the steady-state permeation slopes and the weight loss was calculated as the total cell loss minus the loss through the sealing. LP was calculated taking into account the average film thickness in each case. LP measurements were performed in triplicate.

2.4.7. Statistical Analysis

The optical, thermal, mechanical, and barrier properties were evaluated through analysis of variance (ANOVA) using STATGRAPHICS Centurion XVI v 16.1.03 from StatPoint Technologies, Inc. (Warrenton, VA, USA). Fisher's least significant difference (LSD) was used at the 95% confidence level ($p < 0.05$). Mean values and standard deviations were also reported.

3. Results

3.1. Morphology of RHF Particles

The morphology of the RHF powder was observed by SEM for determining the particle size and shape. In the low-magnification SEM image, shown in Figure 2a, one can see that the particles were not uniform in morphology and their dimension varied broadly. In particular, small particles slightly below 10 µm, in the form of rod-like particles or rectangular junks, coexisted with short fibers, having lengths above 100 µm. The magnified SEM image, included in Figure 2b, illustrates that the outer surface of rice husk was relatively smooth but its inner part was densely covered with orderly bulges. In this regard, it has been reported that the globular structure of rice husk is responsible for its high absorption capacity [61], which can also positively contribute to favoring mechanical interactions with the biopolymer matrix. Similar morphologies were reported for instance by Schneider et al. [62], in which the RHF particles also presented a heterogeneous morphology, namely larger particles of different textures, some smoother, others rough, with also grooves along the structure. The histogram of particle size of RHF is shown in Figure 2c, where one can see that the average fiber length was ~190 μm whereas the diameter corresponding to 90% cumulative (D_{90}) was ~320 μm.

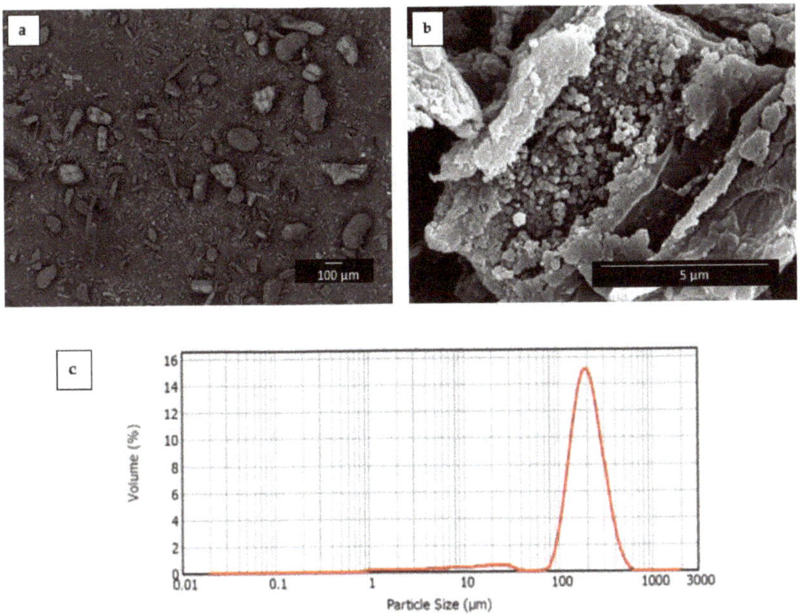

Figure 2. Scanning electron microscopy (SEM) images of rice husk flour (RHF) taken at (**a**) 50× with scale marker of 100 μm and (**b**) 10,000× with scale marker of 5 μm; (**c**) particle size histogram of RHF.

3.2. Morphology of PHB/PHBV/RHF Films

The morphology of the film cross-sections was observed by SEM and the images are gathered in Figure 3. One can observe that all films presented a smooth and featureless fracture surface without much deformation, corresponding to a typical brittle fracture behavior. The presence of the RHF particles can be observed in the cross-sections of all the composite film samples, that is, Figure 3a,c–g, whereas the surface of the PHB/PHBV film in Figure 3b suggests that the matrix was monophasic. Interestingly, the images taken at higher magnifications revealed that RHF fillers were tightly bonded to the polymer as inferred from the absence of gap between the particles and the biopolymer matrix. Moreover, no evidence of filler pull-out or void formation was noticed. At the same time, a rough surface attributed to the matrix deformation can be observed in the fillers surrounding. Additionally, RHF aggregates were not observed but individual particles appeared regularly distributed along the biopolymer matrix indicating that an effective mixing was attained.

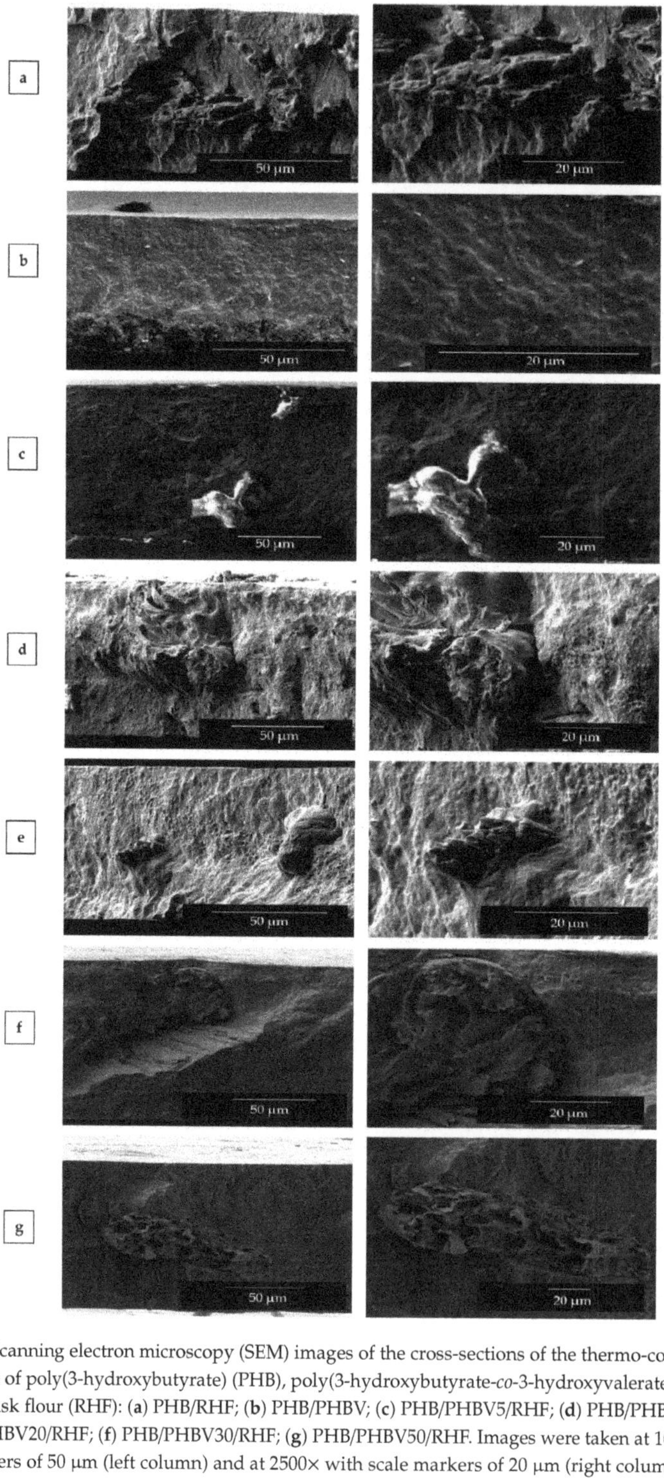

Figure 3. Scanning electron microscopy (SEM) images of the cross-sections of the thermo-compressed films made of poly(3-hydroxybutyrate) (PHB), poly(3-hydroxybutyrate-*co*-3-hydroxyvalerate) (PHBV), and rice husk flour (RHF): (**a**) PHB/RHF; (**b**) PHB/PHBV; (**c**) PHB/PHBV5/RHF; (**d**) PHB/PHBV10/RHF; (**e**) PHB/PHBV20/RHF; (**f**) PHB/PHBV30/RHF; (**g**) PHB/PHBV50/RHF. Images were taken at 1000× with scale markers of 50 μm (left column) and at 2500× with scale markers of 20 μm (right column).

The above observation suggests that a high interfacial adhesion was achieved in the composites due to the combined use of TGIC and DCP during melt processing. Similar morphologies were also described, for instance, by Rosa et al. [63] for PP/RHF composites when MA-modified polypropylene (MAPP) was added as coupling agent. The presence of MAPP successfully reduced the voids sizes and turned the surface more homogeneous, confirming its effect on promoting adhesion in the interfacial region. In relation to TGIC, Hao and Wu [49] recently showed that the addition during melt blending of the isocyanurate additive improved the interfacial adhesion of PLA/sisal fibers (SF) composites. The compatibilization achieved in the green composite was ascribed to the reaction of the –OH groups present at end groups of the biopolyester molecular chains and on the cellulose surface with the epoxy functional groups of TGIC. In the first reaction, ester bonds are known to be formed with the PHA chains by glycidyl esterification of carboxylic acid end groups, which precedes hydroxyl end group etherification [64]. The second reaction generates C–O–C bonds with subsequent hydroxyl side-group formation on the cellulose surface [29]. This reactive compatibilization habitually leads to green composites with enhanced performance properties [65]. For DCP, Wei et al. [52,66] recently described the coupling mechanism of cellulose to PHB and PHBV. Briefly, when the peroxide is exposed to heat during extrusion it decomposes into strong free radicals, which tend to abstract hydrogens (H′) from the biopolymer and cellulose molecular chains and initiate the grafting process between the two phases in the composites. The authors postulated that the grafted copolymers formed on the interfaces of cellulose and PHA coupled the hydrophilic filler to the hydrophobic biopolymer matrix. Therefore, grafting of RHF onto the PHB/PHBV matrix was successfully achieved by the formation of low concentrations of DCP derived radicals at high temperature during extrusion that initiated both the formation both H abstraction and triggered the reaction of the epoxy groups of TGIC with the OH groups of both cellulose and the terminal groups of the biopolyester chains. In relation to the addition of the biowaste derived PHBV, a good mixture between the two PHAs was attained since it was not possible perceive the presence of two phases in the biopolymer matrix, even at the highest PHBV content, that is, 50 wt %. In the film surfaces one can still observe the presence of some remaining impurities, which may be ascribed to organic remnants of small amounts of cell debris or fatty acids from the bioproduction process of PHBV. A similar morphology was recently reported by Martinez-Abad et al. [67] for PHB/unpurified PHBV blends, who also observed a good degree of interaction between the commercial homopolyester and the biosustainably produced copolyester.

3.3. Optical Properties of PHB/PHBV/RHF Films

The visual aspect of the films is displayed in Figure 4 to ascertain their optical properties. Simple naked eye examination of the films indicates that all the samples were slightly opaque but also showed high contact transparency. Additionally, the composite films developed a yellow-to-brown color due to the intrinsic natural color of the RHF powder, which one can observe in previous Figure 1b.

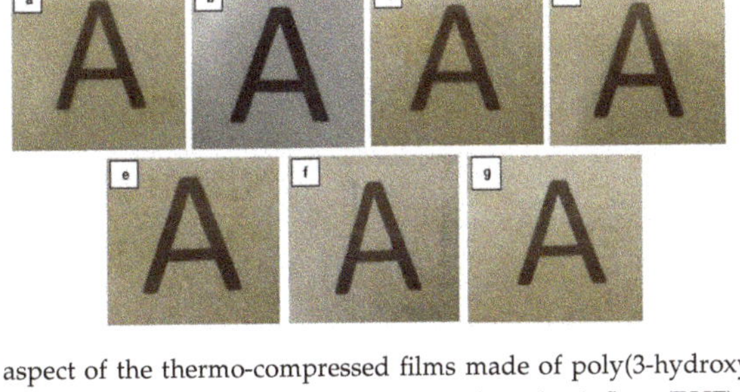

Figure 4. Visual aspect of the thermo-compressed films made of poly(3-hydroxybutyrate) (PHB), poly(3-hydroxybutyrate-*co*-3-hydroxyvalerate) (PHBV), and rice husk flour (RHF): (**a**) PHB/RHF; (**b**) PHB/PHBV; (**c**) PHB/PHBV5/RHF; (**d**) PHB/PHBV10/RHF; (**e**) PHB/PHBV20/RHF; (**f**) PHB/PHBV30/RHF; (**g**) PHB/PHBV50/RHF.

To quantify the color change resulted from the addition of PHBV in PHB/RHF, the color coordinates (a*, b*, L*) and the values of ΔE*, T, and O were determined and reported in Table 2. One can observe that the a* b* coordinates of the PHB/RHF film confirmed the above observed yellow-to-brown color of the sample whereas the PHB/PHBV film presented a natural color. The incorporation of PHBV into the green composite films resulted in a slightly increased the value of a* and, most notably, of b*, confirming the development of a brown tonality. Moreover, the composite films became darker since the L* values were also reduced. In any case, the color differences in the composite films containing different amounts of PHBV were relatively low, that is, ΔE* values remained below 6. One can also observe that the composite films also presented slightly lower transparency and higher opacity with the increasing PHBV contents, having values of T and O around 10–15% and 5% higher, respectively. This haze increase is typically observed in polymer blends due to differences in the refractive index of light diffraction of each polymer [68]. Although the brownish color of the film samples may seem a disadvantage for food packaging, it also offers the capacity to partially block the transmission of ultraviolet and visible (UV–vis) light. This can be a desirable attribute since the films can provide more protection to foodstuff from light, especially UV radiation, which can cause lipid oxidation in the food products [60,69].

Table 2. Optical properties of the thermo-compressed films made of poly(3-hydroxybutyrate) (PHB), poly(3-hydroxybutyrate-*co*-3-hydroxyvalerate) (PHBV), and rice husk flour (RHF).

Film	a*	b*	L*	ΔE*	T	O
PHB/RHF	−0.04 ± 0.01 ^{a}	11.82 ± 0.03 ^{a}	80.20 ± 0.02 ^{a}	-	9.08 ± 0.07 ^{a}	0.107 ± 0.03 ^{a}
PHB/PHBV	−0.25 ± 0.03 ^{b}	1.94 ± 0.02 ^{b}	86.40 ± 0.04 ^{b}	-	8.48 ± 0.03 ^{b}	0.099 ± 0.02 ^{a}
PHB/PHBV5/RHF	0.44 ± 0.05 ^{c}	14.60 ± 0.02 ^{c}	78.05 ± 0.03 ^{c}	3.55 ± 0.05 ^{a}	9.06 ± 0.05 ^{a}	0.111 ± 0.03 ^{a}
PHB/PHBV10/RHF	0.52 ± 0.04 ^{c}	14.84 ± 0.07 ^{c}	77.62 ± 0.03 ^{d}	4.01 ± 0.06 ^{b}	9.91 ± 0.04 ^{c}	0.116 ± 0.05 ^{a}
PHB/PHBV20/RHF	0.56 ± 0.05 ^{c}	15.28 ± 0.06 ^{d}	77.56 ± 0.05 ^{d}	4.39 ± 0.04 ^{c}	10.05 ± 0.03 ^{d}	0.116 ± 0.02 ^{a}
PHB/PHBV30/RHF	0.70 ± 0.07 ^{d}	15.36 ± 0.05 ^{d}	77.29 ± 0.06 ^{e}	4.64 ± 0.03 ^{d}	10.18 ± 0.08 ^{e}	0.117 ± 0.04 ^{a}
PHB/PHBV50/RHF	1.12 ± 0.08 ^{e}	16.17 ± 0.03 ^{e}	76.36 ± 0.04 ^{f}	5.92 ± 0.06 ^{e}	10.17 ± 0.05 ^{e}	0.113 ± 0.04 ^{a}

a*: red/green coordinates (+a red, −a green); b*: yellow/blue coordinates (+b yellow, −b blue); L*: Luminosity (+L luminous, −L dark); ΔE*: color differences; T: transparency; O: opacity. $^{a–f}$ Different letters in the same column indicate a significant difference among the samples ($p < 0.05$).

3.4. *Crystallinity of PHB/PHBV/RHF Films*

Figure 5 displays the DSC thermograms of the film samples obtained during the heating (Figure 5a) and cooling (Figure 5b) scans. Table 3 displays the main thermal transitions obtained from the DSC curves. One can observe that melting in the PHB/RHF film took place in two peaks, which were noted as T_{m1} and T_{m2}, whereas the unfilled PHB/PHBV film melted in a single peak. The double-melting peak phenomenon can be ascribed to the formation of crystalline structures with dissimilar lamellae thicknesses or the presence of crystallite blocks with different degrees of perfection [30]. The first peak originates from the melting of the PHB fraction that crystallized previously during the film formation, while in the second peak contributes the melting of the recrystallized PHB fraction during heating. In this context, other works have reported that melt-extruded films of neat PHB melt in 170–175 °C range [70,71]. Then, one can consider that the presence of the RHF fillers restricted the chain-folding process of PHB during crystallization.

Table 3. Main thermal parameters of the thermo-compressed films made of poly(3-hydroxybutyrate) (PHB), poly(3-hydroxybutyrate-*co*-3-hydroxyvalerate) (PHBV), and rice husk flour (RHF) in terms of: crystallization temperature (T_c), normalized enthalpy of crystallization (ΔH_c), melting temperature (T_m), and normalized melting enthalpy (ΔH_m).

Film	T_c (°C)	ΔH_c (J/g)	T_{m1} (°C)	T_{m2} (°C)	ΔH_m (J/g)
PHB/RHF	113.1 ± 0.6 [a]	70.2 ± 0.2 [a]	160.5 ± 0.8 [a]	167.2 ± 0.9 [a]	71.8 ± 2.3 [a]
PHB/PHBV	113.3 ± 0.3 [a]	73.3 ± 0.4 [b]	165.1 ± 0.3 [b]	-	76.9 ± 1.4 [b]
PHB/PHBV5/RHF	118.9 ± 0.6 [b]	59.8 ± 1.8 [c]	171.9 ± 1.5 [c]	-	63.7 ± 4.8 [c]
PHB/PHBV10/RHF	110.9 ± 0.2 [c]	60.1 ± 3.9 [c]	162.6 ± 1.1 [d]	168.3 ± 0.4 [a]	63.4 ± 1.6 [c]
PHB/PHBV20/RHF	115.4 ± 0.6 [d]	78.0 ± 4.1 [d]	166.0 ± 1.5 [b]	173.3 ± 0.3 [b]	61.4 ± 0.3 [c]
PHB/PHBV30/RHF	113.2 ± 0.6 [a]	72.7 ± 4.7 [d]	162.6 ± 0.7 [d]	173.9 ± 0.1 [c]	59.4 ± 2.2 [c]
PHB/PHBV50/RHF	107.7 ± 0.6 [e]	64.8 ± 0.8 [c]	158.4 ± 1.1 [e]	171.0 ± 0.8 [d]	53.8 ± 0.1 [d]

[a–e] Different letters in the same column indicate a significant difference among the samples ($p < 0.05$).

Interestingly, the incorporation into PHB/RHF of low contents of PHBV, that is, 5 wt %, yielded the film sample with the highest melting peak, that is, ~172 °C, whereas it also suppressed the double-melting peak behavior. This observation points out that the co-addition of 5 wt % PHBV and 10 wt % RHF enhanced the crystallization of PHB molecules, which was further supported by the shift of T_c from 113.1 °C, for the PHB/RHF film, to 118.9 °C, for the PHB/PHBV5/RHF film. The addition of higher contents of PHBV, however, led to films with two melting peaks. The first melting peak can be related to the PHBV-rich fraction, which was seen in the 158–166 °C range, whereas the second one corresponds to PHB melting, observed in 168–174 °C range. This result points out that fully miscibility was only attained in the composite blends containing low amounts of PHBV, that is, 5–10 wt %. At higher PHBV contents, the composite films formed a two-phase system and the T_m values were also reduced. Thus, the films produced with PHBV contents higher than 10 wt % showed a two phase crystallization, having a liquid–liquid separation at temperatures higher than 165 °C. One assumes that in these samples, due to the fact that the HV content in PHBV was relatively high, that is, 20 mol %, phase segregation preceded co-crystallization. One can also observe that the melting enthalpy of the first peak decreased while that of the second peak increased with the increase of the PHBV content in the blend. This observation indicates that crystallization occurred mainly in the PHB-rich regions by which the HV units were partially included into the PHB lattice and also induced some defects in the biopolymer crystals [72]. In this regards, different studies on the crystallization behavior of PHB/PHBV blends have indicated that their degree of miscibility decreases gradually as the HV in the copolyester increases. For instance, blends of PHB with PHBV were fully miscible over up to approximately 60 wt % of copolyester with a HV content of 18.4 mol % [73]. On the contrary, blends consisting of PHB and PHBV of high contents of HV, that is, 76 mol %, showed no depression of the melting point of each PHA, indicating total immiscibility [74]. In another work, Saito et al. [75] studied the competition between co-crystallization and phase segregation in blends of PHB and PHBV with different HV contents. The authors observed almost perfect co-crystallization in blends based on PHB and PHBV with 9 mol % HV, whereas HV contents >15 mol % induced phase segregation, that is, increased the percentage of PHBV that segregates from the growth front of crystals prior to co-crystallization.

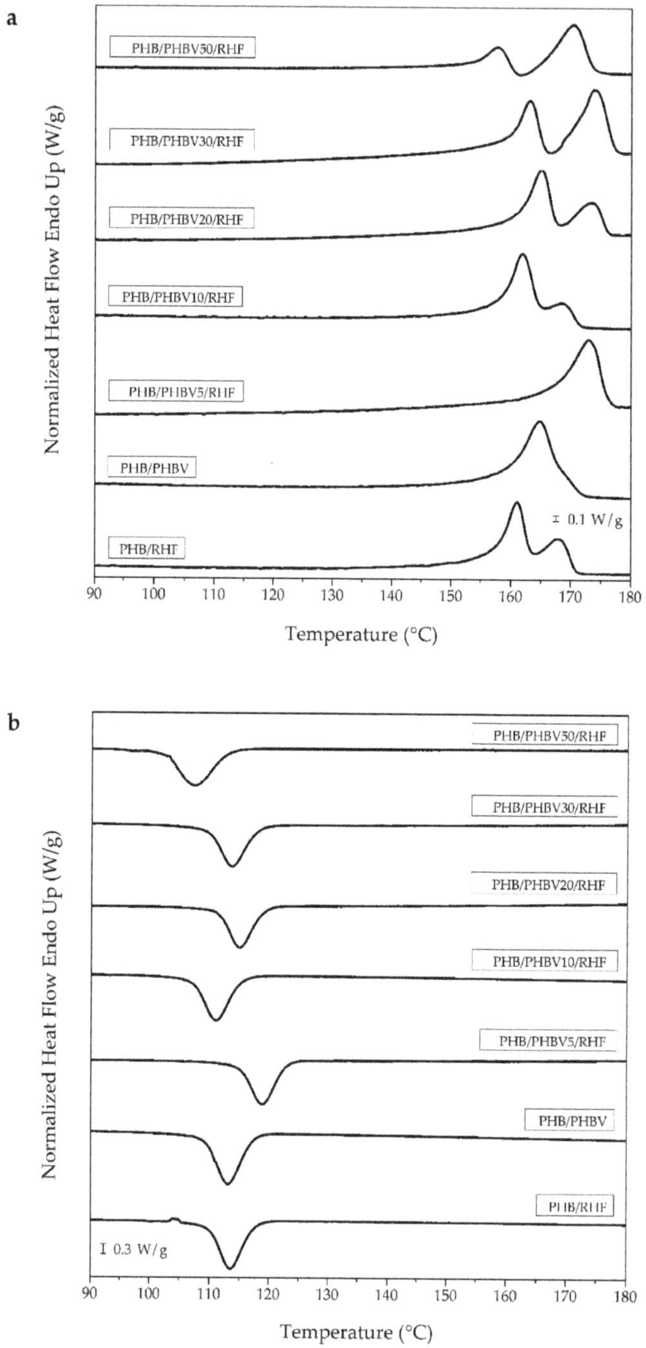

Figure 5. Differential scanning calorimetry (DSC) curves during (**a**) heating and (**b**) cooling of the thermo-compressed films made of poly(3-hydroxybutyrate) (PHB), poly(3-hydroxybutyrate-*co*-3-hydroxyvalerate) (PHBV), and rice husk flour (RHF).

3.5. Thermal Stability of PHB/PHBV/RHF Films

TGA curves are plotted in Figure 6 and the most relevant properties obtained from the curves are listed in Table 4. In relation to RHF, one can observe three main weight losses. The first one occurred around 100 °C, showing a mass loss of ~2%, which corresponds to the moisture released from the lignocellulosic filler. Following the TGA curve of RHF, the second and main degradation peak started at approximately 180 °C and ended at 340 °C with an average mass loss of ~47%. This weight loss includes the degradation of low-M_W components, mainly hemicellulose, followed by cellulose degradation. This zone represents the main devolatilization step of biomass pyrolysis and it is referred as the "active pyrolysis zone" since mass loss rate is high [76]. After this, RHF gradually degraded over a range of temperature from approximately 340 °C to 650 °C with a mass loss of ~16%, which can be seen as a tailing in both TGA and DTG curves. This mass loss is related to the degradation of lignin and it is called "passive pyrolysis zone" since the percentage of mass loss is smaller and the mass loss rate is also much lower compared to that in the second zone [76]. Indeed, when the temperature reached 650 °C, the degradation rates were no longer significant as most volatiles were already pyrolyzed. The rest was converted into char and gases, resulting in a residual mass of nearly 33%, which can be related to the high silicate content in rice husk [47].

Table 4. Main thermal parameters of the thermo-compressed films made of poly(3-hydroxybutyrate) (PHB), poly(3-hydroxybutyrate-*co*-3-hydroxyvalerate) (PHBV), and rice husk flour (RHF) in terms of: onset temperature of degradation ($T_{5\%}$), degradation temperature (T_{deg}), mass loss at T_{deg}, and residual mass at 800 °C.

Film	$T_{5\%}$ (°C)	T_{deg} (°C)	Mass Loss (%)	Residual Mass (%)
RHF	228.9 ± 1.5 ^{a}	335.2 ± 0.7 ^{a}	28.5 ± 0.8 ^{a}	33.1 ± 0.3 ^{a}
PHB/RHF	242.5 ± 1.2 ^{b}	270.0 ± 0.5 ^{b}	67.4 ± 0.7 ^{b}	3.1 ± 1.3 ^{b}
PHB/PHBV	252.6 ± 1.9 ^{c}	282.8 ± 0.4 ^{c}	69.4 ± 1.5 ^{b}	0.6 ± 0.1 ^{c}
PHB/PHBV5/RHF	242.5 ± 1.5 ^{b}	268.2 ± 0.5 ^{d}	65.4 ± 0.3 ^{c}	3.5 ± 0.2 ^{b}
PHB/PHBV10/RHF	247.1 ± 1.3 ^{d}	270.0 ± 0.5 ^{b}	63.9 ± 0.4 ^{d}	3.1 ± 0.8 ^{b}
PHB/PHBV20/RHF	247.5 ± 1.4 ^{d}	274.6 ± 0.6 ^{e}	69.8 ± 1.2 ^{b}	3.3 ± 0.9 ^{b}
PHB/PHBV30/RHF	253.5 ± 1.7 ^{c}	275.5 ± 0.6 ^{e}	61.7 ± 0.2 ^{e}	3.2 ± 0.2 ^{b}
PHB/PHBV50/RHF	257.2 ± 1.6 ^{e}	275.5 ± 0.6 ^{e}	61.5 ± 0.3 ^{e}	3.8 ± 0.4 ^{b}

$^{a–e}$ Different letters in the same column indicate a significant difference among the samples ($p < 0.05$).

Thermal degradation of the unfilled PHB/PHBV film occurred through a single and sharp degradation step that ranged from about 250 °C to 290 °C. PHA degradation typically follows a random chain scission model of the ester linkage that involves a *cis*-elimination reaction of β-CH and a six-member ring transition to form crotonic acid and its oligomers [77]. The presence of RHF reduced both the onset degradation temperature ($T_{5\%}$) and degradation temperature (T_{deg}) by approximately 10 °C and 15 °C, respectively, and also generated a low-intensity second mass centered around 330 °C. This reduction in thermal stability can be mainly ascribed to the lower degradation temperature of the lignocellulosic particles as well as to the presence of some remaining water. One can also observe that all composite films presented a similar thermal degradation profile though certain stability increase was attained for the highest PHBV contents. For instance, the $T_{5\%}$ value increased from 242.5 °C, for the PHB/RHF film, to 257.2 °C, for the PHB/PHBV50/RHF film. Similarly, the values of T_{deg} also increased from 270 °C, for the PHB/RHF film, to 275.5 °C, for the PHB/PHBV50/RHF film. The improvement achieved in the thermal stability can be ascribed to the high purity of the PHBV incorporated in the blend, based on the previous selection of the optimal purification route [58], which has also been reported to be more thermally stable than PHB [78]. Finally, the amount of residual mass was in the 3–4% range, being mainly ascribed to the formed char from RHF.

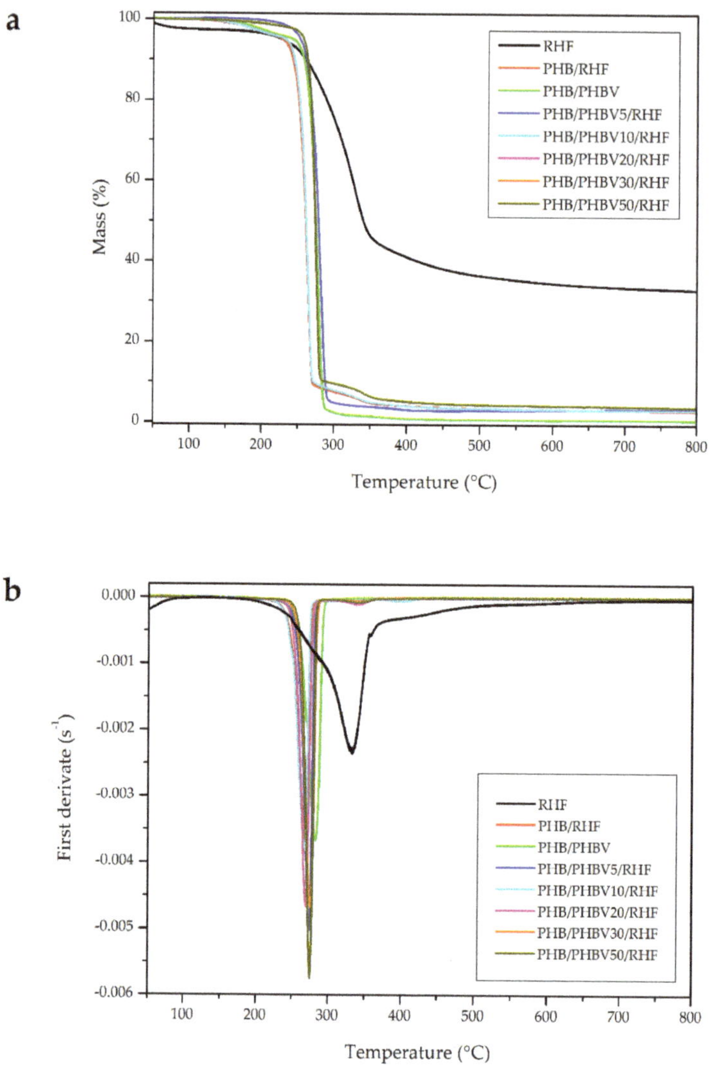

Figure 6. (**a**) Thermogravimetric analysis (TGA) and (**b**) first derivative (DTG) curves of the thermo-compressed films made of poly(3-hydroxybutyrate) (PHB), poly(3-hydroxybutyrate-*co*-3-hydroxyvalerate) (PHBV), and rice husk flour (RHF).

3.6. Mechanical Properties of PHB/PHBV/RHF Films

Figure 7 shows the tensile stress–strain curves, obtained at room temperature, for the thermo-compressed films. The mechanical results, in terms of tensile modulus (E), tensile strength at yield (σ_y), elongation at break (ε_b), and toughness are summarized in Table 5. One can see that all the films were relatively stiff and also brittle due to the intrinsically high crystallinity of PHB, which is derived from secondary crystallization that occurs post-processing with age [79]. This mechanical brittleness habitually represents an obstacle to the practical applications of PHB, for instance in packaging. The presence of RHF further promoted rigidity and brittleness of PHB, showing *E* and ε_b values of 3025 MPa and 1.1%, respectively. One can also observe that the unfilled PHB/PHBV film showed the lowest E value, that is, 1785 MPa, and also the highest values of ε_b and toughness, that is, 8.4% and 1.9 mJ/m^3, respectively. Whereas the mechanical properties of the PHB/RHF and PHB/PHBV5/RHF films were relatively similar, the incorporation of 10 wt % PHBV into PHB/RHF successfully resulted in a film with very balanced performance in terms of mechanical strength and ductility. In particular, the PHB/PHBV10/RHF film showed moderate values of E and σ_y, that is, 2508 MPa and 30.4 MPa, respectively, and a 3-fold increase in ε_b and remarkable higher toughness in comparison to the PHB/RHF film. However, the films produced at higher PHBV contents, that is, 20–50 wt %, showed lower and relatively similar performance. These results indicate that PHBV can successfully produce a toughening effect on the PHB matrix at low or intermediate contents, which can be ascribed to the high solubility achieved in these films samples as described above during the crystallinity analysis. The resultant lower stiffness and higher flexibility of these PHBV-containing films can be attributed to the presence of dislocations, crystal strain, and smaller crystallites in the PHB/PHBV soluble regions due to the insertion of the HV units into the PHB lattice, which acted as defects in the HB crystals [80].

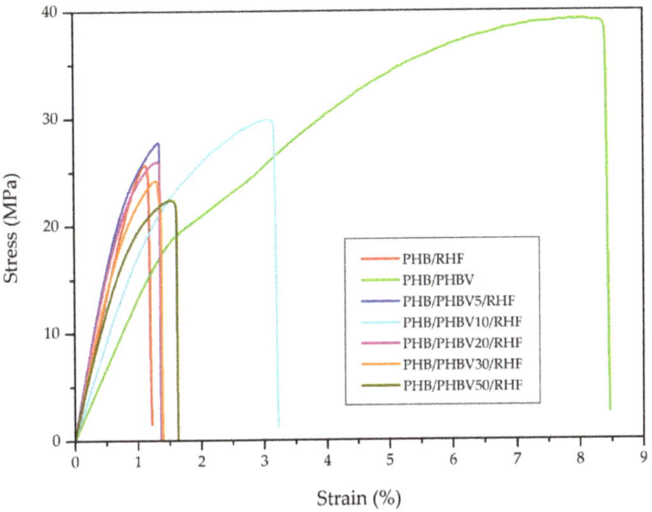

Figure 7. Tensile stress–strain curves of the thermo-compressed films made of poly(3-hydroxybutyrate) (PHB), poly(3-hydroxybutyrate-*co*-3-hydroxyvalerate) (PHBV), and rice husk flour (RHF).

Table 5. Mechanical properties of the thermo-compressed films made of poly(3-hydroxybutyrate) (PHB), poly(3-hydroxybutyrate-*co*-3-hydroxyvalerate) (PHBV), and rice husk flour (RHF) in terms of: tensile modulus (E), tensile strength at yield (σ_y), elongation at break (ε_b), and toughness.

Film	E (MPa)	σ_y (MPa)	ε_b (%)	Toughness (mJ/m^3)
PHB *	~2900 ^{a}	~37 ^{a}	~4 ^{a}	-
PHB/RHF	3025 ± 101 ^{b}	26.7 ± 2.7 ^{b}	1.1 ± 0.1 ^{b}	0.1 ± 0.1 ^{a}
PHB/PHBV	1785 ± 129 ^{c}	38.9 ± 2.0 ^{a}	8.4 ± 1.1 ^{c}	1.9 ± 0.3 ^{b}
PHB/PHBV5/RHF	2985 ± 119 ^{a,b}	27.2 ± 1.1 ^{b}	1.4 ± 0.1 ^{d}	0.3 ± 0.1 ^{a}
PHB/PHBV10/RHF	2508 ± 207 ^{d}	30.4 ± 2.7 ^{b}	3.3 ± 1.3 ^{a}	0.8 ± 0.3 ^{c}
PHB/PHBV20/RHF	2830 ± 193 ^{a,b,d}	26.5 ± 2.5 ^{b}	1.2 ± 0.3 ^{b,d}	0.1 ± 0.1 ^{a}
PHB/PHBV30/RHF	2765 ± 201 ^{a,b,d}	23.9 ± 0.7 ^{b}	1.3 ± 0.2 ^{b,d}	0.2 ± 0.1 ^{a}
PHB/PHBV50/RHF	2649 ± 104 ^{d}	19.5 ± 3.0 ^{c}	1.3 ± 0.4 ^{b,d}	0.2 ± 0.1 ^{a}

* Average mechanical properties for compression-molded films of neat PHB [79]. $^{a-d}$ Different letters in the same column indicate a significant difference among the samples ($p < 0.05$).

3.7. Barrier Performance of PHB/PHBV/RHF Films

Table 6 finally gathers the WVP and LP values of the PHB-based films. The barrier performance to water vapor is one of the main parameters of application interest in packaging, while D-limonene is a commonly used standard compound to test mass transport of aromas. The PHB/RHF film showed somewhat higher WVP values than compression-molded PHB films previously reported by our group, that is, 1.7–1.75 × 10^{-15} kg·m/s·m^2·Pa [81,82]. The lower water vapor barrier attained for the here-prepared green composite films can be related to the high tendency of the lignocellulosic fillers to adsorb water since transport of water vapor molecules is mainly a diffusivity-driven property in PHAs due to their intrinsically low hydrophilic character [83]. The incorporation of biowaste derived PHBV into PHB/RHF increased slightly the WVP values in the films. At low PHBV contents, however, the barrier performance of the composite films was relatively similar since the samples showed variations close to the detection limit of the technique, with deviations oscillating in the 6.0–7.5 × 10^{-16} kg·m/s·m^2·Pa range. Only the films with the highest PHBV contents, that is, 30 wt % and 50 wt %, displayed a significant decrease in the water barrier properties, which was still within the same order of magnitude. The barrier drop observed can be attributed to the higher contribution of the biomass derived PHBV, which results in an overall decrease in molecular order and crystallinity, the presence of defects and discontinuities across the polymer morphology as well as certain plasticization induced by the remaining biomass impurities of PHBV with consequent increase in free volume [67].

Table 6. Water vapor permeability (WVP) and D-limonene permeability (LP) of the thermo-compressed films made of poly(3-hydroxybutyrate) (PHB), poly(3-hydroxybutyrate-*co*-3-hydroxyvalerate) (PHBV), and rice husk flour (RHF).

Film	WVP × 10^{15} (kg·m/m^2·Pa·s)	LP × 10^{15} (kg·m/m^2·Pa·s)
PHB *	1.75 ^{a}	1.95 ^{a}
PHB/RHF	4.52 ± 0.38 ^{b}	2.58 ± 1.35 ^{a}
PHB/PHBV	3.27 ± 0.15 ^{c}	2.04 ± 0.19 ^{a}
PHB/PHBV5/RHF	4.55 ± 0.75 ^{b}	2.16 ± 0.54 ^{a}
PHB/PHBV10/RHF	5.03 ± 0.64 ^{b}	3.10 ± 0.65 ^{a}
PHB/PHBV20/RHF	5.36 ± 0.69 ^{b}	3.38 ± 0.63 ^{a}
PHB/PHBV30/RHF	6.01 ± 0.60 ^{b}	3.72 ± 0.32 ^{a}
PHB/PHBV50/RHF	7.46 ± 0.90 ^{b}	5.04 ± 1.50 ^{a}

* Barrier data for compression-molded PHB films reported in literature [82]. $^{a-c}$ Different letters in the same column indicate a significant difference among the samples ($p < 0.05$).

Similar to WVP, one can observe that the incorporation of PHBV tended to decrease the barrier properties to D-limonene of the PHB/RHF composite films. This aroma compound, as opposed to moisture, is a strong plasticizing component for PHAs, thus, solubility plays a more important role in permeability than diffusion. The LP increase observed suggests that the presence of PHBV favored an increased sorption of D-limonene in the film. Indeed, our previous studies dealing with PHA materials have shown that the LP value of films made of PHBV with 12 mol % HV is two order of magnitude higher, that is, 1.99 × 10^{-13} kg·m/s·m^2·Pa [84], than that of neat PHB films, that is, 1.95 × 10^{-15} kg·m/s·m^2·Pa [82]. In any case, the present results indicate that the incorporation of up to 20 wt % PHBV does not significantly affect the barrier properties of the PHB-based films against water or aroma. In a more packaging oriented application context, the composite films containing low amounts of PHBV present WVP values in the same order of magnitude than those films of petroleum-based PET, that is, 2.30 × 10^{-15} kg·m/s·m^2·Pa [85], which is typically used in medium-barrier applications. In terms of D-limonene, the here-prepared PHB/PHBV/RHF films are two order of magnitude more barrier than compression-molded PET films, that is, 1.17 × 10^{-13} kg·m/s·m^2·Pa [81].

4. Discussion

Results showed an optimum morphology with a regular distribution of RHF, tightly bonded and with absence of gap along the PHB matrix due to the use of reactive compatibilizers. The biowaste derived PHBV also showed a good miscibility with the PHB/RHF composite system, in particular at the lowest contents, that is, 5–10 wt %. At higher concentrations, however, a two-phases system was attained, indicating that crystallization occurred mainly in the PHB-rich regions. The incorporation of PHBV also increased the thermal stability of PHB/RHF, increasing the processing window of the films. With respect to the mechanical properties, contents in the 5–10 wt % range of PHBV yielded films with a more balanced performance in terms of strength and ductility, counteracting the stiffness and fragility induced by RHF. Finally, although the incorporation of PHBV increased the permeability of the films, the water vapor barrier properties of the PHB/PHBV/RHF films still remained in values close to those of PET films, whereas they still presented a high barrier to aroma.

5. Conclusions

The use of natural fillers and biopolymers obtained from agro-food waste currently represents a sustainable alternative to petroleum-based materials. The green composite films prepared herein are potential candidates to be used in rigid packaging for low and medium barrier applications, being processable by current conventional machinery. The valorization of agro-food waste, as well as the relative preservation of physicochemical properties, supports the use of purified biosustainably produced PHBV in the food packaging industry to develop more cost-effective PHA-based articles.

Author Contributions: Conceptualization was devised by L.C. and S.T.-G.; Methodology, B.M.-R. and A.A.; Writing—original draft preparation, B.M.-R.; Writing—review and editing, S.T.-G.; Supervision, L.C., S.T.-G., and J.M.L.; Project Administration, J.M.L.

Funding: This research was supported by the Spanish Ministry of Science, Innovation, and Universities (MICIU) program number AGL2015-63855-C2-1-R and by the EU H2020 projects YPACK (reference number 773872) and ResUrbis (reference number 730349).

Acknowledgments: B.M.-R. and S.T.-G. acknowledge MICIU for her FPI grant (BES-2016-077972) and his Juan de la Cierva-Incorporación contract (IJCI-2016-29675), respectively. The authors also thank the Joint Unit in Polymers Technology between IATA–CSIC and PIMA-Universitat Jaume I.

Conflicts of Interest: The authors declare no conflict of interest.

References

1. Rehm, B.H.A. Polyester synthases: Natural catalysts for plastics. *Biochem. J.* **2003**, *376*, 15–33. [CrossRef]
2. Alaerts, L.; Augustinus, M.; VanAcker, K. Impact of bio-based plastics on current recycling of plastics. *Sustainability* **2018**, *10*, 1487. [CrossRef]
3. Smith, R. *Biodegradable Polymers for Industrial Applications*; Woodhead: Lisle, IL, USA, 2005.
4. Cava, D.; Giménez, E.; Gavara, R.; Lagaron, J.M. Comparative performance and barrier properties of biodegradable thermoplastics and nanobiocomposites versus PET for food packaging applications. *J. Plast. Film. Sheeting* **2006**, *22*, 265–274. [CrossRef]
5. Reis, K.; Pereira, J.; Smith, A.; Carvalho, C.; Wellner, N.; Yakimets, I. Characterization of polyhydroxybutyrate-hydroxyvalerate (PHB-HV)/maize starch blend films. *J. Food Eng.* **2008**, *89*, 361–369. [CrossRef]
6. Nduko, J.M.; Matsumoto, K.; Taguchi, S. Biological lactate-polymers synthesized by one-pot microbial factory: Enzyme and metabolic engineering. *Biobased Monomers Polym. Mater.* **2012**, *1105*, 213–235.
7. Philip, S.; Keshavarz, T.; Roy, I. Polyhydroxyalkanoates: Biodegradable polymers with a range of applications. *J. Chem. Technol. Biotechnol.* **2007**, *82*, 233–247. [CrossRef]
8. Keshavarz, T.; Roy, I. Polyhydroxyalkanoates: Bioplastics with a green agenda. *Curr. Opin. Microbiol.* **2010**, *13*, 321–326. [CrossRef]
9. Blunt, W.; Levin, D.B.; Cicek, N. Bioreactor Operating Strategies for Improved Polyhydroxyalkanoate (PHA) Productivity. *Polymers* **2018**, *10*, 1197. [CrossRef]

10. Kourmentza, C.; Plácido, J.; Venetsaneas, N.; Burniol-Figols, A.; Varrone, C.; Gavala, H.N.; Reis, M.A. Recent Advances and Challenges towards Sustainable Polyhydroxyalkanoate (PHA) Production. *Bioengineering* **2017**, *4*, 55. [CrossRef]
11. Jacquel, N.; Lo, C.W.; Wu, H.S.; Wei, Y.H.; Wang, S.S. Solubility of Solubility of polyhydroxyalkanoates by experiment and thermodynamic correlations. *AIChE J.* **2007**, *53*, 2704–2714. [CrossRef]
12. Domingos, J.M.; Puccio, S.; Martinez, G.A.; Amaral, N.; Reis, M.A.; Bandini, S.; Fava, F.; Bertin, L. Cheese whey integrated valorisation: Production, concentration and exploitation of carboxylic acids for the production of polyhydroxyalkanoates by a fed-batch culture. *Chem. Eng. J.* **2018**, *336*, 47–53. [CrossRef]
13. Samorì, C.; Abbondanzi, F.; Galletti, P.; Giorgini, L.; Mazzocchetti, L.; Torri, C.; Tagliavini, E. Extraction of polyhydroxyalkanoates from mixed microbial cultures: Impact on polymer quality and recovery. *Bioresour. Technol.* **2015**, *189*, 195–202. [CrossRef]
14. Lee, S.Y. Plastic bacteria? Progress and prospects for polyhydroxyalkanoate production in bacteria. *Trends Biotechnol.* **1996**, *14*, 431–438. [CrossRef]
15. Torres-Giner, S.; Montanes, N.; Fombuena, V.; Boronat, T.; Sanchez-Nacher, L. Preparation and characterization of compression-molded green composite sheets made of poly(3-hydroxybutyrate) reinforced with long pita fibers. *Adv. Polym. Technol.* **2018**, *37*, 1305–1315. [CrossRef]
16. Zini, E.; Scandola, M. Green composites: An overview. *Polym. Compos.* **2011**, *32*, 1905–1915. [CrossRef]
17. Saheb, D.N.; Jog, J.P. Natural fiber polymer composites: A review. *Adv. Polym. Technol.* **1999**, *18*, 351–363. [CrossRef]
18. Joshi, S.V.; Drzal, L.T.; Mohanty, A.K.; Arora, S. Are natural fiber composites environmentally superior to glass fiber reinforced composites? *Compos. Part A Appl. Sci. Manuf.* **2004**, *35*, 371–376. [CrossRef]
19. La Mantia, F.P.; Morreale, M. Green composites: A brief review. *Compos. Part A Appl. Sci. Manuf.* **2011**, *42*, 579–588. [CrossRef]
20. Khalil, H.P.S.A.; Bhat, A.H.; Yusra, A.F.I. Green composites from sustainable cellulose nanofibrils: A review. *Carbohydr. Polym.* **2012**, *87*, 963–979. [CrossRef]
21. Ndazi, B.S.; Karlsson, S. Characterization of hydrolytic degradation of polylactic acid/rice hulls composites in water at different temperatures. *Express Polym. Lett.* **2011**, *5*, 119–131. [CrossRef]
22. Yussuf, A.A.; Massoumi, I.; Hassan, A. Comparison of polylactic acid/kenaf and polylactic acid/rice husk composites: The influence of the natural fibers on the mechanical, thermal and biodegradability properties. *J. Polym. Environ.* **2010**, *18*, 422–429. [CrossRef]
23. Quiles-Carrillo, L.; Montanes, N.; Garcia-Garcia, D.; Carbonell-Verdu, A.; Balart, R.; Torres-Giner, S. Effect of different compatibilizers on injection-molded green composite pieces based on polylactide filled with almond shell flour. *Compos. Part B Eng.* **2018**, *147*, 76–85. [CrossRef]
24. Quiles-Carrillo, L.; Montanes, N.; Sammon, C.; Balart, R.; Torres-Giner, S. Compatibilization of highly sustainable polylactide/almond shell flour composites by reactive extrusion with maleinized linseed oil. *Ind. Crop. Prod.* **2018**, *111*, 878–888. [CrossRef]
25. Liminana, P.; García-Sanoguera, D.; Quiles-Carrillo, L.; Balart, R.; Montanes, N. Development and characterization of environmentally friendly composites from poly(butylene succinate) (PBS) and almond shell flour with different compatibilizers. *Compos. Part B Eng.* **2018**, *144*, 153–162. [CrossRef]
26. Montava-Jordà, S.; Quiles-Carrillo, L.; Richart, N.; Torres-Giner, S.; Montanes, N. Enhanced Interfacial Adhesion of Polylactide/Poly(ε-caprolactone)/Walnut Shell Flour Composites by Reactive Extrusion with Maleinized Linseed Oil. *Polymers* **2019**, *11*, 758. [CrossRef]
27. Garcia-Garcia, D.; Carbonell-Verdu, A.; Jordá-Vilaplana, A.; Balart, R.; Garcia-Sanoguera, D.; Garcia-Garcia, D.; Carbonell-Verdu, A.; Jordá-Vilaplana, A.; Garcia-Sanoguera, D. Development and characterization of green composites from bio-based polyethylene and peanut shell. *J. Appl. Polym. Sci.* **2016**, *133*, 43940. [CrossRef]
28. Torres-Giner, S.; Hilliou, L.; Melendez-Rodriguez, B.; Figueroa-Lopez, K.; Madalena, D.; Cabedo, L.; Covas, J.; Vicente, A.; Lagaron, J. Melt processability, characterization, and antibacterial activity of compression-molded green composite sheets made of poly(3-hydroxybutyrate-*co*-3-hydroxyvalerate) reinforced with coconut fibers impregnated with oregano essential oil. *Food Packag. Shelf Life* **2018**, *17*, 39–49. [CrossRef]
29. Quiles-Carrillo, L.; Montanes, N.; Lagaron, J.M.; Balart, R.; Torres-Giner, S. On the use of acrylated epoxidized soybean oil as a reactive compatibilizer in injection-molded compostable pieces consisting of polylactide filled with orange peel flour. *Polym. Int.* **2018**, *67*, 1341–1351. [CrossRef]

30. Montava-Jordà, S.; Torres-Giner, S.; Ferrandiz-Bou, S.; Quiles-Carrillo, L.; Montanes, N. Development of Sustainable and Cost-Competitive Injection-Molded Pieces of Partially Bio-Based Polyethylene Terephthalate through the Valorization of Cotton Textile Waste. *Int. J. Mol. Sci.* **2019**, *20*, 1378. [CrossRef]
31. Ferrero, B.; Fombuena, V.; Fenollar, O.; Boronat, T.; Balart, R. Development of natural fiber-reinforced plastics (NFRP) based on biobased polyethylene and waste fibers from Posidonia oceanica seaweed. *Polym. Compos.* **2015**, *36*, 1378–1385. [CrossRef]
32. Aprianti, E.; Shafigh, P.; Bahri, S.; Farahani, J.N. Supplementary cementitious materials origin from agricultural wastes—A review. *Constr. Build. Mater.* **2015**, *74*, 176–187. [CrossRef]
33. Adam, F.; Appaturi, J.N.; Iqbal, A. The utilization of rice husk silica as a catalyst: Review and recent progress. *Catal. Today* **2012**, *190*, 2–14. [CrossRef]
34. Adam, F.; Kandasamy, K.; Balakrishnan, S. Iron incorporated heterogeneous catalyst from rice husk ash. *J. Colloid Interface Sci.* **2006**, *304*, 137–143. [CrossRef]
35. Zhao, Q.; Zhang, B.; Quan, H.; Yam, R.C.; Yuen, R.K.; Li, R.K.; Yuen, K.K.R. Flame retardancy of rice husk-filled high-density polyethylene ecocomposites. *Compos. Sci. Technol.* **2009**, *69*, 2675–2681. [CrossRef]
36. Panthapulakkal, S.; Law, S.; Sain, M. Enhancement of processability of rice husk filled high-density polyethylene composite profiles. *J. Thermoplast. Compos. Mater.* **2005**, *18*, 445–458. [CrossRef]
37. Nascimento, G.; Cechinel, D.; Piletti, R.; Mendes, E.; Paula, M.; Riella, H.G.; Fiori, M. Effect of Different Concentrations and Sizes of Particles of Rice Husk Ash-RHS in the Mechanical Properties of Polypropylene. *Mater. Sci. Forum* **2010**, *660*, 23–28. [CrossRef]
38. Verheyen, S.; Blaton, N.; Kinget, R.; Kim, H.-S. Thermogravimetric analysis of rice husk flour filled thermoplastic polymer composites. *J. Therm. Anal. Calorim.* **2004**, *76*, 395–404. [CrossRef]
39. Battegazzore, D.; Bocchini, S.; Alongi, J.; Frache, A.; Marino, F. Cellulose extracted from rice husk as filler for poly(lactic acid): Preparation and characterization. *Cellulose* **2014**, *21*, 1813–1821. [CrossRef]
40. Bertini, F.; Canetti, M.; Cacciamani, A.; Elegir, G.; Orlandi, M.; Zoia, L. Effect of ligno-derivatives on thermal properties and degradation behavior of poly(3-hydroxybutyrate)-based biocomposites. *Polym. Degrad. Stab.* **2012**, *97*, 1979–1987. [CrossRef]
41. Boitt, A.P.W.; Barcellos, I.O.; Alberti, L.D.; Bucci, D.Z. Evaluation of the Influence of the Use of Waste from the Processing of Rice in Physicochemical Properties and Biodegradability of PHB in Composites. *Polimeros* **2014**, *24*, 640–645. [CrossRef]
42. Moura, A.; Bolba, C.; Demori, R.; Lima, L.P.F.C.; Santana, R.M. Effect of Rice Husk Treatment with Hot Water on Mechanical Performance in Poly(hydroxybutyrate)/Rice Husk Biocomposite. *J. Polym. Environ.* **2018**, *26*, 2632–2639. [CrossRef]
43. Sánchez-Safont, E.L.; Aldureid, A.; Lagarón, J.M.; Gamez-Perez, J.; Cabedo, L. Biocomposites of different lignocellulosic wastes for sustainable food packaging applications. *Compos. Part B Eng.* **2018**, *145*, 215–225. [CrossRef]
44. Borah, J.S.; Kim, D.S. Recent development in thermoplastic/wood composites and nanocomposites: A review. *Korean J. Chem. Eng.* **2016**, *33*, 3035–3049. [CrossRef]
45. Rowell, R.M.; Young, R.A.; Rowell, J.K.I. *Paper and Composites from Agro-Based Resources*; CRC press: Boca Raton, FL, USA, 1996.
46. Lu, J.Z.; Wu, Q.; McNabb, H.S. Chemical coupling in wood fiber and polymer composites: A review of coupling agents and treatments. *Wood Fiber Sci.* **2000**, *32*, 88–104.
47. George, J.; Sreekala, M.S.; Thomas, S. A review on interface modification and characterization of natural fiber reinforced plastic composites. *Polym. Eng. Sci.* **2001**, *41*, 1471–1485. [CrossRef]
48. Chan, C.M.; Vandi, L.-J.; Pratt, S.; Halley, P.; Richardson, D.; Werker, A.; Laycock, B. Mechanical properties of poly(3-hydroxybutyrate-*co*-3-hydroxyvalerate)/wood flour composites: Effect of interface modifiers. *J. Appl. Polym. Sci.* **2018**, *135*, 46828. [CrossRef]
49. Hao, M.; Wu, H. Effect of in situ reactive interfacial compatibilization on structure and properties of polylactide/sisal fiber biocomposites. *Polym. Compos.* **2018**, *39*, E174–E187. [CrossRef]
50. Dhavalikar, R.; Xanthos, M. Parameters affecting the chain extension and branching of PET in the melt state by polyepoxides. *J. Appl. Polym. Sci.* **2003**, *87*, 643–652. [CrossRef]
51. Quiles-Carrillo, L.; Montanes, N.; Jorda-Vilaplana, A.; Balart, R.; Torres-Giner, S. A comparative study on the effect of different reactive compatibilizers on injection-molded pieces of bio-based high-density polyethylene/polylactide blends. *J. Appl. Polym. Sci.* **2019**, *136*, 47396. [CrossRef]

52. Wei, L.; McDonald, A.G.; Stark, N.M. Grafting of bacterial polyhydroxybutyrate (PHB) onto cellulose via in situ reactive extrusion with dicumyl peroxide. *Biomacromolecules* **2015**, *16*, 1040–1049. [CrossRef]
53. Ahmad, E.E.M.; Luyt, A.S. Effects of organic peroxide and polymer chain structure on mechanical and dynamic mechanical properties of sisal fiber reinforced polyethylene composites. *J. Appl. Polym. Sci.* **2012**, *125*, 2216–2222. [CrossRef]
54. Nogellova, Z.; Kokta, B.V.; Chodak, I. A composite LDPE/wood flour crosslinked by peroxide. *J. Macromol. Sci. Part A* **1998**, *35*, 1069–1077. [CrossRef]
55. Gu, R.; Sain, M.; Kokta, B.V. Evaluation of wood composite additives in the mechanical property changes of PE blends. *Polym. Compos.* **2015**, *36*, 287–293. [CrossRef]
56. Joseph, K.; Thomas, S.; Pavithran, C. Effect of chemical treatment on the tensile properties of short sisal fibre-reinforced polyethylene composites. *Polymer* **1996**, *37*, 5139–5149. [CrossRef]
57. Mokoena, M.A.; Djoković, V.; Luyt, A.S. Composites of linear low density polyethylene and short sisal fibres: The effects of peroxide treatment. *J. Mater. Sci.* **2004**, *39*, 3403–3412. [CrossRef]
58. Melendez-Rodriguez, B.; Castro-Mayorga, J.L.; Reis, M.A.M.; Sammon, C.; Cabedo, L.; Torres-Giner, S.; Lagaron, J.M. Preparation and characterization of electrospun food biopackaging films of poly(3-hydroxybutyrate-*co*-3-hydroxyvalerate) derived from fruit pulp biowaste. *Front. Sustain. Food Syst.* **2018**, *2*, 38. [CrossRef]
59. Figueroa-Lopez, K.; Andrade-Mahecha, M.; Torres-Vargas, O. Development of antimicrobial biocomposite films to preserve the quality of bread. *Molecules* **2018**, *23*, 212. [CrossRef]
60. Kanatt, S.R.; Rao, M.; Chawla, S.; Sharma, A. Active chitosan–polyvinyl alcohol films with natural extracts. *Food Hydrocoll.* **2012**, *29*, 290–297. [CrossRef]
61. Scaglioni, P.T.; Badiale-Furlong, E. Rice husk as an adsorbent: A new analytical approach to determine aflatoxins in milk. *Talanta* **2016**, *152*, 423–431. [CrossRef]
62. Schneider, L.T.; Bonassa, G.; Alves, H.J.; Meier, T.R.W.; Frigo, E.P.; Teleken, J.G. Use of rice husk in waste cooking oil pretreatment. *Environ. Technol.* **2019**, *40*, 594–604. [CrossRef]
63. Rosa, S.M.L.; Santos, E.F.; Ferreira, C.A.; Nachtigall, S.M.B. Studies on the properties of rice-husk-filled-PP composites: Effect of maleated PP. *Mater. Res.* **2009**, *12*, 333–338. [CrossRef]
64. Torres-Giner, S.; Montanes, N.; Boronat, T.; Quiles-Carrillo, L.; Balart, R. Melt grafting of sepiolite nanoclay onto poly(3-hydroxybutyrate-*co*-4-hydroxybutyrate) by reactive extrusion with multi-functional epoxy-based styrene-acrylic oligomer. *Eur. Polym. J.* **2016**, *84*, 693–707. [CrossRef]
65. Formela, K.; Zedler, Ł.; Hejna, A.; Tercjak, A. Reactive extrusion of bio-based polymer blends and composites—Current trends and future developments. *Express Polym. Lett.* **2018**, *12*, 24–57. [CrossRef]
66. Wei, L.; Stark, N.M.; McDonald, A.G. Interfacial improvements in biocomposites based on poly(3-hydroxybutyrate) and poly(3-hydroxybutyrate-*co*-3-hydroxyvalerate) bioplastics reinforced and grafted with α-cellulose fibers. *Green Chem.* **2015**, *17*, 4800–4814. [CrossRef]
67. Martínez-Abad, A.; Cabedo, L.; Oliveira, C.S.; Hilliou, L.; Reis, M.; Lagarón, J.M. Characterization of polyhydroxyalkanoate blends incorporating unpurified biosustainably produced poly(3-hydroxybutyrate-*co*-3-hydroxyvalerate). *J. Appl. Polym. Sci.* **2016**, *133*, 42633. [CrossRef]
68. Maruhashi, Y.; Iida, S. Transparency of polymer blends. *Polym. Eng. Sci.* **2001**, *41*, 1987–1995. [CrossRef]
69. Figueroa-Lopez, K.J.; Vicente, A.A.; Reis, M.A.M.; Torres-Giner, S.; Lagaron, J.M. Antimicrobial and antioxidant performance of various essential oils and natural extracts and their incorporation into biowaste derived poly(3-hydroxybutyrate-*co*-3-hydroxyvalerate) layers made from electrospun ultrathin fibers. *Nanomaterials* **2019**, *9*, 144. [CrossRef]
70. Ollier, R.P.; D'Amico, D.A.; Schroeder, W.F.; Cyras, V.P.; Alvarez, V.A. Effect of clay treatment on the thermal degradation of PHB based nanocomposites. *Appl. Clay Sci.* **2018**, *163*, 146–152. [CrossRef]
71. De Matos Costa, A.R.; Santos, R.M.; Ito, E.N.; de Carvalho, L.H.; Canedo, E.L. Melt and cold crystallization in a poly(3-hydroxybutyrate) poly(butylene adipate-*co*-terephthalate) blend. *J. Therm. Anal. Calorim.* **2019**. [CrossRef]
72. Yoshie, N.; Asaka, A.; Inoue, Y. Cocrystallization and phase segregation in crystalline/crystalline polymer blends of bacterial copolyesters. *Macromolecules* **2004**, *37*, 3770–3779. [CrossRef]
73. Organ, S.J. Phase separation in blends of poly(hydroxybutyrate) with poly(hydroxybutyrate-*co*-hydroxyvalerate): Variation with blend components. *Polymer* **1994**, *35*, 86–92. [CrossRef]

74. Kumagai, Y.; Doi, Y. Enzymatic degradation of poly(3-hydroxybutyrate)-based blends: Poly(3-hydroxybutyrate)/poly(ethylene oxide) blend. *Polym. Degrad. Stab.* **1992**, *35*, 87–93. [CrossRef]
75. Saito, M.; Inoue, Y.; Yoshie, N. Cocrystallization and phase segregation of blends of poly(3-hydroxybutyrate) and poly(3-hydroxybutyrate-*co*-3-hydroxyvalerate). *Polymer* **2001**, *42*, 5573–5580. [CrossRef]
76. Mansaray, K.G.; Ghaly, A.E. Thermogravimetric analysis of rice husks in an air atmosphere. *Energy Sources* **1998**, *20*, 653–663. [CrossRef]
77. Bugnicourt, E. Polyhydroxyalkanoate (PHA): Review of synthesis, characteristics, processing and potential applications in packaging. *Express Polym. Lett.* **2014**, *8*, 791–808. [CrossRef]
78. Li, S.D.; He, J.D.; Yu, P.H.; Cheung, M.K. Thermal degradation of poly(3-hydroxybutyrate) and poly(3-hydroxybutyrate-*co*-3-hydroxyvalerate) as studied by TG, TG–FTIR, and Py–GC/MS. *J. Appl. Polym. Sci.* **2003**, *89*, 1530–1536. [CrossRef]
79. Laycock, B.; Halley, P.; Pratt, S.; Werker, A.; Lant, P. The chemomechanical properties of microbial polyhydroxyalkanoates. *Prog. Polym. Sci.* **2013**, *38*, 536–583. [CrossRef]
80. Orts, W.J.; Marchessault, R.H.; Bluhm, T.L.; Hamer, G.K. Observation of strain-induced β form in poly(β-hydroxyalkanoates). *Macromolecules* **1990**, *23*, 5368–5370. [CrossRef]
81. Sanchez-Garcia, M.D.; Gimenez, E.; Lagaron, J.M. Novel PET nanocomposites of interest in food packaging applications and comparative barrier performance with biopolyester nanocomposites. *J. Plast. Film Sheeting* **2007**, *23*, 133–148. [CrossRef]
82. Cherpinski, A.; Torres-Giner, S.; Cabedo, L.; Lagaron, J.M. Post-processing optimization of electrospun submicron poly(3-hydroxybutyrate) fibers to obtain continuous films of interest in food packaging applications. *Food Addit. Contam. Part A* **2017**, *34*, 1817–1830. [CrossRef]
83. Razumovskii, L.; Iordanskii, A.; Zaikov, G.; Zagreba, E.; McNeill, I. Sorption and diffusion of water and organic solvents in poly(β-hydroxybutyrate) films. *Polym. Degrad. Stab.* **1994**, *44*, 171–175. [CrossRef]
84. Sanchez-Garcia, M.D.; Gimenez, E.; Lagarón, J.M. Morphology and barrier properties of solvent cast composites of thermoplastic biopolymers and purified cellulose fibers. *Carbohydr. Polym.* **2008**, *71*, 235–244. [CrossRef]
85. Lagarón, J.M. *Multifunctional and Nanoreinforced Polymers for Food Packaging*; Woodhead Publishing: Lisle, IL, USA, 2011; pp. 1–28.

materials

MDPI

Article

The Influence of Non-Uniformities on the Mechanical Behavior of Hemp-Reinforced Composite Materials with a Dammar Matrix

Dumitru Bolcu [1,†] and Marius Marinel Stănescu [2,*,†]

1 Department of Mechanics, University of Craiova, 165 Calea Bucureşti, 200620 Craiova, Romania; dbolcu@yahoo.com
2 Department of Applied Mathematics, University of Craiova, 13 A.I. Cuza, 200396 Craiova, Romania
* Correspondence: mamas1967@gmail.com; Tel.: +4-074-035-5079
† These authors contributed equally to this work.

Received: 16 March 2019; Accepted: 9 April 2019; Published: 15 April 2019

Abstract: As a result of manufacture, composite materials can appear to have variations to their properties due to the existence of structural changes. In this paper, we studied the influence of material irregularity on the mechanical behavior of two categories of bars for which we have used hemp fabric as a reinforcing material. The common matrix is a hybrid resin based on Dammar and epoxy resin. We molded two types of bars within each of the previously mentioned categories. The first type, also called "ideal bar", was made of layers in which the volume proportion and the orientation of the reinforcing material was the same in each section. The ideal bar does not show variations of mechanical properties along it. The second type of bar was molded to have one or two layers where, between certain sections, the reinforcing material was interrupted in several segments. We have determined some mechanical properties, the characteristic curves (strain-stress), the tensile strength, and elongation at break for all the sample sets on trial. Moreover, we have studied the influence of the non-uniformities on the mechanical behavior of the composites by entering certain quality factors that have been calculated after experimental determinations.

Keywords: composite materials; hybrid resin; natural reinforcement; non-uniformities; mechanical behavior

1. Introduction

Fiber-reinforced composites are widely used for their advantages over non-reinforced materials. Their applications include construction (see [1]), aerospace (see [2]), medicine (see [3]), and dentistry (see [4]) fields.

The mechanical behavior of composite materials is influenced by environmental factors (humidity, temperature, radiation, chemical agents, see [5–8]), and the mechanical stresses to which they are subjected (the stress type, the stress variation in time, the loading speed, the stress direction, the stress duration, see for example [9,10]).

An important aspect concerning the mechanical properties of composite materials is given by the non-uniformities that appear in the technological process of fabrication. In the case of fiber-reinforced composites the uneven distribution of the matrix-fibers represents the main factor influencing their mechanical behavior. The resin transfer, the structural reactions, and the interface effects are phenomena that can be taken into account when considering the non-uniformity degree (see [10,11]).

Due to the unevenness of the fiber distribution, experimental discrepancies often occur with respect to the physical properties of the studied samples. References [11–13] investigate the influences of non-uniformities on composite material behavior; in these works, the elastic constants and density are assumed to be functions of fiber volume proportion, while the distribution of the fiber volume proportion is given by a polynomial function. Reference [14] studies non-uniformities occurring in composite bars, reinforced with a glass fiber fabric, and introduces a coefficient that estimates the non-uniformities of the composite bars with two areas with different volume proportions of the reinforcement. References [15,16] study the non-uniformities occurring in composite bars reinforced by carbon fabric, and by carbon and Kevlar fabric.

The non-uniformities and discontinuities can be used with a view to obtaining composite materials with controlled properties. The traditional laminated composites, reinforced by fibers, are made of multiple laminae in which the fiber-reinforcing direction is constant. Property control is exercised only by modifying the layer sequence in the laminate. Reference [17] checks on the properties of the composite materials by using so-called "mosaic" composites. Each layer in these composites is made of several pieces, each piece with its own type of fiber orientation, length, and distribution. References [18–20] showed that using this kind of element in regulated assemblies, which are intertwined, led to obtaining composites with a tensile strength up to 90% compared to the composites with continuous reinforcement, although the mere juxtaposition of such layers may bring about a decrease in the tensile strength by 50%. They made composite materials that had improved damage tolerance because there is a mechanism of slowing crack propagation fissure speed by decreasing the cohesion between the adjacent elements. A theory that anticipates crack initiation and propagation is studied in [21] and its effects on fiber-reinforced composites polymerized with light materials are investigated in [22].

The use of hemp fibers as reinforcement for composite materials has risen in the past few years as a result of an increasing demand for the development of biodegradable and recyclable materials. Much attention has been paid in recent decades to composite materials, the components of which, be it matrix or reinforcement, come from nature. The advantages of using green composites lie in both the fibers and the bio-resins being abundantly made by nature, consequently having a low manufacture cost compared to synthetic composites. Moreover, they have relatively good mechanical properties. Hemp fibers can be found in the plant stem, which makes them tough and rigid, an essential requirement for the reinforcement of composite materials. The mechanical properties of hemp fibers are close to those of the glass fibers (see [23–27]). Nevertheless, the biggest disadvantage is the variability of these properties depending on the weather conditions, the harvest area, or other natural factors. In [28], the author explains why the hemp fibers from plants harvested at the beginning of flowering had a stiffness and a tensile strength greater than the fibers of the hemp plant seeds harvested at maturity. In [29] it is shown that the fibers of the middle section of the stock had higher mechanical properties than the fibers of the upper and lower sections of the stem. Studies on the mechanical properties of the hemp fibers are described in detail in [30]. Improvement of the mechanical behavior of hemp fibers, as well as different types of treatments, is shown in [31,32].

Most composite materials studied so far have focused on natural fibers as reinforcing materials in combination with thermoplastic matrices (polypropylene, polyethylene, and vinyl polychloride), or thermo-rigid matrices (phenolic, epoxy, and polyester resins) [33]. The composites made of hemp fibers with thermoplastic, heat resistant, and biodegradable matrices have shown good mechanical properties. In addition, the surface treatments applied to the hemp fibers, with the purpose of improving the bond on the interface between the fibers and the matrix, have led to considerable improvements in mechanical properties. Some mechanical properties which characterize tensile mechanical behavior, torsion, and bending the polypropylene matrix composite material and reinforced with hemp are shown in [34–36].

The synthetic resins have the disadvantage of a processing limit due to high viscosity when melted, a phenomenon that occurs during injection molding. Thus, the final product is hard to recycle. This disadvantage can be eliminated by using thermo-rigid biological matrices, based on vegetable oil resins, since the latter are biodegradable and there is no need for a polymerization process [37–39].

The bio-resins are resins derived from a biological source, and in general they are biodegradable or compostable; therefore, hypothetically, they can be decomposed after use. The natural resins may be fossil (amber), vegetable (Sandarac, Copal, Dammar), or animal (Shellac). Natural resins are insoluble in water but are slightly soluble in oil, alcohol, and, partially, petrol. They form, together with certain organic solvents, solutions used as covering varnishes. Turpentine, rosin, and mastic are products resulting from coniferous resin distillation. An analysis of the chemical composition and the properties of these resins is made in [40] and the applications are presented in [41].

The analysis of these resins focused more on their chemical composition and chemical properties, presented in [42,43], and less on their mechanical properties. For Dammar, which is a gum resin obtained from trees of the family Dipterocarpaceae from India and East Asia, detailed studies on the structure and chemical composition are made by [44].

Reference [45] presents a new silicon and Dammar-modified binder that reduces the use of synthetic binder and that has improved and more ecological properties. The optimal composition for this binder was determined, ensuring the best properties for impact stress, hardness, traction, and adherence. Reference [46] studies how the Dammar addition contributed to improving the rigidity, elasticity modulus, and hardness of a modified silicone.

There are quite a few analyses of the mechanical behavior of natural resins. Reference [47] investigates the mechanical features (tensile strength, percentage elongation and Young's modulus), the water vapor transmission characteristics, and the characteristics of humidity absorption in Dammar films that contain and that do not contain softening agents. The reaction to the compression stress of the oil palm trunk treated with various amounts of Dammar resin was studied in [48]. Also, there are few studies of composite materials with both the matrix and the reinforcement of natural materials. References [49,50] examine the mechanical behavior of some composite materials with a matrix of Dammar-based resin and the reinforcement of cotton, flax, silk, and hemp fabric.

Since natural lakes can form thick resin (see for example references [51,52]), we can conclude that the bio-resins were studied before actual hybrid resins. A hybrid resin is a combination of several constituents, of which at least one is organic and synthetic. It should be noted that most attempts to obtain such resins have been varnishes (see [51–53]). Hybrid resins are alternative environment-friendly compared with synthetic resins. Some practical and useful notes about this problem can be found in the papers [54,55]. Larger investigations about hybridization (in fibers and/or matrices) are available in [56,57].

2. Materials and Methods

2.1. Making the Samples

The first step consisted of casting hybrid resin plates where we used a Dammar volume proportion of 60%. The difference up to 100% consisted of epoxy resin of Resoltech 1050 type together with its associated hardener Resoltech 1055. The casting temperature was 21–23 °C. To realize the Resoltech 1050/Resoltech 1055 combination we respected the manufacturer's instruction. We used a mixture ratio of 7/3 after a given volume. We mixed the epoxy resin obtained with Dammar resin. All samples based on hybrid resin were cut after 10 days.

The synthetic component (the epoxy resin and the reinforcing material) was necessary to generate quick points of activation of the polymerization process. The thermo-mechanical properties of Resoltech 1050 epoxy resin, together with its associated hardener Resoltech 1055, are given by the producer (see [58]).

Figure 1 shows a hybrid resin sample.

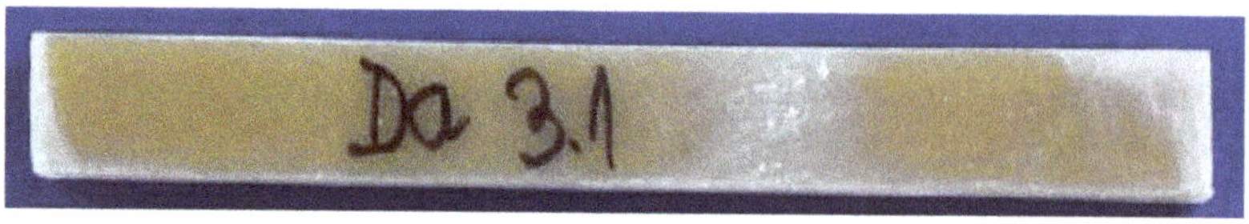

Figure 1. Hybrid resin sample.

The second step consisted of making Dammar-based composite materials by reinforcing them with two types of hemp fabric. The first type of fabric, symbolized by "A" is shown in Figure 2a, the fabric with the characteristic mass of 330 g/cm^3. For this type of fabric, on a surface of 10 cm × 10 cm, we had 50 strands oriented in the direction of the tensile test (longitudinal direction) and 33 strands in transversal direction. The second type of fabric, symbolized by "B", is presented in Figure 2b, the characteristic mass of this fabric being 350 g/cm^3. In this type of fabric, on a surface of 10 cm × 10 cm, we had 62 strands oriented in the direction of the tensile test (longitudinal direction) and 28 strands in transversal direction. The plates obtained had five layers of fabric. In the plates considered as ideal, all five layers are continuous, without interruption. To identify the studied composite materials, we use symbols made of one letter and two digits. The letter may be "A" or "B", depending on the type of fabric used. The first digit is 0, 1, or 2 and represents the number of interrupted layers. For instance, A24 stands for composite material with a reinforcement of the type "A" fabric, with two interrupted layers and the interruption length of the layers of four centimeters.

Figure 2. Hemp fabric: (**a**) type "A"; (**b**) type "B".

The hemp fiber properties are (see [37,50,59]): 1.4–1.5 g/cm^3 density, 30–60 GPa modulus of elasticity, 310–750 MPa breaking load, 2–4% elongation at break.

We produced sets of 10 samples that were 250 mm long and 25 mm wide. These sizes are employed in the tensile tests performed according to ASTM D3039 (see [60]).

In Table 1 we present details about manufactured samples.

Table 1. Details about the manufactured samples.

Abbreviation Sample	The Total Number of Layers of Reinforcement	Number of Layers Interrupted	Length of Interruption [mm]	Number of Samples Tested
A00	5	0	0	10
A10	5	1	0	10
A12	5	1	20	10
A14	5	1	40	10
A20	5	2	0	10
A22	5	2	20	10

Table 1. *Cont.*

Abbreviation Sample	The Total Number of Layers of Reinforcement	Number of Layers Interrupted	Length of Interruption [mm]	Number of Samples Tested
A24	5	2	40	10
B00	5	0	0	10
B10	5	1	0	10
B12	5	1	20	10
B14	5	1	40	10
B20	5	2	0	10
B22	5	2	20	10
B24	5	2	40	10

2.2. Theoretical Considerations

We have a straight bar, with the length l, square section, width b and thickness h, made of n layers of composite materials of constant thickness. A reference system is attached to the bar, with the axis Ox along the bar. The bar is submitted to a tensile test along the Ox axis. We accept that a plane and normal section on the Ox axis before the stress stays plane and normal on this axis throughout the test.

In this hypothesis, the characteristic deformation along the bar depends only on the abscissa x, thus $\varepsilon = \varepsilon(x)$.

The normal tension in layer k in the section of the abscissa x is (see [14]):

$$\sigma_k(x) = E_k(x) \cdot \varepsilon(x), \tag{1}$$

where $E_k(x)$ is the elasticity modulus of the layer k in the section of the abscissa x.

The tension resultant in the section of the abscissa x is the traction force to which the bar is submitted:

$$F = \iint_{(S)} \sigma(x)\, dS = b \sum_{k=1}^{n} h_k E_k(x) \cdot \varepsilon(x), \tag{2}$$

where h_k is the thickness of the layer k.

The average modulus of elasticity of the bar can be determined by the ratio:

$$E_{med} = \frac{F \cdot l}{\Delta l \cdot A} = \frac{l}{h \int_0^l \frac{1}{\sum_{k=1}^{n} h_k E_k(x)} dx}. \tag{3}$$

If in a layer k, in the section of the abscissa x, its breaking load reaches $\sigma_r^{(k)}(x)$, then it breaks and a first bar damage occurs. After the break of the first layer, the efforts are taken over by the other layers, causing a tension change in the layers. Thus, there are two scenarios for the break mechanism. The first presupposes that the breaking load is reached in other layers too, hence their concatenation and the bar break without any increase in the stress force. The second presupposes that after the redistribution of the tensions in the layers the breaking load is not reached in any of the remaining layers. In this case, the stress force may increase, leading to tension increase until the tensile strength is reached in one of the remaining layers, repeating the break phenomenon. The breaking load of the bar, marked by σ_r, is considered to be the average tension in the bar section at the moment when the concatenated break and the bar destruction occur. Since Hooke's law stipulates proportionality between tensions and characteristic deformations, the highest tensions appear in the sections where the characteristic deformations are maximal. These arise in the points where the bar section rigidity is minimal. In practice, the break can be considered to occur at the strands in the area where the bar section rigidity is minimal.

We have two bar types. The first type, called ideal, is made of layers that do not show property variations along the bar. In this type of bar, the volume proportion and the reinforcement orientation, for each layer, is considered to be the same in any section of the bar. The second bar type is considered to have layers where, between certain sections, the reinforcement is removed and the resin remains alone.

The second type of bars can be regarded as bars with different non-uniformities occurring because of the defects that appear during the technological process of bar making. We define the following factors characterizing quality:

- the elasticity factor
$$f_E = \frac{E}{E_{ideal}}, \tag{4}$$
where E is the elasticity modulus of the analyzed sample and E_{ideal} is the elasticity modulus of the material considered to be ideal, without uniformities;
- the resistance factor
$$f_\sigma = \frac{\sigma_r}{\sigma_{r\ ideal}}, \tag{5}$$
where σ_r is the tensile strength of the analyzed sample material and $\sigma_{r\ ideal}$ is the tensile strength of the material considered as ideal, without non-uniformities;
- the uniformity factor
$$f_u = \frac{f_\sigma}{f_E}. \tag{6}$$

These three factors give information on the properties of the analyzed sample material, properties that are compared to those of the reference material, which is considered ideal. The decrease in the elasticity factor shows the decrease in the modulus of elasticity and the decrease of the resistance factor shows the decrease in the tensile strength of the analyzed materials. The decrease in the roughness factor shows the presence of some areas in the material with discontinuities of the mechanical properties.

A particular case is that of the samples with length l, made of n layers, in which at $(n-k)$ layers, which are identical, the reinforcement has the same proportion and orientation along the whole length, and at k layers, between two sections situated at a distance l_0, the reinforcement is removed and replaced by resin. In this case, the elasticity factor and the resistance factor are expressed by:

$$f_E = \left[1 - \frac{l_0}{l} + \frac{l_0}{l}\frac{1}{1 - \frac{k}{n} + \frac{k}{n}\frac{E_r}{E_s}}\right]^{-1}, \tag{7}$$

$$f_\sigma = 1 - \frac{k}{n} + \frac{k}{n}\frac{E_r}{E_s}, \tag{8}$$

where E_s is the elasticity modulus of a layer without discontinuities and E_r is the elasticity modulus of the resin replacing the removed reinforcement.

3. Experimental Determinations

The obtained samples were submitted to a tensile test, which was performed according to the ASTM D3039 provisions (see [60]). We used the testing machine for mechanical trials LLOYD LRX PLUS SERIES (the manufacturer's data [61]) with the following features:

- force range: 5 kN;
- travel: 1 to 735 mm;
- crosshead speed: 0.1 to 1020 mm/min;

- analysis software: NEXYGEN.

Figure 3 presents the installation for a hybrid resin sample tensile test.

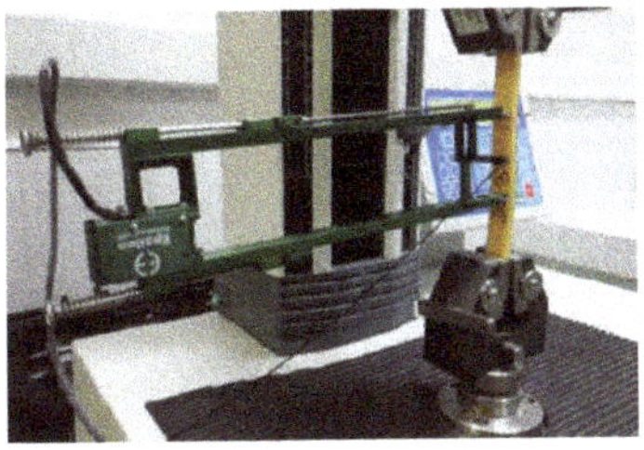

Figure 3. The tensile test assemblage of a sample from the hybrid resin.

Figure 4 shows the characteristic curve of a Dammar-based resin sample. The experimental results for the set of hybrid resin samples are:

- tensile strength: 21–22 MPa;
- elongation at break: 1.95–2.20%;
- modulus of elasticity: 1130–1220 MPa.

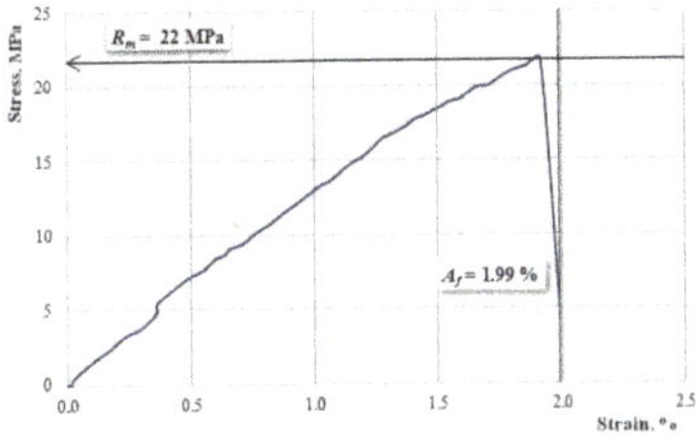

Figure 4. The characteristic curve of a Dammar-based resin sample.

The samples made of hemp-reinforced composite materials were also subject to tensile testing.

Figures 5–7 show the characteristic curves of the composite materials reinforced by type "A" fabric.

Figure 5 presents the characteristic curve for a sample reinforced with hemp fabric of type "A" without reinforcement-interrupted layers.

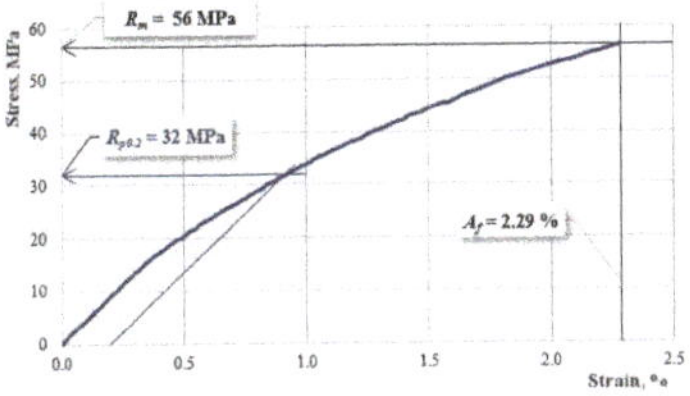

Figure 5. The characteristic curve for a sample A00.

Figure 6 presents the characteristic curves for samples reinforced with hemp fabric of type "A" which have an interrupted layer of reinforcement.

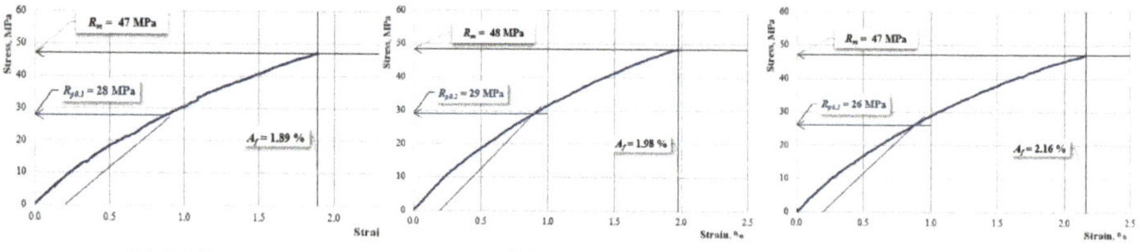

(**a**) A10 (**b**) A12 (**c**) A14

Figure 6. The characteristic curve for a sample: (**a**) A10; (**b**) A12; (**c**) A14.

Figure 7 presents the characteristic curves for samples reinforced with hemp fabric of type "A" which have two interrupted layers of reinforcement.

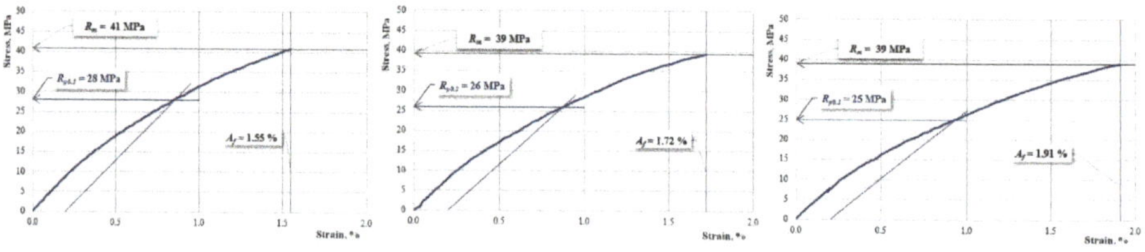

(**a**) A20 (**b**) A22 (**c**) A24

Figure 7. The characteristic curve for a sample: (**a**) A20; (**b**) A22; (**c**) A24.

For every sample reinforced with type "A" hemp fabric, we give the below (Table 2) the lower and upper values (arithmetical average value and deviation value) for the elasticity modulus, tensile strength, and elongation at break. We have not made a statistical analysis for those values simply because all the outcomes are strictly related to samples in the study. The mechanical properties of hemp fibers come under many influences, so a statistical analysis in this case would have not been relevant.

Table 2. Type "A" hemp fabric reinforced samples: lower and upper values (arithmetical average value and deviation value) for the elasticity modulus, tensile strength, and elongation at break.

Abbreviation Sample	Modulus of Elasticity *E* [MPa]	Tensile Strength R_m [MPa]	Elongation at Break *A* [%]
A00	4473–4622 (4547 ± 75)	55–57 (56 ± 1)	2.24–2.29 (2.27 ± 0.03)
A10	4440–4618 (4529 ± 89)	47–48 (47.5 ± 0.5)	1.89–1.95 (1.92 ± 0.03)
A12	4326–4445 (4386 ± 60)	47–48 (47.5 ± 0.5)	1.96–2.02 (1.99 ± 0.03)
A14	4090–4209 (4150 ± 60)	46–48 (47 ± 1)	2.08–2.16 (2.12 ± 0.04)
A20	4421–4573 (4497 ± 76)	40–41 (40.5 ± 0.5)	1.56–1.62 (1.59 ± 0.03)
A22	4062–4199 (4131 ± 68)	39–41 (40 ± 1)	1.71–1.77 (1.74 ± 0.03)
A24	3692–3790 (3741 ± 49)	38–40 (39 ± 1)	1.87–1.93 (1.90 ± 0.03)

Figures 8–10 present the characteristic curves of the composite materials reinforced by type "B" fabric.

The Figure 8 shows the characteristic curve for a sample reinforced with hemp fabric of type "B" without reinforcement-interrupted layers.

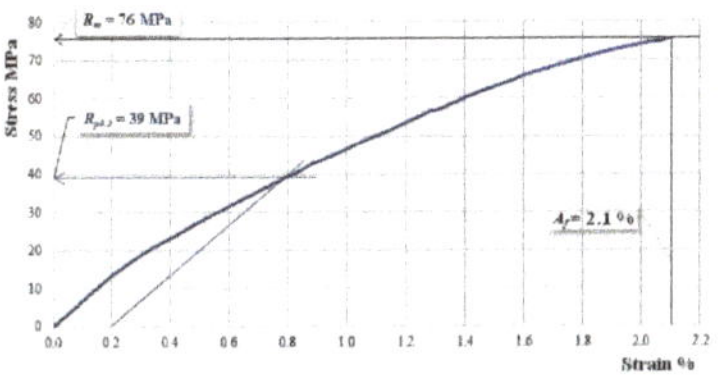

Figure 8. The characteristic curve for a sample B00.

Figure 9 presents the characteristic curves for samples reinforced with hemp fabric of type "B" which have an interrupted layer of reinforcement.

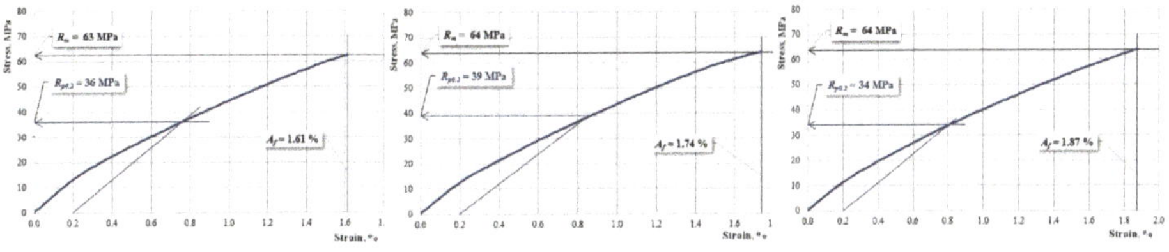

(**a**) B10 (**b**) B12 (**c**) B14

Figure 9. The characteristic curve for a sample: (**a**) B10; (**b**) B12; (**c**) B14.

Figure 10 presents the characteristic curves for samples reinforced with hemp fabric of type "B" which have two interrupted layers of reinforcement.

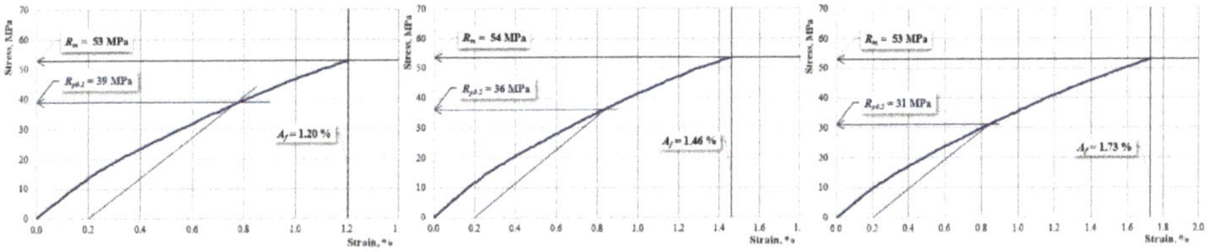

(**a**) B20 (**b**) B22 (**c**) B24

Figure 10. The characteristic curve for a sample: (**a**) B20; (**b**) B22; (**c**) B24.

For every sample reinforced with type "B" hemp fabric, we give, in the below (Table 3), the lower and upper values (arithmetical average value and deviation value) for the elasticity modulus, tensile strength, and elongation at break.

Table 3. Type "B" hemp fabric reinforced samples: lower and upper values (arithmetical average value and deviation value) for the elasticity modulus, tensile strength, and elongation at break.

Abbreviation Sample	Modulus of Elasticity E [MPa]	Tensile Strength R_m [MPa]	Elongation at Break A [%]
B00	6580–6654 (6617 ± 37)	74–76 (75 ± 1)	2.02–2.10 (2.06 ± 0.04)
B10	6528–6610 (6569 ± 41)	62–64 (63 ± 1)	1.60–1.66 (1.63 ± 0.03)
B12	6094–6226 (6160 ± 66)	62–64 (63 ± 1)	1.70–1.74 (1.72 ± 0.02)
B14	5872–5989 (5930 ± 58)	61–64 (62.5 ± 1.5)	1.86–1.92 (1.89 ± 0.03)
B20	6568–6640 (6604 ± 36)	51–53 (52 ± 1)	1.21–1.29 (1.25 ± 0.04)
B22	5770–5899 (5834 ± 64)	50–54 (52 ± 2)	1.44–1.48 (1.46 ± 0.02)
B24	5257–5363 (5310 ± 53)	50–53 (51.5 ± 1.5)	1.72–1.82 (1.77 ± 0.05)

4. Conclusions

The presence of areas with low mechanical properties in composite materials can lead to alteration in the overall properties of the composite materials. Meanwhile, industrial finished products with variable mechanical properties can be obtained that have a mechanically controlled behavior. Specifically, materials can be obtained that should yield to certain applications, and the breakage should occur in areas particularly chosen from the design.

For all the specimens, which have one or two interrupted layers, both the modulus of elasticity and the breaking resistance are proportional to the modulus of elasticity and the tensile strength of the materials considered as ideal. Thus, the ratio of the average modulus of elasticity of the A00-abbreviated specimens and the average modulus of elasticity of the B00-abbreviated specimens is 0.687. Similar values are obtained for the ratios between the modules of elasticity to the composite materials reinforced with hemp weaving "A" and the modules of elasticity to the composite materials reinforced with hemp weaving "B" (ratios are calculated between the modules of elasticity of the specimens with the same number of interrupted layers and the same length of the discontinued area). The ratio of the average breaking strength of the A00-abbreviated specimens and the average breaking strength of the B00-abbreviated specimens is 0.747. Similar values are obtained for the ratios of the tensile stress for composites reinforced with woven hemp type "A" and the tensile stress for composites reinforced with woven hemp type "B" (ratios are calculated between the modules of elasticity of the specimens with the same number of interrupted layers and the same length of the discontinued area).

It is noted that both the elastic modulus and the tensile stress decreased to increase the number of discontinued layers. Tensile strength decreases to increase the number of interrupted layers regardless of length of the interruption zone. The modules of elasticity for the specimens to the length of the discontinued zone is zero (interrupted layers have the reinforced sectioned in place), and are close to the modules of elasticity of the reference specimens that were considered ideal. For the test pieces with interrupted zone length of 20 mm or 40 mm, the modulus of elasticity decreases to increase the number of discontinued layers. In addition, the modulus of elasticity decreases to increase the length of the interrupted area. A particular trend is found for the elongation at breaking. Both for composite materials reinforced with hemp type "A" and those reinforced with hemp type "B" of the reference specimens (indicated A00 and B00) present the highest breaking elongation. Increasing the number of interrupted layers leads to a decrease of breaking elongation. However, for specimens with the same number of interrupted layers, the breaking elongation increases to increase the length of the discontinued area.

Therefore, the studied composite bars properties depend on:

- the properties of the component materials (modulus of elasticity E_s and the tensile strength $\sigma_r^{(s)}$, of the reference material, respectively modulus of elasticity E_r of the matrix);
- number of interrupted layers (the ratio $\frac{k}{n}$ between the number of interrupted layers and the total number of layers);
- the interruption length of the layers (the ratio $\frac{l_0}{l}$ between the interruption length and the bar length).

We believe that the reference values of the elasticity modulus and tensile strength of the materials considered as ideal are the average values of the limits between which these sizes vary in samples with all their layers without discontinuities (see Tables 2 and 3). For type-A00 samples the reference value of the elasticity modulus is 4547 MPa and the tensile strength is 56 MPa. For type-B00 samples the reference value of the elasticity modulus is 6617 MPa and the tensile strength is 75 MPa. We notice that both the elasticity modulus and the tensile strength of the composites reinforced by type "B" fabric have significantly higher values than those of the composites reinforced by type "A" fabric. This can be explained by the different number of strands oriented in the direction of the tensile test (50 for type "A" fabric and 62 for type "B" fabric). The fact that the elasticity modulus of the samples in the reference sets is in correlation with the elasticity modules of the fabrics used for reinforcing, and with the number of fibers oriented in the direction of the stress, shows that the longitudinal modulus of elasticity of the composite material is proportional to the modulus of elasticity of the reinforcing material and its volume proportion, just as in classic mixture theory.

Calculating, in a similar way, the reference values of the elasticity modulus and tensile strength for each of the analyzed sample sets, the presented quality factors can be determined.

Table 4 shows these quality factors for the sample sets reinforced by type "A" fabric and in Table 5 for type "B" fabric.

Table 4. Quality factors for composites reinforced by type "A" fabric.

Sample Type	Elasticity Factor f_E		Resistance Factor f_σ		Uniformity Factor f_u	
	Theoretical	Experimental	Theoretical	Experimental	Theoretical	Experimental
A10	1	0.991	0.853	0.848	0.853	0.855
A12	0.967	0.959	0.853	0.848	0.883	0.884
A14	0.936	0.908	0.853	0.839	0.912	0.924
A20	1	0.984	0.707	0.723	0.707	0.735
A22	0.923	0.903	0.707	0.714	0.765	0.790
A24	0.858	0.818	0.707	0.696	0.824	0.851

Table 5. Quality factors for composites reinforced by type "B" fabric.

Sample Type	Elasticity Factor f_E		Resistance Factor f_σ		Uniformity Factor f_u	
	Theoretical	Experimental	Theoretical	Experimental	Theoretical	Experimental
B10	1	0.993	0.836	0.840	0.836	0.846
B12	0.962	0.931	0.836	0.840	0.869	0.902
B14	0.927	0.896	0.836	0.833	0.902	0.929
B20	1	0.998	0.673	0.693	0.673	0.694
B22	0.911	0.882	0.673	0.693	0.738	0.785
B24	0.837	0.802	0.673	0.687	0.804	0.856

Analyzing Tables 4 and 5 leads us to the following conclusions:

- the elasticity factor decreases if the number of interrupted layers increases; this shows that if the number of layers with interruptions increases, the elasticity modulus of the composite decreases;
- the elasticity factor decreases if the layer interruption length increases; this shows that if the length of the area where resin replaces reinforcement increases, then the elasticity modulus of the composite decreases;
- the resistance factor decreases if the number of interrupted layers increases; therefore, the tensile strength decreases if the number of interrupted layers increases;
- the resistance factor is not influenced by the layer interruption length; this shows that breaking can occur when the fibers in the area where the bar section rigidity minimal break up;
- the uniformity factor decreases if the number of interrupted layers increases;
- the uniformity factor increases if the layer interruption length increases.

5. Discussion

Three factors give information on the properties of the analyzed sample material, properties that are compared to those of the reference material, which is considered to be ideal. Values close to 1 for the three factors point out that the analyzed material has properties that are very close to those of the reference material. A decrease in the values of the three factors indicates the presence of manufacture defects or that the analyzed material is different from the reference material. A value of less than 1 of the resistance factor shows that there are areas in the material where the mechanical properties are inferior to those of the reference material. In the case of composite materials, it indicates that there are areas where either the reinforcement volume proportion is lower than necessary, or the reinforcement orientation does not coincide with that of the reference material. A value less than 1 of the elasticity factor may give information on the dimensions of the area where the mechanical properties are inferior to those of the reference material. The lower the elasticity factor, the larger the faulty area dimensions. If the elasticity factor is 1 and the resistance factor is less than 1, there is a concentrated defect in the analyzed material. The uniformity factor characterizes the uniformity of the analyzed material. Decreased values of this factor point out the existence in the material of certain areas where the mechanical properties differ very much. There might be a case when the resistance factor and the elasticity factor have values less than 1, be very close, or a case in which the uniformity factor is near to the value 1. This shows that the analyzed material is uniform and so different from the reference material that it is actually a different material and not the faulty reference material. In the case of composites, this situation may indicate that the analyzed material has reinforcing material proportions, resin specifically, different from those of the reference material.

Author Contributions: Conceptualization, D.B. and M.M.S.; casting of samples and performed the experiments, D.B. and M.M.S.; analysis and interpretation of experimental data, M.M.S. and D.B.; methodology, M.M.S. and D.B.; writing–original draft, D.B. and M.M.S.; writing–review and editing, M.M.S.

Funding: This research received no external funding.

Conflicts of Interest: The authors declare no conflict of interest.

Nomenclature

f_E	the elasticity factor
f_σ	the resistance factor
f_u	the uniformity factor
E	the elasticity modulus of analyzed sample
E_{ideal}	the elasticity modulus of the material considered to be ideal, without non-uniformities
E_s	the elasticity modulus of a layer without discontinuities
E_r	the elasticity modulus of the resin
$E_k(x)$	the elasticity modulus of the layer k in the section of the abscissa x
σ_r	the tensile strength of the analyzed sample material
$\sigma_r^{(s)}$	the tensile strength of the reference material
$\sigma_k(x)$	the normal tension in layer k in the section of the abscissa x
$\sigma_{r\ ideal}$	the tensile strength of the material considered as ideal, without non-uniformities
$\varepsilon(x)$	the characteristic deformation in layer k in the section of the abscissa x
h_k	the thickness of the layer k
l	the bar length
l_0	the interruption length
b	the bar width
h	the bar thickness.

References

1. Hosseini, A.; Ghafoori, E.; Wellauer, M.; Marzaleh, A.S.; Motavalli, M. Short-term bond behavior and debonding capacity of prestressed CFRP composites to steel substrate. *Eng. Struct.* **2018**, *176*, 935–947. [CrossRef]
2. Soutis, C. Fibre reinforced composites in aircraft construction. *Prog. Aerosp. Sci.* **2005**, *41*, 143–151. [CrossRef]
3. Shahmohammadi, M.; Asgharzadeh Shirazi, H.; Karimi, A.; Navidbakhsh, M. Finite element simulation of an artificial intervertebral disk using fiber reinforced laminated composite model. *Tissue Cell* **2014**, *46*, 299–303. [CrossRef] [PubMed]

4. Scribante, A.; Vallittu, P.K.; Özcan, M. Fiber-reinforced composites for dental applications. *BioMed Res. Int.* **2018**, 4734986. [CrossRef]
5. Zia-ul-Haq; Khan, F.U.; Shah, A.M.H.; Ullah, M.A.; Abdusubhan; Khabir, A.; Ali, F.; Rauf, A.; Khattak, A.K. Effects of different environmental factors on the mechanical properties of different samples of fiber glass. *J. Biodivers. Environ. Sci.* **2017**, *10*, 115–123.
6. Chen, T.T.; Liu, W.D.; Qiu, R.H. Mechanical properties and water absorption of hemp fibers-reinforced unsaturated polyester composites: Effect of fiber surface treatment with a heterofunctional monomer. *BioResources* **2013**, *8*, 2780–2791. [CrossRef]
7. Derrien, K.; Gilormini, P. The effect of moisture-induced swelling on the absorption capacity of transversely isotropic elastic polymer-matrix composites. *Int. J. Solids Struct.* **2009**, *46*, 1547–1553. [CrossRef]
8. Dhakal, H.N.; Zhang, Z.Y.; Richardson, M.O.W. Effect of water absorption on the mechanical properties of hemp fibre reinforced unsaturated polyester composites. *Compos. Sci. Technol.* **2007**, *67*, 1674–1683. [CrossRef]
9. Huang, Z.; Wang, N.; Zhang, Y.; Hu, H.; Luo, Y. Effect of mechanical activation pretreatment on the properties of sugarcane bagasse/poly(vinyl chloride) composites. *Compos. Part A Appl. Sci. Manuf.* **2012**, *43*, 114–120. [CrossRef]
10. Summerscales, J.; Grove, S. Manufacturing methods for natural fibre composites. In *Natural Fibre Composites: Materials, Processes and Properties*; Hodzic, A., Shanks, R., Eds.; Taylor and Francis: Boca Raton, FL, USA, 2014; pp. 157–186.
11. Chatterjee, A. Non-uniform fiber networks and fiber-based composites: Pore size distributions and elastic moduli. *J. Appl. Phys.* **2010**, *108*, 063513. [CrossRef]
12. Chi, S.H.; Chung, Y.L. Mechanical behavior of functionally graded material plates under transverse load, I: Analysis. *Int. J. Solids Struct.* **2006**, *43*, 3657–3674. [CrossRef]
13. Tanaka, M.; Hojo, M.; Hobbiebrunken, T.; Ochiai, S.; Hirosawa, Y.; Fujita, K. Influence of non-uniform fiber arrangement on tensile fracture behavior of unidirectional fiber/epoxy model composites. *Compos. Interface* **2005**, *12*, 365–378. [CrossRef]
14. Bolcu, D.; Stanescu, M.M.; Ciuca, I.; Dumitru, S.; Sava, M. The non-uniformity from the composite materials reinforced with fiber glass fabric. *Mater. Plast.* **2014**, *51*, 97–100.
15. Stanescu, M.M.; Bolcu, D.; Ciuca, I.; Dinita, A. Non uniformity of composite materials reinforced with carbon and carbon-kevlar fibers fabric. *Mater. Plast.* **2014**, *51*, 355–358.
16. Bolcu, D.; Sava, M.; Dinita, A.; Miritoiu, C.M.; Baciu, F. The influence of discontinuities on elastic and mechanical properties of composite materials reinforced with woven carbon, carbon-kevlar and kevlar. *Mater. Plast.* **2016**, *53*, 23–28.
17. Jarve, D.; Kim, R. Strength prediction and measurement in model multilayered discontinuous tow reinforced composites. *J. Compos. Mater.* **2004**, *38*, 5–18.
18. Baucom, J.N.; Thomas, J.P.; Pogele, W.R. Tiled composite laminates. *J. Compos. Mater.* **2010**, *44*, 3115–3132. [CrossRef]
19. Dyskin, A.V.; Estrin, Y.; Pasternak, E. Topological interloking of platonic solids: A way to new materials and structures. *Philos. Mag. Lett.* **2003**, *83*, 197–203. [CrossRef]
20. Dyskin, A.V.; Pasternak, E.; Estrin, Y.; Pasternak, E.; Kanel-Belov, A.J. A new principle in design of composite materials: Reinforcement by interloked elements. *Compos. Sci. Technol.* **2003**, *63*, 483–491. [CrossRef]
21. Ghafoori, E.; Motavalli, M. A retrofit theory to prevent fatigue crack initiation in aging riveted bridges using carbon fiber-reinforced polymer materials. *Polymers* **2016**, *8*, 308. [CrossRef]
22. Scribante, A.; Massironi, S.; Pieraccini, G.; Vallittu, P.; Lassila, L.; Sfondrini, M.F.; Gandini, P. Effects of nanofillers on mechanical properties of fiber-reinforced composites polymerized with light-curing and additional postcuring. *J. Appl. Biomater. Funct. Mater.* **2015**, *13*, e296–e299. [CrossRef]
23. Bogoeva-Gaceva, G.; Avella, M.; Malinconico, M.; Buzaovska, A.; Grozdanov, A.; Gentile, G.; Errico, M.E. Naturalfiber eco-composites. *Polym. Compos.* **2007**, *28*, 98–107. [CrossRef]
24. Brouwer, W.D. Natural fiber composites—From uphols-tery to structural components. In Proceedings of the Natural Fibres for Automotive Industry Conference, Manchester Conference Centre, Manchester, UK, 28 November 2000.

25. Ivens, J.; Bos, H.; Verpoest, I. The applicability of natural fibers as reinforcement for polymer composites. In Proceedings of the Symposium on Renewable Bioproducts—Industrial Outlets and Research for 21st Century, International Agriculture Centre, Wegeningen, The Netherlands, 24–25 June 1997.
26. Mougin, G. Natural fiber composites—Problems and solutions. *JEC Compos. Mag.* **2006**, *25*, 32–35. Available online: http://www.jeccomposites.com/news/composites-news/natural-fibre-composites-problemsand-solutions (accessed on 25 October 2011).
27. Mueller, D.H.; Krobjilowski, A. New discovery in the properties of composites reinforced with natural fibers. *J. Ind. Text.* **2003**, *33*, 111–130. [CrossRef]
28. Liu, M.; Fernando, D.; Daniel, G.; Madsen, B.; Meyer, A.; Ale, M.; Thygesen, A. Effect of harvest time and field retting duration on the chemical composition, morphology and mechanical properties of hemp fibers. *Ind. Crop. Prod.* **2015**, *69*, 29–39. [CrossRef]
29. Liu, M.; Fernando, D.; Meyer, A.S.; Madsen, B.; Daniel, G.; Thygesen, A. Characterization and biological depectinization of hemp fibers originating from different stem sections. *Ind. Crop. Prod.* **2015**, *76*, 880–891. [CrossRef]
30. Shahzad, A. Hemp fiber and its composites—A review. *J. Compos. Mater.* **2012**, *46*, 973–986. [CrossRef]
31. Liu, M.; Thygesen, A.; Summerscales, J.; Meyer, A. Targeted pre-treatment of hemp bast fibres for optimal performance in biocomposite materials: A review. *Ind. Crop. Prod.* **2017**, *108*, 660–683. [CrossRef]
32. Hoque, M.B.; Alam, A.B.M.; Mahmud, H.; Nobi, A. Mechanical, degradation and water uptake properties of fabric reinforced polypropylene based composites: Effect of alkali on composites. *Fibers* **2018**, *6*, 94. [CrossRef]
33. Malkapuram, R.; Kumar, V.; Yuvraj, S.N. Recent development in natural fibre reinforced polypropylene composites. *J. Reinf. Plast. Compos.* **2008**, *28*, 1169–1189. [CrossRef]
34. Hargitai, H.; Racz, I.; Anandjiwala, R.D. Developmentof hemp fiber reinforced polypropylene composites. *J. Thermoplast. Compos. Mater.* **2008**, *21*, 165–174. [CrossRef]
35. Mechraoui, A.; Riedl, B.; Rodrigue, D. The effect of fiber and coupling agent content on the mechanical properties of hemp/polypropylene composites. *Compos. Interfaces* **2007**, *14*, 837–848. [CrossRef]
36. Sain, M.; Suhara, P.; Law, S.; Bouillox, A. Interface mod-ification and mechanical properties of natural fiber-poly-olefin composite products. *J. Reinf. Plast. Compos.* **2005**, *24*, 121–130. [CrossRef]
37. Mohanty, A.K.; Misra, M.; Hinrichsen, G. Biofibers, biodegradable polymers and biocomposites: An overview. *Macromol. Mater. Eng.* **2000**, *276–277*, 1–24. [CrossRef]
38. Shogren, R.L.; Petrovic, Z.; Liu, Z.S.; Erhan, S.Z. Biodegradation behavior of some vegetable oil-based polymers. *J. Polym. Environ.* **2004**, *12*, 173–178. [CrossRef]
39. Uyama, H.; Kuwabara, M.; Tsujimoto, T.; Kobayashi, S. Enzymatic synthesis and curing of biodegradable epoxide-containing polyesters from renewable resources. *Biomacromolecules* **2003**, *4*, 211–215. [CrossRef]
40. Prati, S.; Sciutto, G.; Mazzeo, R.; Torri, C.; Fabbri, D. Application of ATR-far-infrared spectroscopy to the analysis of natural resins. *Anal. Bioanal. Chem.* **2011**, *399*, 3081–3091. [CrossRef]
41. Suprakas, S.R.; Mosto, B. Biodegradable polymers and their layered silicate nanocomposites: In greening the 21st century materials world. *Prog. Mater. Sci.* **2005**, *50*, 962–1079.
42. Azemard, C.; Menager, M.; Vieillescazes, C. On the tracks of sandarac, review and chemical analysis. *Environ. Sci. Pollut. Res.* **2017**, *24*, 27746–27754. [CrossRef]
43. Kononenko, I.; Viguerie, L.; Rochut, S.; Walter, P. Qualitative and quantitative studies of chemical composition of sandarac resin by GC-MS. *Environ. Sci. Pollut. Res.* **2017**, *24*, 2160–2165. [CrossRef]
44. Hidayat, A.T.; Farabi, K.; Harneti, D.; Maharani, R.; Mayanti, T.; Setiawan, A.S.; Supratman, U.; Shiono, Y. Cytotoxicity and structure activity relationship of Dammarane-type triterpenoids from the bark of aglaia elliptica against P-388 murine leukemia cells. *Nat. Prod. Sci.* **2017**, *23*, 291–298. [CrossRef]
45. Zakaria, R.; Ahmad, A.H. Adhesion and hardness evaluation of modified silicone-Dammar as natural coating materials. *J. Appl. Sci.* **2012**, *9*, 890–893. [CrossRef]
46. Zakaria, R.; Ahmad, A.H. The performance of modified silicone-Dammar resin in nanoindentation test. *Int. J. Adv. Sci. Technol.* **2012**, *42*, 33–44.
47. Pethe, A.M.; Joshi, S.B. Mechanical and film forming studies of novel biomaterial. *Int. J. Pharm. Sci. Res.* **2013**, *4*, 2761–2769.
48. Nurfajriani, L.W.; Gea, S.; Thamrin, B.W. Mechanical properties of oil palm trunk by reactive compregnation methode with Dammar resin. *Int. J. PharmTech Res.* **2015**, *8*, 74–79.

49. Ciuca, I.; Bolcu, A.; Stanescu, M.M. A study of some mechanical properties of bio-composite materials with a Dammar-based matrix. *Environ. Eng. Manag. J.* **2017**, *16*, 2851–2856.
50. Stanescu, M.M.; Bolcu, D. A study of some mechanical properties of a category of composites with a hybrid matrix and natural reinforcements. *Polymers* **2019**, *11*, 478. [CrossRef]
51. Kanehashi, S.; Oyagi, H.; Lu, R.; Miyakoshi, T. Developement of bio-based hybrid resin, from natural lacquer. *Prog. Org. Coat.* **2010**, *77*, 24–29. [CrossRef]
52. Ishimura, T.; Lu, R.; Yamasaki, K.; Miyakoshi, T. Development of an eco-friendly hybrid lacquer based on kurome lacquer sap. *Prog. Org. Coat.* **2010**, *69*, 12–15. [CrossRef]
53. Drisko, G.L.; Sanchez, C. Hybridization in materials science—Evolution, current state, and future aspirations. *Eur. J. Inorg. Chem.* **2012**, *32*, 5097–5105. [CrossRef]
54. De Paola, S.; Minak, G.; Fragassa, C.; Pavlovic, A. Green Composites: A Review of State of Art. In Proceedings of the 30th Danubia Adria Symposium on Advanced Mechanics, Primosten, Croatia, 25–28 September 2013; Croatian Society of Mechanics, Ed.; pp. 77–78.
55. Hyseni, A.; De Paola, S.; Minak, G.; Fragassa, C. Mechanical characterization of ecocomposites. In Proceedings of the 30th Danubia Adria Symposium on Advanced Mechanics, Primosten, Croatia, 25–28 September 2013; pp. 175–176.
56. Zivkovic, I.; Pavlovic, A.; Fragassa, C.; Brugo, T. Influence of moisture absorption on the impact properties of flax, basalt and hybrid flax/basalt fiber reinforced green composites. *Compos. Part B: Eng.* **2017**, *111*, 148–164. [CrossRef]
57. Fragassa, C.; Pavlovic, A.; Santulli, C. Mechanical and impact characterisation of flax and basalt fibre bio-vinylester composites and their hybrids. *Compos. Part B Eng.* **2018**, *137*, 247–259. [CrossRef]
58. Resoltech 1050, hardeners 1053 to 1059. Structural Lamination Epoxy System. Available online: www.scabro.com/images/.../1/.../Resoltech%201050/DS-1050.pdf (accessed on 9 January 2019).
59. Singh, A.A.; Afrin, S.; Karim, Z.L. Green composites: Versatile material for future, chapter Green Biocomposites. In *Series Green Energy and Technology*; Springer International Publishing: Berlin, Germany, 2017; pp. 29–44, doi:10.1007/978-3-319-49382-4_2.
60. ASTM D3039, Standard Test Method for Tensile Properties of Polymer Matrix Composite Materials. Available online: https://www.astm.org/Standards/D3039 (accessed on 9 January 2019).
61. LLOYD LRX PLUS SERIES, Materials Testing Machine. Available online: http://www.elis.it/lloyd-pdf/LRXPlus.pdf (accessed on 28 March 2019).

materials

MDPI

Article

Fully Biodegradable Composites: Thermal, Flammability, Moisture Absorption and Mechanical Properties of Natural Fibre-Reinforced Composites with Nano-Hydroxyapatite

Pooria Khalili [1,2], Xiaoling LIU [1,*], Zirui ZHAO [1] and Brina Blinzler [2]

[1] Ningbo Nottingham New Materials Research Institute, University of Nottingham Ningbo China (UNNC), Ningbo 315100, China; pooria.khalili@gmail.com (P.K.); zy18541@nottingham.edu.cn (Z.Z.)

[2] Department of Industrial and Materials Science, Chalmers University of Technology, 412 96 Gothenburg, Sweden; brina.blinzler@chalmers.se

* Correspondence: Xiaoling.Liu@nottingham.edu.cn; Tel.: +86-574-8818-0000 (ext. 8057)

Received: 7 March 2019; Accepted: 1 April 2019; Published: 3 April 2019

Abstract: Natural fibre-reinforced poly(lactic acid) (PLA) laminates were prepared by a conventional film stacking method from PLA films and natural fabrics with a cross ply layup of [0/90/0/90/0/90], followed by hot compression. Natural fibre (NF) nano-hydroxyapatite (nHA) filled composites were produced by the same manufacturing technique with matrix films that had varying concentrations of nHA in the PLA. Their flammability, thermal, moisture absorption and mechanical properties were analysed in terms of the amount of nHA. The flame behavior of neat PLA and composites evaluated by the UL-94 test demonstrated that only the composite containing the highest quantity of nHA (i.e., 40 wt% nHA in matrix) was found to achieve an FH-1 rating and exhibited no recorded burn rate, whereas other composites obtained only an FH-3. The thermal degradation temperature and mass residue were also observed, via thermogravimetric analysis, to increase when increasing concentrations of nHA were added to the NF composite. The tensile strength, tensile modulus and flexural modulus of the neat resin were found to increase significantly with the introduction of flax fibre. Conversely, moisture absorption was found to increase and mechanical properties to decrease with both the presence of NF and increasing concentrations of nHA, and subsequent mechanical properties experienced an obvious reduction.

Keywords: poly(lactic acid) (PLA); natural fibre (NF); nano-hydroxyapatite (nHA); flammability; mechanical properties

1. Introduction

Polymeric thermosets such as vinyl ester, unsaturated polyester and epoxy are chosen for most composites due to their resistance to chemicals, hydrophobicity and suitable mechanical performances. However, thermoplastic matrices have recently gained significant attention in production industries such as construction, automotive and packaging. Bio-based polymers such as thermoplastic starch, poly(lactic acid) (PLA) and poly(3-hydroxybutyrate) (PHB) can be substituted for polyethylene, polystyrene or polypropylene. Petroleum based polymers can be replaced with some bio-based counterparts to deal with disposal issues and problems of diminishing fossil fuel stocks. For many applications where mechanical recycling is an issue, biodegradable polymers are of great interest, as they can be made to be biodegradable and compostable [1]. PLA, amongst other bio-based polymers, possesses relatively high crystallinity, melting point and stiffness, which demonstrates a greater commercial potential [2,3]. However, for engineering applications, in order to obtain the required mechanical performance, this type of polymer must be reinforced.

Recent regulations on the recyclability of materials and environmental requirements compel manufacturers and research institutions to develop composites from renewable sources. Plant fibres (PF) i.e., flax, ramie, jute, hemp and sisal are biodegradable, renewable and economical to use, have high specific modulus and strength, and have low density [4–6]. PFs possess approximately 40% lower density than glass fibres, which leads to the manufacturing of lighter components than polymeric parts reinforced with glass fibres [7,8]. This is crucial for applications related to transportation resulting in reduced emissions and enhanced fuel efficiency [9]. Therefore, the eco-advantages and the reduced mass of these materials make them competitive with synthetic fibre-reinforced composites. However, biocomposites are considered flammable when compared with the traditional fibre-reinforced plastics e.g., carbon and glass fibre composites. Therefore, bio-based composites are more flammable than the traditional composites. This limits the usage of biocomposites in applications where the fire regulations are stringent, for instance the aviation and railway industries [10]. The enhancement of fire and thermal resistivity of biocomposites (e.g., PLA-based biocomposites) is required even for less stringent industries like automotive and packaging.

Inorganic fillers such as silica nanoparticles at high concentration were used to impart high thermal stability to the cellulose-based composites, as obtained from thermogravimetric analysis (TGA) [11]. Nano-clay was also used in a different study to improve the thermal stability and flame retardancy of natural fibre polymer composites [12]. In another investigation, the incorporation of halloysite nanotubes into natural fibre polymeric composites was reported [13]. The thermal degradation temperature was observed to enhance with the incorporation of these nano-fillers. Except for the nano-inorganic additives mentioned, the thermal properties and flame retardancy of natural fibre-reinforced plastics has been improved by the inclusion of various flame-retardants (FRs) into polymers. For instance, the introduction of 3–7 wt% of expandable graphite (EG) [10] and 5–15 wt% ammonium polyphosphate (APP) [6] into 20 wt% natural fibre-reinforced polymer provided 0 s drip flame time from a vertical Bunsen burner test, reduced gross heat of combustion as obtained from bomb calorimeter and increased the mass residue from thermogravimetric analysis. However, the addition of these fillers was observed to reduce the mechanical performance of the FR-filled composites. This is because high concentrations of inorganic additive, especially compared to organic fillers, are necessary to impart fire resistivity into composites [14]. The high concentrations lead to the embrittlement and deterioration of the mechanical performance of composites.

Nano hydroxyapatite (nHA), $Ca_{10}(PO_4)_6(OH)_2$, is an inorganic bio-filler that has medical applications and is the main calcium phosphate phase present in bone. nHA does not present high strength, which makes it unfit for load-bearing applications [15]. Akindoyo et al. [15] showed that addition of 10 wt% nHA into PLA enhanced the content of mass residue from 0.35% to 6.17% at 750 °C from thermogravimetric analysis. This demonstrates higher thermal performance of nHA than PLA as inclusion of nHA can result in enhanced char residue.

To the best of the authors' knowledge, few literature studies have reported the improved thermal performance of PLA with the addition of nHA [5,16], and no investigation has highlighted the enhanced flame resistivity and char formation of natural fibre PLA laminates with the inclusion of nHA particles. The water absorption and mechanical behavior of natural fibre-reinforced PLA composites containing nHA additives have also not been reported thus far. The composites were made of fully green materials i.e., bio-based polymer, natural fibre and bio-filler to produce an environmentally friendly composite. In this research, 100% biodegradable composites were developed using various loadings of nHA and these nano-composites were compared with neat PLA and flax fibre-reinforced laminate. The aim of this study was to utilize the good thermal properties of nHA to develop flax fabric PLA composites with improved flame retardancy and thermal resistivity. The formulations produced were investigated based on flammability, thermal, mechanical and water absorption behaviors.

2. Materials and Methods

2.1. Materials

Unidirectional flax fabrics were purchased from Easy Composites Ltd (Stoke-on-Trent UK) and had the mass over area and density of 150 g/m^2 and 1.5 g/m^3, respectively. Nano-hydroxyapatite (nHA), $Ca_{10}(PO_4)_6(OH)_2 \geq 99.5\%$, was supplied by Yunduan New Materials, Weifang, China. The particle size was less than 40 nm, had the density of 3.16 g/cm^3 and had pH of 7.41. The mass loss after drying was 0.59% and mass loss after burning was 2.59%. The poly (lactic acid) (PLA, Ingeo^{TM} bio-based polymer PLA3251D) was provided by NatureWorks LLC (Minnetonka, MN, USA) and had the specific gravity of 1.24 g/cm^3.

2.2. Processing

The nanocomposites were compounded using an internal mixer. The PLA along with nHA was compounded at 190 °C for 20 min at a mixing speed of 200 rpm. The quantity of nHA was prepared at 20 wt%, 30 wt% and 40 wt% in PLA matrix. The compounding materials (PLA/nHA) and pure PLA were pressed using a hot press machine to produce films to be used as the matrix. The films with a thickness of roughly 0.3 mm were produced using an XLB50 (flat vulcanization press (YueQing TOPS machinery CO., Ltd, Shanghai, China) at a pressure of 3 MPa for the period of 10 min at the temperature of 190 °C. The PLA and PLA/nHA films were cold pressed for 2 min, immediately after removal from the hot press. It is worth noting that the PLA, nHA and films were kept in a convectional laboratory oven for 24 h at 50 °C before any stage of processing and the flax fabrics were kept in an oven at 80 °C prior to the hot compression.

Composites were produced by a conventional film-stacking method, which is constructed by adding alternating layers of matrix film and plant fabric. The cross-ply of [0/90/0/90/0/90] was utilized to produce the laminates. The prepared films and flax fabric were hot pressed into 280 × 160 × 2 mm^3 plates using the same hydraulic hot press set at a pressure of 5 bar for 10 min at 190 °C for consolidation, and subsequently the composite plates were cold pressed to room temperature for 2 min. The prepared composites were named CPLA (control), CN20, CN30 and CN40. CPLA was made of flax fabric and PLA, and the number in front of CN20, CN30 and CN40 (nano-composites) denotes the amount of nHA in the matrix. The CPLA and nano-composites contained 30% fibre volume fraction. Neat PLA was also fabricated for comparison purposes.

2.3. Flammability Test

The flame behaviours of the PLA, control and composite laminates were examined using the UL-94 test in accordance with ISO 1210 [17]. The samples were placed horizontally and the burn rate of each sample in the horizontal orientation was reported. The sample is classified as FH-1 if the combustion front does not pass the 25 mm mark. It is categorized as FH-2 if the combustion front passes the 25 mm mark, but does not pass the 100 mm mark. The burnt length is added to the classification designation. It is classified as FH-3 if combustion front passes the 100 mm mark and the rate of burning does not exceed 75 mm/min for samples measuring a thickness < 3 mm. If the rate of burning surpasses the mentioned value, it is categorized as FH-4. If the flame does not continue after the initial mark, which is 25 mm from the end of specimen, the burn rate is denoted "not applicable" (NA).

2.4. Thermogravimetric Analysis (TGA)

Thermogravimetric analysis (TGA) was conducted on a TA instrument (SDT Q600) (New Castle, DE, USA, 2016) to investigate the thermal stability of the composites. The samples of approximately 20 mg were placed in a platinum crucible and were heated from 30 °C to 600 °C in a nitrogen environment. The flow rate and ramping rate were set to 50 mL/min and 20 °C/min, respectively. The corresponding TGA and differential thermal gravimetry (DTG) were obtained

2.5. *Differential Scanning Calorimeter (DSC) Test*

The differential scanning calorimeter (DSC) test was conducted on a TA machine. (SDT Q20) (New Castle, DE, USA, 2016). Samples weighing 17–19 mg were heated from room temperature (25 °C) to 200 °C at a ramping rate of 10 °C/min with a nitrogen flow rate of 50 mL/min. The same heating rate and flow rate were used to cool down the samples to the room temperature. The crystallinity of PLA based matrix was obtained using the Equation (1) [18] below:

$$X_{DSC}\% = \frac{\Delta H_m - \Delta H_c}{\Delta H_m^{\circ}} \times \frac{100}{w} \quad (1)$$

where ΔH_m is the melting enthalpy, ΔH_c is the enthalpy of cold crystallization, ΔH_m° is 93.7 J/g for pure crystalline PLA and w is the mass fraction of the PLA-based matrix.

2.6. *Scanning Electron Microscopy (SEM)*

Scanning electron microscopy (SEM, ZEISS Sigma/VP SEM) (ZEISS IGMA/VF, Jena, Germany, 2012) was used to investigate the morphological structure of fractured surfaces of composites after tensile experiments with an acceleration voltage of 5 kV. A Leica EM SCD 500 high vacuum Sputter Coater (Lecia Microsystems, Prague, Czech Republic, 2012) was employed to gold coat the fracture surfaces with the plasma exposition of 60 s prior to scanning.

2.7. *Tensile Test*

The tensile test was carried out on dog bone specimens prepared according to EN ISO 527-4 [19], using an MTS EXCEED E45 universal tester (MTS Systems Corporation, Shanghai, China 2016). The specimens were conditioned at 50% relative humidity and 23 °C prior to the testing at the crosshead speed of 1 mm/min. The extensometers were used in the middle of gauge length. The average results of five specimens were recorded for the tensile strength, tensile modulus and elongation at break.

2.8. *Flexural Test*

The bending test was conducted according to EN ISO 14125 [20] on an MTS ECCEED E42 universal tester (MTS Systems Corporation, Shanghai, China) at a speed of 0.5 mm/min. The specimens measured 60 mm × 15 mm × 2 mm with the span length of 40 mm and were preconditioned the environmental method mentioned in the previous test. The flexural strength and modulus were recorded as an average of the five specimens.

2.9. *Water Absorption Test*

The water absorption test was conducted for all the formulations in accordance with EN ISO 62:2008 [21]. Prior to the test, the samples were conditioned at room temperature with 50% relative humidity after being dried in an oven for 72 h at 40 °C. The specimens were then soaked in two different containers filled with distilled water at the temperatures of 25 °C and 60 °C. The amount of moisture absorption was measured per 24 h for 14 days. m_0 is the mass of test sample after initial drying and before immersion, and m is the mass of test sample after immersion and final drying. The average mass gain of three test samples for each formulation was calculated and reported. The mass gain (c) is the percentage change in mass relative to the initial mass and is calculated using the Equation (2):

$$c = \frac{m - m_0}{m_0} \times 100\% \quad (2)$$

3. Results

3.1. Flammability of Poly(lactic Acid) (PLA), Control and Nano-Hydroxyapatite (nHA)-Filled Composites

The UL-94 results are displayed based on the rate of burning and rating in Table 1. The PLA sample burned continuously until its burn length reached 31 mm and sample drips fell on the chamber bed. For the Control, CN20 and CN30, the UL-94 rating was FH-3, indicating the complete combustion of these composites. The burn rates were recorded 20, 19.6 and 17.7 mm/min, respectively, and no dripping was observed. This demonstrates that increasing volumes of nHA enhanced the flame retardancy. Further addition of nHA (CN40) improved the fire resistivity significantly by providing self-extinguishing performance and no burn rate. This is due to the formation of a more thermally stable char in the respective sample, thereby acting as an effective barrier, shielding the underlying materials from the flame zone and heat. PLA undergoes thermal decomposition via a hydroxyl end-initiated ester interchange process and chain hemolysis creating lactide, oligomers, carbon dioxide, carbon monoxide and acetaldehyde [19], and fillers that can act as flame-retardants are capable of transforming its decomposition pathway to produce char and decrease the formation of combustible products [22,23].

Table 1. UL-94 results of poly(lactic acid) (PLA), Control and nano-hydroxyapatite (nHA) filled composite laminates.

Sample	Burn Rate (mm/min)	UL 94 Rating
PLA	NA	FH-2-31 mm
CPLA	20	FH-3
CN20	19.6	FH-3
CN30	17.7	FH-3
CN40	NA	FH-1

3.2. Thermal Studies

The thermogravimetric analysis (TGA) and differential thermogravimetric analysis (DTGA) curves of the composite samples are shown in Figures 1 and 2. The degradation temperatures, mass residue and mass loss rate of the samples were determined. The TGA curve displayed one decomposition step for all formulations and that the main decomposition took place between 300 °C and 400 °C. It is worth noting that bio-fibres consist of hemicellulose, lignin and cellulose, and their pyrolysis takes place at different temperature ranges of approximately 160–900 °C, 220–315 °C and 315–400 °C for lignin, hemicellulose and cellulose, respectively [24,25]. The temperatures at 5% mass loss ($T_{5\%}$) of PLA, CPLA and nanocomposites were approximately 328 °C, 304 °C and 290 °C, and the degradation temperatures (T_d), obtained from the peak of DTGA curves, were about 373 °C, 361 °C and 372 °C, respectively. This demonstrates that the $T_{5\%}$ and T_d of CPLA and nHA filled composites are lower than those of PLA, which can be explained by a higher destabilization of PLA in the composites. Amongst the composites, the T_d improved with the addition of nHA relative to that of the Control and revealed nearly the same value as neat PLA. It was reported that electrostatic attraction between the polymeric carboxylate group and CA^{2+} of nHA ions affect the interfacial bonding in composites containing nHA [26]. The maximum rates of thermal degradation (Figure 2) were measured 2.88, 1.94, 1.91, 1.7 and 1.67 %/°C for PLA, CPLA, CN20, CN30 and CN40, respectively, suggesting that both the natural fibre and nHA contributed to the reduction in the decomposition rate. The mass residues at 600 °C (Figure 1) were 1.85%, 8.14%, 16.97%, 22.70% and 30.84% for PLA, CPLA, CN20, CN30 and CN40, respectively. The inclusion of natural fibres was observed to improve the mass reside by about 340 % relative to that of neat PLA, and introduction of 40 wt% nHA into the matrix was found to increase the residue by 279% compared to that of CPLA. This enhancement in the increased char residue is in accordance with flammability test, which resulted in the formation of thermally resistive char on the surface of CN40 and subsequent protection of the bulk of the substrate.

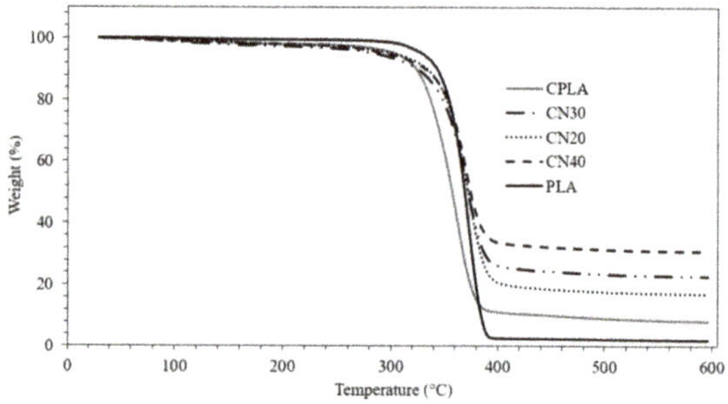

Figure 1. Thermogravimetric analysis (TGA) curves of PLA, control and nanocomposites.

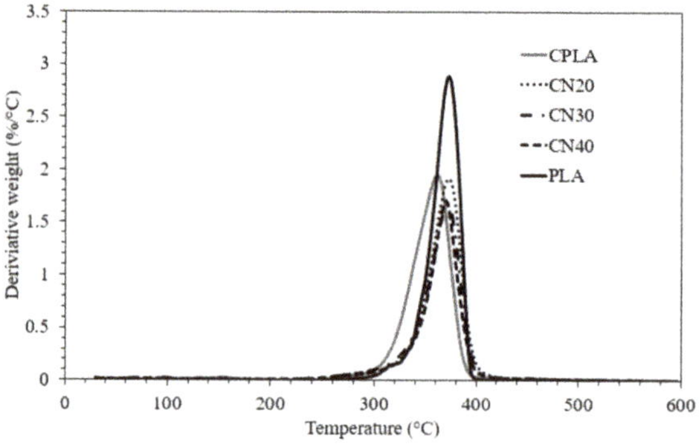

Figure 2. Differential thermogravimetric analysis (DTGA) curves of PLA, control and nanocomposites.

3.3. Crystallization and Melting Properties

The DSC thermograms of CPLA, PLA and CN40 from the heating run are displayed in Figure 3. The glass transition temperature (T_g), crystallization temperature (T_c), melting temperature (T_m), crystallization enthalpy (ΔH_c), melting enthalpy (ΔH_m) and degree of crystallinity (X_{DSC}) obtained are shown in Table 2. The addition of natural fibres was observed to show an increase in T_g compared to that of PLA. This observation displays that an increased T_g consequently indicates a change from flexible and soft behaviors to tough and hard properties [27]. The inclusion of nHA slightly decreased the T_g, indicating improved polymer chain mobility. The crystallinity and T_c of CPLA decreased by 12 °C after incorporation of fibres, which implies that the natural fibres hinder the diffusion and migration of molecular chains of PLA to the nucleus surface in CPLA. As expected, the addition of nHA was observed to further reduce the T_c by approximately 13%, signifying a faster crystallization of the nanocomposites. This can be ascribed to nHA acting as sites of nucleation, leading to heterogeneous nucleation within the PLA [28]. A similar increase in T_g (by approximately. 2 °C) and reduction in T_c (by about 14 °C) was obtained after the addition of talc fibre into the PLA [29]. For the last transition, the formation of a smaller second peak is due to the presence of NFs in the PLA which influenced the overall melting behavior of the composites. This suggests the presence of two different types of crystal [30]. Inducing heterogeneous nucleation due to the introduction of nHA can also contribute to the higher T_m. This is attributed to the creation of less perfect crystals, which would usually melt at greater temperatures than more perfect crystals [15].

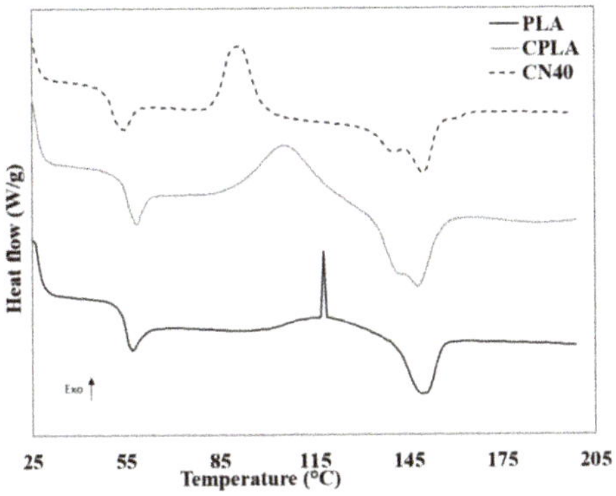

Figure 3. Differential scanning calorimeter (DSC) thermograms of PLA, control (CPLA) and CN40.

Table 2. The T_g, T_c, T_m, ΔH_c, ΔH_m and X_{DSC} obtained from DSC test.

Samples	T_g (°C)	T_c (°C)	T_m (°C)	ΔH_c (J/g)	ΔH_m (J/g)	X_{DSC} (%)
PLA	57	118	150	6.02	9.04	4.95
CPLA	59	106	148	17.47	18.41	1.54
CN40	56	92	151	13.83	15.04	1.98

3.4. Morphological Properties

The fracture surface of the control and composites containing nHA particles were scanned after tensile tests were performed, as illustrated in Figure 4. As observed from Figure 4a, no fibre pull-out or obvious evidence of poor interfacial bonding in the flax fibre-reinforced PLA can be seen. The red and yellow arrows highlight the flax fibres and PLA matrix in the composite, respectively. The impact of the inclusion of nHA fillers in the composite laminates (CN20, CN30 and CN40) can be observed in their breaking behavior, as shown in Figure 4b–d. A smooth fracture surface was observed in the matrix of CPLA, whereas a rough fracture surface was obtained in the matrix of the nanocomposites. Incorporation of 20 wt% nHA into the matrix was detected to be sufficient to provide even dispersion of particles in the PLA-based matrix, as depicted in the red circles Figure 4b. However, some fibre pull-out and fibre debonding were observed, indicating poor fibre/matrix interfacial adhesion. As the amount of filler increased in the matrix, clear agglomeration was found in the composite laminates ((Figure 4c,d), which is displayed in yellow circles. The agglomerated spots can be points where stress concentration occurs, which can cause premature failure of the composites. It has been stated [28] that well-distributed nHA filler in the PLA matrix can result in improved mechanical performances.

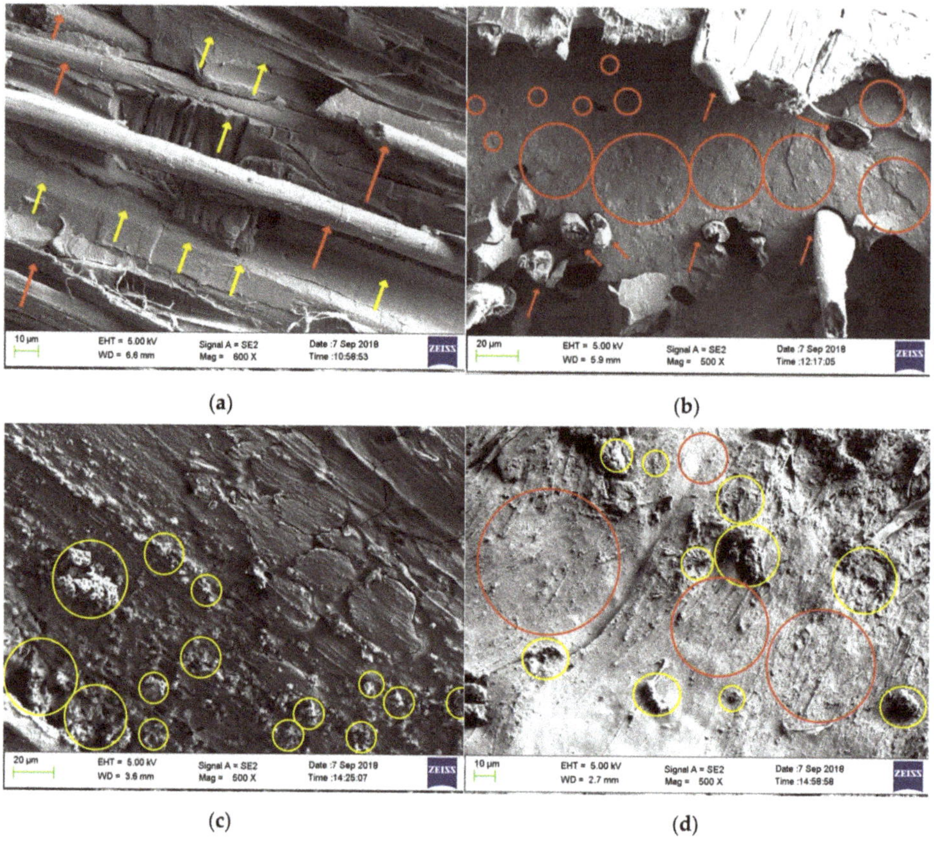

Figure 4. Scanning electron microscope (SEM) micrographs of the fracture surface of tensile tested samples at 500× magnification. (**a**) CPLA, (**b**) CN20, (**c**) CN30 and (**d**) CN40.

3.5. Tensile Properties

The tensile properties of the PLA, control and nHA loaded laminates were studied and the results are illustrated in Figure 5. The Young's moduli of the PLA and Control were 5.1 GPa and 12.6 GPa, the tensile strengths, 45 MPa and 50.7 MPa, and the elongation at break 4.5% and 2%, respectively. As expected, reinforcing the PLA with 6 layers of flax fabric enhanced the tensile strength and modulus by 13% and 146% compared to neat PLA, respectively, which is a significant improvement. This is because tensile performance is primarily fibre-dependent and flax fibre has greater tensile properties than those of neat PLA. Moreover, the observed enhancements in mechanical performance are affected by good interfacial adhesion between matrix and fibre, as demonstrated in Figure 4a, leading to better load transfer between fibre and matrix, and thereby improved mechanical performance. However, after the addition of fabrics, the elongation at break experienced a reduction. As compared to the control, the tensile properties decreased with the inclusion of nHA due to fibre pull-out and fibre debonding. The reduction was substantial upon the addition of 30 wt% and 40 wt% nHA in the matrix, which can be attributed to the low strength of nHA as revealed previously [26,31]. Its low strength can be associated with the elimination of organic compounds during synthesis [15]. Another reason could be the agglomeration of nHA in the matrix, as revealed by the SEM micrographs (Figure 4c,d), which triggers an early failure.

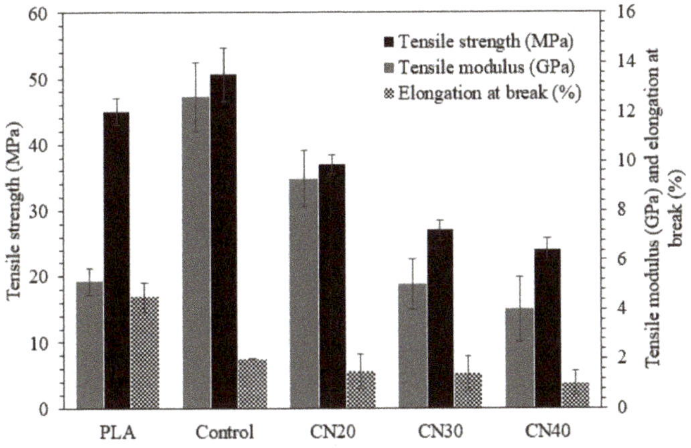

Figure 5. Tensile properties of PLA, control and nHA filled composites.

3.6. Flexural Properties

The flexural performances of the PLA, control and nHA loaded laminates were investigated, and the results are displayed in Figure 6. Neat PLA had a flexural strength and modulus of 57.5 MPa and 2.9 GPa, respectively, whereas the control showed a slightly lower bending strength of 54 MPa and a significantly higher modulus of 9.1 GPa than those of neat PLA, indicating 212% improvement in flexural modulus. As observed in Figure 6, both the flexural strength and the modulus reduced with the increased nHA as compared to those of the control. For instance, for the highest amount of nHA (CN40), the reduction in strength and modulus was 67% and 35%, respectively, relative to those of the control. As mentioned in Section 3.5, the drop can be related to the low strength of nHA, which occurred during its synthesis because of elimination of organic components. The second reason for these reductions is the issue of nHA dispersion in the PLA matrix. Poor dispersion results in agglomeration (Figure 4c,d) which affects the flexural properties as they are matrix dependent. Significant agglomeration can result in unwanted premature failure at the interface of PLA and nHA as well as ineffective load transfer.

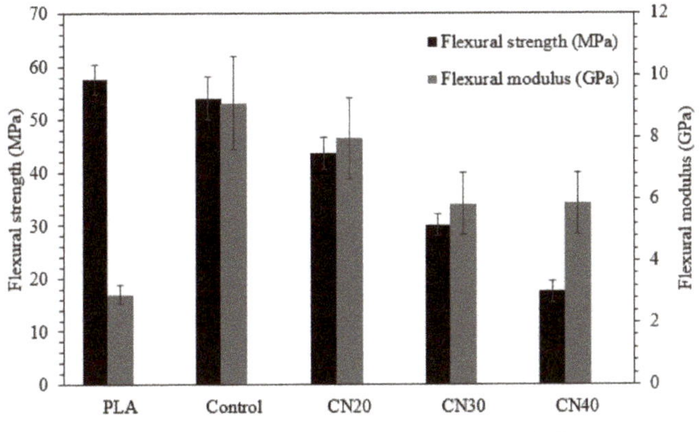

Figure 6. Flexural properties of PLA, control and nHA filled composites.

3.7. Water Absorption Behaviors

The mass gain as a function of the immersion time (day) at 25 °C and 60 °C for the produced composites is displayed in Figures 7 and 8. For PLA at room temperature (Figure 7), the mass gain occurred for two consecutive days and reached approximately 0.6% at the saturation point. The mass remained almost constant for the following days. With the addition of natural fabric into the PLA (CPLA composite), the mass gain increased up to 12 days resulting in a higher saturation value of 8.9%. This is attributed to the hydrophilic nature of flax fibre due to the polar groups e.g., carboxyl and hydroxyl groups. The results obtained are in agreement with another report [32], suggesting that hydrophobic PLA demonstrates a lower tendency to absorb water than lignocellulosic fabrics [33]. For CN20 and CN30, the mass gain continued for 10 days and reached 16.3% and 16.7%, respectively. The saturation time for CN40 was 6 days with the mass gain of 19.8%. This proves that addition of nHA fillers into the matrix increased the water absorption of flax fibre composites, indicating poor interfacial adhesion between the fibres and matrix due to the inclusion of nano fillers, as detected in SEM images (Figure 4), and demonstrated in mechanical analysis. This is because increasing the concentration of nano fillers reduces the amount of PLA in the composite. As a result, less matrix is available to adhere to the natural fibre, thereby exposing the fibres to the environment. Therefore, the natural fibres absorbed more water due to their lack of adhesion to the PLA/nHA matrix. In addition, nHA is a polar (hydrophilic) filler, which further increases the water absorption tendency.

The water absorption behavior of composites immersed in water at 60 °C is displayed in Figure 8. The PLA reached the peak of its mass gain with the value of ca. 0.8% and it became saturated rapidly after about one day. This shows higher amount of mass gain and shorter saturation time as compared to those of PLA at room temperature. For CPLA, the mass gain increased for around 11 days and its value was recorded at 10.4%. These results also show a higher mass gain and shorter saturation relative to those of CPLA soaked at room temperature. After 11 days, the mass of the samples started reducing slightly, which was due to the separation of tiny PLA pieces as a result of disintegration of the water-soluble materials [32]. The CN20, CN30 and CN40 samples saturated rapidly, after 4, 2 and 2 days, when compared to the saturation time obtained at room temperature, respectively. The mass gain was higher for the samples with greater concentration of nHA. These composites experienced a continuous mass loss after the peak of mass gain due to peeling and degradation of the nHA/PLA samples. This could be due to the poor interfacial adhesion between the nHA/PLA matrix and the flax fibre, thereby creating a higher surface area for degradation of the matrix material. It is worth highlighting that more cracks were visibly be observed on the surface of composites reinforced with flax fibre than neat PLA, and the cracks were more obvious for composites containing higher concentrations of nHA.

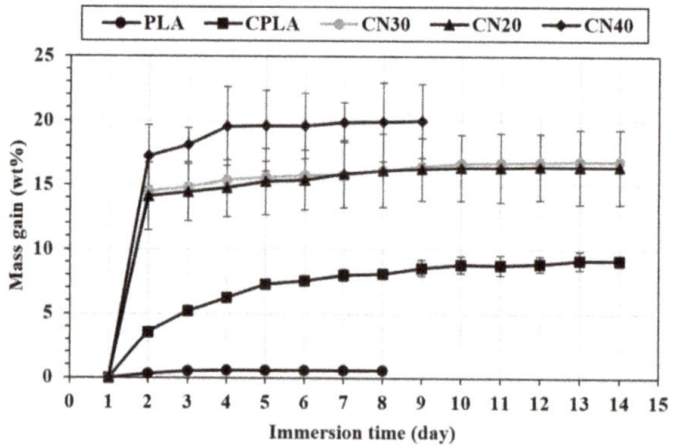

Figure 7. Water absorption behavior of PLA, CPLA and nanocomposites at 23 °C.

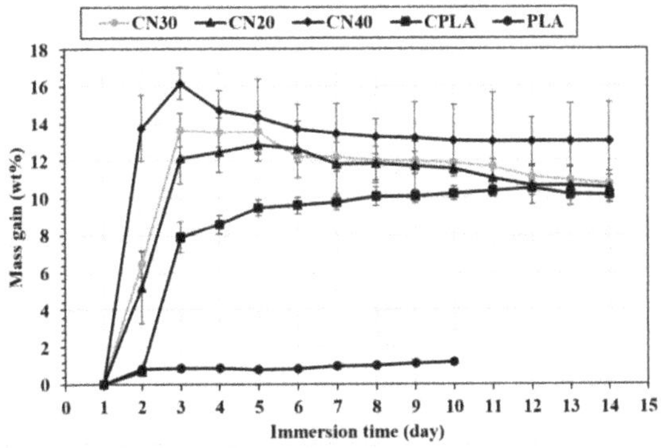

Figure 8. Water absorption behavior of PLA, CPLA and nanocomposites at 60 °C.

4. Conclusions

This research studied environmentally friendly biocomposites which were prepared using hot compression with the incorporation of fully bio-sourced constituents (i.e., bio-filler, NF and bio-thermoplastic). The study demonstrates the flame resistivity and thermal improvement of nHA as an additive by preventing flame development in the composite specimen (CN40) and reducing the burn rate as obtained via the UL-94 test study. It should also be noted that the mass residue was enhanced by 279%, thermal decomposition temperature was improved slightly, and the mass loss rate was reduced by approximately 14% upon the addition of the nanofillers determined by the initial TGA investigation. The efficiencies were more pronounced in terms of flame retardancy and thermal resistivity at higher concentrations of nHA particles. From the DSC analysis, it was found that the degree of crystallinity reduced in CPLA when compared to that of neat PLA, and with the presence of nHA additives in the composite, X_{DSC} experienced an increase as compared to its value of CPLA. The nHA particles, while imparting good thermal resistivity to the composite laminates, induced lower mechanical properties at higher concentrations. This reduction could be attributed to the agglomeration of the nHA particles within the matrix, in particular at higher concentrations, as well as increased fibre pull-out/debonding upon nHA inclusion, based on morphological observations. The water absorption increased with the addition of NF as well as nHA particles at both room temperature and 60 °C, which is ascribed to the poor adhesion at the interface between nHA/PLA matrix and flax fibres, as validated in the SEM micrograph, thereby enabling easier moisture absorption of NFs in the composites. This is because there is less matrix available to coat the natural fibres, which results in easy exposure to the environment. Furthermore, water absorption tendency increases in the presence of inorganic fillers such as nHA. At higher temperature, the saturation time was shortened remarkably for all the composites. In addition, the matrixof composite specimens started to disintegrate throughout the test due to the hydrolytic degradation. Therefore, there must be a compromise between the flame retardancy/thermal resistivity and mechanical/moisture absorption behavior for these types of biocomposites.

Author Contributions: Conceptualization, P.K.; Funding acquisition, X.L.; Investigation, P.K.; Methodology, P.K. and Z.Z.; Project administration, X.L.; Resources, X.L.; Supervision, P.K.; Validation, X.L. and B.B.; Writing original draft, P.K.; Review and editing, X.L. and B.B.

Funding: This work was done in "the ACC TECH-UNNC joint laboratory in Sustainable Composite Materials". The authors would like to acknowledge the financial support by Ningbo S&T bureau collaboration project (project code: 2017D10030) and the Ningbo 3315 Innovation Team, Scheme of "Marine Composites Development and Manufacturing for Sustainable Environment.

Conflicts of Interest: The authors declare no conflict of interest.

References

1. Hapuarachchi, T.D.; Peijs, T. Multiwalled carbon nanotubes and sepiolite nanoclays as flame retardants for polylactide and its natural fibre reinforced composites. *Compos. Part A Appl. Sci. Manuf.* **2010**, *41*, 954–963. [CrossRef]
2. Plackett, D.; Løgstrup Andersen, T.; Batsberg Pedersen, W.; Nielsen, L. Biodegradable composites based on l-polylactide and jute fibres. *Compos. Sci. Technol.* **2003**, *63*, 1287–1296. [CrossRef]
3. Barkoula, N.M.; Garkhail, S.K.; Peijs, T. Biodegradable composites based on flax/polyhydroxybutyrate and its copolymer with hydroxyvalerate. *Ind. Crops A Prod.* **2010**, *31*, 34–42. [CrossRef]
4. Gurunathan, T.; Mohanty, S.; Nayak, S.K. A review of the recent developments in biocomposites based on natural fibres and their application perspectives. *Compos. Part A Appl. Sci. Manuf.* **2015**, *77*, 1–25. [CrossRef]
5. Väisänen, T.; Das, O.; Tomppo, L. A review on new bio-based constituents for natural fiber-polymer composites. *J. Clean. Prod.* **2017**, *149*, 582–596. [CrossRef]
6. Khalili, P.; Tshai, K.; Hui, D.; Kong, I. Synergistic of ammonium polyphosphate and alumina trihydrate as fire retardants for natural fiber reinforced epoxy composite. *Compos. Part B Eng.* **2017**, *114*, 101–110. [CrossRef]
7. Shen, L.; Patel, M.K. Life Cycle Assessment of Polysaccharide Materials: A Review. *J. Polym. Environ.* **2008**, *16*, 154. [CrossRef]
8. Corbière-Nicollier, T.; Gfeller Laban, B.; Lundquist, L.; Leterrier, Y.; Månson, J.A.E.; Jolliet, O. Life cycle assessment of biofibres replacing glass fibres as reinforcement in plastics. *Res. Conserv. Recycl.* **2001**, *33*, 267–287. [CrossRef]
9. John, M.J.; Thomas, S. Biofibres and biocomposites. *Carbohydr. Polym.* **2008**, *71*, 343–364. [CrossRef]
10. Khalili, P.; Tshai, K.Y.; Kong, I. Natural fibre reinforced expandable graphite filled composites: Evaluation of the flame retardancy, thermal and mechanical performances. *Compos. Part A Appl. Sci. Manuf.* **2017**, *100*, 194–205. [CrossRef]
11. Rodríguez-Robledo, M.C.; González-Lozano, M.A.; Ponce-Peña, P.; Quintana Owen, P.; Aguilar-González, M.A.; Nieto-Castañeda, G.; Bazán-Mora, E.; López-Martínez, R.; Ramírez-Galicia, G.; Poisot, M. Cellulose-Silica Nanocomposites of High Reinforcing Content with Fungi Decay Resistance by One-Pot Synthesis. *Materials* **2018**, *11*, 575. [CrossRef]
12. Biswal, M.; Mohanty, S.; Nayak, S.K. Thermal stability and flammability of banana-fiber-reinforced polypropylene nanocomposites. *J. Appl. Polym. Sci.* **2012**, *125*, E432–E443. [CrossRef]
13. Subasinghe, A.; Das, R.; Bhattacharyya, D. Materials, N. Study of thermal, flammability and mechanical properties of intumescent flame retardant PP/kenaf nanocomposites. *Int. J. Smart Nano Mater.* **2016**, *7*, 202–220. [CrossRef]
14. Hapuarachchi, T.; Ren, G.; Fan, M.; Hogg, P.; Peijs, T. Fire retardancy of natural fibre reinforced sheet moulding compound. *Appl. Compos. Mater.* **2007**, *14*, 251–264. [CrossRef]
15. Akindoyo, J.O.; Beg, M.D.; Ghazali, S.; Heim, H.P.; Feldmann, M. Effects of surface modification on dispersion, mechanical, thermal and dynamic mechanical properties of injection molded PLA-hydroxyapatite composites. *Compos. Part A Appl. Sci. Manuf.* **2017**, *103*, 96–105. [CrossRef]
16. Akindoyo, J.O.; Beg, M.D.; Ghazali, S.; Heim, H.P.; Feldmann, M. Impact modified PLA-hydroxyapatite composites–Thermo-mechanical properties. *Compos. Part A Appl. Sci. Manuf.* **2018**, *107*, 326–333. [CrossRef]
17. ISO. Thermal Properties. In *Method 140A: Determination of the Burning Behaviour of Horizontal and Vertical Specimens in Contact with a Small-Flame Ignition Source*; International Organization for Standardization: Geneva, Switzerland, 1992; Vol. BS 2782-1: Method 140A: 1992 ISO 1210:1992.
18. Pilla, S.; Gong, S.; O'Neill, E.; Rowell, R.M.; Krzysik, A.M. Polylactide-pine wood flour composites. *Polym. Eng. Sci.* **2008**, *48*, 578–587. [CrossRef]
19. ISO. Plastics—Determination of Tensile Properties. In *Part 4: Test Conditions for Isotropic and Orthotropic Fibre-Reinforced Plastic Composites*; International Organization for Standardization: Geneva, Switzerland, 1997; Vol. BS EN ISO BS EN ISO 527-4:1997 BS 2782-3: Method 326F:1997.
20. *Fibre-Reinforced Plastic Composites—Determination of Flexural Properties*; International Organization for Standardization: Geneva, Switzerland, 2011; British Standard: 2011; Vol. BS EN ISO 14125:1998 +A1:2011.
21. ISO. *Plastics—Determination of Water Absorption*; International Organization for Standardization: Geneva, Switzerland, 2008; Vol. BS EN ISO 62:2008.

22. Bourbigot, S.; Fontaine, G. Flame retardancy of polylactide: An overview. *Polym. Chem.* **2010**, *1*, 1413–1422. [CrossRef]
23. Bocz, K.; Szolnoki, B.; Marosi, A.; Tábi, T.; Wladyka-Przybylak, M.; Marosi, G. Flax fibre reinforced PLA/TPS biocomposites flame retarded with multifunctional additive system. *Polym. Degrad. Stab.* **2014**, *106*, 63–73. [CrossRef]
24. Yang, H.; Yan, R.; Chen, H.; Lee, D.H.; Zheng, C. Characteristics of hemicellulose, cellulose and lignin pyrolysis. *Fuel* **2007**, *86*, 1781–1788. [CrossRef]
25. Alvarez, V.; Rodriguez, E.; Vázquez, A. Thermaldegradation and decomposition of jute/vinylester composites. *J. Therm. Anal. Calorim.* **2006**, *85*, 383–389. [CrossRef]
26. Cucuruz, A.T.; Andronescu, E.; Ficai, A.; Ilie, A.; Iordache, F. Synthesis and characterization of new composite materials based on poly(methacrylic acid) and hydroxyapatite with applications in dentistry. *Int. J. Pharm.* **2016**, *510*, 516–523. [CrossRef]
27. Tesoro, G. *Textbook of Polymer Science*, 3rd ed; Billmeyer, F.W., Jr., Ed.; Wiley-Interscience: New York, NY, USA, 1984; 578p, No price given. [CrossRef]
28. Liuyun, J.; Chengdong, X.; Lixin, J.; Dongliang, C.; Qing, L. Effect of n-HA content on the isothermal crystallization, morphology and mechanical property of n-HA/PLGA composites. *Mater. Res. Bull.* **2013**, *48*, 1233–1238. [CrossRef]
29. Huda, M.S.; Drzal, L.T.; Mohanty, A.K.; Misra, M. The effect of silane treated- and untreated-talc on the mechanical and physico-mechanical properties of poly(lactic acid)/newspaper fibers/talc hybrid composites. *Compos. Part B Eng.* **2007**, *38*, 367–379. [CrossRef]
30. Aydın, M.; Tozlu, H.; Kemaloglu, S.; Aytac, A.; Ozkoc, G. Effects of Alkali Treatment on the Properties of Short Flax Fiber–Poly(Lactic Acid) Eco-Composites. *J. Polym. Environ.* **2011**, *19*, 11–17. [CrossRef]
31. Wang, X.; Li, Y.; Wei, J.; de Groot, K. Development of biomimetic nano-hydroxyapatite/poly(hexamethylene adipamide) composites. *Biomaterials* **2002**, *23*, 4787–4791. [CrossRef]
32. Baghaei, B.; Skrifvars, M.; Salehi, M.; Bashir, T.; Rissanen, M.; Nousiainen, P. Novel aligned hemp fibre reinforcement for structural biocomposites: Porosity, water absorption, mechanical performances and viscoelastic behaviour. *Compos. Part A Appl. Sci. Manuf.* **2014**, *61*, 1–12. [CrossRef]
33. Mofokeng, J.P.; Luyt, A.S.; Tábi, T.; Kovács, J. Comparison of injection moulded, natural fibre-reinforced composites with PP and PLA as matrices. *J. Thermoplast. Compos. Mater.* **2011**, *25*, 927–948. [CrossRef]

materials

MDPI

Article

Influence of Process Parameters in Graphene Oxide Obtention on the Properties of Mechanically Strong Alginate Nanocomposites

Izaskun Larraza [1], Lorena Ugarte [2], Aintzane Fayanas [1], Nagore Gabilondo [1], Aitor Arbelaiz [1], Maria Angeles Corcuera [1,*] and Arantxa Eceiza [1,*]

1 'Materials+Technologies' Research Group (GMT) and Department of Chemical and Environmental Engineering, Faculty of Engineering, Gipuzkoa, University of the Basque Country (UPV/EHU), Pza Europa 1, 20018 Donostia-San Sebastian, Gipuzkoa, Spain; izaskun.larraza@ehu.eus (I.L.); aintzane.fa@gmail.com (A.F.); nagore.gabilondo@ehu.eus (N.G.); aitor.arbelaiz@ehu.eus (A.A.)

2 'Materials+Technologies' Research Group and Department of Engineering Design and Project Management, Faculty of Engineering, Gipuzkoa-Eibar Section, University of the Basque Country (UPV/EHU), Otaola Hiribidea 29, 20600 Eibar, Gipuzkoa, Spain; lorena.ugarte@ehu.eus

* Correspondence: marian.corcuera@ehu.eus (M.A.C.); arantxa.eceiza@ehu.eus (A.E.)

Received: 23 December 2019; Accepted: 26 February 2020; Published: 28 February 2020

Abstract: Sodium alginate, a biopolymer extracted from brown algae, has shown great potential for many applications, mainly due to its remarkable biocompatibility and biodegradability. To broaden its fields of applications and improve material characteristics, the use of nanoreinforcements to prepare nanocomposites with enhanced properties, such as carbonaceous structures which could improve thermal and mechanical behavior and confer new functionalities, is being studied. In this work, graphene oxide was obtained from graphite by using modified Hummers' method and exfoliation was assisted by sonication and centrifugation, and it was later used to prepare sodium alginate/graphene oxide nanocomposites. The effect that different variables, during preparation of graphene oxide, have on the final properties has been studied. Longer oxidation times showed higher degrees of oxidation and thus larger amount of oxygen-containing groups in the structure, whereas longer sonication times and higher centrifugation rates showed more exfoliated graphene sheets with lower sizes. The addition of graphene oxide to a biopolymeric matrix was also studied, considering the effect of processing and content of reinforcement on the material. Materials with reinforcement size-dependent properties were observed, showing nanocomposites with large flake sizes, better thermal stability, and more enhanced mechanical properties, reaching an improvement of 65.3% and 83.3% for tensile strength and Young's modulus, respectively, for a composite containing 8 wt % of graphene oxide.

Keywords: graphene oxide; size selection; sodium alginate; mechanical properties; thermal stability

1. Introduction

In recent years, there is a growing interest in renewably sourced materials in the polymeric materials field, due to the fluctuations in oil price and the increase in environmental concern. Biopolymers, extracted from renewable sources, can be classified into three main groups: extracted from biomass, such as polysaccharides and proteins; synthesized from bioderived monomers, such as polylactic acid; and those produced by organisms or bacteria, such as bacterial cellulose [1,2]. Sodium alginate is an algae-based anionic and hydrophilic linear polysaccharide, characterized by its excellent biocompatibility, biodegradability, non-toxicity, and low cost [3,4]. Sodium alginate is a salt of alginic acid and is composed of (1→4)-β-D-mannuronic acid (M) and (1→4)-α-L-guluronic acid (G) units in the form of homopolymeric (MM or GG blocks) and heteropolymeric sequences (MG or GM blocks) depending on the source. It is extracted from brown algae (*Macrocystis pyrifera*) and can also be synthesized from microorganisms [5,6]. Alginate M and G units contain lateral hydroxyl and carboxyl groups that serve as reactive sites for several modifications [7]. Alginate is most commonly used in a water-soluble sodium alginate powder form, allowing physical cross-linking usually by the addition of divalent cations, such as Ca^{2+}, Ba^{2+}, and Zn^{2+} [7]. These divalent cations form a chelated structure, according to the commonly called egg-box model, preferentially with the GG blocks, thereby creating a stable three-dimensional network [8]. Alginate has potential in applications such as biomedicine, pharmaceutics, food industry, textiles, and additive manufacturing [9].

When compared to commercial synthetic polymers, the main drawbacks of alginate and biopolymers, in general, are their inferior thermal and mechanical properties [10], as well as their strong hydrophilic character. Currently, much effort has been made to improve the performance of alginates. In this sense, the addition of inorganic nanoreinforcements has been demonstrated to be an effective strategy to overcome the disadvantages of alginate [11–13], as well as to provide new functionalities. Gholizadeh et al. [14] incorporated different contents of hydroxyapatite nanoparticles to alginate and obtained an improvement in physical and mechanical properties, as well as antimicrobial activity, compared to alginate [14]. The addition of CuO nanoparticles to alginate improved the thermal properties and also conferred antifungal activity, which is valuable in biomedical applications [15]. The incorporation of silicon dioxide to alginate further reduced the water vapor permeability and swelling degree and significantly increased the mechanical properties, important parameters for packaging industries [16].

Similarly, when carbonaceous structures, such as graphene and graphene derivatives, are incorporated to alginate, not only do the mechanical and thermal behaviors improve and the hydrophilicity decreases, but nanocomposites with electrical conductivity and antibacterial properties are also achieved. Graphene is a 2D allotropic variety of carbon, constituted by a one-atom-thick layer of sp2-hybridized carbon atoms. It has been reported that a single-layered graphene exhibits a Young's modulus of ~1100 GPa, a tensile strength of 130 GPa, and electrical and thermal conductivities up to 6000 S/cm^{-1} and 5000 $W\ m^{-1}\ K^{-1}$, respectively [17,18]. Graphene has the disadvantage of often being difficult to disperse in some biopolymeric matrices. Graphene derivatives, such as graphene oxide (GO), are more compatible due to the presence of hydroxyl, carboxylic, and epoxy groups on their surface and are consequently easier to disperse. GO is usually obtained by graphite oxide exfoliation through strong sonication treatments. Graphite oxide has been commonly synthesized by means of oxidative treatments and three main methods have been reported in the literature: Brodie's method [19], Staudenmaier's method [20], and Hummers' method [21]. Among all of them, Hummers' method and its modified method, are the most used since they are less toxic. Moreover, due to the hydrophilic character and to the presence of oxygenated groups at the surface of graphene oxide, it can be easily exfoliated and dispersed in aqueous media, being it compatible with most of aqueous soluble or dispersible polymers, this also has the benefit of being an environmentally. Liang et al. added 0.7 wt % GO to a poly(vinyl alcohol) matrix and observed an improvement of 76% and 62% in tensile strength and Young's modulus, respectively [22]. Si et al. [23] biosynthesized a bacterial cellulose/graphene oxide nanocomposite (BC/GO) by adding GO to a culture medium. The BC/GO nanocomposite containing 0.48 wt % of GO showed an increase of 38% and 120% in tensile strength and Young's modulus, respectively [23]. GO has also been incorporated into chitosan to prepare membranes for the removal of heavy metals and for evaluating their effect with respect to traditional chitosan and toxic glutaraldehyde membranes. An improvement in heavy metal adsorption has been observed with the addition of 5 wt % GO [24].

It has been observed that the properties of GO-containing nanocomposites are strongly dependent on the physical-chemical properties and morphology of GO, which are mainly influenced by the oxidation treatment and exfoliation degree. Long oxidation times lead to high contents of oxygenated groups and to high exfoliation degrees, which result in GO flakes with small size (several hundred nanometers) and thickness usually formed by a low number of layers (less than 2 nm) [25–27]. Thus, the interfacial interactions between GO and the matrix and hence, the final properties of the nanocomposite, will be influenced by the characteristics of GO and therefore, its preparation method.

There are many papers in the literature on the effect of graphene-obtaining treatment on the final properties of graphene, and quite a number of studies have assessed the effect of graphene content on the final properties of nanocomposites; however, not many studies are known, in which the effect of different types of graphene, with different characteristics depending on the treatment used to obtain them, on the final properties of nanocomposites is studied, and even less in the case of nanocomposites in which the matrix is a biopolymer. May et al. [28] concluded that the size of graphene flakes plays an important role in reinforcing polymers. They observed that when using large flakes and maintaining the thickness constant, the elastic modulus and tensile strength values of polyvinyl alcohol-graphene nanocomposites were significantly higher than that obtained by using smaller graphene flakes. Nawaz et al. [29] also observed that large flake sizes enhanced the elastic modulus and tensile strength, while elongation at break values diminished in polyacrylonitrile-graphene nanocomposites. Moreover, Szparaga et al. [30] observed a clear correlation between composite mechanical behavior and altered crystallinity in the structure, where mechanical properties of calcium alginate/GO composites improved with changes in crystallinity and average crystal area.

Therefore, this work focuses on obtaining graphene oxide with different characteristics for its subsequent incorporation (with varying content) to an alginate polymer matrix. In this sense, an extensive study of how oxidative treatment and graphene oxide isolation, assisted by centrifugation and sonication, affect its physical-chemical characteristics and morphology has been carried out. The effect of different sizes as well as thicknesses of graphene oxide flakes, determined by morphological studies, on the mechanical properties of nanocomposites has been analyzed.

2. Experimental Section

2.1. Materials and Methods

Graphite flakes were purchased from Aldrich. Sulfuric acid (H_2SO_4, 96%), sodium nitrate ($NaNO_3$, 99%), potassium permanganate ($KMnO_4$, 99%), hydrogen peroxide (H_2O_2, 30% *w/v*) and hydrochloric acid (HCl, 37%) were supplied by Panreac (Barcelona, Spain). Medium viscosity alginic acid sodium salt from brown algae (SA powder, 4 Pa.s and $M_v = 2.4 \times 10^5$ g mol^{-1}, determined by viscosity measurements) was purchased from Sigma-Aldrich (St. Louis, MO, USA).

2.1.1. Oxidation of Graphite

The oxidation process of graphite was carried out according to modified Hummers' method [21]. Graphite flakes (1 g) were mixed with 0.5 g $NaNO_3$ and 23 mL H_2SO_4 in an iced cooled bath at 0 °C for 30 min under continuous magnetic agitation. Then, 3 g $KMnO_4$ was added to obtain a green-colored mixture. The mixture was kept at 0 °C for 2 h under magnetic agitation, until a purple color was achieved. Straightaway, it was heated at 35 °C in an oil bath for 30 min. Then, 46 mL of deionized water was slowly added. By this addition, the temperature of the mixture reached 98 °C. The mixture was kept at 98 °C for different times, 15 min and 30 min, and the samples thus obtained were designated as GO15 and GO30, respectively. The bath heater was shut down and 10 mL of H_2O_2 was added. The mixture was kept in the oil bath until the formation of bubbles stopped and it reached room temperature. Once at room temperature, 150 mL of deionized water was added. The resulting supernatant was discarded and a yellow-like mixture was obtained. The mixture was washed by centrifugation, using HCl (5 wt %) at 4500 rpm for 20 min. This step was repeated 5 times. Then, the same centrifugation procedure was repeated using deionized water until neutral pH was achieved. Finally, the mixture was filtered through polyamide filters (0.2 µm pore size, Sartorius, Göttingen, Germany) and was vacuum-dried at 50 °C for 48 h. Graphite oxide films, GO15 and GO30, were obtained.

2.1.2. Exfoliation and Size Selection of Graphene Oxide

The obtained graphite oxide was exfoliated in water (0.5 mg mL^{-1}), assisted by ultrasonication (Vibracell 75043, Sonics & Materials, Newton, MA, USA 30% amplitude). Ultrasonication times of 3 h and 4 h were applied for samples oxidized for 30 min, and these fractions were designated as GO30S and GO30L, respectively. The graphene oxide flakes, thus obtained, were size selected by centrifugation. Firstly, they were centrifuged at 4000 and/or 3000 rpm for 30 min and the supernatant fractions (ca. 80%) were collected. The fractions were named according to the applied ultrasonication and centrifugation treatments as GO30L-4000, GO30L-3000, and GO30S-4000. The remaining sediments of GO30L-4000 and GO30L-3000 fractions were collected and redispersed in water for 15 min using an ultrasonic bath. Centrifugation and redispersion steps were repeated for 2000 and 1000 rpm centrifugation rates to obtain size-selected graphene oxide flakes [31], denoted as GO30L-2000 and GO30L-1000, respectively.

The supernatant fractions were filtered through polyamide filters (Sartorius, 0.2 μm pore size) and dried for 48 h at room temperature. The designation of GO fractions and the applied oxidation and exfoliation procedures are summarized in Table 1.

Table 1. Designation of graphene oxide (GO) fractions.

Sample	Oxidation Time (min)	Ultrasonication Time (h)	Centrifugation Rate (rpm)
GO15	15	-	-
GO30	30	-	-
GO30L	30	4	-
GO30S	30	3	-
GO30L-4000	30	4	4000
GO30L-3000	30	4	3000
GO30L-2000	30	4	2000
GO30L-1000	30	4	1000
GO30S-4000	30	3	4000

2.1.3. Preparation of Nanocomposites

First, dispersions of GO30L-4000 and GO30S-4000 fractions in water (5 mg mL^{-1}) were prepared by ultrasonication for 1 h. Afterwards, sodium alginate/GO films were obtained. For this, sodium alginate in water at 2 wt % was used to incorporate different volumes of the GO dispersions previously prepared, so that the amount of GO was 1, 4, 6, and 8 wt % in the nanocomposites. The films were obtained by solvent casting and evaporating the water for 7 days at room temperature, and the samples were designated as SA-GO30L-4000-x% and SA-GO30S-4000-x%, where x is the percentage of GO in the nanocomposites. The samples were stored in a desiccator until their characterization.

2.2. *Characterization Techniques*

2.2.1. Fourier Transform Infrared Spectroscopy

Fourier transform infrared (FTIR) spectroscopy was used to identify the characteristic functional groups of graphite, graphite oxide, and nanocomposites. Measurements were performed with a Nicolet Nexus FTIR spectrometer (Thermofisher Scientific, Waltham, MA, USA). For carbonaceous nanostructure characterization, KBr pellets (0.0025 mg sample g^{-1} KBr) were employed for the analysis. Single-beam spectra of the samples were obtained after averaging 32 scans in the range of 4000 to 400 cm^{-1}, with a resolution of 2 cm^{-1}. For composites, an MKII Golden Gate accessory (Specac) with a diamond crystal at a nominal incidence angle of 45° and ZnSe lens were used. Spectra were recorded in attenuated total reflection (ATR) mode between 4000 and 650 cm^{-1} with a resolution of 4 cm^{-1} and 32 scans.

2.2.2. Ultraviolet-Visible Spectrophotometry

The absorbance of graphite and graphene oxide was measured by ultraviolet-visible spectrophotometry (UV-Vis), using open-top quartz cells. For sample preparation, low concentration dispersions were prepared (0.5 g sample mL^{-1} solvent) by using ethanol and deionized water for graphite and graphene oxide, respectively. The spectra were obtained in a UV-3600 UV-VIS-NIR spectrophotometer (Shidmazu, Kioto, Japan) in the wavelength range of 200 to 600 nm.

2.2.3. Raman Spectroscopy

Raman spectra of graphite and graphene oxide were obtained using a Reninshaw InVia (Renishaw, Wotton-under-Edge, UK) spectrometer, coupled to a Leica DMLM microscope (50x), with a laser of 514 nm wavelength (ModuLaser) operating at 5% of nominal potency. Data were collected in the range of 150 to 3200 cm^{-1}. Values of exposure time and accumulations were set at 20 s and 5 respectively.

2.2.4. X-Ray Diffraction

X-Ray diffraction (XRD) analyses were performed in a Philips X'Per PRO (Malvern Panalytical, Malvern, UK) automatic diffractometer, operating at 40 kV and 40 mA in theta-theta configuration. A secondary monochromator with radiation Cu-Kα (λ = 0, 154 nm) and the solid state detector PIXCEL (active length in 2θ: 3.347°) were used. Data were collected in the 2θ range of 5° to 70° (step size: 0.026 and time between steps: 60 s) in continuous mode.

The interplanar distance in the different samples was analyzed according to Bragg's law [32,33]:

$$n\lambda = 2d\ sin\theta \quad (1)$$

where n is a natural number between 1 and ∞ (in this case, n = 1), λ is the wavelength of the X-rays used in the analysis (in this case, λ = 0.154 nm), d is the interplanar distance in crystal structure, and θ is the angle between the incident rays and the dispersion planes.

2.2.5. Atomic Force Microscopy

The morphology of graphene oxide flakes was analyzed by atomic force microscopy (AFM). Height images were obtained in a Dimension Icon (Bruker, Billerica, MA, USA). scanning probe microscope equipped with a Nanoscope V controller (Bruker). Tapping mode was employed in air, using an integrated tip/cantilever (125 µm length with ca. 300 kHz resonant frequency).

For sample preparation, GO fractions were dispersed in water (0.1 mg mL^{-1}) using an ultrasonic tip for 1 h. A droplet of graphene oxide dispersion was put on a prewashed silicon wafer substrate and water was eliminated by spin coating at 1200 rpm for 120 s. Prior to analysis, samples were kept at room temperature for 48 h.

2.2.6. Thermogravimetric Analysis

Thermal degradation of graphene oxide flakes and nanocomposites was assessed by thermogravimetric analysis (TGA) performed in a TGA/STDA 851 (Mettler Toledo, Columbus, OH, USA) equipment. Samples of around 5 mg were heated from room temperature to 700 °C at a heating rate of 10 °C min^{-1} under nitrogen atmosphere.

2.2.7. Mechanical Properties

Young modulus, tensile strength, and elongation at break of nanocomposites were analyzed in an Instron 5697 equipment (Instron, Norwood, MA, USA) using a load cell of 500 N in tensile mode. Tests were carried out at a crosshead speed of 2 mm min^{-1} and with an initial grip separation of 10 mm. Rectangular samples of 70 mm × 5 mm × 0.05 mm (length × width × thickness) were used. Tensile modulus (E), stress at yield (σ_y), stress at break (σ_b), and elongation at break (ε_b) were determined from stress-strain curves and averaged from five specimens.

3. Results and Discussion

3.1. Oxidation Process

In order to study the different types of functional groups formed in the oxidation process of graphite as well as the effects of the different oxidation times, FTIR analysis was used. The FTIR spectra of graphite and GO are shown in Figure 1.

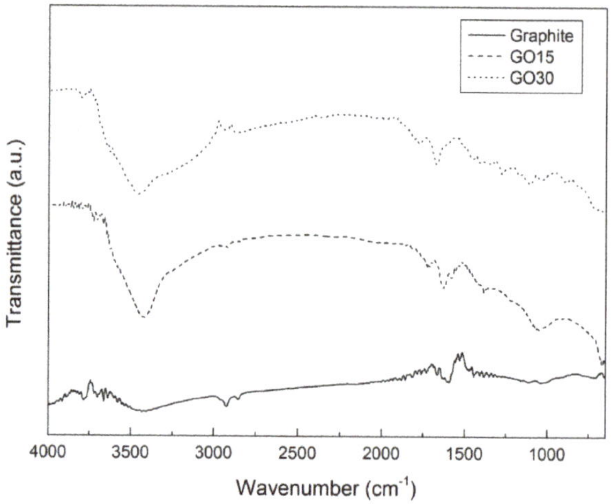

Figure 1. Fourier transform infrared (FTIR) spectra of graphite, GO15, and GO30 samples (y-axis of the curves were translated in order to avoid overlapping and to improve the visibility of the characteristic bands).

The peak corresponding to the stretching vibration of C=C bonds was clearly observed at around 1590 cm^{-1} [34] for graphite, while this peak was not so clear and could be overlapped in the case of oxidized graphite. Moreover, GO15 and GO30 samples showed additional peaks. The pronounced peak at around 3400 cm^{-1} was assigned to the stretching vibration of OH groups, derived from hydroxide and carboxylic acid groups, as well as from some moisture traces [35]. Moreover, a peak attributed to O–H bending can be seen at 1630 cm^{-1}. The peak at 1735 cm^{-1} was assigned to the stretching vibration of C=O bonds in carboxylic acid [36]. Finally, a peak was also observed at 1050 cm^{-1} in graphite oxide samples, which was related to the stretching vibration of C–O–C [25]. These results confirmed the presence of oxygen-containing functional groups in graphite oxide, indicating that the oxidation process was carried out satisfactorily. No significant differences were observed for different oxidation grades of GO15 and GO30 samples.

The differences in oxidation grades of graphite oxide samples were analyzed by UV-Vis spectroscopy. The obtained spectra are shown in Figure 2.

The graphite sample did not show remarkable peaks. The spectra of graphite oxide samples showed two absorption maximums at around 230 nm and 300 nm. The peak around 230 nm was attributed to the π–π^* transitions of aromatic C–C bonds, while the peak at 300 nm corresponded to the n–π^* transitions of carbonyl (C=O) groups [25,37] and both can be bathochromically shifted by conjugation [38]. Both peaks were characteristic of graphite oxide, indicating that the oxidative process was effective, in good agreement with the FTIR analysis.

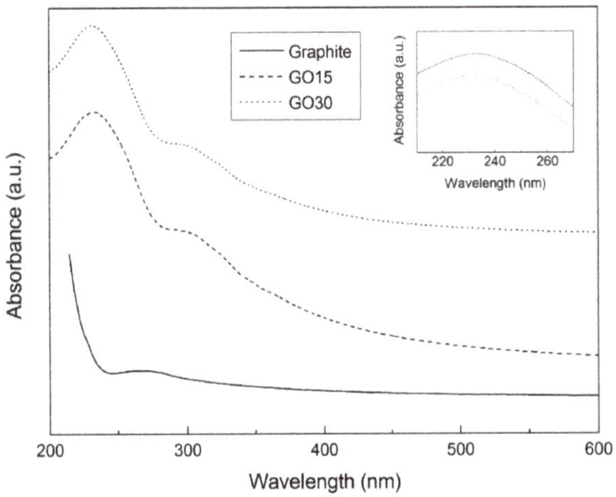

Figure 2. Ultraviolet-visible (UV-Vis) spectra of graphite, GO15, and GO30 samples.

The absorption peak of π–π^* (C–C) transitions was studied in more detail (Figure 2, inset). It was observed that the peak appeared at 233 nm (wavenumber) in sample GO15, while it appeared at 231 nm in sample GO30. This shift in the wavenumber suggested that higher oxidation grades (GO30) resulted in a higher disruption of the structure of sp^2 domain, thereby reducing the concentration of π electrons. As a consequence, more energy is needed for π–π^* transitions [25,39].

Results of Raman spectroscopy analyses of graphite and graphite oxide are shown in Figure 3. All spectra showed typical G, D, and 2D bands associated with carbon materials [39]. The G band is assigned to the in-plane vibration mode due to the bond stretching of sp^2 carbon pairs and the 2D band is related to the second order of zone-boundary phonons [39,40]. The D band is associated with flake edges since it needs a defect for activation [41,42]. In the case of graphite (Figure 3a), G, 2D, and D bands were observed at 1570, 2700, and 1354 cm^{-1}, respectively.

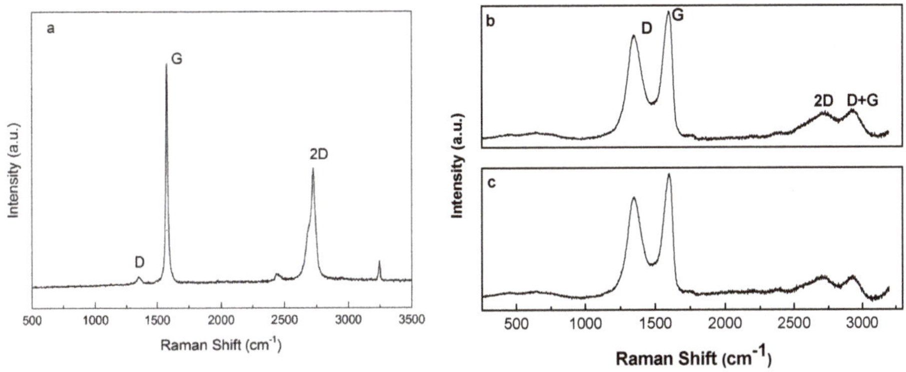

Figure 3. Raman spectra of (**a**) graphite, (**b**) GO15, and (**c**) GO30 samples.

In graphite oxide samples (Figure 3b), a shift to higher wavenumbers was observed for G peak compared to graphite. The maximum of the peak was observed at 1596 cm^{-1} and 1600 cm^{-1} wavenumber values for GO15 and GO30 samples, respectively. This shift to higher wavenumbers suggested a reduction of the in-plane sp^2 domains as a result of the oxidation of graphite [43]. In the same fashion, an increase in the wavenumber of D band was observed in the GO30 sample (1348 cm^{-1} and 1351 cm^{-1} for GO15 and GO30, respectively). This may indicate the presence of more defects and disorders caused by hetero-atoms, grain boundaries, aliphatic chains, etc. as a consequence of stronger oxidation [35]. As shown in spectra b and c of Figure 3, beside the 2D band, a band located around 2920 cm^{-1}, denoted as D + G band and related to defects, was also noted in the spectra of GO. For the 2D band, a slight decrease in intensity was observed in the GO30 sample, which can be explained by the breaking of the stacking order of graphene sheets along the z-axis due to oxidation [44]. These results confirmed stronger oxidation of the GO30 sample.

The relative intensity of D and G bands can be taken as indicative of crystallite size, according to the equation proposed by Cancado et al. [45]:

$$\mathbf{L_a} = \left[\left(2,4 * 10^{-10}\right)(\lambda_1)^4\right] / \left[\mathbf{I}_{(\mathbf{D})} / \mathbf{I}_{(\mathbf{G})}\right] \tag{2}$$

where L_a is the average size of sp^2 domain crystals, λ_1 is the input laser energy, I_D is the intensity of D band, and I_G is the intensity of G band.

The I_D/I_G ratios and L_a values obtained for graphite, GO15, and GO30 samples are shown in Table 2. It was observed that, as the oxidation degree increased, the I_D/I_G ratio increased, while L_a values decreased. This indicated that higher oxidation degrees resulted in smaller crystallites, the formation of defects, sp^3 hybridizations, and changes in crystallinity [35,46]. In general, Raman results suggested that the structure of graphite was modified by oxidation. Variations due to different oxidation degrees were also observed, in good agreement with the UV-Vis analysis.

Table 2. I_D/I_G ratios and L_a values for graphite, GO15, and GO30 samples.

Sample	I_D/I_G Ratio	L_a (nm)
Graphite	0.063	264.4
GO15	0.81	20.7
GO30	0.85	19.8

The effect of different oxidation times on the interplanar distance of graphite was analyzed by XRD analysis. XRD patterns of graphite, GO15, and GO30 samples are shown in Figure 4. The XRD pattern of graphite structure showed a pronounced peak at 2θ = 26.63°, corresponding to the (002) plane of graphite [46] and taken as indicative of pure graphite [37]. This peak, although with lower intensity, also appeared in GO15 and GO30 samples, suggesting that total oxidation was not achieved. Furthermore, in oxidized samples, a new peak was observed at 2θ = 10.63° for GO15 and at 2θ = 10.53° for GO30, associated with a higher interlayer spacing, owing to the formation of more oxygen-containing functional groups on GO.

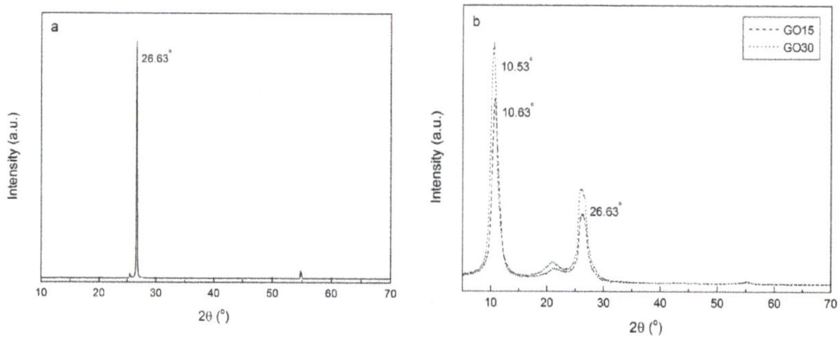

Figure 4. XRD patterns obtained for (**a**) graphite and (**b**) GO15 and GO30 oxidized samples.

The distance between planes for the three systems was calculated using Equation (1) and the values obtained from the diffractograms (θ = 13.315°, θ = 5.315°, and θ = 5.265° for graphite, GO15, and GO30, respectively). Results are shown in Table 3. An increase in the interplanar distance was observed when comparing pure graphite with oxidized samples. These results suggested that the interplanar distance increased as a consequence of the insertion of functional groups containing oxygen and water molecules between the graphene oxide layers [34,37].

Table 3. Interplanar distance (*d*) values calculated for graphite, GO15, and GO30.

Sample	2θ (°)	*d* (nm)
Graphite	26.63	0.335
GO15	10.63	0.828
GO30	10.53	0.837

According to the results, it was concluded that the oxidation process was carried out satisfactorily and that the GO30 sample presented a higher degree of oxidation. Graphite residues were observed in the XRD patterns of GO15 and GO30 samples, indicating that a fraction of graphite was not oxidized.

In view of these results, the sample GO30 was subjected to an exfoliation and posterior centrifugation process to eliminate the residual graphite fraction and obtain small-thickness graphene oxide flakes, according to the previously described procedure. The GO30 sample was selected due to the higher content of oxygen-containing hydrophilic groups that make graphite oxide easier to exfoliate in a polar medium.

3.2. *Exfoliation and Size Selection of Graphene Oxide*

To analyze the effect of sonication times (L: 4 h and S: 3 h) and centrifugation rates (4000, 2000, and 1000 rpm) on the characteristics of GO30 flakes, GO30L-4000, GO30L-2000, GO30L-1000, and GO30S-4000 samples were analyzed by XRD analysis and AFM.

XRD spectra of GO30 samples are shown in Figure 5. It was observed that, as centrifugation rate increased, the residual graphite content decreased. The intensity of the peak at 26.63° was the highest for the sample centrifuged at 1000 rpm and almost disappeared in sample GO30L-4000. This decrease in the intensity of the main peak corresponding to the plane (002) is related to a high level of exfoliation and disorder between GO flakes [37]. The distance between graphene oxide flakes, calculated according to Equation (2), increased with the centrifugation rate (Table 4). This indicated a higher level of exfoliation of the fraction GO30L-4000 [35]. For the effect of sonication times, it was observed that a shorter sonication time resulted in smaller distance between graphene oxide flakes, together with a higher fraction of graphite, in good agreement with a shorter exfoliation time.

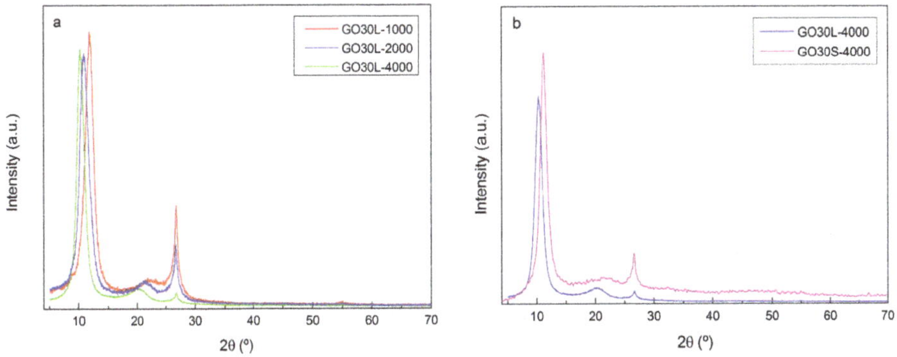

Figure 5. XRD analysis of (**a**) GO30L series and (**b**) GO30-4000 series.

Table 4. Interplanar distance (*d*) values for GO30L-1000, GO30L-2000, GO30L-4000, and GO30S-4000 samples.

Sample	2θ (°)	*d* (nm)
GO30L-1000	12.03	0.733
GO30L-2000	10.90	0.811
GO30L-4000	10.27	0.865
GO30S-4000	11.12	0.795

The morphology of the GO flakes was observed by AFM and the results are shown in Figure 6. Concerning flake size, in GO30L-1000 and GO30L-2000 samples, irregular flakes of sizes between 300 and 500 nm were observed. In the GO30L-4000 sample, homogeneous flakes of size at around 250 nm were observed. A similar effect was observed in a previous work on centrifugation-based size selection of graphene [47]. Flake thickness was analyzed by cross-sectional profiles, and the GO30L-1000 and GO30L-2000 samples showed a heterogeneous distribution with values ranging from 4 to 10 nm. Flake thickness determined by AFM and the number of layers of graphene (*N*) can be related by Equation (3) [47,48]:

$$N = \frac{t_{AFM} - 0.4}{d_{spacing}} \tag{3}$$

where t_{AFM} is the thickness measured by AFM, *0.4* is the factor that takes into account substrate–graphene and graphene–tip interactions, and $d_{spacing}$ corresponds to interplane spacing in each sample. When this equation was applied to graphene oxide, measured thickness values for GO30L-1000 and GO30L-2000 fractions corresponded to multilayer GO flakes. For fraction GO30L-4000, thickness values at around 2 nm were observed, which may be related to few-layer graphene oxide flakes.

Comparing the effect of ultrasonication time on flake morphology in GO30L-4000 and GO30S-4000 fractions (parts c and d of Figure 6, respectively), flakes of bigger size and similar thickness were observed when decreasing the ultrasonication time. Specifically, thickness values of around 3 nm and an average flake size of 450 nm were observed for GO30S-4000, whereas the thickness and size values were around 2 nm and 250 nm, respectively, for GO30L-4000. This suggested that ultrasonication time was related to the breaking of GO flakes, while thickness was more dependent on the centrifugation rate.

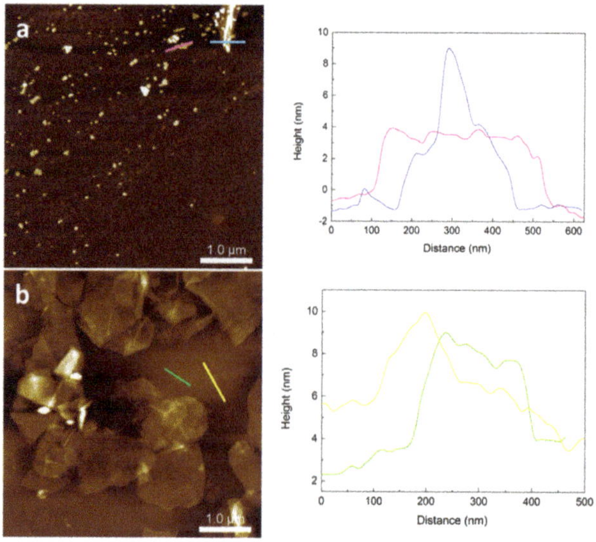

Figure 6. *Cont.*

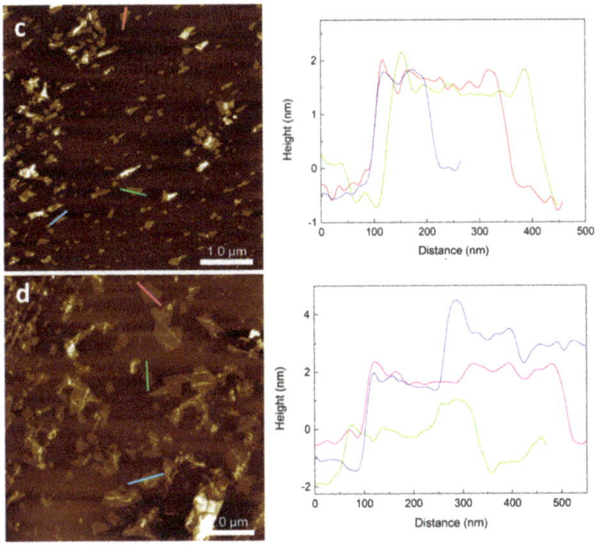

Figure 6. Atomic force microscopy (AFM) height images (left) and cross-sectional profiles (right) of (**a**) GO30L-1000, (**b**) GO30L-2000, (**c**) GO30L-4000, and (**d**) GO30S-4000 GO fractions.

3.3. Nanocomposites

Considering the results obtained in graphene oxide characterization, sodium alginate-based nanocomposites were prepared with GO30L-4000 and GO30S-4000 graphene oxide samples. The most oxidized fraction (GO30) was chosen since the higher presence of oxygen-containing groups could favor the dispersibility of graphene oxide in water, as well as the interaction with the matrix. A centrifugation rate of 4000 rpm was chosen due to the presence of better exfoliated graphene flakes.

FTIR spectra of the nanocomposites and pure SA matrix are shown in Figure 7. Figure 7a shows the spectra of SA-GO30L-4000-1% and SA-GO30L-4000-8%, as well as the SA matrix. No significant differences were observed when comparing the nanocomposites and the matrix. With the incorporation of graphene oxide, a slight broadening and a shift to lower wavenumbers were observed in the peak corresponding to the stretching vibration of the O–H bond. These changes would be an evidence of hydrogen-bonding interactions occurring between SA and GO. A similar behavior was observed when comparing SA-GO30L-4000-8% and SA-GO30S-4000-8% samples with the SA matrix (Figure 7b). No differences were observed with relation to different sonication times.

Thermogravimetric (TG) and derivative thermogravimetric (DTG) results for the SA matrix and GO30L-4000, as well as SA-GO30L-4000-8% and SA-GO30S-4000-8% nanocomposites, are shown in parts a and b of Figure 8, respectively. The TG curve of GO showed that the thermal degradation process was carried out in three steps. In the first step, occurring between 25 °C and 100 °C, the mass loss was associated with the evaporation of water trapped between GO flakes [36]. The second step occurred between 230 °C and 260 °C and was related to the decomposition of the less stable oxygen-containing functional groups [36]. Finally, the slow mass loss observed from 260 °C was related to the more stable functional groups present in the graphene oxide, giving place to a high amount of residue [49]. The TG curves of nanocomposites showed a two-step degradation process. The first step, occurring at around 100 °C, was associated with the evaporation of water absorbed by the nanocomposite films. The second step, occurring between 200 °C and 300 °C, was related to the thermal decomposition of SA [26]. A higher amount of residue was observed for nanocomposites when compared with the matrix, which could be attributed to the excellent thermal stability of GO [50]. The sample SA-GO30S-4000-8% showed a higher amount of residue when compared with the SA-GO30L-4000-8% sample, suggesting that the bigger size of flakes obtained by shorter sonication times favored the thermal stability of final nanocomposites.

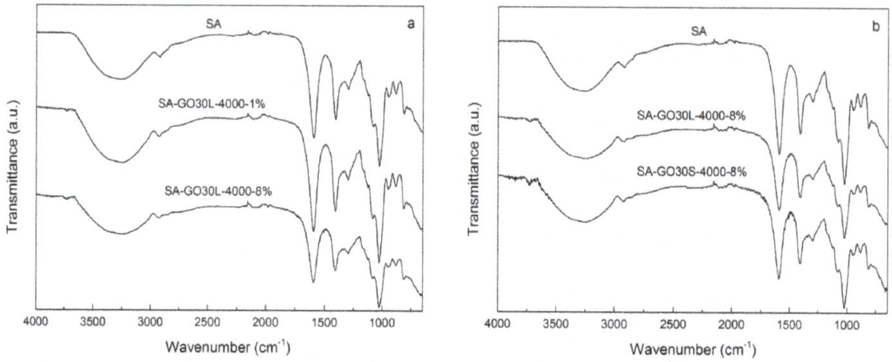

Figure 7. FTIR spectra of (**a**) SA (sodium alginate) matrix, SA-GO30L-4000-1%, and SA-GO30L-4000-8% nanocomposites and (**b**) SA matrix, SA-GO30L-4000-8%, and SA-GO30S-4000-8% nanocomposites (y-axis of the curves was translated in order to avoid overlapping and to improve the visibility of the characteristic bands).

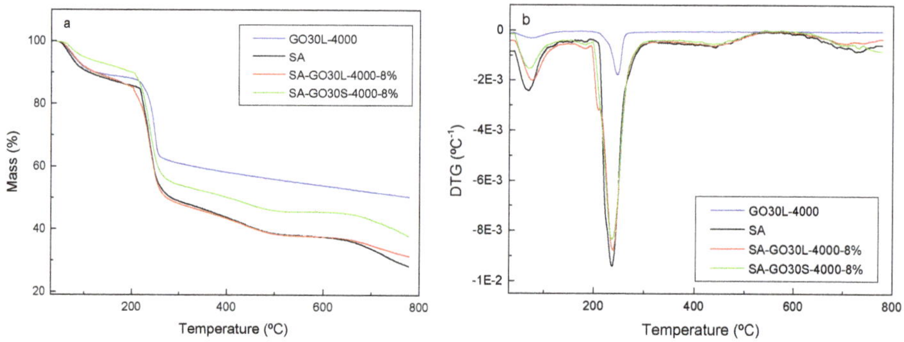

Figure 8. (**a**) Thermogravimetric (TG) and (**b**) derivative thermogravimetric (DTG) curves of GO30L-4000, SA matrix, GO30L-4000-8%, and GO30S-4000-8% nanocomposites.

Concerning DTG curves, shown in Figure 8b, it was observed that the temperature of maximum degradation rate slightly increased for nanocomposites with respect to the pure SA matrix, suggesting that the incorporation of GO enhanced the thermal stability and delayed the pyrolysis of nanocomposite films [51]. This increase in thermal stability in the presence of GO could be a result of interactions occurring between GO and SA. The presence of GO could hinder the mobility of SA molecular chains, increasing the energy required for thermal decomposition of nanocomposite films.

The effect of GO content in the mechanical properties of nanocomposites was assessed by tensile tests. Figure 9 shows the stress-strain curves of the SA matrix and SA-GO30L-4000 nanocomposite series, and their Young's modulus, tensile strength, and elongation at break values are shown in Table 5.

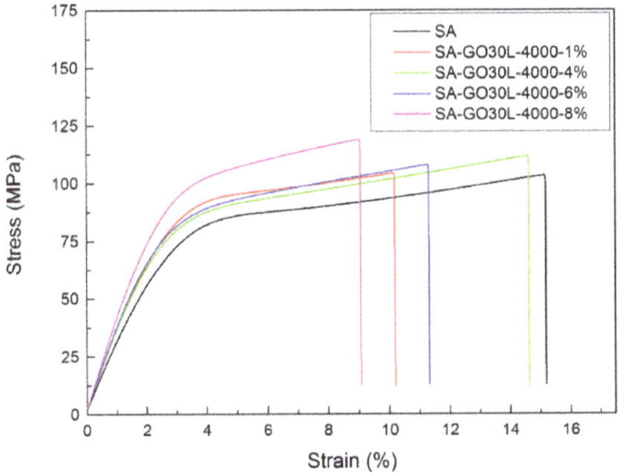

Figure 9. Stress-strain curves of SA-GO30L-4000 nanocomposite series with different GO30L-4000 content and pure SA matrix.

Table 5. Tensile strength, Young's modulus, and elongation at break of SA-GO30L-4000 nanocomposite series and pure SA matrix.

Sample	Tensile Strength (MPa)	Young's Modulus (GPa)	Elongation at Break (%)
SA	101.4 ± 4.8	3.0 ± 0.1	12.8 ± 3.9
SA-GO30L-4000-1%	104.5 ± 9.1	3.3 ± 0.3	10.2 ± 0.9
SA-GO30L-4000-4%	111.9 ± 15.1	3.2 ± 0.5	14.6 ± 2.9
SA-GO30L-4000-6%	108.3 ± 8.9	3.6 ± 0.3	11.3 ± 1.6
SA-GO30L-4000-8%	119.1 ± 6.2	3.7 ± 0.3	9.0 ± 3.0

Young's modulus values of nanocomposites showed a slight increase with respect to the SA matrix. In general, an increase in tensile strength was observed with the incorporation of GO. Pure SA matrix showed a tensile strength of 101.4 MPa, while the maximum value obtained was 119.1 MPa for the nanocomposite with 8% of GO content. This resulted in an increase of 17.4% with respect to the matrix. The rest of the nanocomposites also showed improved tensile strength with respect to pure SA matrix. The improvement of mechanical properties was attributed to the uniform dispersion of GO in the SA matrix, as well as to the effective interfacial interactions, resulting in a good transference of stress from the SA matrix to rigid GO nanoreinforcements [26,50,51]. The elongation at break values obtained for the nanocomposites were lower than values obtained for the matrix. These results suggested that the addition of GO as a nanoreinforcement improved the strength and the stiffness of the films at the expense of flexibility.

To analyze the effect of GO flake size on mechanical properties, SA-GO30L-4000 and SA-GO30S-4000 nanocomposites series were compared, with a GO content of 6 wt % and 8 wt %. In AFM analysis, GO30L-4000 fraction showed thickness and flake size values around of 2 nm and 250 nm, respectively, while the thickness and flake size values for the GO30S-4000 fraction were around 3 nm and 450 nm, respectively. The stress-strain curves obtained for these samples are shown in Figure 10. Young's modulus, tensile strength, and elongation at break values calculated from the stress-strain curves are listed in Table 6.

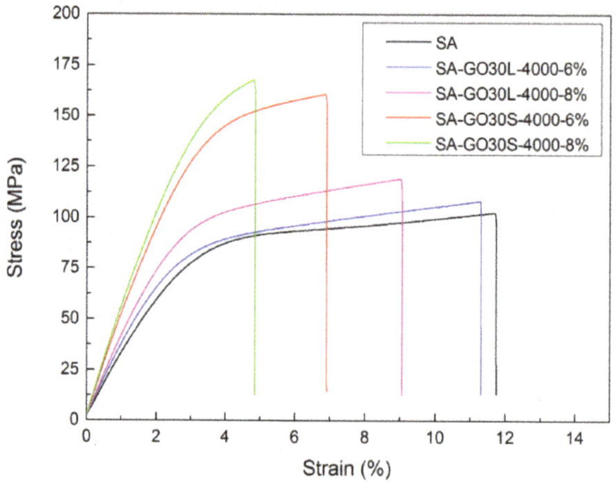

Figure 10. Stress-strain curves of SA-GO30L-4000 and SA-GO30S-4000 nanocomposites series as well as pure SA matrix.

Table 6. Tensile strength, Young's modulus, and elongation at break of SA-GO30L-4000 and SA-GO30S-4000 nanocomposites series and pure SA matrix.

Sample	Tensile Strength (MPa)	Young's Modulus (GPa)	Elongation at Break (%)
SA	101.4 ± 4.8	3.0 ± 0.1	12.8 ± 3.9
SA-GO30L-4000-6%	108.3 ± 8.9	3.6 ± 0.3	11.3 ± 1.6
SA-GO30S-4000-6%	160.8 ± 17.2	5.1 ± 0.7	6.9 ± 1.2
SA-GO30L-4000-8%	119.1 ± 6.2	3.7 ± 0.3	9.0 ± 3.0
SA-GO30S-4000-8%	167.7 ± 17.4	5.5 ± 0.3	4.8 ± 0.8

A significant increase was observed in Young's modulus and tensile strength when the GO30S-4000 fraction was used as a nanoreinforcement. Concerning tensile strength, an improvement of 65.3% was observed for sample SA-GO30S-4000-8% when compared with the matrix (167.7 MPa vs. 101.4 MPa). For the SA-GO30S-4000-6% sample, an improvement of 58.8% was observed with respect to the matrix (160.8 MPa vs. 101.4 MPa), while for the SA-GO30L-4000-6% sample, an improvement of 6.7% was observed with respect to the SA matrix. Young's modulus also increased significantly when the GO30S-4000 fraction was used. The values obtained for SA-GO30S-4000-6% and SA-GO30S-4000-8% nanocomposites were 5.1 GPa and 5.5 GPa, respectively, while for SA-GO30L-4000-6% and SA-GO30L-4000-8% nanocomposites, values of 3.6 GPa and 3.7 GPa were obtained, respectively. The elongation at break values diminished in the SA-GO30S-4000 series with respect to the SA-GO30L-4000 series. These results suggested that the effect of flake size is an important factor influencing the reinforcement effect when flakes of similar thickness are employed. Similar results were reported with other matrices by other authors [28,29].

In order to assess the effect of the addition of GO and its size on the structure of alginate and, thus, in final composite properties, X-ray analyses were carried out. The resulting diffractograms are shown in Figure 11. The alginate diffractogram showed a very intense crystalline peak centered around $2\theta = 13°$, attributed to the (110), and a peak at $2\theta = 22°$, corresponding to the (200) plane. Regarding the amorphous zones, a broad peak can be seen in the diffractogram centered around $2 = 40°$ [52].

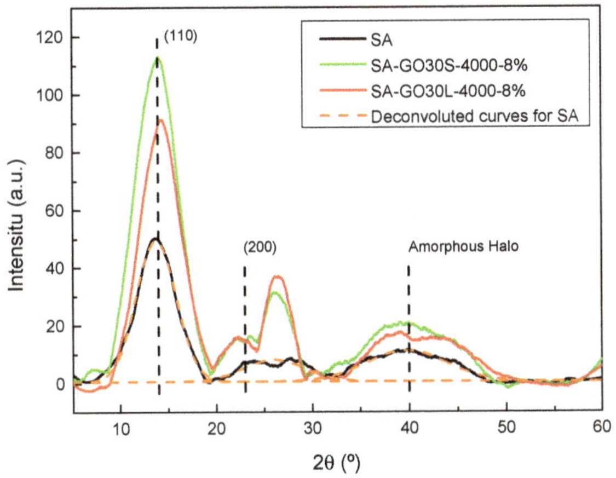

Figure 11. X-ray diffractograms for SA matrix, SA-GO30L-4000-8% and SA-GO30S-4000-8%, and deconvoluted curves for SA.

For composites, a clear increase in the intensity of the crystalline zones took place. A new peak centered around 2θ = 26° was also present, attributed to the (002) plane of the remaining graphitic structure of the carbonaceous reinforcements. In order to further analyze the system structure and crystallinity, numerical analyses were carried out to determine the crystalline phase content of each system. For that, diffractograms were deconvoluted with originPro9 using Gauss function and Equation (4) was used to calculate each value.

$$X_c(\%) = \frac{A_C}{A_C + A_A} x\ 100 \quad (4)$$

where A_C is the sum of areas under the crystalline peaks and A_A is the area under the amorphous halo.

The crystallinity values for SA, SA-30GOL-4000-8%, and SA-30GOS-4000-8% were 74%, 78%, and 87%, respectively. The addition of GO resulted in more crystalline materials, which was greatly affected by the size of the flakes. Larger flakes, produced by shorter sonication times, resulted in systems with higher crystallinity degrees; this dependence on reinforcement shape has been previously reported by Szparaga et al. [30].

The crystallinity values were in agreement with mechanical behaviors shown by the samples. The higher crystallinity degree shown by reinforced composites, specially SA-30GO3-4000-8%, could explain the lower elongation at break values shown by these systems. This higher crystallinity could also add to the reinforcement supplied by the GO flakes in increasing the strength and Young's modulus.

4. Conclusions

When graphite was subjected to the described oxidative process, it was observed that oxygen-containing groups were satisfactorily introduced into the graphitic structure. As the time of oxidative treatment increased (30 min vs. 15 min), a higher degree of oxidation, as well as a higher disruption of the graphitic structure, was observed.

During the sonication and centrifugation processes to obtain graphene oxide, it was observed that higher centrifugation rates resulted in a graphene oxide fraction with lower amount of residual graphite. In the same fashion, graphene oxide flakes of lower size and thickness were isolated as centrifugation rate increased. According to AFM results, it was observed that, when using a centrifugation rate of 4000 rpm, few-layer graphene oxide flakes were obtained. When sonication time was decreased from 4 h to 3 h and final centrifugation rate was maintained at 4000 rpm, an increase in flake size was observed, while flake thickness values remained unchanged. In view of these results, it was concluded that ultrasonication time is related to the breaking of graphene oxide flakes, while flake thickness is more dependent on the centrifugation rate.

The nanocomposites showed evidences of hydrogen-bonding interactions between the SA matrix and graphene oxide, because of the oxygen-containing groups introduced during the oxidative process. The incorporation of GO increased the resistance to thermal degradation of the nanocomposites, probably due to the restrictions in SA chains motion as a consequence of interactions with graphene oxide. Higher flake sizes resulted in an improvement in the resistance to thermal degradation. In general, the incorporation of graphene oxide improved the tensile strength and Young's modulus of the SA matrix. It was observed that, at similar graphene oxide thickness values, the increase in flake size significantly improved the mechanical properties.

Author Contributions: M.A.C. and A.E. conceived the project and designed research. I.L., L.U., A.F., N.G., A.A., carried out all experimental measurements and preparations including graphite oxidation, graphene oxide exfoliation and composites preparation, and their corresponding characterization. I.L., L.U., M.A.C. and A.E. performed data analysis and wrote the manuscript and All authors have read and agreed to the published version of the manuscript.

Funding: This research was funded by Spanish Ministry of Science, Innovation and Universities in the frame of MAT2016-76294-R project, the Basque Government for PIBA 2019-44 project and the Gipuzkoa Council in the frame of Programa de Red Gipuzkoana de Ciencia, Tecnología e Innovación 2019.

Acknowledgments: The Universidad del País Vasco/Euskal Herriko Unibertsitatea (UPV/EHU) (GIU18/216 research group) and the Macrobehavior-Mesostructure-Nanotechnology SGIker unit of UPV/EHU are gratefully acknowledged.

Conflicts of Interest: The authors declare no conflict of interest.

References

1. Rouf, T.B.; Kokini, J.L. Biodegradable biopolymer–graphene nanocomposites. *J. Mater. Sci.* **2016**, *51*, 9915–9945. [CrossRef]
2. Halley, P.J.; Dorgan, J.R. Next-generation biopolymers: Advanced functionality and improved sustainability. *MRS Bull.* **2011**, *36*, 687–691. [CrossRef]
3. Mørch, Ý.A.; Holtan, S.; Donati, I.; Strand, B.L.; Skjåk-Bræk, G. Mechanical properties of C-5 epimerized alginates. *Biomacromolecules* **2008**, *9*, 2360–2368. [CrossRef] [PubMed]
4. Draget, K.I. *Handbook of Hydrocolloids;* Woodhead Publishing Limited: Cambridge, UK, 2000.
5. Olivas, G.I.; Barbosa-Cánovas, G.V. Alginate-calcium films: Water vapor permeability and mechanical properties as affected by plasticizer and relative humidity. *LWT Food Sci. Technol.* **2008**, *41*, 359–366. [CrossRef]
6. Vauchel, P.; Kaas, R.; Arhaliass, A.; Baron, R.; Legrand, J. A new process for extracting alginates from Laminaria digitata: Reactive extrusion. *Food Bioprocess Technol.* **2008**, *1*, 297–300. [CrossRef]
7. Lee, K.Y.; Mooney, D.J. Alginate: Properties and biomedical applications. *Prog. Polym. Sci.* **2012**, *37*, 106–126. [CrossRef]
8. Vicini, S.; Mauri, M.; Wichert, J.; Castellano, M. Alginate gelling process: Use of bivalent ions rich microspheres. *Polym. Eng. Sci.* **2017**, *57*, 531–536. [CrossRef]
9. Dodero, A.; Pianella, L.; Vicini, S.; Alloisio, M.; Ottonelli, M.; Castellano, M. Alginate-based hydrogels prepared via ionic gelation: An experimental design approach to predict the crosslinking degree. *Eur. Polym. J.* **2019**, *118*, 586–594. [CrossRef]
10. Shchipunov, Y. Bionanocomposites: Green sustainable materials for the near future. *Pure Appl. Chem.* **2012**, *84*, 2579–2607. [CrossRef]
11. Coleman, J.N.; Khan, U.; Blau, W.J.; Gun'ko, Y.K. Small but strong: A review of the mechanical properties of carbon nanotube-polymer composites. *Carbon* **2006**, *44*, 1624–1652. [CrossRef]
12. Liu, C.; Liu, H.; Xiong, T.; Xu, A.; Pan, B.; Tang, K. Graphene oxide reinforced alginate/PVA double network hydrogels for efficient dye removal. *Polymers* **2018**, *10*, 835. [CrossRef] [PubMed]
13. Huang, H.; Liu, C.; Wu, Y.; Fan, S. Aligned carbon nanotube composite films for thermal management. *Adv. Mater.* **2005**, *17*, 1652–1656. [CrossRef]

14. Gholizadeh, B.S.; Buazar, F.; Hosseini, S.M.; Mousavi, S.M. Enhanced antibacterial activity, mechanical and physical properties of alginate/hydroxyapatite bionanocomposite film. *Int. J. Biol. Macromol.* **2018**, *116*, 786–792. [CrossRef]
15. Safaei, M.; Taran, M.; Imani, M.M. Preparation, structural characterization, thermal properties and antifungal activity of alginate-CuO bionanocomposite. *Mater. Sci. Eng. C* **2019**, *101*, 323–329. [CrossRef] [PubMed]
16. Yang, M.; Xia, Y.; Wang, Y.; Zhao, X.; Xue, Z.; Quan, F.; Geng, C.; Zhao, Z. Preparation and property investigation of crosslinked alginate/silicon dioxide nanocomposite films. *J. Appl. Polym. Sci.* **2016**, *133*, 1–9. [CrossRef]
17. Du, X.; Skachko, I.; Barker, A.; Andrei, E.Y. Approaching ballistic transport in suspended graphene. *Nat. Nanotechnol.* **2008**, *3*, 491–495. [CrossRef] [PubMed]
18. Lee, C.; Wei, X.; Kysar, J.W.; Hone, J. Measurement of the elastic properties and intrinsic strength of monolayer graphene. *Science* **2008**, *321*, 385–388. [CrossRef]
19. Brodie, B.C. XIII. On the atomic weight of graphite. *Philos. Trans. R. Soc. Lond.* **1859**, *149*, 249–259. [CrossRef]
20. Staudenmaier, L. Method for the preparation of the graphite acid. *Eur. J. Inorg. Chem.* **1898**, *31*, 1481–1487.
21. Hummers, W.S.; Offeman, R.E. Preparation of graphitic oxide. *J. Am. Chem. Soc.* **1958**, *80*, 1339. [CrossRef]
22. Liang, J.; Huang, Y.; Zhang, L.; Wang, Y.; Ma, Y.; Cuo, T.; Chen, Y. Molecular-level dispersion of graphene into poly(vinyl alcohol) and effective reinforcement of their nanocomposites. *Adv. Funct. Mater.* **2009**, *19*, 2297–2302. [CrossRef]
23. Si, H.; Luo, H.; Xiong, G.; Yang, Z.; Raman, S.R.; Guo, R.; Wan, Y. One-step in situ biosynthesis of graphene oxide-bacterial cellulose nanocomposite hydrogels. *Macromol. Rapid Commun.* **2014**, *35*, 1706–1711. [CrossRef] [PubMed]
24. Liu, L.; Li, C.; Bao, C.; Jia, Q.; Xiao, P.; Liu, X.; Zhang, Q. Preparation and characterization of chitosan/graphene oxide composites for the adsorption of Au(III) and Pd(II). *Talanta* **2012**, *93*, 350–357. [CrossRef] [PubMed]
25. Chen, Y.; Yin, Q.; Zhang, X.; Jia, H.; Ji, Q.; Xu, Z. Impact of various oxidation degrees of graphene oxide on the performance of styrene–butadiene rubber nanocomposites. *Polym. Eng. Sci.* **2018**, *58*, 1409–1418. [CrossRef]
26. Nie, L.; Liu, C.; Wang, J.; Shuai, Y.; Cui, X.; Liu, L. Effects of surface functionalized graphene oxide on the behavior of sodium alginate. *Carbohydr. Polym.* **2015**, *117*, 616–623. [CrossRef]
27. Appel, A.K.; Thomann, R.; Mülhaupt, R. Polyurethane nanocomposites prepared from solvent-free stable dispersions of functionalized graphene nanosheets in polyols. *Polymer* **2012**, *53*, 4931–4939. [CrossRef]
28. May, P.; Khan, U.; O'Neill, A.; Coleman, J.N. Approaching the theoretical limit for reinforcing polymers with graphene. *J. Mater. Chem.* **2012**, *22*, 1278–1282. [CrossRef]
29. Nawaz, K.; Ayub, M.; Ul-Haq, N.; Khan, M.B.; Niazi, M.B.K.; Hussain, A. Effects of selected size of graphene nanosheets on the mechanical properties of polyacrylonitrile polymer. *Fibers Polym.* **2014**, *15*, 2040–2044. [CrossRef]
30. Szparaga, G.; Brzezińska, M.; Pabjańczyk-Wlazło, E.; Puchalski, M.; Sztajnowski, S.; Krucińska, I. Structure-Property of Wet-Spun Alginate-Based Precursor Fibers Modified with Nanocarbons. *Autex Res. J.* **2019**, 1–11. [CrossRef]
31. Khan, U.; O'Neill, A.; Porwal, H.; May, P.; Nawaz, K.; Coleman, J.N. Size selection of dispersed, exfoliated graphene flakes by controlled centrifugation. *Carbon* **2012**, *50*, 470–475. [CrossRef]
32. Zachariasen, W.H. A general theory of X-ray diffraction in crystals. *Acta Crystallogr.* **1967**, *23*, 558–564. [CrossRef]
33. Bragg, W.L. The diffraction of short electromagnetic Waves by a Crystal. *Proc. Camb. Philol. Soc.* **1913**, *17*, 43–57.
34. Dinari, M.; Salehi, E.; Abdolmaleki, A. Thermal and morphological properties of nanocomposite materials based on graphene oxide and L-leucine containing poly (benzimidazole-amide) prepared by ultrasonic irradiation. *Ultrason. Sonochem.* **2018**, *41*, 59–66. [CrossRef] [PubMed]
35. Krishnamoorthy, K.; Veerapandian, M.; Yun, K.; Kim, S. The chemical and structural analysis of graphene oxide with different degrees of oxidation. *Carbon* **2012**, *53*, 38–49. [CrossRef]
36. Sharifi-Bonab, M.; Rad, F.A.; Mehrabad, J.T. Preparation of laccase-graphene oxide nanosheet/alginate composite: Application for the removal of cetirizine from aqueous solution. *J. Environ. Chem. Eng.* **2016**, *4*, 3013–3020. [CrossRef]
37. Todorova, N.; Giannakopoulou, T.; Boukos, N.; Vermisoglou, E.; Lekakou, C.; Trapalis, C. Self-propagating solar light reduction of graphite oxide in water. *Appl. Surf. Sci.* **2017**, *391*, 601–608. [CrossRef]

38. Paredes, J.I.; Villar-Rodil, S.; Martínez-Alonso, A.; Tascón, J.M.D. Graphene oxide dispersions in organic solvents. *Langmuir* **2008**, *24*, 10560–10564. [CrossRef]
39. Ferrari, A.C. Raman spectroscopy of graphene and graphite: Disorder, electron—Phonon coupling, doping and nonadiabatic effects. *Solid State Commun.* **2007**, *143*, 47–57. [CrossRef]
40. Gayathri, S.; Jayabal, P.; Kottaisamy, M.; Ramakrishnan, V. Synthesis of few layer graphene by direct exfoliation of graphite and a Raman spectroscopic study. *AIP Adv.* **2014**, 4. [CrossRef]
41. Ferrari, A.C.; Meyer, J.C.; Scardaci, V.; Casiraghi, C.; Lazzeri, M.; Mauri, F.; Piscanec, S.; Jiang, D.; Novoselov, K.S.; Roth, S.; et al. Raman Spectrum of Graphene and Graphene Layers. *Phys. Rev. Lett.* **2006**, *97*, 187401. [CrossRef]
42. Dresselhaus, M.S.; Jorio, A.; Saito, R. Characterizing Graphene, Graphite, and Carbon Nanotubes by Raman Spectroscopy. *Annu. Rev. Condens. Matter Phys.* **2010**, *1*, 89–108. [CrossRef]
43. Sharma, S.; Susan, D.; Kothiyal, N.C.; Kaur, R. Graphene oxide prepared from mechanically milled graphite: Effect on strength of novel fly-ash based cementitious matrix. *Constr. Build. Mater.* **2018**, *177*, 10–22. [CrossRef]
44. Lespade, P.; Al-Jishi, R.; Dresselhaus, M.S. Model for Raman scattering from incompletely graphitized carbons. *Carbon* **1982**, *20*, 427–431. [CrossRef]
45. Cançado, L.G.; Takai, K.; Enoki, T.; Endo, M.; Kim, Y.A.; Mizusaki, H.; Jorio, A.; Coelho, L.N.; Magalhães-Paniago, R.; Pimenta, M.A. General equation for the determination of the crystallite size la of nanographite by Raman spectroscopy. *Appl. Phys. Lett.* **2006**, *88*, 163106. [CrossRef]
46. Guerrero-contreras, J.; Guerrero-contreras, J. Graphene oxide powders with different oxidation degree, prepared by synthesis variations of the Hummers method Graphene oxide powders with different oxidation degree, prepared by synthesis variations of the Hummers method. *Mater. Chem. Phys.* **2015**, *153*, 209–220. [CrossRef]
47. Ugarte, L.; Gómez-Fernández, S.; Tercjak, A.; Martínez-Amesti, A.; Corcuera, M.A.; Eceiza, A. Strain sensitive conductive polyurethane foam/graphene nanocomposites prepared by impregnation method. *Eur. Polym. J.* **2017**, *90*, 323–333. [CrossRef]
48. Shearer, C.J.; Slattery, A.D.; Stapleton, A.J.; Shapter, J.G.; Gibson, C.T. Accurate thickness measurement of graphene. *Nanotechnology* **2016**, *27*, 125704. [CrossRef]
49. Sohail, M.; Saleem, M.; Ullah, S.; Saeed, N.; Afridi, A.; Khan, M.; Arif, M. Modified and improved Hummer's synthesis of graphene oxide for capacitors applications. *Mod. Electron. Mater.* **2017**, *3*, 110–116. [CrossRef]
50. Ionita, M.; Pandele, M.A.; Iovu, H. Sodium alginate/graphene oxide composite films with enhanced thermal and mechanical properties. *Carbohydr. Polym.* **2013**, *94*, 339–344. [CrossRef]
51. Hu, X.; Zhang, X.; Tian, M.; Qu, L.; Zhu, S.; Han, G. Robust ultraviolet shielding and enhanced mechanical properties of graphene oxide/sodium alginate composite films. *J. Compos. Mater.* **2016**, *50*, 2365–2374. [CrossRef]
52. Sundarrajan, P.; Eswaran, P.; Marimuthu, A.; Subhadra, L.B.; Kannaiyan, P. One pot synthesis and characterization of alginate stabilized semiconductor nanoparticles. *Bull. Korean Chem. Soc.* **2012**, *33*, 3218–3224. [CrossRef]

materials

MDPI

Article

Improved Toughness in Lignin/Natural Fiber Composites Plasticized with Epoxidized and Maleinized Linseed Oils

Franco Dominici [1,*], María Dolores Samper [2], Alfredo Carbonell-Verdu [2], Francesca Luzi [1], Juan López-Martínez [2], Luigi Torre [1] and Debora Puglia [1]

1 Civil and Environmental Engineering Department, University of Perugia, UdR INSTM, Strada di Pentima 4, Terni 05100, Italy; francesca.luzi@unipg.it (F.L.); luigi.torre@unipg.it (L.T.); debora.puglia@unipg.it (D.P.)
2 Instituto de Tecnología de Materiales, Universitat Politècnica de València, Plaza Ferrandiz y Carbonell, 03801 Alcoy-Alicante, Spain; masammad@upvnet.upv.es (M.D.S.); alcarve1@epsa.upv.es (A.C.-V.); jlopezm@mcm.upv.es (J.L.-M.)
* Correspondence: franco.dominici@unipg.it; Tel.: +39-0744-492910

Received: 31 December 2019; Accepted: 27 January 2020; Published: 28 January 2020

Abstract: The use of maleinized (MLO) and epoxidized (ELO) linseed oils as potential biobased plasticizers for lignin/natural fiber composites formulations with improved toughness was evaluated. Arboform®, a lignin/natural fiber commercial composite, was used as a reference matrix for the formulations. The plasticizer content varied in the range 0–15 wt % and mechanical, thermal and morphological characterizations were used to assess the potential of these environmentally friendly modifiers. Results from impact tests show a general increase in the impact-absorbed energy for all the samples modified with bio-oils. The addition of 2.5 wt % of ELO to Arboform (5.4 kJ/m^2) was able to double the quantity of absorbed energy (11.1 kJ/m^2) and this value slightly decreased for samples containing 5 and 10 wt %. A similar result was obtained with the addition of MLO at 5 wt %, with an improvement of 118%. The results of tensile and flexural tests also show that ELO and MLO addition increased the tensile strength as the percentage of both oils increased, even if higher values were obtained with lower percentages of maleinized oil due to the possible presence of ester bonds formed between multiple maleic groups present in MLO and the hydroxyl groups of the matrix. Thermal characterization confirmed that the mobility of polymer chains was easier in the presence of ELO molecules. On the other hand, MLO presence delayed the crystallization event, predominantly acting as an anti-nucleating agent, interrupting the folding or packing process. Both chemically modified vegetable oils also efficiently improved the thermal stability of the neat matrix.

Keywords: Arboform; epoxidized oil; maleinized linseed oil; toughness; thermal stability

1. Introduction

At the present time, the use of eco-sustainable materials is becoming a required condition of worldwide interest. It is known that polymeric materials are derived from fossil fuels and these limited resources can be preserved if, from a sustainable perspective, biobased materials containing the maximum amount of renewable biomass derivatives will be considered. The main concern is related to the possible substitution of materials and products traditionally made from petroleum resources with biobased plastics and composites.

To antedate this need, the German Fraunhofer Institute for Chemical Technology together with Tecnaro GmbH Company studied and developed a new material based on wood components that can be processed as a thermoplastic polymeric material. Arboform® is composed of natural fibers, lignin and additives and can be obtained by using different types of lignin, typically at 30% wt, natural fibers (60 wt %, flax, hemp, sisal, wood) and 10% natural additives (softeners, pigments, processing agents, etc.) that make it a fully biodegradable bioplastic composite known as "liquid wood [1]". The material properties—biodegradability and recyclability up to ten times without modifications of its features—recommended it to be the near future alternative to various traditional plastic materials [2]. Depending on the quantity of the mixed components, Arboform can be commercially found in three different options: LV3 Nature, F45 Nature and LV5 Nature. These composite materials can be extruded and injected by using the same technologies applied to engineer polymers based on lignin as a matrix and reinforced with natural fibers [3–5].

The appearance of Arboform samples is typical of a woody-like material, while the mechanical behavior properties are in the range of the engineering thermoplastics, such as polyamides. However, even if it is characterized by high values for strength and elastic modulus, this composite material shows a drawback in a modest toughness, which restricts its applications as an impact-resistant material [6,7]. The plasticization of a lignin biopolymer has been studied by Bouajila et al. [8], who clearly indicated that mechanisms of lignin plasticization are totally altered in dry and in wet conditions. In detail, they demonstrated that unsaturated lignin (dry lignin, plasticized by low amounts of plasticizers) is better plasticized by molecules that can be involved in H bonds, while hydrated lignin is better plasticized by aromatic molecules with a structure similar to that of monolignols (e.g., vanillin) and by molecules with a solubility parameter similar to that of the matrix [9]. The plasticization of lignin with organic plasticizers was originally reported in 1975 by Sakata and Senju [10], who studied the thermal softening temperatures of lignin plasticized by a variety of plasticizers, including esters of phthalic acid, phosphoric acid and aliphatic acids. Previous examples of combined use of vegetable oils and lignin can be found in Antonsson et al. [11], where low-molecular-weight lignin was used together with a vegetable oil to produce a new hydrophobic lignin derivative similar to suberin, demonstrating the potential use in paper-coating applications (due to its capability to interact well with wood fibers and make paper hydrophobic). However, to the best of the authors' knowledge, there are no examples of natural additives considered as toughening agents for lignin biopolymer. In the present study, linseed oil was selected as a natural additive able to improve the properties of the composite without compromising its biopolymeric nature. In order to improve compatibility with the matrix, an epoxidized linseed oil (ELO) and a linseed oil with maleic anhydride modification (MLO) were chosen.

2. Materials and Methods

A commercial composite material named Arboform® grade L (Tecnaro GmbH, Ilsfeld, Germany), V3 Nature was supplied by Tecnaro GmbH (Ilsfeld, Germany). This thermoplastic material, obtained by biorenewable resources, is characterized by a density of 1.29 g/cm^3 and a melt volume rate (MVR) (at 190 °C/2.16 Kg, 15 cm^3/10 min). Epoxidized linseed oil (ELO), supplied by Traquisa S.A. (Barcelona, Spain) with a molecular weight 1.037–1.039 and an epoxy equivalent weight (EEW) of 178 g equiv^{-1}, and maleinized linseed oil (MLO) supplied as Veomer Lin by Vandeputte (Mouscron, Belgium) with a viscosity of 10 dPa s at 20 °C and an acid value of 105–130 mg KOH g^{-1} were used as plasticizers.

2.1. *Processing of Plasticized Compounds*

Formulations based on Arboform L, V3 added with ELO and MLO are reported in Table 1. Due to the saturation limit of the matrix with epoxidized oil (miscibility problems between the matrix and this quantity of epoxidized oil during the blending process), it was not possible to produce the formulation with 15 wt % of ELO. Four formulations of Arboform L, 3V plasticized with a content of ELO between 1 wt % and 10 wt % and five using MLO at a percentage between 1 wt % and 15 wt % were produced to compare their properties with the neat matrix.

Table 1. Composition and labeling of the formulations with different modified oils

Reference	Arboform L, V3, wt %	ELO, wt %	MLO, wt %
ARBOFORM	100	0	0
ARB_1ELO	99	1	0
ARB_2.5ELO	97.5	2.5	0
ARB_5ELO	95	5	0
ARB_10ELO	90	10	0
ARB_1MLO	99	0	1
ARB_2.5MLO	97.5	0	2.5
ARB_5MLO	95	0	5
ARB_10MLO	90	0	10
ARB_15MLO	85	0	15

After drying at 50 °C for 24 h, Arboform® L, V3 was mixed with the modified oils in a co-rotating twin-screw extruder (D = 30 mm; L/D = 20:1) by DUPRA (Alicante, Spain) at a rotation speed of 25 rpm with a temperature profile of 165–170–173–175 °C. After cooling to room temperature, the materials were pelletized and dried at 50 °C for 24 h. All 10 formulations were processed in a Meteor 270/75 injection molding machine (Mateu & Solé, Barcelona, Spain) with a temperature profile of 160–165–170–175 °C and an injection pressure P_{inj} = 60 MPa. Samples for tensile, flexural, impact, heat deflection and dynamic mechanical thermal analysis (DTMA) tests have been manufactured. Standard samples for tensile tests and rectangular samples sizing 80 × 10 × 4 mm^3 were obtained for the characterization.

2.2. Measurements of Mechanical Properties

The effects of ELO and MLO content on mechanical properties were studied by impact tests. All specimens were conditioned, according to ISO 291 [12], at 23 °C and 50% RH before testing and tested in the same conditions. Impact-absorbed energy was measured with a 6 J Charpy pendulum from Metrotec S.A. (San Sebastián, Spain) on unnotched samples according to ISO 179 standard testing [13]. At least five samples for each material were tested.

Flexural and tensile tests were carried out by a universal test machine Ibertest Elib 30 (Ibertest S.A.E., Madrid, Spain) at room temperature. A minimum of five different samples was tested using a 5 kN load cell. The crosshead speed for the tensile tests was set at 10 mm min^{-1} as recommended by ISO 527 standard [14]. An axial extensometer by Ibertest was used to give accurate values of Young's modulus for each material. Flexural characterization was performed setting the crosshead speed to 5 mm min^{-1} for three points bending test, as suggested by ISO 178 standard [15].

2.3. Thermo-Mechanical Characterization

Thermo-mechanical behavior of the mixtures was studied by heat deflection temperature (HDT) tests and by dynamic mechanical thermal analysis (DMTA). The HDT was determined by the A method according to ISO 75 standard [16] that recommends a load of 1.8 MPa and a heating rate of 120 °C min h^{-1}. Tests were carried out in a VICAT/HDT station DEFLEX 687-A2 by Metrotec S.A. (San Sebastian, Spain).

DMTA in torsion mode of Arboform L, V3 and its formulations, plasticized with ELO and MLO, was carried out with an oscillatory rheometer AR G2 by TA Instruments (New Castle, DE, USA), equipped with accessory clamps for solid samples. Rectangular samples sizing 40 × 10 × 4 mm^3 were subjected to a temperature ramp from 30 °C to 100 °C, setting the heating rate at 2 °C min^{-1}. The maximum deformation (γ) was at 0.1% and all samples were tested at a constant frequency of 1 Hz. The curves of the storage moduli (G′), the loss moduli (G″) and their ratio G″/G′ as Tan Delta (tan(δ)) versus the temperature were determined.

Differential scanning calorimetry (DSC) tests were carried out with a TA Instruments Mod. Q200 (TA Instrument, New Castle, DE, USA) calorimeter to determine the effect of ELO and MLO content on the thermal properties of the blends based on Arboform L, V3. For DSC analysis, approximately 10 mg of each sample was placed in a hermetically sealed sample pan after the calibration of the instrument with indium standard. Tests were performed in a three-step cycle: heating, cooling and heating scans, from 25 °C to 180 °C at 10 °C min^{-1}. Glass transition temperature (T_g), crystallization and melting phenomena of the blends were determined.

Thermal degradation behavior of the formulations was evaluated by thermogravimetric analysis (TGA, Seiko Exstar 6300, Tokyo, Japan); around 5 mg samples were used to perform dynamic tests in a nitrogen atmosphere (200 mL min^{-1}) from 30 to 600 °C at 10 °C min^{-1}. Thermal degradation curves for the Arboform L, V3 and formulations with 5 wt % and 10 wt % modified with each oil were evaluated.

2.4. Morphological Characterization

The morphology and the microstructure of the blend samples were observed. Fractured surfaces from impact tests were captured by field emission scanning electron microscopy (FESEM) in a Zeiss Ultra microscope 55 (Oxford Instruments, Oxfordshire, UK) with an accelerating voltage of 2 kV. Fractured surfaces were previously coated with a thin platinum layer in a sputter coater EM MED020 (Leica Microsystems, Wetzlar, Germania).

3. Results and Discussion

The study of the effect of the addition of linseed oils modified with epoxidation and maleinization treatment to the commercial biopolymer Arboform grade L, V3 Nature was carried out by a wide characterization of the formulated materials. Since modest ability to absorb shocks is a critical property of Arboform L, V3, it was tried to increase its toughness by adding plasticizers.

The results of impact tests represent the combination of two effects: on the one hand, fracture resistance associated with the mechanical strength; on the other hand, deformation capability that is directly related to the ductile mechanical behavior. The impact-absorbed energy depends on several factors, such as crack sizes and growth speed, presence of stress concentrators, phase separation, etc. All these factors can modify the overall deformation capacity and, subsequently, the total energy absorbed during deformation and fracture. Impact test results in Figure 1 show a general increase in the impact-absorbed energy for all the samples modified with bio-oils. Absorbed energy values of neat Arboform L, V3 (5.4 kJ/m^2) increased to 11.1 kJ/m^2 with an improvement of 105% with ARB_2.5MLO. Formulations ARB_5MLO and ARB_10MLO show values of impact-adsorbed energy of 10.7 kJ/m^2 and 10.8 kJ/m^2, respectively, corresponding to a percentage increase of 98% and 100% with the addition of 5 and 10 wt % of MLO. The addition of 2.5% wt. of ELO to Arboform is enough to double the quantity of absorbed energy and this value slightly decreased for samples containing 5 and 10 wt % of ELO. A similar result is also obtained with the addition of MLO, but in this case, 1 wt % was enough to appreciate an improvement of the toughness of the materials. In the case of MLO-modified samples, the absorbed energy values result doubled respect to the matrix for all the formulations. The best result of impact resistance is appreciated in the material with 5 wt % of MLO, with an improvement of 118%, after which a slight decrease is observed. Similar behavior was also reported by others [2,17,18].

It was found that the addition of synthetic compatibilizers, such as polymeric methylene diphenyl diisocyanate (PMDI) at low concentrations (1 wt %), produces an increase of the absorbed energy up to 92% and then progressively decreases, increasing the PMDI content to 2 wt %. Similarly, to what was established for PMDI, high quantities of modified oils can produce a phase separation that does not contribute to the improvement of impact resistance [19,20].

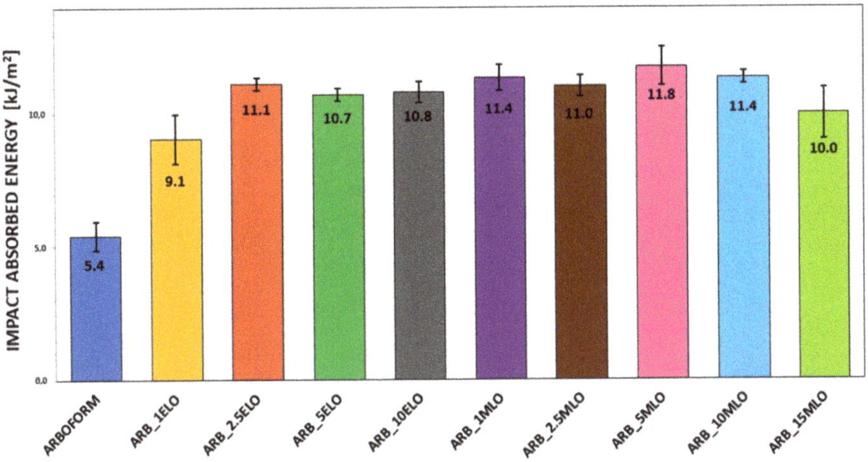

Figure 1. Impact-adsorbed energy for epoxidized (ELO)- and maleinized (MLO)-modified samples based on Arboform L, V3 matrix.

The results of the flexural tests (Figure 2) show that the addition of ELO produces a progressive increase in tensile strength as the percentage of oil increases; simultaneously, a reduction in the flexural modulus is observed when the epoxidized oil content increases. The effectiveness of ELO as a coupling agent, as well as a plasticizer, with biopolymers and lignin-based fillers, was already demonstrated [21–23]. The progressive increase in strength with the ELO content is attributable to the good compatibility between lignin and fibrous reinforcement contained in the Arboform. Above 2.5 wt % of ELO, the plasticizing effect of the epoxidized linseed oil molecules enables polymer chain mobility so that the modulus decreases [24].

The addition of maleinized oil produces a significant improvement in flexural strength, which exceeds the strength of the matrix for all formulations added with MLO. The ARB_2.5MLO formulation shows a flexural strength of 65.2 MPa compared to 41.6 MPa of the Arboform, with an increase of 56.7% [19,25,26].

Formulations with an MLO content of 5 wt % and above show a slight decrease in the flexural strength, essentially due to an anti-plasticizing effect. Some authors have described this anti-plasticizing effect with a potential saturation of the plasticizer [20]. Gutierrez-Villareal et al. experimentally found this phenomenon when citrate esters were used to plasticize poly (methyl methacrylate) (PMMA) [27] with a low concentration of plasticizer. Vidotti et al. have guessed that the anti-plasticizing effect may be due to a reduction in the free volume [28] so that when the free volume is filled by the plasticizer, the phenomenon may appear. The increase in strength, associated with the anti-plasticization phenomenon, can also be explicated by considering the trend in crystallinity as the plasticizer enhances chain mobility, thus the crystallization tendency is favored. The anti-plasticizing effect depends on the molecular weight, the concentration of the plasticizer and the characteristics of the polymer matrix, and it is specific for each polymer–plasticizer system [29]. Elastic modulus increased for the formulations added with 1, 2.5 and 5 wt % of MLO. The drop of the flexural modulus for the formulations with higher percentages of modified oils is essentially related to an excess of oil molecules, which does not bind directly to the polymeric matrix. Plasticizer excess can have a negative effect on homogeneity by forming a dispersed phase in the main matrix. A high plasticizer content produces intense plasticizer–plasticizer interactions, leading to a phase separation [30]. Chieng et al. concluded that, due to the high plasticizer content, only a small part is directly in contact between the interface area, while the excess is dispersed in the polymer matrix [31,32].

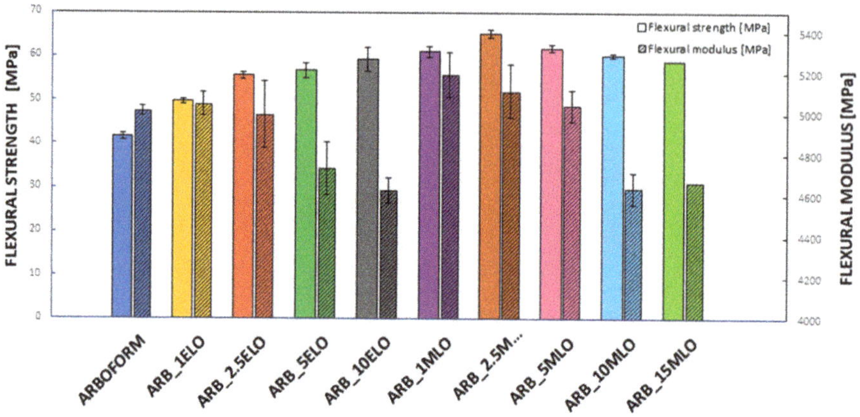

Figure 2. Comparative bar plot of flexural strength and flexural modulus for neat Arboform L, V3 and its formulations with ELO and MLO.

The results of the tensile tests (Table 2) show a slight decrease in the values of the elastic modulus, which progresses with the increase in the content of both modified oils. A general increase in tensile strength is obtained for all the formulations produced: this increase is significant for the formulations with the higher epoxidized oil contents (5, 10 wt % ELO), while the higher values of the tensile strength are obtained with lower percentages of maleinized oil (1, 2.5, 5 wt % MLO).

Table 2. Result of the tensile test made on ELO- and MLO-modified materials.

Reference	Tensile Modulus [MPa]	Tensile Strength [MPa]	Strain at Break [%]
ARBOFORM	5520 ± 270	23 ± 2	2.1 ± 0.3
ARB_1ELO	5260 ± 400	27 ± 2	2.4 ± 0.2
ARB_2.5ELO	5200 ± 370	28 ± 3	2.6 ± 0.3
ARB_5ELO	4860 ± 410	32 ± 2	2.9 ± 0.2
ARB_10ELO	4730 ± 310	31 ± 2	3.1 ± 0.3
ARB_1MLO	4900 ± 660	32 ± 4	3.0 ± 0.3
ARB_2.5MLO	4760 ± 220	32 ± 2	3.3 ± 0.4
ARB_5MLO	4660 ± 180	33 ± 1	4.2 ± 0.7
ARB_10MLO	4320 ± 310	30 ± 3	3.8 ± 0.5
ARB_15MLO	4500 ± 350	27 ± 1	3.6 ± 0.3

The effect of high ELO contents compared to low quantities of MLO is justified by the different operating principles. Lignin is considered a potential substitute for bisphenol A in the synthesis of epoxy resins due to the presence of hydroxyl groups (in particular of phenolic hydroxyl group) in the lignin structure. However, structure and steric hindrance limit epoxidation reactions. The reduced reactivity with the epoxy groups requires high quantities of epoxy oil for a binding effect to be detected [33]. Furthermore, for high quantities of ELO, a self-polymerization phenomenon of linseed oil cannot be excluded [34]. Essentially, the best results are obtained for formulations with higher ELO content. It was observed that ester bonds can be potentially formed between multiple maleic groups present in MLO and the hydroxyl groups of green composites, as in the case of Arboform. This reaction could induce an effective stress transfer between lignin and fibrous components of the composite material, improving their mechanical properties [25].

Moving the composition from 5 to 10 wt % of MLO, a reduction in both strength and modulus was observed. This negative effect can be related to the phase separation caused by an excessive concentration of MLO in the composite. Excess MLO molecules, which do not form direct bonds with the composite molecules, are placed between the biopolymer chains and act as a lubricant with a greater influence on the mobility of the chain. This result, together with previous mechanical tests, suggests that optimal and mechanically balanced performance is obtained for MLO contents of about 5 wt % The best results of tensile strength are obtained with the formulations containing 5 wt % of each oil. The values of the strains at the break of the materials modified with epoxidized oil show modest improvement, while a slight increase in strain values at the break was found for materials modified with MLO. The formulations with the content of each oil between 5 wt % and 10 wt % show the best performance in terms of tensile elongation at break.

HDT measurements: The bubble chart in Figure 3 represents the deflection temperature distributions of the materials added with the modified linseed oils. The center of the bubble indicates the average value of the HDT, and the diameter size is its standard deviation. The initial HDT value for the Arboform composite is 52.2 °C, which shows that moderate temperatures lead to material softening. The addition of 1% by weight of ELO produces a slight increase at 52.4 °C in HDT, which is negligible when considering the standard deviation, compared to the Arboform L, V3 matrix. Higher percentages of epoxidized oil cause a gradual decrease in characteristic deflection temperatures. In relation to the evolution of HDT in terms of ELO wt %, the plasticization provided by ELO leads to softer and more flexible materials, therefore the values of the deflection temperatures decrease with increasing ELO content up to values of 51.3 °C, 50.0 °C and 49.5 °C for an ELO content of 2.5, 5 and 10% by weight, respectively. Epoxidized linseed oil acts as a plasticizer, producing a visible softening effect, which is more evident by the application of external loads low-temperature values. Mobility of polymer chains is easier in the presence of ELO molecules, since these reduce the intermolecular attraction forces between polymeric macromolecules [35]. The stability of HDT for the ARB_1ELO formulation can be explained, in addition to the low oil content, by a slight increase of crystallinity with ELO content, as the plasticizer allows more intense polymer chain motion. The presence of ELO plasticizer allows chain mobility, and this has a positive effect on crystallinity because polymer chains can rearrange to better ordered/packed structures. For ELO contents greater than 1%, the plasticizing effect prevails on crystallization and the benefit on HDT becomes irrelevant [24].

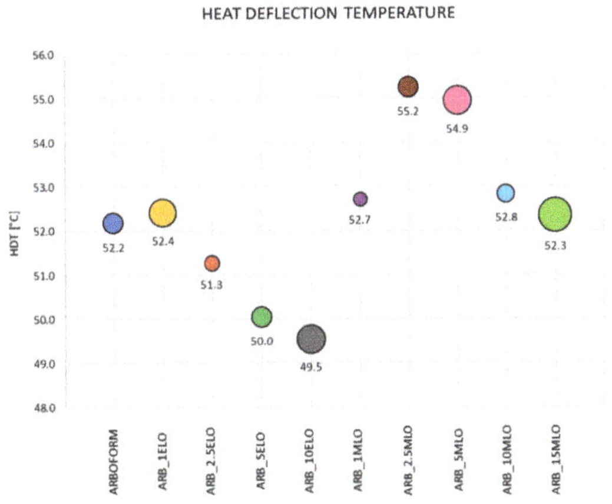

Figure 3. Bubble chart of the Heat Deflection Temperature test.

The formulation containing 1 wt % of MLO shows only a slight increase in HDT with a narrow dispersion of results. The use of maleinized oil produces an increase in the deflection temperature at the percentages between 2.5 and 5 wt %, to 55.2 °C and 54.9 °C, respectively, while higher percentages cause the drop of HDT to values close to the Arboform matrix deflection temperature. This behavior can be explained by the compatibilizing effect of MLO, which, thanks to the presence of maleic groups, binds easily to the hydroxyl groups of the lignin and the Arboform natural fiber reinforcements, forming bonds that stabilize the composite; this effect, together with the anti-plasticizing effect, is effective for MLO contents up to 5 wt %. For higher quantities of maleinized oil (10–15 wt %), the plasticizing effect becomes predominant by exerting a lubricating action by distancing the polymer chains, as the MLO molecules are positioned between the molecules of the matrix weakening the polymer–polymer interactions (hydrogen bonds, van der Waals or ionic forces), thus increasing the mobility of the chains. This effect produces greater deformability, which results in a decrease of HDT for high MLO contents [25,34,36].

DMTA analysis: Figure 4a shows the results of DMTA analysis for the Arboform material modified with ELO. Curves of the storage modulus (G′) show a deflection between 50 °C and 60 °C at the glass transition temperature of the reference matrix. In general, the trend for G′ is similar, except for the G′ curve of ARB_10ELO formulation, which shows slightly lower values (in accordance with the results of the moduli obtained from the tensile tests). On the other hand, the higher value for G′ was found for the of the ARB_1ELO formulation, confirming the results of the HDT tests. The crve of ARB_10ELO loss modulus (G″) highlights a shift toward lower temperatures, while the G″ curve of ARBOFORM shows the best thermal stability of the set. Peak values of the tan curves (not reported here) show the tendency to progressively decrease in intensity with increasing ELO content [37,38].

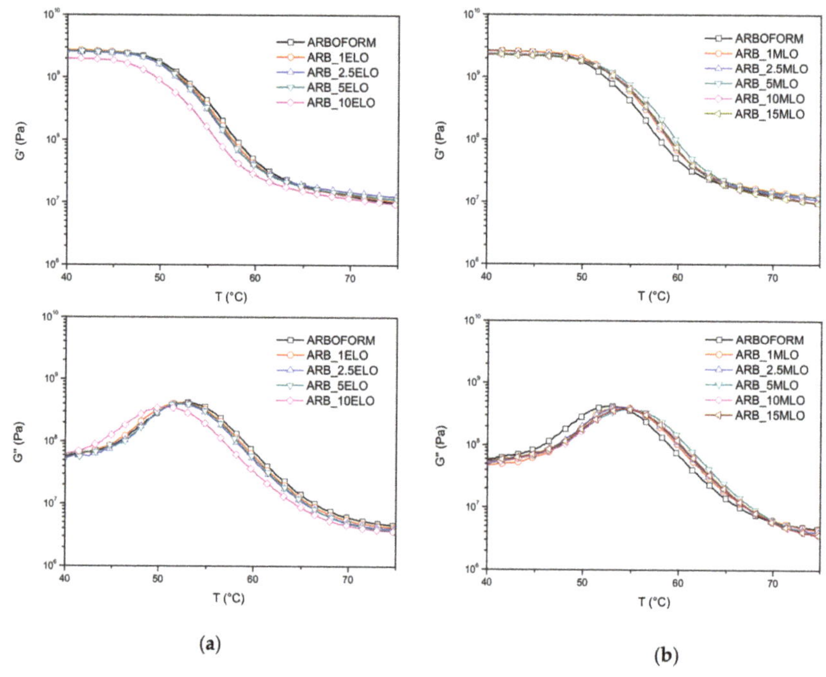

Figure 4. Curves of storage (G′) and loss (G″) moduli of (**a**) ELO-added and (**b**) MLO-added formulations.

In Figure 4b, the results of the DMTA analysis carried out for the materials with increasing content of MLO show that all the formulations have comparable storage moduli in the temperature range between 40 °C and 50 °C; above this range, the modified materials show improved G′ values when compared to the reference matrix. The formulation ARB_5MLO shows the highest modulus value of this material set. All the G″ curves of the MLO-added materials show a shift toward higher temperatures. The improvement of the thermal mechanical performance of the materials containing MLO is also confirmed by the increase in T_g. The T_g values, calculated as the maximum peak of the tan (δ) curves (Table 3), show an increase in the glass transition temperatures of all the compounds modified with MLO. The highest value is obtained for the formulation with 5 wt % of MLO, in accordance with HDT tests.

Table 3. Values of glass transition temperature, calculated at the maximum of the tan(δ) curve.

Reference	T_g (°C)
ARBOFORM	59.1
ARB_1MLO	58.8
ARB_2.5MLO	58.4
ARB_5MLO	58.4
ARB_10MLO	57.5
ARB_1MLO	60.1
ARB_2.5MLO	60.4
ARB_5MLO	61.4
ARB_10MLO	61.0
ARB_15MLO	60.8

DSC analysis: Results of thermal characterization for Arboform formulations containing different amounts of ELO- and MLO-modified oils are reported in Figure 5. Other than glass transition events, two main peaks appeared on the DSC thermograms due to an exothermic cold crystallization and an endothermic melting transformation [6,39]. Both of these phenomena can be attributed, in our opinion, to a polylactic acid (PLA) fraction (probably added to improve the workability of liquid wood). The presence of PLA is unequivocally confirmed by the melting peak observed in the neat sample of Arboform, which falls in the melting range of pure PLA [40].

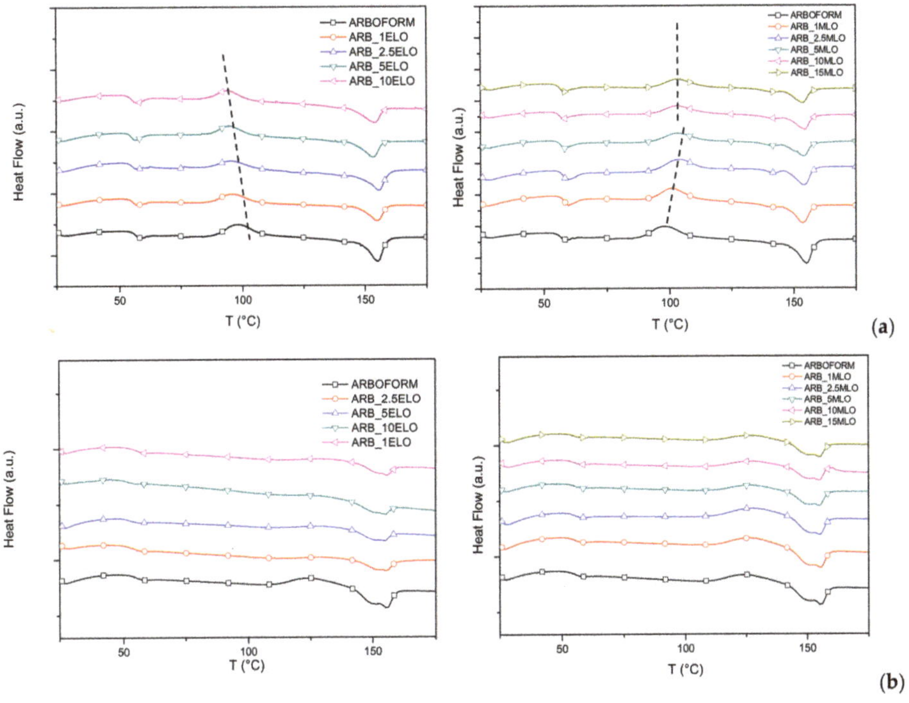

Figure 5. DSC curves (1st heating run (**a**) and 2nd heating run (**b**)) of ELO (left) and MLO (right) added formulations.

Table 4 shows the results of T_g, T_{cc} and T_m, for the studied materials. The literature reports that the MLO has an effect on the reduction of the glass transition temperature for PLA based formulation [41]: in our case, we noticed that T_g values remain constant and comparable to the reference material, with no significant differences between MLO and ELO modified formulations. A double melt peak was identified in the formulations (more evident in the second heating scan) due to the formation of non-perfect small crystals that melt at lower temperatures, causing the formation of a small peak [42]. Regarding the cold crystallization temperature (T_{cc}), a clear effect was indeed found in every composite containing the modified oils. As seen in Figure 5 and the values in Table 4, the addition of ELO provides a decrease in T_{cc} of Arboform formulations. In particular, ELO favors crystallization due to increased chain mobility, owing to a plasticization effect that allows crystallization to occur with lower energy content. On the other hand, MLO presence delayed the crystallization event: it could be that chain extension, branching or cross-linking of polylactic chains can be promoted, showing a moderate increase in elongation at break in combination to a significant improvement of the mechanical resistant properties (both tensile modulus and strength). This macromolecular change can be based on the fact that the cold crystallization peak was broader (or even disappeared) as the MLO content increased and the melting peak slightly shifted to lower temperatures. In particular, the T_{cc} of Arboform formulations with 1 wt % and 2.5 wt % MLO moved up to 100.7 °C and 103.8 °C, respectively, while the neat matrix showed a T_{cc} of 97.2 °C. A similar T_{cc} was observed for formulations with 5, 10 and 15 wt % MLO with values of 102–103 °C [43]. This suggests a break of the crystalline structure in the polymeric phase, so the shift in the cold crystallization process can be related to the network formation of PLA molecules of higher molecular weight that inhibits chain motion during packing and rearrangement. While one can consider that ELO acts as a plasticizer in the green composites, typically increasing volume and reducing polymer–polymer interactions, MLO particles predominantly acted as an anti-nucleating agent, interrupting the folding or packing process of the predominant PLA chains.

Table 4. Thermal parameters from DSC analysis of ELO- and MLO-modified Arboform formulations.

	1st Heating						Cooling	2nd Heating					
	T_g [°C]	T_{cc} [°C]	ΔH_{cc} [Jg^{-1}]	$T_{m'}$ [°C]	$T_{m''}$ [°C]	ΔH_m [Jg^{-1}]	T_g [°C]	T_g [°C]	T_{cc} [°C]	ΔH_{cc} [Jg^{-1}]	T_m [°C]	$T_{m''}$ [°C]	ΔH_m [Jg^{-1}]
ARBOFORM	55.5	97.2	16.0	140.5	154.3	22.7	46.4	55.3	124.9	7.3	148.6	154.9	13.9
ARB_1ELO	55.1	97.3	18.7	141.1	154.7	27.3	47.8	54.7	130.1	3.7	150.4	155.4	9.4
ARB_2.5ELO	54.7	93.7	18.7	140.5	154.2	27.5	46.4	54.0	132.1	3.8	150.7	155.6	6.6
ARB_5ELO	54.4	96.7	18.5	140.2	154.4	28.1	46.4	54.0	130.0	4.0	151.4	154.9	6.8
ARB_10ELO	54.3	95.5	19.0	140.0	154.9	29.0	47.3	52.0	131.1	4.1	151.6	155.2	7.0
ARB_1MLO	57.7	100.7	18.5	140.7	153.6	25.2	47.3	55.5	126.6	12.6	151.4	155.4	14.0
ARB_2.5MLO	57.6	103.8	21.2	142.1	154.0	22.9	47.7	55.6	127.5	12.6	152.2	155.0	13.3
ARB_5MLO	57.2	102.6	21.8	142.8	154.3	24.8	48.4	55.8	127.1	11.6	152.4	155.5	14.1
ARB_10MLO	56.6	102.8	21.6	143.0	154.4	26.9	47.2	55.4	127.0	13.9	150.9	155.0	13.8
ARB_15MLO	56.2	103.2	22.8	142.0	153.4	30.5	46.5	55.3	127.2	15.2	148.6	155.0	14.2

Thermogravimetric analysis: Results of TGA analysis carried out on materials added with 5 wt % and 10 wt % of the two modified oils (Figure 6) shows that the epoxidized oil confers good thermal stability in the temperature range between 160 °C and 240 °C in comparison with a neat matrix. An increase in thermal stability in materials modified with MLO above 260 °C is to be noted. The derivative curves of thermogravimetric test show an increase in the temperatures of the main peaks of the modified materials with respect to the reference matrix. The highest values for each oil are obtained with 5% by weight: the peak of the matrix at 341 °C moves to 347 °C with the epoxidized oil and it rises to 353 °C with the oil with maleic modification. This result indicates an effective bond of modified oils with the matrix. It is well known that, owing to their high thermal stability, vegetable oils and their chemically modified derivatives can be used as thermal stabilizers in polymers, i.e., poly (vinyl chloride), in addition, it has been also recently reported that EVOs and other vegetable oil-derived materials can efficiently thermally stabilize polylactic based formulations [44]. Linear chain extension, branching and/or, even more intensely, cross-linking effects of MLO are responsible for the achieved thermal stabilization. This is related to the fact that these phenomena are considered to counteract chain scission, potentially leading to a more branched structure.

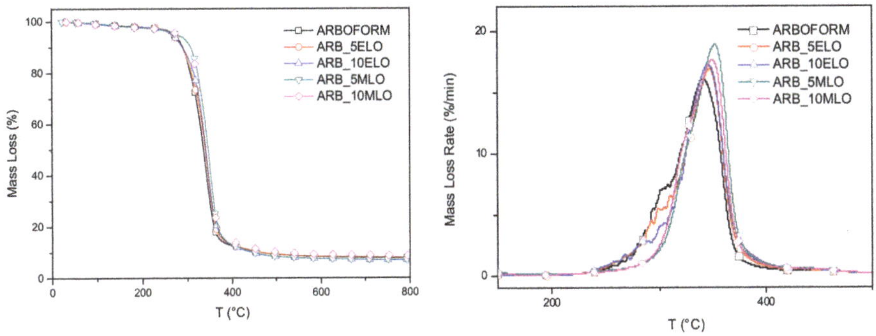

(a) (b)

Figure 6. TG (**a**), DTG (**b**), DSC curves of ELO- and MLO-modified Arboform formulations.

Morphological analysis: Scanning electron microscopy (SEM) micrographs of fractured surfaces for the Arboform formulations are shown in Figure 7.

It can be observed that phase separation has already occurred at low oil contents, highlighting surfaces with submicrometric sized cavities formed upon solidification of the matrix due to the presence of micro-drops of oil. Arboform surface shows a brittle fracture with some radial cracks typical of materials characterized by low deformation and poor toughness. The neat matrix surface appears smooth and relatively flat surface with low roughness, while a considerable increase in the number of round shape irregularities and/or microvoids on the surface fractures appeared in the case of modified formulations [45]. The addition of increasing amounts of the different vegetable oils (VOs) provides some differences: in particular, the Arboform blend compatibilized with 1 wt % ELO shows a smooth fracture surface similar to that of the neat matrix, while Arboform compatibilized with 1 wt % of MLO shows enhanced presence of cavities/voids.

Phase separation has been reported for vegetable oil-derived additives over 5 wt % as small spherical domains due to excess plasticizer/compatibilizer [46]. The fractured surface of the ELO compatibilized Arboform blend shows strong heterogeneity, which can be related to phase separation, while the MLO compatibilized blends offer a more homogeneous fracture surface, which could be responsible for the maximum achieved elongation at break (see Table 2). Improved miscibility for MLO-modified Arboform formulations can be related to a similar solubility parameter between MLO and the biopolyester contained in the Arboform matrix, thus allowing interactions between them. It has been reported that chemically modified vegetable oils have solubility parameters close to PLA, specifically epoxidized compounds derived from soybean oil achieved a solubility parameter of 16.70 MPa$^{1/2}$ [47], while maleinized oil showed a value of 19 MPa$^{1/2}$, quite similar to the solubility parameter of PLA, which has been reported to be in the range 19.5-22.0 MPa$^{1/2}$ [48,49].

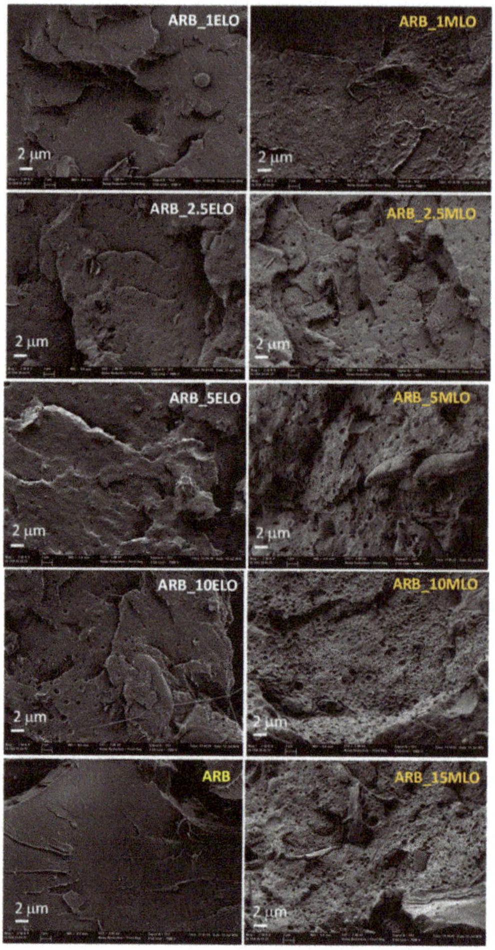

Figure 7. FESEM micrographs of fractured surfaces of ELO- and MLO-modified Arboform formulations.

4. Conclusions

The addition of environmentally friendly plasticizers derived from vegetable oils, epoxidized (ELO) and maleinized linseed oil (MLO) at relatively low contents (0–15 wt %) led to a significant increase in mechanical properties, such as impact-absorbed energy, tensile and flexural strength, proving that these attractive additives can provide plasticization to brittle polymers and also improve compatibility in immiscible or partially miscible polymer blends. In detail, ARB_5ELO and ARB_5MLO showed an increase, respectively, of +98% and 118% of impact-absorbed energy, +39% and +43% of tensile strength and +37% and +49% of flexural strength. The best results in terms of impact-absorbed energy were obtained with ARB_5MLO (+118%), also tensile (+43%) and flexural strength (+49%), which were resultantly positively affected. In addition, the natural origin of the vegetable oils represents an environmentally effective solution to progress in the preparation and commercialization of industrial formulations based on biopolymers and biopolymeric blends.

Author Contributions: Conceptualization, F.D. and M.D.S.; investigation, F.D., M.D.S and A.C.-V.; writing—original draft preparation, F.D., D.P. and F.L.; writing—review and editing, F.D., D.P. and J.L.-M.; supervision, J.L.-M. and L.T. All authors have read and agreed to the published version of the manuscript.

Funding: This research received no external funding.

Conflicts of Interest: The authors declare no conflicts of interest.

References

1. Barhalescu, M.L.; Jarcau, M.; Poroch-Seritan, M.; Costescu, E.; Vaideanu, D.; Petrescu, T.C. *The Behaviour of "Liquid Wood" When Exposed to Some Physico-Chemical Factors*; IOP Publishing: Bristol, UK; p. 012034.
2. Plăvănescu, S. Biodegradable Composite Materials—Arboform: A Review. *Int. J. Mod. Manuf. Technol.* **2014**, *6*, 63–84.
3. Nägele, H.; Pfitzer, J.; Nägele, E.; Inone, E.R.; Eisenreich, N.; Eckl, W.; Eyerer, P. Arboform®-A Thermoplastic, Processable Material from Lignin and Natural Fibers. In *Chemical Modification, Properties, and Usage of Lignin*; Hu, T.Q., Ed.; Springer US: Boston, MA, USA, 2002; pp. 101–119.
4. Nagele, H.; Pfitzer, J.; Ziegler, L.; Inone-Kauffmann, E.R.; Eckl, W.; Eisenreich, N. Lignin Matrix Composites from Natural Resources-ARBOFORM®. In *Bio-Based Plastics: Materials and Applications*; Kabasci, S., Ed.; John Wiley & Sons, Ltd.: Hoboken, NJ, USA, 2013; pp. 89–115. [CrossRef]
5. Mokhena, T.C.; Mochane, M.J.; Sadiku, E.R.; Agboola, O.; John, M.J. Opportunities for PLA and Its Blends in Various Applications. In *Green Biopolymers and their Nanocomposites*; Materials Horizons: From Nature to Nanomaterials; Gnanasekaran, D., Ed.; Springer: Singapore. [CrossRef]
6. Nedelcu, D.; Lohan, N.M.; Volf, I.; Comaneci, R. Thermal behaviour and stability of the Arboform® LV3 Nature liquid wood. *Compos. Part B Eng.* **2016**, *103*, 84–89. [CrossRef]
7. Nedelcu, D.; Santo, L.; Santos, A.; Plavanescu, S. Mechanical Behaviour Evaluation of Arboform Material Samples by Bending Deflection Test. *Mat. Plast.* **2015**, *52*, 423–426.
8. Bouajila, J.; Dole, P.; Joly, C.; Limare, A. Some laws of a lignin plasticization. *J. Appl. Polym. Sci.* **2006**, *102*, 1445–1451. [CrossRef]
9. Wang, C.; Kelley, S.S.; Venditti, R.A. Lignin-based thermoplastic materials. *Chem. Sus. Chem.* **2016**, *9*, 770–783. [CrossRef]
10. Sakata, I.; Senju, R. Thermoplastic behavior of lignin with various synthetic plasticizers. *J. Appl. Polym. Sci.* **1975**, *19*, 2799–2810. [CrossRef]
11. Antonsson, S.; Henriksson, G.; Johansson, M.; Lindström, M.E. Low Mw-lignin fractions together with vegetable oils as available oligomers for novel paper-coating applications as hydrophobic barrier. *Ind. Crop. Prod.* **2008**, *27*, 98–103. [CrossRef]
12. *Plastics — Standard atmospheres for conditioning and testing*; ISO 291; International Organization for Standardization: Geneva, Switzerland, 2008.
13. *Plastics — Determination of Charpy impact properties — Part 1: Non-instrumented impact test*; ISO 179; International Organization for Standardization: Geneva, Switzerland, 2008.
14. *Plastics — Determination of tensile properties — Part 1: General principles*; ISO 527; International Organization for Standardization: Geneva, Switzerland, 2019.
15. *Plastics — Determination of flexural properties*; ISO 178; International Organization for Standardization: Geneva, Switzerland, 2019.
16. *Plastics — Determination of temperature of deflection under load — Part 2: Plastics and ebonite*; ISO 75; International Organization for Standardization: Geneva, Switzerland, 2013.
17. Jiang, L.; Chen, F.; Qian, J.; Huang, J.; Wolcott, M.; Liu, L.; Zhang, J. Reinforcing and toughening effects of bamboo pulp fiber on poly (3-hydroxybutyrate-co-3-hydroxyvalerate) fiber composites. *Ind Eng. Chem. Res.* **2009**, *49*, 572–577. [CrossRef]
18. Sahoo, S.; Misra, M.; Mohanty, A.K. Enhanced properties of lignin-based biodegradable polymer composites using injection moulding process. *Compo. Part A Appl. Sci. Manuf.* **2011**, *42*, 1710–1718. [CrossRef]
19. Ferri, J.M.; Garcia-Garcia, D.; Sánchez-Nacher, L.; Fenollar, O.; Balart, R. The effect of maleinized linseed oil (MLO) on mechanical performance of poly (lactic acid)-thermoplastic starch (PLA-TPS) blends. *Carbohydr. Polym.* **2016**, *147*, 60–68. [CrossRef]
20. Mikus, P.Y.; Alix, S.; Soulestin, J.; Lacrampe, M.F.; Krawczak, P.; Coqueret, X.; Dole, P. Deformation mechanisms of plasticized starch materials. *Carbohydr. Polym.* **2014**, *114*, 450–457. [CrossRef] [PubMed]

21. Oliveira de Castro, D.; Frollini, E.; Ruvolo-Filho, A.; Dufresne, A. "Green polyethylene" and curauá cellulose nanocrystal based nanocomposites: Effect of vegetable oils as coupling agent and processing technique. *J. Polym. Sci. Part B Polym. Phys.* **2015**, *53*, 1010–1019. [CrossRef]
22. Coles, S. Bioplastics from Lipids. In *Bio-Based Plastics: Materials and Applications*; Kabasci, S., Ed.; John Wiley & Sons, Ltd.: Hoboken, NJ, USA, 2013; pp. 117–135. [CrossRef]
23. Samper, M.-D.; Ferri, J.M.; Carbonell-Verdu, A.; Balart, R.; Fenollar, O. Properties of biobased epoxy resins from epoxidized linseed oil (ELO) crosslinked with a mixture of cyclic anhydride and maleinized linseed oil. *EXPRESS Polym. Lett.* **2019**, *13*, 407–418. [CrossRef]
24. Balart, J.F.; Fombuena, V.; Fenollar, O.; Boronat, T.; Sánchez-Nacher, L. Processing and characterization of high environmental efficiency composites based on PLA and hazelnut shell flour (HSF) with biobased plasticizers derived from epoxidized linseed oil (ELO). *Compos. Part B Eng.* **2016**, *86*, 168–177. [CrossRef]
25. Ferri, J.M.; Garcia-Garcia, D.; Montanes, N.; Fenollar, O.; Balart, R. The effect of maleinized linseed oil as biobased plasticizer in poly (lactic acid)-based formulations. *Polym. Int.* **2017**, *66*, 882–891. [CrossRef]
26. Carbonell-Verdu, A.; Garcia-Garcia, D.; Dominici, F.; Torre, L.; Sanchez-Nacher, L.; Balart, R. PLA films with improved flexibility properties by using maleinized cottonseed oil. *Eur. Polym. J.* **2017**, *91*, 248–259. [CrossRef]
27. Gutierrez-Villarreal, M.H.; Rodríguez-Velazquez, J. The effect of citrate esters as plasticizers on the thermal and mechanical properties of poly (methyl methacrylate). *J. Appl. Polym. Sci.* **2007**, *105*, 2370–2375. [CrossRef]
28. Vidotti, S.E.; Chinellato, A.C.; Hu, G.H.; Pessan, L.A. Effects of low molar mass additives on the molecular mobility and transport properties of polysulfone. *Journal of applied polymer science* **2006**, *101*, 825–832. [CrossRef]
29. Moraru, C.I.; Lee, T.C.; Karwe, M.V.; Kokini, J.L. Plasticizing and Antiplasticizing Effects of Water and Polyols on a Meat-Starch Extruded Matri. *J. Food Sci* **2002**, *67*, 3396–3401. [CrossRef]
30. Silverajah, V.S.; Ibrahim, N.A.; Yunus, W.M.Z.W.; Hassan, H.A.; Woei, C.B. A comparative study on the mechanical, thermal and morphological characterization of poly (lactic acid)/epoxidized palm oil blend. *Int. J. Mol. Sci.* **2012**, *13*, 5878–5898. [CrossRef]
31. Chieng, B.; Ibrahim, N.; Then, Y.; Loo, Y. Epoxidized vegetable oils plasticized poly (lactic acid) biocomposites: Mechanical, thermal and morphology properties. *Molecules* **2014**, *19*, 16024–16038. [CrossRef]
32. Chieng, B.W.; Ibrahim, N.A.; Yunus, W.M.Z.W.; Hussein, M.Z. Plasticized poly (lactic acid) with low molecular weight poly (ethylene glycol): Mechanical, thermal, and morphology properties. *J. Appl. Polym. Sci.* **2013**, *130*, 4576–4580. [CrossRef]
33. Lau, A.K.-t.; Hung, A.P.Y. *Natural Fiber-Reinforced Biodegradable and Bioresorbable Polymer Composites*; Woodhead Publishing, imprint of Elsevier Ltd.: Duxford, UK, 2017.
34. Quiles-Carrillo, L.; Montanes, N.; Sammon, C.; Balart, R.; Torres-Giner, S. Compatibilization of highly sustainable polylactide/almond shell flour composites by reactive extrusion with maleinized linseed oil. *Ind. Crop. Prod.* **2018**, *111*, 878–888. [CrossRef]
35. Alam, J.; Alam, M.; Raja, M.; Abduljaleel, Z.; Dass, L. MWCNTs-reinforced epoxidized linseed oil plasticized polylactic acid nanocomposite and its electroactive shape memory behavior. *Int. J. Mol. Sci.* **2014**, *15*, 19924–19937. [CrossRef] [PubMed]
36. Garcia-Campo, M.; Quiles-Carrillo, L.; Masia, J.; Reig-Pérez, M.; Montanes, N.; Balart, R. Environmentally friendly compatibilizers from soybean oil for ternary blends of poly (lactic acid)-PLA, poly (ε-caprolactone)-PCL and poly (3-hydroxybutyrate)-PH. *Materials* **2017**, *10*, 1339. [CrossRef] [PubMed]
37. Mazurchevici, S.; Quadrini, F.; Nedelcu, D. The liquid wood heat flow and material properties as a function of temperature. *Mat. Res. Express* **2018**, *5*, 035303. [CrossRef]
38. Plavanescu, S.; Carausu, C.; Comaneci, R.; Nedelcu, D. *The Influence of Technological Parameters on the Dynamic Behavior of "Liquid Wood" Samples Obtained by Injection Molding*; AIP Publishing: Melville, NY, USA, 2017; p. 030038.
39. Nedelcu, D.; Ciofu, C.; Lohan, N.M. Microindentation and differential scanning calorimetry of "liquid wood". *Compos. Part B Eng.* **2013**, *55*, 11–15. [CrossRef]
40. Cicala, G.; Tosto, C.; Latteri, A.; La Rosa, D.A.; Blanco, I.; Elsabbagh, A.; Russo, P.; Ziegmann, G. Green Composites Based on Blends of Polypropylene with Liquid Wood Reinforced with Hemp Fibers: Thermomechanical Properties and the Effect of Recycling Cycles. *Materials* **2017**, *10*, 998. [CrossRef]

41. Pawlak, F.; Aldas, M.; López-Martínez, J.; Samper, D.M. Effect of Different Compatibilizers on Injection-Molded Green Fiber-Reinforced Polymers Based on Poly(lactic acid)-Maleinized Linseed Oil System and Sheep Wool. *Polymers* **2019**, *11*, 1514. [CrossRef]
42. Magoń, A.; Pyda, M. Study of crystalline and amorphous phases of biodegradable poly(lactic acid) by advanced thermal analysis. *Polymer* **2009**, *50*, 3967–3973. [CrossRef]
43. Quiles-Carrillo, L.; Blanes-Martínez, M.M.; Montanes, N.; Fenollar, O.; Torres-Giner, S.; Balart, R. Reactive toughening of injection-molded polylactide pieces using maleinized hemp seed oil. *Eur. Polym. J.* **2018**, *98*, 402–410. [CrossRef]
44. Chieng, B.W.; Ibrahim, N.A.; Then, Y.Y.; Loo, Y.Y. Epoxidized Jatropha Oil as a Sustainable Plasticizer to Poly(lactic Acid). *Polymers* **2017**, *9*, 204. [CrossRef] [PubMed]
45. Carbonell-Verdu, A.; Ferri, J.M.; Dominici, F.; Boronat, T.; Sanchez-Nacher, L.; Balart, R.; Torre, L. Manufacturing and compatibilization of PLA/PBAT binary blends by cottonseed oil-based derivatives. *Express Polym. Lett.* **2018**, *12*, 808–823. [CrossRef]
46. Garcia-Garcia, D.; Fenollar, O.; Fombuena, V.; Lopez-Martinez, J.; Balart, R. Improvement of Mechanical Ductile Properties of Poly(3-hydroxybutyrate) by Using Vegetable Oil Derivatives. *Macromol. Mat. Eng.* **2017**, *302*, 1600330. [CrossRef]
47. Wang, Q.; Chen, Y.; Tang, J.; Zhang, Z. Determination of the Solubility Parameter of Epoxidized Soybean Oil by Inverse Gas Chromatography. *J. Macromol. Sci. Part B* **2013**, *52*, 1405–1413. [CrossRef]
48. Domingues, R.C.C.; Pereira, C.C.; Borges, C.P. Morphological control and properties of poly(lactic acid) hollow fibers for biomedical applications. *J. Appl. Polym. Sci.* **2017**, *134*, 45494. [CrossRef]
49. Abbott, S. Chemical compatibility of poly (lactic acid): A practical framework using Hansen solubility parameters. In *Poly (lactic acid): Synthesis, Structures, Properties, Processing, and Applications*; Auras, R.A., Lim, L.-T., Selke, S.E.M., Tsuji, H., Eds.; John Wiley & Sons: Hoboken, NJ, USA; pp. 83–95. [CrossRef]

materials

MDPI

Article

Optimization of the Loading of an Environmentally Friendly Compatibilizer Derived from Linseed Oil in Poly(Lactic Acid)/Diatomaceous Earth Composites

Lucia Gonzalez [1], Angel Agüero [1], Luis Quiles-Carrillo [1,*], Diego Lascano [2] and Nestor Montanes [1]

1 Technological Institute of Materials (ITM), Universitat Politècnica de València (UPV), Plaza Ferrándiz y Carbonell 1, 03801 Alcoy, Spain; lugona@epsa.upv.es (L.G.); guanche.ar@gmail.com (A.A.); nesmonmu@upvnet.upv.es (N.M.)

2 Escuela Politécnica Nacional, Quito 17-01-2759, Ecuador; diegol-zn@hotmail.com

* Correspondence: luiquic1@epsa.upv.es; Tel.: +34-966-528-433

Received: 25 April 2019; Accepted: 13 May 2019; Published: 17 May 2019

Abstract: Maleinized linseed oil (MLO) has been successfully used as biobased compatibilizer in polyester blends. Its efficiency as compatibilizer in polymer composites with organic and inorganic fillers, compared to other traditional fillers, has also been proved. The goal of this work is to optimize the amount of MLO on poly(lactic acid)/diatomaceous earth (PLA/DE) composites to open new potential to these materials in the active packaging industry without compromising the environmental efficiency of these composites. The amount of DE remains constant at 10 wt% and MLO varies from 1 to 15 phr (weight parts of MLO per 100 g of PLA/DE composite). The effect of MLO on mechanical, thermal, thermomechanical and morphological properties is described in this work. The obtained results show a clear embrittlement of the uncompatibilized PLA/DE composites, which is progressively reduced by the addition of MLO. MLO shows good miscibility at low concentrations (lower than 5 phr) while above 5 phr, a clear phase separation phenomenon can be detected, with the formation of rounded microvoids and shapes which have a positive effect on impact strength.

Keywords: maleinized linseed oil MLO; poly(lactic acid); diatomaceous earth; biocomposites; active containers

1. Introduction

Natural oils and, in particular, vegetable oils, are currently being widely investigated as they could be a source of a wide variety of new environmentally friendly materials from renewable resources that could positively contribute to sustainable development. In addition, some of these natural vegetable oils cannot be used in the food industry because of regulation restrictions due to their composition and other components. For this reason, some of these vegetable oils are obtained as by-products from other industries, and this contributes to their high worldwide availability, together with their cost-effective price. Recently, selectively modified vegetable oils have been proposed as interesting materials for compatibilization of polymer blends. Other applications of these modified vegetable oils include partially biobased thermosetting resins as an alternative to petroleum-derived resins such as epoxies, which can also be used as matrices in high environmental efficiency green composites. In addition to this, modified vegetable oils are widely used as secondary plasticizers in poly(vinyl chloride)—PVC—to provide increased thermal stability [1–8]. To tailor the desired functionality of a vegetable oil, different chemical modifications have been proposed, including epoxidation, maleinization, acrylation, and hydroxylation, among others.

Vegetable oils are interesting from a chemical point of view because of their triglyceride structure, which consists of a glycerol basic structure which is chemically bonded to different fatty acids through ester bonds. Fatty acids can be saturated as stearic acid (C18:0, which means a chain length of 18 carbon atoms without any unsaturation), palmitic acid (C16:0) or margaric acid (C17:0). These saturated fatty acids are not interesting for chemical modification. Nevertheless, some fatty acids can contain one, two or more unsaturations, thus leading to unsaturated fatty acids such as palmitoleic acid (C16:1), oleic acid (C18:1), linoleic acid (C18:2) or linolenic acid (C18:3), and others. Figure 1 shows a schematic representation of the chemical structure of an unsaturated vegetable oil.

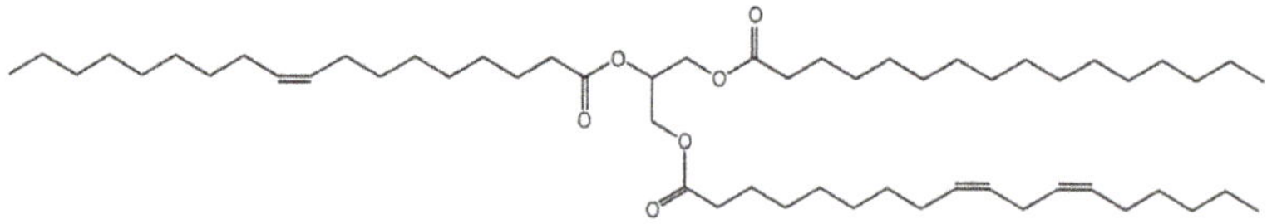

Figure 1. Schematic representation of a triglyceride structure which is the base of vegetable oils, showing different (saturated and unsaturated) fatty acids bonded to a glycerol basic structure through ester bonds.

Unsaturations are highly reactive points, such that they represent the base for a chemical modification to provide the desired functionality. Epoxidation is one of the most investigated chemical modification of a vegetable oil. By a simple epoxidation process with peroxoacids derived from, for example, hydrogen peroxide and acetic acid, unsaturations can be converted into oxirane rings [5,9–11]. These oxirane rings allow crosslinking in a similar way to a petroleum-derived epoxy resins with different hardener systems [12]. Moreover, oxirane rings increase the polarity of the triglyceride, and this provides good plasticization properties to different polymers. In particular, epoxidized soybean oil (ESBO) and epoxidized linseed oil (ELO) are commercially available as secondary plasticizers for polyvinyl chloride (PVC) but their use has been extended to other polymers such as poly(lactic acid), poly(hydroxybutyrate), and so on [13–18]. Other interesting chemical modification of vegetable oils is acrylation, which is carried out on previously epoxidized vegetable oils by reaction with acrylic monomers (acrylic acid, methyl methacrylate, and so on). These acrylic monomers react with the oxirane rings thus leading to increased reactivity to give interesting plasticizer materials or thermosetting resins with a similar behavior to vinyl-ester resins [12,15,19]. On the other hand, vegetable oils can be subjected to a maleinization process, which increases reactivity as well as the above-mentioned methods [20–23]. This modification is based on the reaction of maleic anhydride with unsaturations, thus leading to anchorage of maleic anhydride in the triglyceride structure [21,24]. Some recent investigations have revealed the interesting plasticization effect of maleinized vegetable oils in some polymers such as poly(lactic acid)—PLA—to increase chain mobility [1,25], but in general, there are few research works on the potential of these biobased materials as additive in plastic formulations. In addition to the plasticization effect of maleinized linseed oil (MLO) on PLA, it has been corroborated the coupling/compatibilizing effect of MLO on PLA composites with diatomaceous earth [26]. For this reason, MLO stands out as an alternative to traditional compatibilizers in the food packaging industry and, in particular, in active packaging, as diatomaceous earth particles are highly porous structures that can be loaded with antioxidants that can be released in a controlled way to increase the shelf life of a product. This approach has given interesting results. For example, Tornuk et al. and Brandelli et al. have reported the use of montmorillonite (MMT) and halloysite nanotubes (HNTs) as carriers for different active principles for active packaging [27–32]. Diatomaceous earth (DE) represent an interesting alternative to other clays/nanoclays. From a chemical point of view, DE is composed of amorphous silica, $SiO_2{\cdot}n\,H_2O$. From a morphological point of view, it consists hierarchical micro/nanoporous structure. It is this hierarchical porosity that allows the use of DE as carriers for active principles in active packaging. DE is composed of micro-shells of marine unicellular eukaryote organisms in phytoplankton and formed a sediment millions of years ago. Diatom fossilization led to formation of huge diatomaceous earth deposits; therefore, it is an abundant cost-effective product. The main properties of DE have qualities of very low density, porous structure, abrasive, chemical inertness, biocompatible, high absorption capacity, low thermal conductivity, high resistance to acids, and permeability, among others. Currently, diatomaceous earth is widely used as filtration media, for absorption, as a natural insecticide, as functional additives, dental fillings, membranes, and chemical sensors, among other things. When used as natural fillers, DE can provide two different effects: on the one hand, they can provide some reinforcement effect, and on the other hand, they can act as carriers for the controlled release of active principles [33–37]. In a previous work [26], we reported the exceptional performance of maleinized linseed oil as a compatibilizer in PLA/DE composites, compared to other conventional compatibilizers. In this work, the main goal is to optimize the DE/MLO ratio to obtain the best balanced properties on poly(lactic acid)—PLA/diatomaceous earth—DE composites.

2. Experimental

2.1. Materials

The polymer matrix used in this study was a commercial grade of poly(lactic acid) manufactured and distributed by Nature Works LLC (Minnetonka, MN, USA). This commercial grade was Ingeo Biopolymer 6201D with a melt flow index in the 15–30 g/(10 min) range at 210 °C, which makes it suitable for injection moulding and melt spinning of fibres as well. It is a lightweight material with a typical density of 1.24 g cm^{-3}. Regarding diatomaceous earth (DE), this was supplied by ECO-Tierra de diatomeas (Granada, Spain). Table 1 summarizes the composition of this DE.

Table 1. Composition of diatomaceous earth used in PLA/DE composites.

Component	Weight Percentage (wt%)
SiO_2	89.00
$Na_2O + K_2O$	1.88
CaO	6.73
Al_2O_3	1.00
Fe_2O_3	0.46

This DE shows different particle sizes and shapes, but triangular shapes with rounded angles are predominant, as can be seen in Figure 2. The average particle size is between 4 and 7 μm. It is worth noting the highly porous structure of these DE particles.

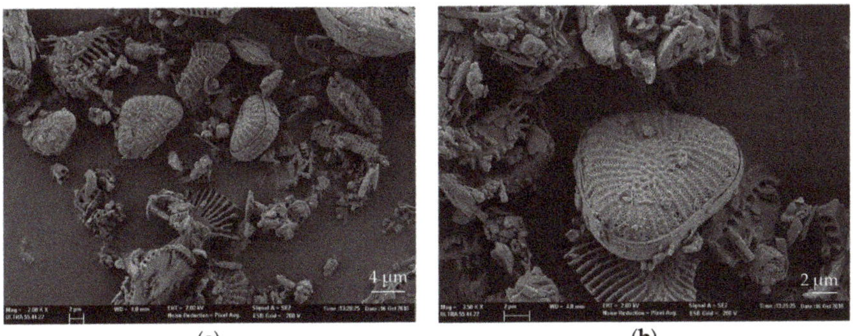

Figure 2. FESEM images of diatomaceous earth at different magnification (**a**) 2000× and (**b**) triangular shape detail at 3500×.

The compatibilizer used in this study was maleinized linseed oil (MLO), which is obtained from the reaction of maleic anhydride (MA) with the unsaturations contained in linseed oil (oleic acid-C18:1, linoleic acid-C18:2 and linolenic-C18:3). This MLO was a commercial-grade VEOMER LIN supplied by Vandeputte (Mouscron, Belgium). Some of its features included a viscosity of 10 dPa s measured at 20 °C and an acid value in the 105–130 mg KOH g^{-1} range.

2.2. *Manufacturing of PLA/DE Composites Compatibilized with MLO*

Table 2 summarizes the compositions and coding of the developed formulations. Prior to any processing, PLA and diatomaceous earth were dried at 60 °C for 18 h to remove the residual moisture.

Table 2. Composition and labelling of PLA/DE composites with different amounts of MLO compatibilizer.

Code	PLA (wt%)	DE (wt%)	MLO (phr) *
PLA	100	0	0
PLA+DE	90	10	0
PLA+DE+1MLO	90	10	1
PLA+DE+2.5MLO	90	10	2.5
PLA+DE+5MLO	90	10	5
PLA+DE+10MLO	90	10	10
PLA+DE+15MLO	90	10	15

* phr represents the weight parts of MLO per 100 weight parts of PLA/DE composites.

The above-mentioned compositions were subjected to an initial compounding stage in a co-rotating twin screw extruder from DUPRA S.L. (Alicante, Spain). Different temperatures were selected for the extrusion process by taking into account that the melt peak temperature of PLA is close to 170 °C; therefore, the initial heating stage close to the hopper was set to 165 °C and progressively increased up to 180 °C in the extrusion die. A rotating speed of 20–25 rpm was used. After the compounding stage in a co-rotating twin screw extruder a continuous filament (4 mm diameter) was obtained. This filament was cooled down in air to room temperature (to avoid hydrolysis) and dropped into a shredder manufactured by Mayper (Valencia, Spain) which gave an average pellet size 3 mm in diameter and 2–2.5 mm in height. The pellet of the different composites obtained was then processed by injection moulding process using a Meteor 270/75 from Mateu & Solé (Barcelona, Spain) injection moulding machine. The temperature profile, from the feeding zone to the injection nozzle, was set to: 170–180–190–200 °C. The material was processed with a holding pressure of 75 bar, with an injection time of 8 s in mold and 20 s as cooling time in the open mould.

2.3. *Characterization and Testing*

2.3.1. Thermal and Thermo-Mechanical Characterization

DE can affect thermal behaviour of PLA matrix. For this reason, the effects of DE addition and MLO on thermal transitions of PLA/DE composites were obtained using differential scanning calorimetry (DSC) using a Mettler Toledo 821 calorimeter (Schwerzenbach, Switzerland). A typical procedure is based on the use of a sample weight of about 7–10 mg. Samples were accurately weighed and placed into standard aluminium sealed pans (40 μL). The thermal program was divided into three different stages. The first stage was programmed from 30 °C to 200 °C at a heating rate of 10 °C min^{-1}. This stage was applied to remove the previous thermal history which is particularly important in semicrystalline polymers. After this, a cooling stage from 200 °C down to 0 °C at a constant cooling rate of 10 °C min^{-1} was applied. With this stage, samples are subjected to a controlled cooling process which allows further comparisons. Finally, a new heating cycle from 0 °C to 300 °C at a heating rate of 10 °C min^{-1} was applied, and all thermal transitions were obtained in this second heating cycle. To avoid undesired oxidations a nitrogen inert atmosphere (66 mL min^{-1}) was used. An important parameter in semicrystalline polymers is the degree of crystallinity (χ_c) which represents the ratio between the crystalline areas contained in the polymer and the total volume. The degree of crystallinity (χ_c) was calculated by using the following expression:

$$\chi_c = \frac{\Delta H_m - \Delta H_{cc}}{\Delta H_m^0 \cdot (1-w)} \cdot 100(\%) \quad (1)$$

In this equation, ΔH_m and ΔH_{cc} (J g^{-1}) represent the melt and cold crystallization enthalpies, respectively, while ΔH_m^0 corresponds to the theoretical melt enthalpy of a fully crystalline PLA, and was taken as 93.0 J g^{-1}, as reported in literature [38]. Finally, (1-w) stands for the weight fraction of PLA in the sample without DE or MLO.

Complementary to the characterization of thermal transitions, the thermal stability was evaluated by means of thermogravimetric analysis (TGA) using a TGA/SDTA 851 thermobalance from Mettler Toledo Inc. thermobalance (Schwerzenbach, Switzerland). The selected thermal program was a dynamic heating from 30 °C to 700 °C at a constant heating rate of 20 °C min^{-1} using air atmosphere to simulate more aggressive conditions than using inert atmosphere. The sample weight mass varied in the 8–10 mg range and all the samples had similar dimensions to obtain comparable and reproducible results. Standard alumina crucibles (70 mL) were employed for TGA characterization.

Dynamic mechanical behaviour of PLA/DE composites with different MLO loadings was used to follow the evolution of the storage modulus (G') and the dynamic damping factor (*tan* δ) as a function of increasing temperature. To this end, an AR-G2 oscillatory rheometer from TA Instruments (New Castle, PA, USA), equipped with an environmental test chamber (ETC) and a special clamp device for solid samples, was using in torsion/shear mode. Rectangular samples with dimensions of 40 × 10 mm^2 and an average thickness of 4 mm were subjected to a temperature sweep from 30 °C up to 140 °C at a heating rate of 2 °C min^{-1}. This temperature range was selected because the main thermal transitions of PLA in the solid state, i.e., the glass transition temperature (T_g) and the cold crystallization occur in this range. Other characteristics of this experiment were defined by a maximum shear deformation (%γ) of 0.1% and a frequency of 1 Hz.

2.3.2. Mechanical Characterization

Mechanical properties of PLA/DE composites with varying MLO loading were obtained from tensile tests following ISO 527-1 in a universal test machine ELIB 30 from S.A.E. Ibertest (Madrid, Spain). The selected conditions for these tests were: load cell of 5 kN, crosshead speed of 10 mm min^{-1}. Different tensile properties were obtained and averaged from five different tests for each sample, i.e., tensile strength (σ_t), tensile modulus (E_t) and elongation at break ($\%\varepsilon_b$).

Mechanical response of PLA/DE composites in impact conditions were obtained using a Charpy pendulum with a total energy of 1 J from Metrotec S.A. (San Sebastián, Spain) following the guidelines of ISO 179. Five unnotched samples were tested for each formulation, and the impact strength was calculated in kJ m^{-2} by taking into account the cross section of samples.

In addition to the above-mentioned characterization techniques, Shore D hardness was obtained in a durometer 673-D from J. Bot S.A. (Barcelona, Spain), as indicated in ISO 868. In a similar way, hardness was measured in five different samples, and the average values were collected.

2.3.3. Microscopic Characterization

The internal morphology of PLA/DE composites was studied from fractured samples on impact tests. A field emission scanning electron microscope (FESEM) from Oxford Instruments (Abingdon, United Kingdom) working at an acceleration voltage of 2 kV was used. To provide conducting properties to samples and avoid sample charge, all fractured samples were covered with an ultrathin gold-palladium alloy in a Quorum Technologies Ltd. EMITECH model SC7620 sputter coater (East Sussex, UK).

3. Results and Discussion

3.1. Thermal Properties of PLA/DE Composites with Varying MLO Loading

A comparative plot of the DSC thermograms of neat PLA and PLA/DE composites with varying MLO content is gathered in Figure 3. A first thermal transition can be seen at around 60 °C that corresponds to the glass transition temperature (T_g) of PLA. As PLA is highly sensitive to the cooling process, which affects the degree of crystallinity, a cold crystallization peak can be observed with a peak maximum of 119 °C, while this characteristic peak moves down to lower values in PLA/DE composites. The cold crystallization process occurs at lower temperatures with 10 wt% DE. In particular, the peak maximum is displaced to 112 °C. This slight change in the cold crystallization characteristic temperatures is directly related to the fact that DE particles can act as nucleants for crystallization, thus favouring crystallite formation [39–42]. At higher temperatures, close to 170 °C, an endothermic peak can be observed which is attributed to the melt process of the crystalline fraction in PLA. Table 3 shows the main thermal results obtained from DSC characterization.

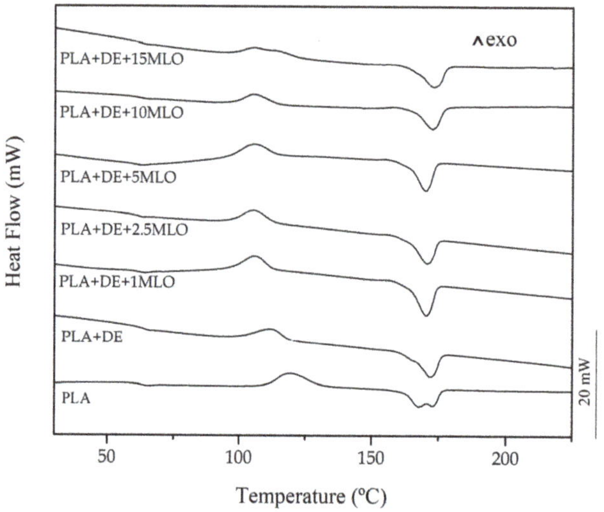

Figure 3. DSC thermograms of neat PLA and PLA+DE composites with different MLO loading (expressed in phr).

Table 3. Main thermal parameters of PLA and PLA+DE composites with different MLO loading (expressed in phr) obtained by differential scanning calorimetry (DSC).

Sample	T_g (°C)	T_{cc} (°C)	ΔH_{cc} (J g^{-1})	T_m (°C)	ΔH_m (J g^{-1})	χ_c (%)
PLA	63.0	119.5	27.5	169.9	36.5	9.7
PLA-DE	63.8	111.8	19.5	171.3	32.7	15.7
PLA-DE-1MLO	61.8	105.7	21.6	169.7	33.85	14.7
PLA-DE-2.5MLO	61.3	105.2	21.5	170.0	33.6	15.0
PLA-DE-5MLO	60.2	105.5	21.2	169.5	31.2	12.5
PLA-DE-10MLO	62.1	105.5	16.9	172.0	26.9	13.0
PLA-DE-15MLO	61.5	105.6	18.7	172.4	28.1	13.0

Addition of MLO provides a slight decrease in T_g of PLA/DE composites. In particular, the maximum decrease is obtained for a MLO loading of 5 phr which gives a T_g value of 60.2 °C (3.6 °C lower than PLA/DE composites). This slight decrease in T_g is representative for poor plasticization effects, as observed in other polymer systems [1,43,44]. Nevertheless, with regard to the cold crystallization process, a clear decrease in the peak temperature (T_{cc}) can be seen from 112 °C (PLA/DE composite) down to values of 105 °C for almost all composites, independently of the MLO loading. MLO favors crystallization due to increased chain mobility. On the other hand, the melt peak temperature of the obtained materials does not change in a remarkable way, with values of about 170 °C, even with increasing MLO content. With regard to normalized enthalpies related to the cold crystallization and melting processes, it is worth noting that they are very useful for making an estimation of the degree of crystallinity (χ) of the PLA/DE composites with increasing MLO content. Neat PLA shows a degree of crystallinity of 9.7%, while the addition of 10 wt% DE leads to increased crystallinity up to values of 15.7% due to the nucleant effect of DE. This is also consistent with the decrease in the cold crystallization peak temperature, as stable crystallites can be obtained at lower temperatures. By adding low MLO loads in the 1–2 phr range, the degree of crystallinity remains almost constant but high MLO loading in the 5–15 phr range, favour the stability of the amorphous PLA domains which is detectable by a decrease in the degree of crystallinity to values of 13%.

With regard to the thermal stability of PLA/DE composites with varying MLO content, Figure 4 shows the thermogravimetric TGA degradation profiles of neat PLA, PLA/DE composite and compatibilized PLA/DE composite with different MLO loadings. As can be observed in Figure 4, all the developed materials in this study show a one-step degradation process. It is important to take into account that the overall thermal stability is related to chemical and physical interactions between the base polymer matrix, PLA, inorganic particles (DE) and MLO [45]. Table 4 shows the main thermal results obtained from TGA analysis.

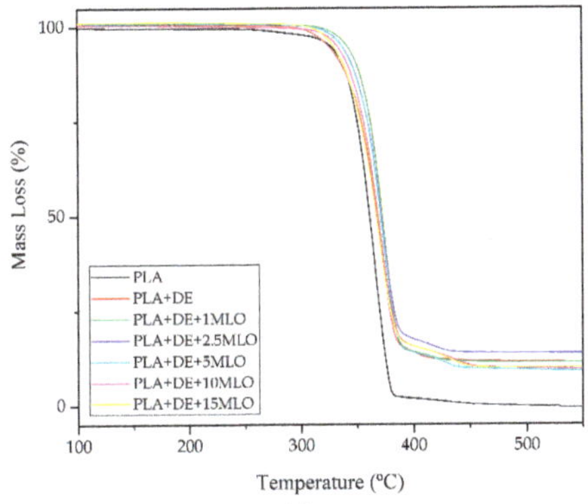

(a)

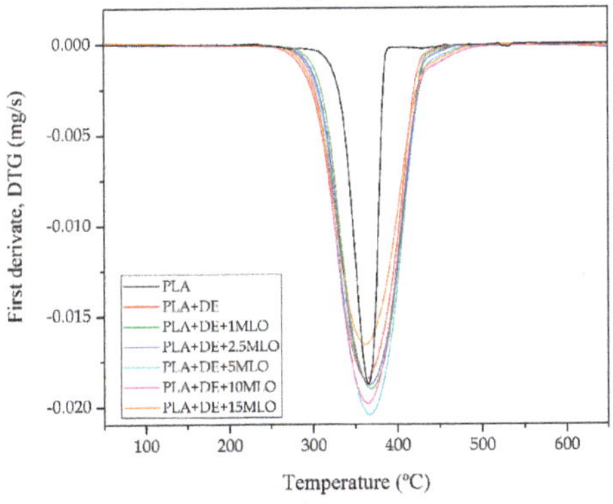

(b)

Figure 4. Comparative thermogravimetric (TGA) plots corresponding to PLA/DE composites with varying MLO content: (**a**) TGA thermograms curves and (**b**) first derivative, DTG curves.

Table 4. Main thermal parameters of the thermal degradation of PLA/DE composites with varying MLO content obtained by thermogravimetric analysis, TGA.

Sample	T_{onset} (°C)	T_{max} (°C)	Residual Mass (wt. %)
PLA	264.1	366.3	0.16
PLA-DE	294.3	364.3	11.6
PLA-DE-1MLO	316.7	369.6	11.5
PLA-DE-2.5MLO	316.2	367.4	13.8
PLA-DE-5MLO	315.6	367.3	10.0
PLA-DE-10MLO	302.0	364.6	10.3
PLA-DE-15MLO	309.3	361.6	10.2

The simple addition of DE increases the thermal stability of neat PLA. In fact, the onset degradation temperature of PLA (Tonset) changes from 264 °C for neat PLA to 294.3 °C for the PLA/DE composites with 10 wt% DE. These results are in agreement with the work of Carrasco et al., which suggests that addition of small amounts of inorganic materials into a polymer matrix provides increased thermal stability [46]. Moreover, addition of MLO also provides increased thermal stability up to onset degradation temperature values of 315–316 °C for MLO loading in the 1–5 phr range [1]. In fact, the onset degradation temperature for composites with 1–5 phr MLO provides an increase of about 22 °C with regard to the uncompatibilized PLA/DE composite [44]. This phenomenon could be somewhat related to interactions between the PLA chains and the modified vegetable oil [45]. With regard to the residual mass, PLA is almost fully decomposed, while all its composites with DE show residual mass values close to 10% which is in total accordance with the amount of DE filler (10 wt%). DE is composed of inorganic siliceous particles which do not undergo degradation in the temperature range comprised between 30 and 700 °C.

3.2. Thermomechanical Properties of PLA/DE Composites with Varying MLO Loading

Figure 5 gathers dynamic mechanical thermal analysis (DMTA) curves corresponding to the evolution of the storage modulus, G' (Figure 5a) and dynamic damping factor, *tan* δ (Figure 5b) with temperature. Regarding the storage modulus, below the glass transition temperature (T_g), all materials show a typical elastic-glassy behaviour with high G' values. The uncompatibilized PLA/DE composite shows a G' value of 1095 MPa at 40 °C. This G' value is remarkably higher than that of the neat PLA (565 MPa at the same temperature). Addition of MLO leads to a decrease in stiffness (lower G' values). Thus, the PLA/DE composite with 1 phr MLO shows a G' value of 832 MPa at 40 °C. As the MLO content increases, G' values show a decreasing tendency as it can be seen in Table 5. This behaviour indicates some plasticization effect of MLO. In fact, some recent research works have suggested that modified triglyceride molecules are placed between polymer chains and, therefore, an increase in chain mobility is achieved, leading to decreased G' values. In addition to this internal lubrication effect, MLO molecules increase the free volume this reducing the intermolecular attraction forces between adjacent PLA chains, all this having a positive effect on overall chain mobility [25,47,48]. Above the T_g, a clear softening occurs, and G' values decrease in a remarkable way. As can be seen in Figure 5a, addition of MLO moves the corresponding curves to lower temperatures as MLO enables chain mobility [49,50]. The cold crystallization process can be detected as an increase in G' values above the T_g. As certain temperature is reached, some amorphous areas of PLA tend to re-arrange to form packed structures or crystallites and this increases the density, which is directly related to the stiffness and, consequently, the G' is increased [51,52].

Table 5. Summary of some dynamic-mechanical thermal properties of PLA/DE with varying MLO loading.

Sample	T_g (°C) *	Storage Modulus, G' (MPa) at 40 °C
PLA	68.1	564.9
PLA-DE	66.2	1095.5
PLA-DE-1MLO	65.9	831.9
PLA-DE-2.5MLO	65.8	624.3
PLA-DE-5MLO	64.5	588.8
PLA-DE-10MLO	64.2	518.4
PLA-DE-15MLO	63.2	545.9

* The T_g value has been obtained by using the peak maximum criterium for *tan* δ.

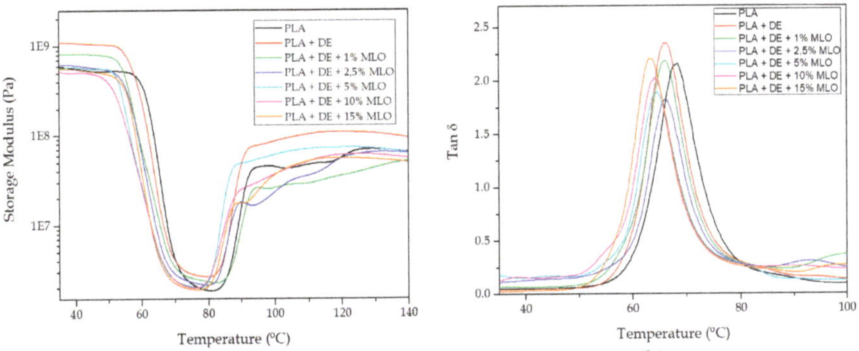

(a) (b)

Figure 5. Comparative plot of dynamic mechanical thermal properties of PLA/DE with varying MLO loading as a function of temperature: (**a**) Storage modulus, G' and (**b**) dynamic damping factor (*tan* δ).

Figure 5b shows the damping factor, *tan* δ, for PLA/DE composites with increasing MLO loading. As it can be seen the damping factor is slightly moved towards lower temperatures, compared to neat PLA. Despite it being possible to evaluate the glass transition temperature by using different criteria (G', G'', *tan* δ), one of the most widely used is the peak maximum for *tan* δ. By using this criterion, neat PLA shows a T_g of 68.1 °C and PLA/DE composite offers a T_g of 66.2 °C. As MLO loading increases, a slight decrease in T_g can be observed which is in total accordance with previous results obtained by DSC. These T_g values are summarized in Table 5.

3.3. Mechanical Properties of PLA/DE Composites with Varying MLO Loading

Table 6 summarizes the main mechanical properties of PLA/DE composites with MLO. The only addition of DE (10 wt%) promotes an increase in tensile modulus from 900 MPa (neat PLA) up to 1344 MPa (PLA/DE composite with 10 wt% DE). Nevertheless, both the tensile strength (σ_t) and the elongation at break (ε_b) decrease. The tensile strength changes from 65 MPa to 53 MPa with 10 wt% DE and the elongation at break is decreased up to half the value of neat PLA. Dispersion of DE particles leads to a clear embrittlement, as these embedded particles can exert a stress concentration effect, with a subsequent decrease in elongation at break [31].

Table 6. Summary of mechanical properties of PLA/DE composites with varying MLO loading, obtained from tensile, impact and hardness tests.

Sample	Tensile Modulus, E_t (MPa)	Tensile Strength, σ_b (MPa)	Elongation at break, ε_b (%)	Impact Strength (kJ m^{-2})	Hardness Shore D
PLA	900 ± 75	65 ± 0.5	6.3 ± 0.9	28 ± 3.4	80.5 ± 3.5
PLA-DE	1344 ± 54	53 ± 2.0	3.5 ± 0.3	12.4 ± 2.1	82.3 ± 2.2
PLA-DE-1MLO	1329 ± 70	50 ± 0.5	4.7 ± 0.2	13.4 ± 2.3	82.3 ± 2.1
PLA-DE-2.5MLO	1322 ± 20	46 ± 2.2	4.9 ± 0.3	14.6 ± 2.9	81.9 ± 1.6
PLA-DE-5MLO	1275 ± 48	40 ± 0.6	5.6 ± 0.4	15.6 ± 1.6	82.0 ± 2.2
PLA-DE-10MLO	1161 ± 117	33 ± 0.6	19.5 ± 2	18.4 ± 3.8	81.7 ± 1.6
PLA-DE-15MLO	1075 ± 166	33 ± 0.9	22.8 ± 1.9	21.7 ± 3.6	79.8 ± 2.9

Addition of MLO on PLA/DE composites provides a decrease in rigidity, as expected. For low MLO loadings in the 1–2 phr range, the tensile modulus is almost constant. Although the average value is slightly lower, if we take into consideration the standard deviation, the change is not significant. Nevertheless, over 5 phr MLO, a clear change in mechanical behaviour can be detected. In particular, the tensile modulus is remarkably decreased. It is worth noting that PLA/DE composites with 15 phr MLO show a tensile modulus of 1075 MPa, which represents a 20% decrease regarding the PLA/DE composite without MLO. The same tendency can be detected for tensile strength. Addition of 5 phr produces a decrease of 10% in tensile strength while above 5 phr, the percentage decrease in tensile strength is comprised between 25–38%. This pronounced change in mechanical resistant properties is inversely related to elongation at break which increases with increasing MLO loading. Uncompatibilized PLA/DE composite is extremely brittle, with an elongation at break of 3.5%. This is slightly increased to values of 5% for low MLO loading but, as expected, above 10 phr MLO, it is possible to obtain a noticeable increase in elongation at break, with values of close to 20% in this range. Some previous works have demonstrated different effects of MLO on PLA and other polyesters. These effects include plasticization, chain extension, branching and, in some cases, some crosslinking [43,47,53]. The plasticization effect has been corroborated by a slight decrease in the glass transition temperature. MLO exerts a lubricant effect on PLA polymer chains and this is responsible for the observed increase in ductility. Changes in stiffness are also related to the ability of the materials to absorb energy. It is important to note that impact strength is related to both ductile and resistant properties. Neat PLA is characterized by an impact strength of 28 kJ m^{-2}; due to the stress concentration phenomenon of DE, the PLA/DE composite with 10 wt% DE shows a remarkable decrease in impact strength down to values of 12.4 kJ m^{-2}. These results are in total agreement with the dramatic decrease in both tensile strength and elongation at break. As the MLO loading increases, the impact strength improves, and it is worth noting that PLA/DE composites containing 15 phr MLO reach an impact strength of about 22 kJ m^{-2}. This is lower than neat PLA, but it is remarkably superior to the uncompatibilized PLA/DE composite. Above 5 phr, MLO can counteract the negative effect of DE on impact strength. Finally, with regard to Shore D hardness, it is worthy to note that although the average values show a slight increase in hardness with DE addition and a decrease with MLO loading, the standard deviation of these values shows that Shore D hardness values remain almost unchanged centred in 80.

3.4. Morphological Characterization of PLA/DE Composites with Varying MLO Loading

Figure 6 shows a FESEM image corresponding to the fractured surface from the impact test of PLA/DE sample at 2500×. This morphology is characterized by a matrix phase consisting on PLA and a dispersed phase constituted by embedded DE particles. The polymer matrix shows a uniform topography with small steps and, subsequently, low roughness. These are clear evidence of a brittle fracture (or a fracture with very short plastic deformation). In addition, it is possible to identify individual DE particles and the surrounding of these particles show a small gab regarding the polymer matrix. These features are in total agreement with the previous mechanical properties with a decrease in elongation at break and a decrease in tensile strength due to the stress concentration phenomenon provided by uncompatibilized DE particles.

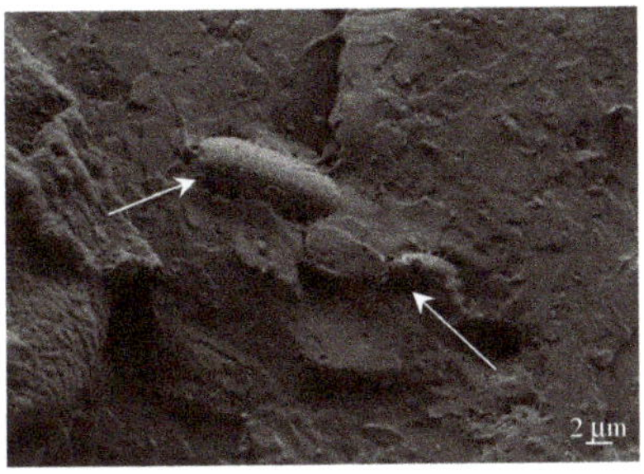

Figure 6. FESEM image (2500×) of the surface of a PLA/DE sample obtained after impact test.

Figure 7 shows FESEM images corresponding to PLA/DE composites compatibilized with different MLO loadings. From these images, the PLA matrix and the embedded DE particles (disperse phase) are clearly distinguishable. For low MLO content (1–5 phr), the PLA matrix shows a homogeneous topography, absolutely uniform and smooth. The only difference with the uncompatibilized PLA/DE composite (Figure 6) is that the steps and roughness are rounded and less pronounced, as can be seen in Figure 7a–c. Moreover, it is possible to observe formation of somewhat interphase between PLA and DE particles. In this sense, MLO can act as a compatibilizer of PLA matrix and DE particles due to the tendency of the maleic anhydride group to react with hydroxyl groups in both PLA (terminal groups) and siliceous surface of DE particles. These improvements in the interface phenomena between PLA and DE are responsible for an improvement in ductile properties and, subsequently, on the impact strength, as previously indicated. With respect to PLA/DE composites with high MLO content (10–15 phr), the obtained morphology is somewhat different. The polymeric matrix is less uniform, with a high density of rounded microvoids. In addition, some phase separation can be observed. This indicates that PLA is not completely miscible with MLO. This immiscibility is much more evident with an excess of MLO (see white circles in Figure 7d,e). This restricted miscibility promotes formation of microvoids (see white arrows in Figure 7d,e), as reported in previous works [12,20,25,49,50,54]. In addition to this, it is possible to observe some evidence of plastic deformation (filaments) of the PLA matrix [1]. Therefore, it is possible to conclude that the increase in impact strength is directly related to somewhat interaction between PLA matrix and DE particles.

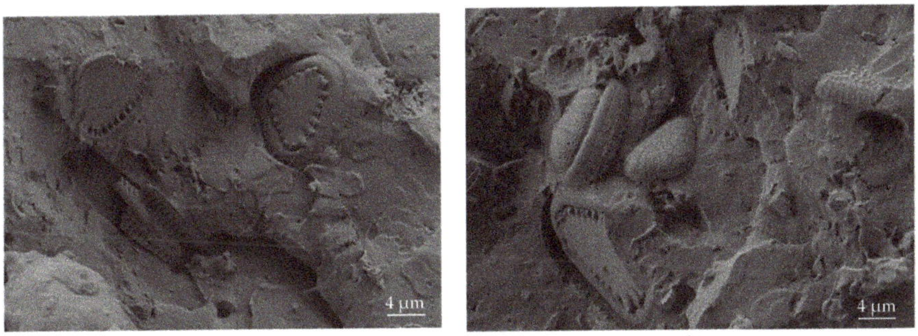

(a) (b)

Figure 7. *Cont.*

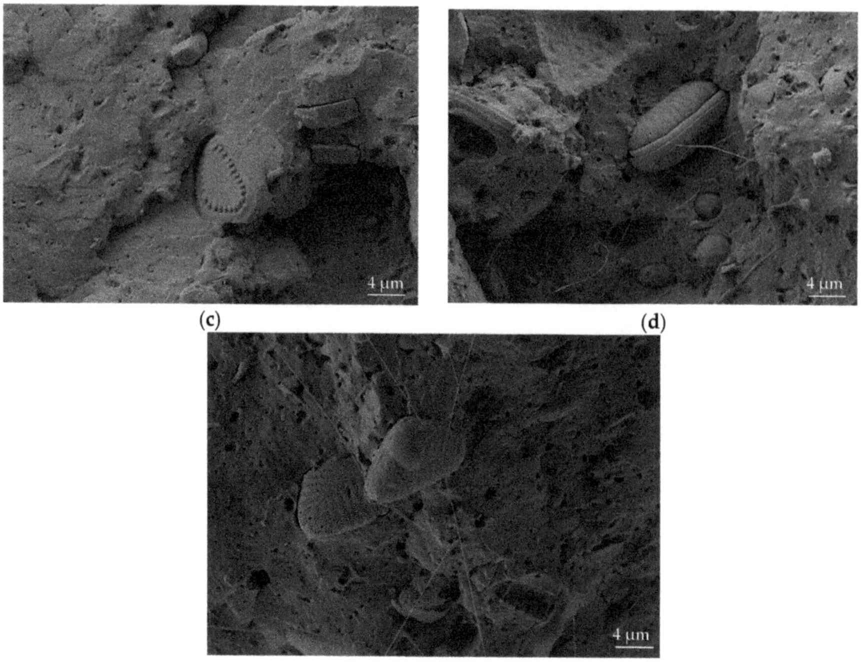

(**c**) (**d**)

(**e**)

Figure 7. FESEM images (2500×) of the surface of a PLA/DE sample obtained after impact test, with varying MLO loading (in phr): (**a**) 1, (**b**) 2.5, (**c**) 5, (**d**) 10 and (**e**) 15.

4. Conclusions

This work reports the efficiency of maleinized linseed oil as a biobased compatibilizer in poly(lactic acid)—PLA—composites with diatomaceous earth (DE) at a constant loading of 10 wt%. In particular, the work focuses on the optimization of MLO loading to obtain the most balanced properties on PLA/DE composites. The obtained results show a clear embrittlement of PLA/DE composites without any compatibilizer. The elongation at break is reduced to half the value of neat PLA. On the contrary, the tensile modulus increases as expected. On the other hand, the impact strength of uncompatibilized PLA/DE composites goes down to a value of 12.4 kJ m^{-2}, which is remarkably lower than neat PLA (28 kJ m^{-2}). Two different levels of effect can be seen depending on the MLO loading. For MLO loadings in the 1–5 phr range, a slight increase in ductile properties can be detected, with a slight decrease in the glass transition temperature (T_g). Nevertheless, above 5 phr MLO, ductile properties are remarkably improved, and the impact strength increases to values close to 22 kJ m^{-2} which is almost double the value of uncompatibilized PLA/DE composite. The morphology of these composites shows that MLO exerts a compatibilizing effect, bridging the PLA matrix and the DE particles. Therefore, it is possible to conclude that MLO loading in the 10–15 phr range gives optimum and balanced properties for PLA/DE composites without compromising the ecoefficiency of the developed composites.

Author Contributions: Conceptualization, N.M. and L.G.; methodology, A.A. and D.L.; validation, L.Q.-C., L.G. and N.M.; data analysis, L.G. and D.L.; original draft preparation, L.G. and A.A.; review and editing, A.A. and L.Q.-C.; supervision, N.M.; project administration, N.M.

Funding: This research was funded by the Ministry of Science, Innovation, and Universities (MICIU) project number MAT2017-84909-C2-2-R. L. Quiles-Carrillo is recipient of a FPU grant (FPU15/03812) from the Spanish Ministry of Education, Culture, and Sports (MECD). D. Lascano acknowledges UPV for the grant received though the PAID-01-18 program. N. Montanes acknowledges the project "Development and production of new material from revalued industrial wastes for technological sector applications" for partially funding this research.

Conflicts of Interest: The authors declare no conflict of interest

References

1. Quiles-Carrillo, L.; Blanes-Martinez, M.M.; Montanes, N.; Fenollar, O.; Torres-Giner, S.; Balart, R. Reactive toughening of injection-molded polylactide pieces using maleinized hemp seed oil. *Eur. Polym. J.* **2018**, *98*, 402–410. [CrossRef]
2. Islam, M.R.; Beg, M.D.H.; Jamari, S.S. Development of Vegetable-Oil-Based Polymers. *J. Appl. Polym. Sci.* **2014**, *131*. [CrossRef]
3. Lu, Y.; Larock, R.C. Novel Polymeric Materials from Vegetable Oils and Vinyl Monomers: Preparation, Properties, and Applications. *ChemSusChem* **2009**, *2*, 136–147. [CrossRef]
4. Sharma, V.; Kundu, P.P. Addition polymers from natural oils—A review. *Prog. Polym. Sci.* **2006**, *31*, 983–1008. [CrossRef]
5. Miao, S.; Wang, P.; Su, Z.; Zhang, S. Vegetable-oil-based polymers as future polymeric biomaterials. *Acta Biomater.* **2014**, *10*, 1692–1704. [CrossRef]
6. Petrovic, Z.S.; Guo, A.; Javni, I.; Cvetkovic, I.; Hong, D.P. Polyurethane networks from polyols obtained by hydroformylation of soybean oil. *Polym. Int.* **2008**, *57*, 275–281. [CrossRef]
7. Xia, Y.; Larock, R.C. Vegetable oil-based polymeric materials: Synthesis, properties, and applications. *Green Chem.* **2010**, *12*, 1893–1909. [CrossRef]
8. Xia, Y.; Quirino, R.L.; Larock, R.C. Bio-based Thermosetting Polymers from Vegetable Oils. *J. Renew. Mater.* **2013**, *1*, 3–27. [CrossRef]
9. Malarczyk, K.; Milchert, E. Methods for epoxidation of vegetable oils on heterogenous catalysts. *Przem. Chem.* **2015**, *94*, 412–415.
10. Milchert, E.; Malarczyk, K.; Klos, M. Technological Aspects of Chemoenzymatic Epoxidation of Fatty Acids, Fatty Acid Esters and Vegetable Oils: A Review. *Molecules* **2015**, *20*, 21481–21493. [CrossRef] [PubMed]
11. Tan, S.G.; Chow, W.S. Biobased Epoxidized Vegetable Oils and Its Greener Epoxy Blends: A Review. *Polym. Plast. Technol. Eng.* **2010**, *49*, 1581–1590. [CrossRef]
12. Carbonell-Verdu, A.; Bernardi, L.; Garcia-Garcia, D.; Sanchez-Nacher, L.; Balart, R. Development of environmentally friendly composite matrices from epoxidized cottonseed oil. *Eur. Polym. J.* **2015**, *63*, 1–10. [CrossRef]
13. Alam, J.; Alam, M.; Raja, M.; Abduljaleel, Z.; Dass, L.A. MWCNTs-Reinforced Epoxidized Linseed Oil Plasticized Polylactic Acid Nanocomposite and Its Electroactive Shape Memory Behaviour. *Int. J. Mol. Sci.* **2014**, *15*, 19924–19937. [CrossRef]
14. Fenollar, O.; Garcia-Sanoguera, D.; Sanchez-Nacher, L.; Lopez, J.; Balart, R. Effect of the epoxidized linseed oil concentration as natural plasticizer in vinyl plastisols. *J. Mater. Sci.* **2010**, *45*, 4406–4413. [CrossRef]
15. Ray, D.; Sain, S. Thermosetting bioresins as matrix for biocomposites. In *Biocomposites for High-Performance Applications*; Elsevier: Amsterdam, The Netherlands, 2017; pp. 57–80.
16. Xing, C.; Matuana, L.M. Epoxidized soybean oil-plasticized poly(lactic acid) films performance as impacted by storage. *J. Appl. Polym. Sci.* **2016**, *133*. [CrossRef]
17. Carbonell-Verdu, A.; Garcia-Sanoguera, D.; Jorda-Vilaplana, A.; Sanchez-Nacher, L.; Balart, R. A new biobased plasticizer for poly(vinyl chloride) based on epoxidized cottonseed oil (vol 33, 43642, 2016). *J. Appl. Polym. Sci.* **2016**, *133*. [CrossRef]
18. Vijayarajan, S.; Selke, S.E.M.; Matuana, L.M. Continuous Blending Approach in the Manufacture of Epoxidized SoybeanPlasticized Poly(lactic acid) Sheets and Films. *Macromol. Mater. Eng.* **2014**, *299*, 622–630. [CrossRef]
19. Sotoodeh-Nia, Z.; Hohmann, A.; Buss, A.; Williams, R.C.; Cochran, E.W. Rheological and physical characterization of pressure sensitive adhesives from bio-derived block copolymers. *J. Appl. Polym. Sci.* **2018**, *135*. [CrossRef]
20. Carbonell-Verdu, A.; Dolores Samper, M.; Garcia-Garcia, D.; Sanchez-Nacher, L.; Balart, R. Plasticization effect of epoxidized cottonseed oil (ECSO) on poly(lactic acid). *Ind. Crop. Prod.* **2017**, *104*, 278–286. [CrossRef]
21. Carbonell-Verdu, A.; Garcia-Garcia, D.; Dominici, F.; Torre, L.; Sanchez-Nacher, L.; Balart, R. PLA films with improved flexibility properties by using maleinized cottonseed oil. *Eur. Polym. J.* **2017**, *91*, 248–259. [CrossRef]

22. Ernzen, J.R.; Bondan, F.; Luvison, C.; Wanke, C.H.; Martins, J.D.N.; Fiorio, R.; Bianchi, O. Structure and properties relationship of melt reacted polyamide 6/malenized soybean oil. *J. Appl. Polym. Sci.* **2016**, *133*. [CrossRef]
23. Mauck, S.C.; Wang, S.; Ding, W.; Rohde, B.J.; Fortune, C.K.; Yang, G.; Ahn, S.-K.; Robertson, M.L. Biorenewable Tough Blends of Polylactide and Acrylated Epoxidized Soybean Oil Compatibilized by a Polylactide Star Polymer. *Macromolecules* **2016**, *49*, 1605–1615. [CrossRef]
24. Rosu, D.; Mustata, F.; Tudorachi, N.; Musteata, V.E.; Rosu, L.; Varganici, C.D. Novel bio-based flexible epoxy resin from diglycidyl ether of bisphenol A cured with castor oil maleate. *RSC Adv.* **2015**, *5*, 45679–45687. [CrossRef]
25. Ferri, J.M.; Garcia-Garcia, D.; Montanes, N.; Fenollar, O.; Balart, R. The effect of maleinized linseed oil as biobased plasticizer in poly (lactic acid)-based formulations. *Polym. Int.* **2017**, *66*, 882–891. [CrossRef]
26. Agüero, A.; Quiles-Carrillo, L.; Jorda-Vilaplana, A.; Fenollar, O.; Montanes, N. Effect of different compatibilizers on environmentally friendly composites from poly (lactic acid) and diatomaceous earth. *Polym. Int.* **2019**, *68*, 893–903. [CrossRef]
27. Brandelli, A.; Wentz Brum, L.F.; Zimnoch dos Santos, J.H. Nanostructured bioactive compounds for ecological food packaging. *Environ. Chem. Lett.* **2017**, *15*, 193–204. [CrossRef]
28. Gorrasi, G.; Senatore, V.; Vigliotta, G.; Belviso, S.; Pucciariello, R. PET-halloysite nanotubes composites for packaging application: Preparation, characterization and analysis of physical properties. *Eur. Polym. J.* **2014**, *61*, 145–156. [CrossRef]
29. Kumar, N.; Kaur, P.; Bhatia, S. Advances in bio-nanocomposite materials for food packaging: A review. *Nutr. Food Sci.* **2017**, *47*, 591–606. [CrossRef]
30. Kuswandi, B. Environmental friendly food nano-packaging. *Environ. Chem. Lett.* **2017**, *15*, 205–221. [CrossRef]
31. Rhim, J.-W.; Park, H.-M.; Ha, C.-S. Bio-nanocomposites for food packaging applications. *Prog. Polym. Sci.* **2013**, *38*, 1629–1652. [CrossRef]
32. Tornuk, F.; Hancer, M.; Sagdic, O.; Yetim, H. LLDPE based food packaging incorporated with nanoclays grafted with bioactive compounds to extend shelf life of some meat products. *LWT-Food Sci. Technol.* **2015**, *64*, 540–546. [CrossRef]
33. Aw, M.S.; Simovic, S.; Yu, Y.; Addai-Mensah, J.; Losic, D. Porous silica microshells from diatoms as biocarrier for drug delivery applications. *Powder Technol.* **2012**, *223*, 52–58. [CrossRef]
34. Cacciotti, I.; Mori, S.; Cherubini, V.; Nanni, F. Eco-sustainable systems based on poly(lactic acid), diatomite and coffee grounds extract for food packaging. *Int. J. Biol. Macromol.* **2018**, *112*, 567–575. [CrossRef] [PubMed]
35. Davoudizadeh, S.; Ghasemi, M.; Khezri, K.; Bahadorikhalili, S. Poly(styrene-co-butyl acrylate)/mesoporous diatomaceous earth mineral nanocomposites by in situ AGET ATRP. *J. Therm. Anal. Calorim.* **2018**, *131*, 2513–2521. [CrossRef]
36. Medarevic, D.P.; Losic, D.; Ibric, S.R. Diatoms—nature materials with great potential for bioapplications. *Hem. Ind.* **2016**, *70*, 613–627. [CrossRef]
37. Ozen, I.; Simsek, S.; Okyay, G. Manipulating surface wettability and oil absorbency of diatomite depending on processing and ambient conditions. *Appl. Surf. Sci.* **2015**, *332*, 22–31. [CrossRef]
38. Saeidlou, S.; Huneault, M.A.; Li, H.; Park, C.B. Poly(lactic acid) crystallization. *Prog. Polym. Sci.* **2012**, *37*, 1657–1677. [CrossRef]
39. Liu, M.; Zhang, Y.; Zhou, C. Nanocomposites of halloysite and polylactide. *Appl. Clay Sci.* **2013**, *75*, 52–59. [CrossRef]
40. Tham, W.L.; Poh, B.T.; Ishak, Z.A.M.; Chow, W.S. Thermal behaviors and mechanical properties of halloysite nanotube-reinforced poly (lactic acid) nanocomposites. *J. Therm. Anal. Calorim.* **2014**, *118*, 1639–1647. [CrossRef]
41. Prashantha, K.; Lecouvet, B.; Sclavons, M.; Lacrampe, M.F.; Krawczak, P. Poly (lactic acid)/halloysite nanotubes nanocomposites: Structure, thermal, and mechanical properties as a function of halloysite treatment. *J. Appl. Polym. Sci.* **2013**, *128*, 1895–1903. [CrossRef]
42. De Silva, R.T.; Soheilmoghaddam, M.; Goh, K.L.; Wahit, M.U.; Bee, S.A.H.; Chai, S.P.; Pasbakhsh, P. Influence of the processing methods on the properties of poly (lactic acid)/halloysite nanocomposites. *Polym. Compos.* **2016**, *37*, 861–869. [CrossRef]

43. Carbonell-Verdu, A.; Ferri, J.M.; Dominici, F.; Boronat, T.; Sanchez-Nacher, L.; Balart, R.; Torre, L. Manufacturing and compatibilization of PLA/PBAT binary blends by cottonseed oil-based derivatives. *Express Polym. Lett.* **2018**, *12*, 808–823. [CrossRef]
44. Li, M.; Li, S.; Xia, J.; Ding, C.; Wang, M.; Xu, L.; Yang, X.; Huang, K. Tung oil based plasticizer and auxiliary stabilizer for poly(vinyl chloride). *Mater. Des.* **2017**, *122*, 366–375. [CrossRef]
45. Prempeh, N.; Li, J.; Liu, D.; Das, K.; Maiti, S.; Zhang, Y. Plasticizing Effects of Epoxidized Sun Flower Oil on Biodegradable Polylactide Films: A Comparative Study. *Polym. Sci. Ser. A* **2014**, *56*, 856–863. [CrossRef]
46. Carrasco, F.; Gamez-Perez, J.; Santana, O.O.; Maspoch, M.L. Processing of poly(lactic acid)/organomontmorillonite nanocomposites: Microstructure, thermal stability and kinetics of the thermal decomposition. *Chem. Eng. J.* **2011**, *178*, 451–460. [CrossRef]
47. Quiles-Carrillo, L.; Montanes, N.; Sammon, C.; Balart, R.; Torres-Giner, S. Compatibilization of highly sustainable polylactide/almond shell flour composites by reactive extrusion with maleinized linseed oil. *Ind. Crop. Prod.* **2018**, *111*, 878–888. [CrossRef]
48. Bocqué, M.; Voirin, C.; Lapinte, V.; Caillol, S.; Robin, J.J. Petro-based and bio-based plasticizers: Chemical structures to plasticizing properties. *J. Polym. Sci. Part A Polym. Chem.* **2016**, *54*, 11–33. [CrossRef]
49. Silverajah, V.S.G.; Ibrahim, N.A.; Zainuddin, N.; Yunus, W.M.Z.W.; Abu Hassan, H. Mechanical, Thermal and Morphological Properties of Poly(lactic acid)/Epoxidized Palm Olein Blend. *Molecules* **2012**, *17*, 11729–11747. [CrossRef]
50. Ali, F.; Chang, Y.-W.; Kang, S.C.; Yoon, J.Y. Thermal, mechanical and rheological properties of poly (lactic acid)/epoxidized soybean oil blends. *Polym. Bull.* **2009**, *62*, 91–98. [CrossRef]
51. Shah, D.U. Developing plant fibre composites for structural applications by optimising composite parameters: A critical review. *J. Mater. Sci.* **2013**, *48*, 6083–6107. [CrossRef]
52. Yu, Y.; Cheng, Y.; Ren, J.; Cao, E.; Fu, X.; Guo, W. Plasticizing effect of poly (ethylene glycol) s with different molecular weights in poly (lactic acid)/starch blends. *J. Appl. Polym. Sci.* **2015**, *132*. [CrossRef]
53. Quiles-Carrillo, L.; Montanes, N.; Garcia-Garcia, D.; Carbonell-Verdu, A.; Balart, R.; Torres-Giner, S. Effect of different compatibilizers on injection-molded green composite pieces based on polylactide filled with almond shell flour. *Compos. Part B Eng.* **2018**, *147*, 76–85. [CrossRef]
54. Silverajah, V.S.G.; Ibrahim, N.A.; Yunus, W.M.Z.W.; Abu Hassan, H.; Woei, C.B. A Comparative Study on the Mechanical, Thermal and Morphological Characterization of Poly(lactic acid)/Epoxidized Palm Oil Blend. *Int. J. Mol. Sci.* **2012**, *13*, 5878–5898. [CrossRef] [PubMed]

materials

MDPI

Article

The Effect of Varying Almond Shell Flour (ASF) Loading in Composites with Poly(Butylene Succinate (PBS) Matrix Compatibilized with Maleinized Linseed Oil (MLO)

Patricia Liminana, Luis Quiles-Carrillo *, Teodomiro Boronat, Rafael Balart and Nestor Montanes

Technological Institute of Materials (ITM), Universitat Politècnica de València (UPV), Plaza Ferrándiz y Carbonell 1, 03801 Alcoy, Spain; patligre@mcm.upv.es (P.L.); tboronat@dimm.upv.es (T.B.); rbalart@mcm.upv.es (R.B.); nesmonmu@upvnet.upv.es (N.M.)

* Correspondence: luiquic1@epsa.upv.es; Tel.: +34-966-528-433

Received: 19 October 2018; Accepted: 1 November 2018; Published: 3 November 2018

Abstract: In this work poly(butylene succinate) (PBS) composites with varying loads of almond shell flour (ASF) in the 10–50 wt % were manufactured by extrusion and subsequent injection molding thus showing the feasibility of these combined manufacturing processes for composites up to 50 wt % ASF. A vegetable oil-derived compatibilizer, maleinized linseed oil (MLO), was used in PBS/ASF composites with a constant ASF to MLO (wt/wt) ratio of 10.0:1.5. Mechanical properties of PBS/ASF/MLO composites were obtained by standard tensile, hardness, and impact tests. The morphology of these composites was studied by field emission scanning electron microscopy—FESEM) and the main thermal properties were obtained by differential scanning calorimetry (DSC), dynamical mechanical-thermal analysis (DMTA), thermomechanical analysis (TMA), and thermogravimetry (TGA). As the ASF loading increased, a decrease in maximum tensile strength could be detected due to the presence of ASF filler and a plasticization effect provided by MLO which also provided a compatibilization effect due to the interaction of succinic anhydride polar groups contained in MLO with hydroxyl groups in both PBS (hydroxyl terminal groups) and ASF (hydroxyl groups in cellulose). FESEM study reveals a positive contribution of MLO to embed ASF particles into the PBS matrix, thus leading to balanced mechanical properties. Varying ASF loading on PBS composites represents an environmentally-friendly solution to broaden PBS uses at the industrial level while the use of MLO contributes to overcome or minimize the lack of interaction between the hydrophobic PBS matrix and the highly hydrophilic ASF filler.

Keywords: green composites; natural fillers; poly(butylene succinate) (PBS); almond shell flour (ASF)

1. Introduction

Over the last years, research on new polymer materials has attracted much research with the aim of minimizing the environmental impact of petroleum-derived polymers. These new polymers, also known as biopolymers, have demonstrated a clear contribution to decrease the carbon footprint in comparison to conventional plastics [1,2]. High environmentally-friendly polymers can be obtained from renewable resources and can potentially find interesting engineering applications. These biobased polymers include polysaccharides (cellulose, starch, chitosan, and so on), protein polymers (gluten, ovalbumin, soy protein, collagen, among others), and bacterial polymers such as poly(3-hydroxybutyrate), PHB, and other polymers obtained from biomass fermentation by different microorganisms [3–7]. Some polymers can be obtained from petroleum resources, but they show high environmental efficiency at the end-of-life as they can undergo full disintegration under certain conditions (compost). Aliphatic polyesters such as poly(butylene succinate) (PBS), poly(glycolic acid) (PGA), poly(ε-caprolactone) (PCL), poly(butylene succinate-*co*-adipate) (PBSA), and some aliphatic-aromatic copolyesters, i.e., poly(butylene succinate-*co*-terephthalate) (PBAT), poly(butylene succinate-*co*-terephthalate) (PBST), among others, belong to these petroleum-based, disintegrable (biodegradable) polymers [8,9].

Among these polyesters PBS is gaining relevance due to its high flexibility which allows its use in the packaging industry. Poly(butylene succinate) can be obtained from polycondensation of succinic acid and 1,4-butanediol (BDO). Although the most common route to obtain PBS is from petroleum-derived monomers (in fact, the first PBS commercial grades were petroleum-derived), currently it is possible to obtain both starting monomers from renewable resources such as starch, glucose or cellulose by bacterial fermentation [10,11], and this will open a new age in the development of biopolyesters from renewable resources. Obviously, bio-derived PBS is a high environmentally-friendly material, from both points of view: origin (bio-derived) and end-of-life (disintegrable in controlled compost soil). Nevertheless, petroleum-derived PBS lacks the "bio" origin, but it does not generate problematics at the end-of-life since, as other aliphatic polyesters, it can undergo disintegration in controlled compost soil [12]. So, currently, petroleum-based PBS and derivatives are interesting alternatives to other non-biodegradable plastics. Recently, Puchalski et al. have reported the degradation of petroleum-derived PBS and PBSA subjected to different environmental conditions. In addition they reported the change in physical and mechanical properties of PBS and PBSA during the degradation processes [13].

Poly(butylene succinate) possesses comparable properties to those of some commodities such as poly(ethylene) (PE) or poly(propylene) (PP) [14]. In addition, processing of PBS can be carried out at moderate temperatures, with a pre-drying stage to remove moisture, which is responsible for hydrolysis. Its main uses include film/sheet for the packaging and agricultural industries [15,16]; despite this, its use is increasing in the automotive industry and medical devices as well [17–19]. The main drawback of PBS is its high cost compared to commodity and some engineering plastics.

One way to partially overcome this drawback without compromising its biodegradability is by blending it with less expensive biodegradable polymers such as poly(lactic acid) (PLA), PCL, among others [20,21]. Another approach is by using lignocellulosic fillers to give the so-called natural polymer composites (NPCs) with PBS matrix. Natural polymer composites can positively contribute to give sustainable materials with balanced properties (mechanical, thermal, barrier, physical, and so on), similar to commodities [22–24]. An interesting approach to these fillers is the use of industrial or agroforestry by-products to act as reinforcing fillers in NPCs. It has been widely reported the potential of industrial wastes from the food industry (fruit shells, stalks, fruit skins, seeds, among others) in NPCs [18,25–27].

Poly(butylene succinate) has been successfully used as matrix with a wide variety of natural fibers such as sisal [28], hemp [29], kenaf [30], and so on. Regarding PBS composites with agricultural wastes, it is worthy to note the work by Tserki et al. [31], in which, lignocellulosic waste flours (spruce, olive husk and paper) were used as fillers into PBS matrices. Yeng et al. [32] reported the use of fillers from wheat bran into modified PBS matrices. They reported the positive effect of grafting maleic anhydride into PBS chains to give poly(butylene succinate-*g*-maleic anhydride) which showed increased interactions with the cellulosic components contained in the wheat bran.

Among the wide variety of potential lignocellulosic wastes, almond shell is an abundant waste in countries such as Spain, which stands as the third major producer of almonds just after the USA and Australia. Almond shell has been employed as lignocellulosic fillers in several NPCs. In particular, it has been added to commodities such as PP [33,34], but moreover, several studies involving almond shell wastes and some biopolymers, PLA [35,36] or PCL [37], have been reported in the last years. In a previous work [38], our group reported the need of compatibilizer agents to provide increased interactions between the highly hydrophobic PBS matrix and the highly hydrophilic almond shell flour (ASF) filler as observed in other polymer/lignocellulosic filler composites [39,40]. Several solutions to overcome the low polymer/filler interactions have been proposed such as silane treatments, acetylation, plasma treatments, and so on [41–43]. Excellent results have been obtained by using maleic anhydride grafted copolymers in both polymer/lignocellulosic fillers and binary/ternary blends with immiscible polymers [44,45]. Recently, vegetable oils have been proposed as environmentally-friendly compatibilizers as an alternative to conventional petroleum-based ones [46,47]. Epoxidized vegetable oils (EVOs), such as epoxidized soybean oil (ESBO), epoxidized linseed oil (ELO) [48,49], epoxidized palm oil (EPO) [50], and so on, have been successfully used as compatibilizers in NPCs. Another vegetable oil derivative, namely maleinized linseed oil (MLO) has been used as compatibilizers in polymer/lignocellulosic composites [51,52]. Ferri et al. [53] reported a clear improvement on processability, mechanical ductile properties and thermal stability on PLA composites with low MLO loading content. In our previous work, focused on PBS/ASF composites with a constant ASF content of 30 wt %, several compatibilizer families based on different reactive groups, i.e., epoxy, maleic anhydride and acrylic were used. Maleinized linseed oil—MLO—gave the best results in terms of balanced properties due to the reaction of the succinic anhydride group attached to the triglyceride molecule, towards the hydroxyl groups contained in both PBS (end-chain groups) and ASF (cellulose and hemicellulose) [38].

The aim of this work is to expand the potential of PBS/ASF composites by varying the ASF wt % loading up to 50 wt % and using MLO as reactive compatibilizer.

2. Materials and Methods

2.1. Materials

Poly(butylene succinate)-based composites were manufactured with a PBS commercial grade Bionolle 1020MD supplied by Showa Denko Europe GmbH (Munich, Germany). This PBS grade is petroleum-derived but fully disintegrable in controlled compost soil. This PBS possesses a melt flow index—MFI comprised between 20 and 34 g/10 min and a density of 1.26 g cm^{-3}. The lignocellulosic filler was almond shell supplied by JESOL Materias Primas (Valencia, Spain). The almond shell was grinded and sieved in a CISA®RP09 sieve shaker (CISA Cedacería Industrial, Barecelona, Spain) to an average particle size of 150 µm. The selected compatibilizer was maleinized linseed oil—MLO, VEOMER LIN supplied by Vandeputte (Mouscron, Belgium). This modified vegetable oil is characterized by a viscosity of 10 dPa s at 20 °C and an acid value comprised in the 105–130 mg KOH g^{-1} range.

2.2. Manufacturing of PBS/ASF/MLO Composites

As polyesters are very sensitive to hydrolysis, PBS was previously dried at 50 °C for 24 h to avoid degradation during processing. The ASF was also dried in the same conditions as PBS in a dehumidifying dryer MDEO from Industrial Marsé (Barcelona, Spain). The MLO was heated at 40 °C for 30 min to reduce its viscosity and enhance mixing with both PBS and ASF.

Table 1 shows the formulations developed in this study. The appropriate amounts of each component were weighed and mechanically pre-mixed in a zipper bag for 5 min. Then, the mixtures were compounded in a twin-screw co-rotating extruder from Construcciones Mecánicas DUPRA, S.L. (Alicante, Spain). The rotation speed was set to 40 rpm and the temperature of the four heated barrels was programmed to 120 °C (hopper), 125 °C, 130 °C, and 130 °C (dye). The screws had a diameter of 25 mm and a length-L to diameter-D ratio of 24. After extrusion, the obtained compounds were pelletized and dried again at 50 °C for 24 h before further processing in a Meteor 270/75 injection moulding machine from Mateu&Sole (Barcelona, Spain). The temperature profile was set to (from the hopper to the injection nozzle): 110 °C, 115 °C, 130 °C, and 125 °C.

Table 1. Formulations and sample coding of poly(butylene succinate)/almond shell flour/ maleinized linseed oil (PBS/ASF/MLO) composites.

Sample Code	PBS (wt %)	ASF (wt %)	MLO (wt %)
PBS	100	0	0
PBS + 10 ASF + 1.5 MLO	88.5	10	1.5
PBS + 20 ASF + 3 MLO	77	20	3
PBS + 30 ASF + 4.5 MLO	65.5	30	4.5
PBS + 40 ASF + 6 MLO	54	40	6
PBS + 50 ASF + 7.5 MLO	42.5	50	7.5

2.3. *Mechanical Characterization*

PBS/ASF/MLO composites were characterized by standard tensile tests following ISO 527-1:2012 in a universal test machine ELIB 50 from S.A.E. Ibertest (Madrid, Spain). The load cell was 5 kN and the crosshead speed was set to 10 mm min^{-1}. In addition, Shore D hardness values of PBS/ASF/MLO composites were obtained in a durometer model 676-D from J. Bot Instruments (Barcelona, Spain) as indicated in ISO 868:2003. Finally, the impact strength was obtained using the Charpy pendulum (with an energy of 1 J) on notched samples ("V" type notch with a radius of 0.25 mm), according to ISO 179-1:2010. At least five different samples were used in each mechanical test and the average values of the corresponding parameters were obtained. In particular, the elongation at break—ε_b, maximum tensile strength—σ_t and the tensile modulus—E_t, were obtained from tensile tests. All the tests were conducted at room temperature.

2.4. *Thermal and Thermomechanical Characterization*

Differential scanning calorimetry (DSC) was used to study the main thermal transitions of PBS/ASF/MLO composites in a model DSC 821 calorimeter from Mettler-Toledo (Schwerzenbach, Switzerland). The average sample weight was comprised between 5 and 7 mg and standard aluminium crucibles with a volume of 40 μL were used. All the samples were subjected to a thermal program consisting on three stages. The first stage consisted on a first heating cycle from 25 °C up to 200 °C. Then, a cooling stage down to −50 °C was applied, and finally, a second heating stage from −50 °C up to 300 °C was scheduled. The heating/cooling rate was 10 °C min^{-1} for all three stages and a constant nitrogen flow of 66 mL min^{-1} was used. All DSC tests were run in triplicate to obtain reliable results. The degree of crystallinity (χ_c) was calculated using the following equation:

$$\%\chi_c = \left[\frac{\Delta H_m}{\Delta H_m^0 \cdot (1-w)}\right] \cdot 100 \quad (1)$$

where ΔH_m corresponds to the measured melt enthalpy of PBS. ΔH_m^0 (J g^{-1}) stands for the theoretical melt enthalpy of a fully crystalline PBS, which was taken as 110.3 J g^{-1} for PBS as previously reported [54]. Regarding w, it represents the total weight percent of all components (ASF and MLO) added to PBS matrix.

Thermal stability of PBS/ASF/MLO composites was studied by thermogravimetry in a TGA/SDTA 851 thermobalance from Mettler-Toledo (Schwerzenbach, Switzerland). The average sample weight for TGA characterization was in the 7–10 mg range and standard alumina crucibles with a total volume of 70 μL were used. The scheduled thermal program was a dynamic heating from 30 °C up to 700 °C at a heating rate of 20 °C min^{-1} in air atmosphere.

Thermomechanical characterization was carried out by dynamic mechanical thermal analysis (DMTA) in a Mettler-Tolledo DMA1 (Schwerzenbach, Switzerland). Samples were subjected to a dynamic flexural test in single cantilever at a frequency of 1 Hz. The thermal program consisted on a heating sweep from −50 °C up to 80 °C at a constant heating rate of 2 °C min^{-1}. The maximum flexural deformation was set to 10 μm. On the other hand, the thermal/dimensional stability was determined in a thermomechanical analyzer—TMA Q400 from TA Instruments (Delaware, USA). Squared samples with parallel faces sizing 4 × 10 × 10 mm^3 were subjected to a constant load of 0.02 N and subsequently subjected to a heating program from −50 °C to 80 °C at a heating rate of 2 °C min^{-1}. The coefficient of linear thermal expansion—CLTE was calculated below and over the glass transition temperature, T_g.

2.5. Morphology Characterization

The morphology of PBS/ASF/MLO composites was studied by field emission scanning electron microscopy (FESEM) in a FESEM model ZEISS ULTRA 55 from Oxford Instruments (Abingdon, UK). Fractured samples from impact tests were subjected to a sputtering process with an aurum-palladium alloy inside a sputter coater model EMITECH SC7620 from Quorum Technologies (East Sussex, UK).

3. Results and Discussion

3.1. Appearance and Mechanical Properties of PBS/ASF/MLO Composites.

As observed in Figure 1, neat PBS is white due to its semicrystalline nature [13,55]. As the ASF loading increases, we can observe a slight change in color but in general, their appearance is like other wood plastic composites.

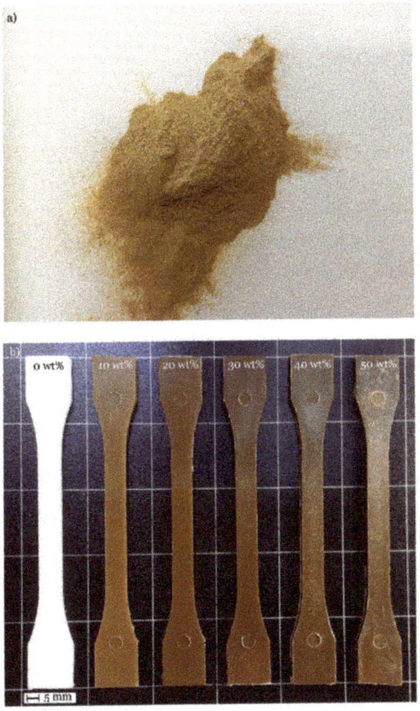

Figure 1. (**a**) Powdered almond shell flour (ASF) and (**b**) injection moulded PBS/ASF/MLO composites with varying ASF content in wt %.

As the ASF increases, the material becomes darker and this can be followed by the evolution of the luminance (L^*) of the samples as observed in Figure 2a. In addition, the colour coordinates a^* and b^* offer the real changes in brownish colour. The a^* coordinate changes from negative values (green) to positive values (red) while the b^* coordinate provides a measurement of the change in colour from blue (negative values) to yellow (positive values). As it can be seen in Figure 2b, all samples are placed in the $a^* > 0$ and $b^* > 0$ quadrant. In particular, both the a^* and b^* coordinates decrease as the ASF loading increases. Figure 2b also shows the colour coordinates (a^*b^*) of several commercial woods [56,57]. Poly(butylene succinate)/ASF/MLO composites with 10 wt % show similar colour coordinates to those of eucalypt [58] and teak woods [59]. As the ASF content increases, the yellow content decreases and the obtained materials are browner.

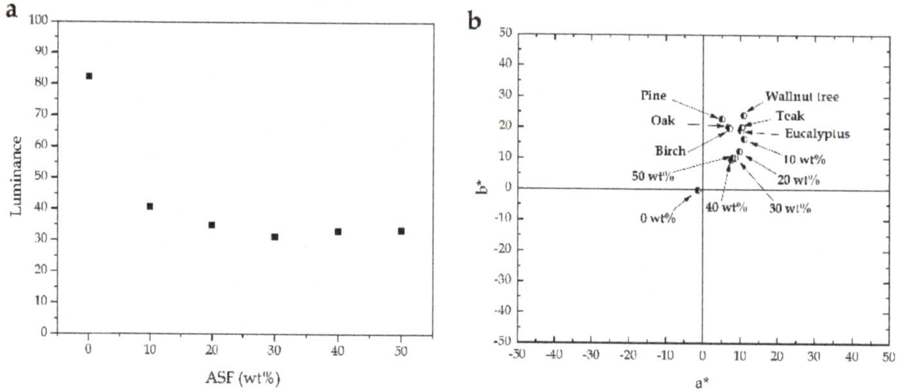

Figure 2. (**a**) Variation of the luminance (L^*) and (**b**) colour coordinates (a^*b^*) of PBS/ASF/MLO composites with varying ASF loading.

Typically, the addition of an uncompatibilized lignocellulosic filler into a polymer matrix, leads to a decrease in mechanical properties due to the low polymer/particle interactions. This lack (or very poor) interactions results in a decrease in material's cohesion which gives decreased elongation at break—ε_b and maximum tensile strength—σ_t as they are very sensitive to cohesion. In contrast, the material usually becomes stiffer with increased modulus as this represents the stress to strain ratio in the linear region [60]. Individual MLO additive into a thermoplastic polyester matrix, acts with several overlapping effects. The first one is a typical plasticization effect which gives a slight decrease in T_g with the subsequent decrease in all mechanical resistant properties (maximum tensile strength and modulus), while the elongation at break is increased. A second effect of MLO is chain extension due to reactions of the attached succinic anhydride group with hydroxyl groups located on polyester end-chains which gives increased elongation at break [52,53]. In this work, addition of both ASF and MLO should lead to overlapping all the above-mentioned phenomena in a complex way. Figure S1 in Supplementary Information shows a plot comparison of the typical stress (σ)—strain (ε) curves for PBS/ASF/MLO composites with different ASF and MLO loading. Table 2 summarizes the main mechanical properties of PBS/ASF/MLO composites. As it can be seen, a progressive decrease in maximum tensile strength from neat PBS (31.5 MPa) down to very low values of 7.1 MPa (50 wt % ASF + 7.5 MLO) are obtained. This dramatic decrease in tensile strength is directly related to the large filler content (in fact, the PBS polymer matrix only represents 42.5 wt % in composites with 50 wt % ASF). With regard to elongation at break—ε_b, the behavior is quite interesting. PBS is a very flexible polymer with a ε_b of 215.6%. The only addition of uncompatibilized 30 wt % ASF leads to a dramatic decrease in ε_b down to 6.3% as reported in our previous work [38], and this value was remarkably improved up to 25.8% by compatibilization with 4.5 wt % MLO. A decrease in ε_b is typical of uncompatibilized particle-filled polymers. The lack (or poor) polymer-particle interactions are responsible for this high decrease. It is important to remark that lignocellulosic fillers are high hydrophilic due to their cellulose and hemicellulose content while, on the other hand, most polymers are hydrophobic. These extremely high difference in polarity is responsible for poor polymer-particle interactions and this leads to poor material cohesion. As ε_b is highly sensitive to cohesion, uncompatibilized composites with a polymer matrix and a lignocellulosic filler, show a dramatic decrease in ε_b values. Despite some petroleum-derived compatibilizers (mainly copolymers) have been widely used to increase polymer-particle filler interactions, ε_b is not remarkably improved but in this work, the use of a flexible molecule derived from a natural triglyceride, allows overcoming or minimizing this poor polymer-particle interaction thus allowing a moderate increase in ε_b, with regard to uncompatibilized PBS/ASF composites [61–64]. It seems evident that MLO addition provides improved ductile properties to PBS/ASF composite with 30 wt % ASF. In fact, the ε_b is higher than 16% for all composites, depending on the ASF and MLO content. For low ASF loading (up to 30 wt %) an increasing tendency in ε_b can be observed. As indicated previously, the particle filler provides a remarkable decrease in ε_b while individual MLO gives increased elongation due to different phenomena. The positive compatibilizing effect of MLO is evident for these concentrations, with increasing ε_b as MLO content increases from 1.5 up to 4.5 wt %. In fact, the compatibilized composite with 30 wt % ASF and 4.5 wt % MLO gives the highest ε_b value. It seems that there is an ASF threshold at 30 wt %. Below this threshold, MLO can effectively compatibilize PBS/ASF/MLO composites in a complex process which involves several MLO mechanisms. Over 30 wt % ASF, a decrease in ε_b down to 16% is obtained. This could mean that once the compatibilization threshold has been overpassed, compatibilization does not occur in a correct way (maybe due to the high filler content). As indicated previously, MLO could be responsible for some overlapping phenomena and, moreover, as it is not fully miscible with polyesters, plasticizer saturation could occur as reported previously with polymers and composites with modified vegetable oils [65]. The evolution of the tensile modulus also suggests that there is an ASF threshold at about 30 wt % above which, compatibilization does not occur in an appropriate way.

Table 2. Summary of the main mechanical properties of PBS/ASF/MLO composites obtained by tensile, hardness and impact tests.

Sample Code	Maximum Tensile Strength, σ_t (MPa)	Tensile Modulus, E_t (MPa)	Elongation at Break, ε_b (%)	Shore D Hardness	Impact Strength (J m^{-2})
PBS	31.5 ± 0.9	417 ± 21	215.6 ± 16.5	60.1 ± 0.5	16.5 ± 0.8
PBS + 30 ASF [38]	14.8 ± 0.5	790 ± 56	6.3 ± 0.9	71.2 ± 0.3	1.8 ± 0.3
PBS + 10 ASF + 1.5 MLO	24.6 ± 0.2	561 ± 29	17.0 ± 0.6	66.7 ± 0.7	5.4 ± 0.4
PBS + 20 ASF + 3 MLO	18.6 ± 0.4	601 ± 63	20.7 ± 1.3	66.9 ± 0.9	3.9 ± 0.9
PBS + 30 ASF + 4.5 MLO	13.8 ± 0.3	535 ± 51	25.8 ± 1.0	67.2 ± 0.2	3.8 ± 0.5
PBS + 40 ASF + 6 MLO	9.3 ± 0.4	465 ± 76	16.3 ± 0.7	65.3 ± 0.5	2.6 ± 0.2
PBS + 50 ASF + 7.5 MLO	7.1 ± 0.2	364 ± 47	16.4 ± 1.0	63.3 ± 0.5	2.6 ± 0.1

With regard to Shore D hardness values, addition of ASF offers the same tendency than that observed for elongation at break but the changes are not significant. Regarding the impact strength, PBS/ASF/MLO composites show interesting behavior. As reported in our previous work, the addition of 30 wt % ASF to PBS without any compatibilizer gives an impact energy of 1.8 kJ m^{-2}, which is increased up to double by the addition of 4.5 wt % MLO (3.8 kJ m^{-2}) [38]. It is important to remark that the impact strength is directly related to mechanical resistant properties and ductile properties as well. So that, both an increase in tensile strength and an increase in elongation at break are representative for improved impact strength. So, by taking into account this fact, a decreasing tendency can be observed for impact strength with values of 5.4 kJ m^{-2} for PBS/ASF/MLO composites with 10 wt % ASF down to values of 2.6 kJ m^{-2} for the composite with the highest ASF content (50 wt %). It is worthy to note that even for this high ASF content, the impact strength is remarkably higher than that of the uncompatibilized composite consisting on PBS and 30 wt % ASF.

This particular mechanical response can be understood by evaluating the fracture surfaces from impact tests. Figure 3 shows the typical shape of almond shell flour (ASF) microparticles. As one can see, it is possible to find the typical spotted surface of almond shell on isolated micro-particles (Figure 3a). These particles provide a porous structure that can be positive for polymer/particle interactions. In addition to these shapes, it is possible to find rounded particles with smoothed surface a lower porosity (Figure 3b).

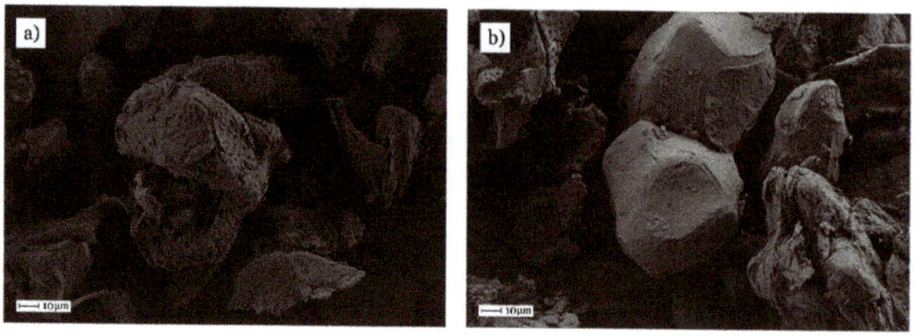

Figure 3. Field emission scanning electron microscopy (FESEM) images (500×) corresponding to almond shell flour (ASF) particles, (**a**) with spotted surfaces and porous structure and (**b**) with a smooth surface.

Figure 4 gathers the field emission scanning electron microscopy (FESEM) images for PBS/ASF/MLO composites with increasing ASF content. Figure 4a shows the fractured surface of neat PBS under impact conditions. Obviously, this surface appears to be rough due to deformation during impact. As it has been previously described, a threshold at 30 wt % ASF can be detected. Below 30 wt % ASF, the compatibilizing effect of MLO seems to be clear, thus leading to a progressive increase in mechanical ductile properties. Over 30 wt %, the MLO content is not enough to provide good compatibilization and this suggested poor polymer/particle cohesion. This different behavior can be explained by the following FESEM study. Figure 4b shows the fracture surface of the PBS/ASF/MLO composite with 10 wt % ASF (and the corresponding MLO content, i.e., 1.5 wt %). Both the matrix and the dispersed particles can be clearly identified. A typical spotted surface of an ASF particle is surrounded by the PBS matrix. As it can be seen (with white ellipses), the gap between the particle and the surrounding matrix is almost non-existent thus indicating the positive effect of MLO on establishing polymer/particle interactions through the reaction of succinic anhydride pendant groups with hydroxyl groups in both PBS (end-chains) and cellulose in ASF [66]. Another phenomenon can be observed in this composite. The PBS matrix shows scattered spherical shapes (white arrows) corresponding to MLO [67]. Similar situation can be found for PBS/ASF/MLO composites with 20 wt % (Figure 4c). In this case, a small gap (white ellipse) in the range of several hundred nanometers can be found between the ASF particle and the surrounding matrix, which also shows some more spherical shapes (white arrows) due to MLO with a typical diameter in the nanoscale range. Figure 4d shows the fracture surface of the PBS/ASF/MLO composite with 30 wt % and 4.5 wt % MLO which shows the best balanced mechanical properties (ductile and resistant). ASF are fully embedded inside the PBS matrix and clear evidences of plastic deformation (filaments) can be seen. Some small ASF particles can be seen with a small gap of 100–200 nm (white ellipse). Nevertheless, composites with 40 wt % ASF show a clear difference with regard to the previous composites with 30 wt % ASF or less. If we observe Figure 4e, the gap between the ASF particle and the surrounding matrix (white ellipse) is noticeably higher of 1–2 μm. This phenomenon indicates an excess ASF particles and MLO is not enough to provide a homogeneous interface, thus leading to poor cohesion, which in turn, is responsible for a decrease in ductile properties. This same behavior can be observed for PBS/ASF/MLO composites with 50 wt % ASF (Figure 4f. It is worthy to note that the PBS content in these composites is only 42.5 wt % so that it is evident the poor cohesion among the surface. The PBS matrix also shows the spherical shapes corresponding to MLO. So that, the FESEM study is in total agreement with the previous mechanical properties thus giving evidence of an ASF threshold which determined if MLO compatibilization is effective or not. For 30 wt % ASF or less, MLOs can provide a homogeneous interface and contribute to compatibilize ASF with the PBS matrix. Over 30 wt % ASF, the amount of filler seems to be extremely high and MLO content is not enough to compatibilize ASF with PBS thue leading to poor cohesion is detected.

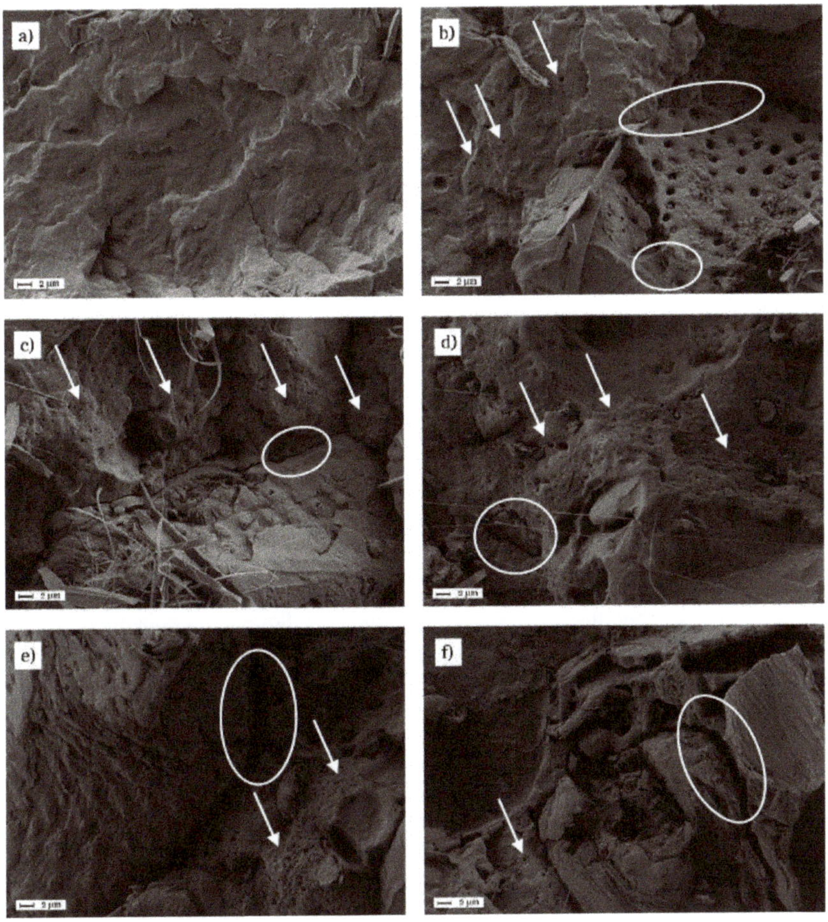

Figure 4. Field emission scanning electron microscopy (FESEM) images (2000×) of PBS/ASF/MLO composites with varying ASF content: (**a**) neat PBS; (**b**) 10 wt % ASF, 1.5 wt % MLO; (**c**) 20 wt % ASF, 3.0 wt %; (**d**) 30 wt % ASF, 4.5 wt %; (**e**) 40 wt % ASF, 6.0 wt %, and (**f**) 50 wt % ASF, 7.5 wt %.

3.2. Thermal Properties of PBS/ASF/MLO Composites

Differential scanning calorimetry was used to obtain the main thermal transitions of PBS/ASF/MLO composites. Figure S2 in Supplementary Materials shows a comparison of the DSC thermograms of PBS/ASF/MLO composites with different ASF and MLO content. Table 3 gathers the main thermal parameters obtained through DSC characterization. As it can be observed, a slight decrease in the melt peak temperature (T_m) from 119.1 °C down to 113–114 °C can be observed thus indicating that ASF favours crystallization as the crystal structure of cellulose in ASF acts as nucleant during crystallization [68]. Obviously, the normalized melt enthalpy (ΔH_m) decreases with the increasing ASF and MLO content. Nevertheless, taking into account the actual PBS mass in each PBS/ASF/MLO composite, calculation of the degree of crystallinity ($\%\chi_c$) leads to slightly higher values with increasing ASF content. So that, in addition to the nucleant effect of ASF, it allows developing higher percentage of crystallinity in combination with MLO, which provides increased chain mobility. For this reason, the $\%\chi_c$ changes from 57.7% for neat PBS up to values of 62–66% with different ASF content. Similar tendency has been found in some biopolyesters such as PLA and other semicrystalline polymers [34,48]. In particular, similar findings were reported by Calabia et al. for PBS composites with varying cotton fiber loading in the 0–40 wt % range. They reported a clear increasing tendency on crystallinity with increasing cotton fiber loading [68].

Table 3. Main thermal properties of PBS/ASF/MLO composites with varying ASF content, obtained by differential scanning calorimetry (DSC) analysis.

Sample Code	Melt Enthalpy, ΔH_m (J g^{-1})	Melt Peak Temperature, T_m (°C)	χ_c (%)
PBS	65.1 ± 1.7	119.6 ± 0.9	57.7 ± 1.7
PBS + 10 ASF + 1.5 MLO	61.0 ± 2.4	113.8 ± 1.2	62.4 ± 2.5
PBS + 20 ASF + 3 MLO	52.5 ± 1.2	113.8 ± 2.1	61.8 ± 1.4
PBS + 30 ASF + 4.5 MLO	47.8 ± 2.8	113.9 ± 1.9	66.2 ± 3.9
PBS + 40 ASF + 6 MLO	37.2 ± 0.9	113.8 ± 0.9	62.5 ± 1.5
PBS + 50 ASF + 7.5 MLO	31.0 ± 1.6	114.1 ± 1.7	66.1 ± 3.4

With regard to thermal degradation at high temperatures, thermogravimetric analysis—TGA gave the main degradation parameters, i.e., $T_{5\%}$ and T_{max} which correspond to the temperature for a mass loss of 5% and the temperature for a maximum mass loss rate, respectively. Figure 5a shows the TGA degradation curves of PBS/ASF/MLO composites with increasing ASF content and Figure 5b shows the first derivative that allows identifying the temperature for the maximum mass loss rate. Poly(butylene succinate) degrades in a single step process and its $T_{5\%}$ is close to 338.1 °C, thus indicating high thermal stability. Almond shell flour, as other lignocellulosic fillers degrades in a complex process with several overlapping stages. The first stage is residual moisture removal at a temperature range of 80–100 °C. Over 250–270 °C, hemicellulose starts its degradation reactions followed by the more thermally stable cellulose domains. Degradation of cellulose and hemicellulose involves complex reactions comprised in the temperature range of 250–370 °C. Regarding lignin, it is worthy to note that its degradation occurs in a wider temperature range from 250 °C up to 450–500 °C [69,70]. In general, as ASF shows lower thermal stability than PBS matrix, the typical TGA curves are moved towards lower temperatures with increasing the ASF content as Figure 5a shows. As shown in Table 4, the $T_{5\%}$ changes progressively from 338.1 °C for neat PBS down to values of 256.3 °C for the PBS/ASF/MLO composite containing 50 wt % ASF. The same tendency can be found for the temperature corresponding to the maximum mass loss rate (T_{max}), which is represented in Figure 5b as the peak minimum which is moved towards lower temperature values as it can be quantified in Table 4. In particular, the T_{max} changes from 414.7 °C for neat PBS to values of 378.7 °C for the PBS/ASF/MLO composite with the highest ASF content (50 wt %). All composites (even that with the highest ASF content of 50 wt %) are thermally stable up to 250 °C which indicates processing can be carried out in a wide temperature window since the melt process of PBS is comprised between 100–120 °C. Poly(butylene succinate) shows high thermal stability as other polyesters but lignocellulosic particles degrade at lower temperatures. Nevertheless, as the melt process of PBS is moderate, it is usually processed at temperatures in the 130–140 °C. As it has been shown by TGA analysis, all composites show thermal stability up to 250 °C. So that, although ASF addition leads to decreased thermal stability, it does not compromise processing and applications of PBS/ASF composites which can find interesting applications in the automotive industry (interior panels), construction and building (fencing, gates, panels, railings, and so on), outdoor furniture parts, etc. [71–73].

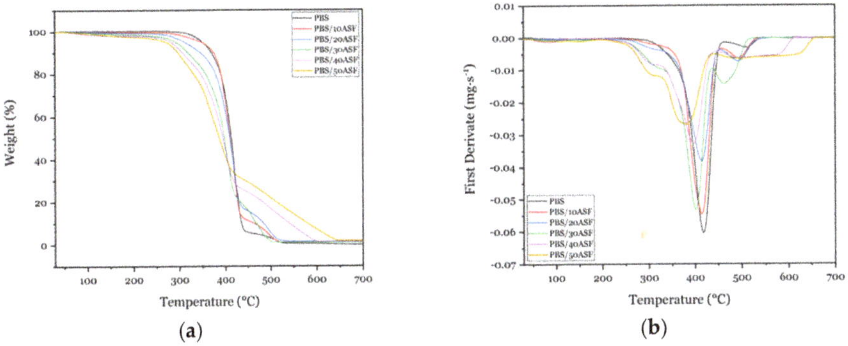

Figure 5. (**a**) thermogravimetric (TGA) degradation curves and (**b**) first derivative (DTG) of PBS/ASF/MLO composites with varying ASF loading.

Table 4. Main thermal degradation parameters of PBS/ASF/MLO composites with varying ASF content, obtained by thermogravimetric analysis (TGA).

Sample Code	$T_{5\%}$ (°C)	T_{max} (°C)	Residual Weight (%)
PBS	338.1	414.7	0.39
PBS + 10 ASF + 1.5 MLO	338.0	414.9	1.14
PBS + 20 ASF + 3 MLO	305.3	414.2	1.58
PBS + 30 ASF + 4.5 MLO	295.8	407.6	0.68
PBS + 40 ASF + 6 MLO	272.7	395. 7	1.39
PBS + 50 ASF + 7.5 MLO	256.3	378.7	1.86

3.3. Thermomechanical Properties of PBS/ASF/MLO Composites.

Dynamic mechanical thermal characterization—DMTA was used to evaluate the influence of temperature on mechanical behavior of PBS/ASF/MLO composites. Figure 6a shows the plot evolution of the storage modulus (E') as a function of temperature for PBS/ASF/MLO composites with increasing ASF loading. At low temperatures of −50 °C, all composites show similar storage modulus and the difference in behavior can be observed over the glass transition temperature, T_g (located in the −40/−10 °C range). At room temperature the material with the lowest stiffness is neat PBS and, as the ASF increases, the characteristic DMTA curve is moved to higher E' values. This behavior is typical in dynamic tests with other polymer/natural filler composites [74]. PBS is a viscoelastic polymer and as the damping factor represents the ratio between the loss modulus (E'') to the storage modulus (E'), it is possible to see in Figure 6b that over the glass transition process, the loss modulus increases with increasing ASF due to internal friction between PBS polymer chains and ASF particles. For this reason, above the glass transition process, the damping factor also increases with the ASF content. Regarding the damping factor (*tan δ*), (Figure 6c), the maximum damping factor corresponds to PBS and decreases as the ASF content increases. The peak maximum corresponding to the damping factor, can be assigned to the glass transition temperature (T_g) of the PBS-rich phase. Neat PBS shows a T_g value of about −23 °C and this is slightly moved to −20 °C by the addition of MLO to PBS/ASF/MLO composites with up to 30 wt % PBS. This indicates somewhat interaction as observed by FESEM. Nevertheless, the change in T_g is not remarkable.

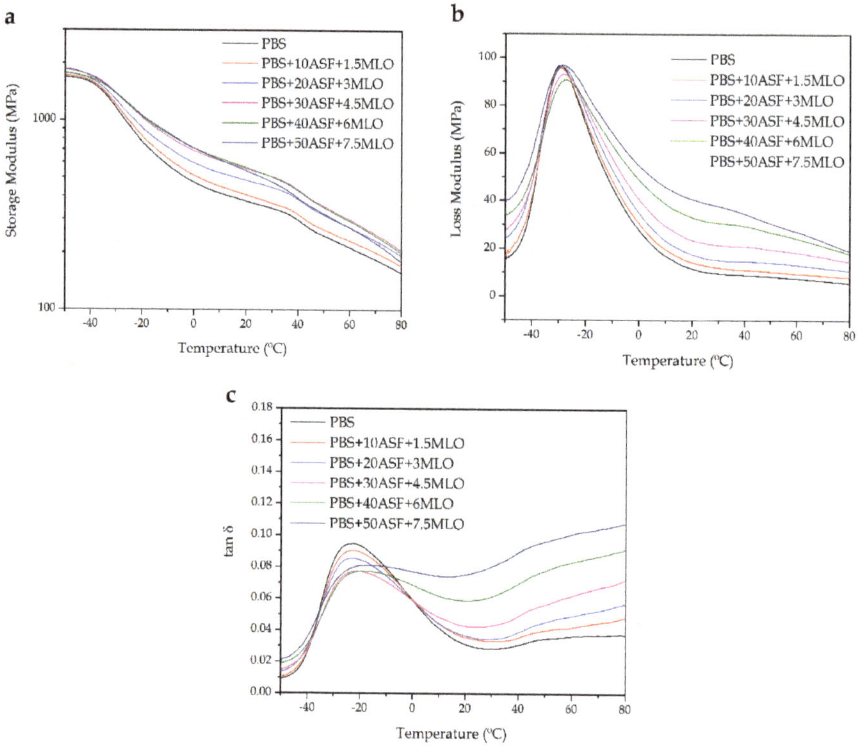

Figure 6. Plot evolution of (**a**) storage modulus (E'), (**b**) loss modulus (E''), and (**c**) damping factor (*tan δ*) of PBS/ASF/MLO composites with varying ASF loading.

In addition to dynamic mechanical thermal characterization (DMTA), thermomechanical analysis was used to study the thermal stability of PBS/ASF/MLO composites. Table 5 summarizes the values of the coefficient of linear thermal expansion (CLTE) both below and above the T_g. Obviously, the *CLTE* values are lower below the T_g since the material behaves as rigid and temperature does not affect in a great extent to a change in dimension. On the other hand, the *CLTE* values measured above the T_g, are remarkably higher as the material behaves as a softened/plastic material. In particular, neat PBS possesses a *CLTE* of 84.4 μm m^{-1} $°C^{-1}$ and 220.3 μm m^{-1} $°C^{-1}$ below and above the T_g, respectively. As it can be expected, the addition of a lignocellulosic filler leads to improved dimensional stability. Thus, for CLTE values measured below the T_g, all PBS/ASF/MLO composites show lower values compared to neat PBS. Nevertheless, the dimensional stabilization that ASF can provide to PBS/ASF/MLO composites is much evidence by analysing the CLTE values measured above the T_g that changes from 220.3 μm m^{-1} $°C^{-1}$ (neat PBS) down to values of 149.8 μm m^{-1} $°C^{-1}$ for composites with 50 wt % ASF [75].

Table 5. Variation of the coefficient of linear thermal expansion (*CLTE*) below and above the glass transition temperature (T_g) of PBS/ASF/MLO composites with varying ASF loading.

Sample Code	*CLTE* (μm m^{-1} °C^{-1}) Obtained by TMA		Thermal Parameters Obtained by DTMA	
	Below T_g	Above T_g	T_g (°C)	*tan δ*
PBS	84.4	222.3	−22.9	0.094
PBS + 10 ASF + 1.5 MLO	85.11	193.6	−22.8	0.090
PBS + 20 ASF + 3 MLO	81.8	166.9	−23.0	0.085
PBS + 30 ASF + 4.5 MLO	74.9	167.1	−21.6	0.077
PBS + 40 ASF + 6 MLO	83.5	168.0	−18.2	0.077
PBS + 50 ASF + 7.5 MLO	81.7	149.8	−17.9	0.081

4. Conclusions

In this work, high environmentally-friendly composites with a PBS matrix and a lignocellulosic waste from almond shell, were successfully manufactured by extrusion/compounding followed by injection moulding. Almond shell waste, in the form of powder (ASF), was added in the 0–50 wt % range and, to improve polymer/particle interactions, a vegetable oil-derived compatibilizer, namely maleinized linseed oil, was used with a constant ratio regarding the ASF content. These composites offer a wood like color and can be used as wood plastic composites. Composites containing 30 wt % ASF and 4.5 wt % MLO, offer the best balanced properties. In particular, it shows the maximum impact strength with balanced modulus and elongation at break. The obtained results suggest an ASF threshold of 30 wt %. Below this threshold MLO can provide improved PBS/ASF interactions as confirmed by field emission scanning electron microscopy. Over 30 wt % ASF, composites are more brittle and interface phenomena are less intense thus leading to poor material cohesion, therefore indicating poor compatibilization of MLO due to high ASF content and MLO saturation. As PBS is a very flexible material, the PBS/ASF/MLO composites obtained in this work, represent an interesting technical solution to low-medium mechanical properties composites with potential uses as wood plastic composites.

Supplementary Materials: The following are available online at http://www.mdpi.com/1996-1944/11/11/2179/s1, Figure S1: Comparative plot of the stress (σ)–strain (ε) curves of PBS/ASF/MLO composites with varying ASF content., Figure S2: Comparative plot of the DSC thermograms of PBS/ASF/MLO composites with varying ASF content.

Author Contributions: Conceptualization was devised by N.M., L.Q.-C. and R.B.; Methodology, Validation, and Formal Analysis was carried out by P.L., L.Q.-C., N.M., T.B., and R.B.; Investigation, Resources, Data Curation, and Writing-Original Draft Preparation was performed by P.L. and L.Q.-C.; Writing-Review & Editing, P.L. and R.B.; Supervision, N.M. and R.B.; Project Administration, R.B.

Funding: This research was supported by the Ministry of Economy, Industry and Competitiveness (MINECO) program number MAT2017-84909-C2-2-R.

Acknowledgments: L.Q.-C. wants to thank Generalitat Valenciana for his FPI grant (ACIF/2016/182) and the Spanish Ministry of Education, Culture, and Sports (MECD) for his FPU grant (FPU15/03812).

Conflicts of Interest: The authors declare no conflict of interest.

References

1. Hottle, T.A.; Bilec, M.M.; Landis, A.E. Biopolymer production and end of life comparisons using life cycle assessment. *Resour. Conserv. Recyc.* **2017**, *122*, 295–306. [CrossRef]
2. Niaounakis, M. *Biopolymers: Reuse, Recycling, and Disposal*; Elsevier: Amsterdam, The Netherlands, 2013; pp. 1–413.
3. Zhu, Y.; Romain, C.; Williams, C.K. Sustainable polymers from renewable resources. *Nature* **2016**, *540*, 354–362. [CrossRef] [PubMed]
4. Gandini, A.; Lacerda, T.M. From monomers to polymers from renewable resources: Recent advances. *Prog. Polym. Sci.* **2015**, *48*, 1–39. [CrossRef]

5. Eichhorn, S.J.; Gandini, A. Materials from Renewable Resources. *MRS Bull.* **2010**, *35*, 187–190. [CrossRef]
6. Fombuena, V.; Sanchez-Nacher, L.; Samper, M.D.; Juarez, D.; Balart, R. Study of the Properties of Thermoset Materials Derived from Epoxidized Soybean Oil and Protein Fillers. *J. Am. Oil Chem. Soc.* **2013**, *90*, 449–457. [CrossRef]
7. Ferrero, B.; Boronat, T.; Moriana, R.; Fenollar, O.; Balart, R. Green Composites Based on Wheat Gluten Matrix and Posidonia Oceanica Waste Fibers as Reinforcements. *Polym. Compos.* **2013**, *34*, 1663–1669. [CrossRef]
8. Kondratowicz, F.; Ukielski, R. Synthesis and hydrolytic degradation of poly (ethylene succinate) and poly (ethylene terephthalate) copolymers. *Polym. Degrad. Stab.* **2009**, *94*, 375–382. [CrossRef]
9. Mochizuki, M.; Hirami, M. Structural effects on the biodegradation of aliphatic polyesters. *Polym. Adv. Technol.* **1997**, *8*, 203–209. [CrossRef]
10. Debuissy, T.; Pollet, E.; Averous, L. Synthesis of potentially biobased copolyesters based on adipic acid and butanediols: Kinetic study between 1,4-and 2,3-butanediol and their influence on crystallization and thermal properties. *Polymer* **2016**, *99*, 204–213. [CrossRef]
11. Patel, M.K.; Bechu, A.; Villegas, J.D.; Bergez-Lacoste, M.; Yeung, K.; Murphy, R.; Woods, J.; Mwabonje, O.N.; Ni, Y.Z.; Patel, A.D.; Gallagher, J.; Bryant, D. Second-generation bio-based plastics are becoming a reality—Non-renewable energy and greenhouse gas (GHG) balance of succinic acid-based plastic end products made from lignocellulosic biomass. *Biofuels Bioprod. Biorefin. Biofpr.* **2018**, *12*, 426–441. [CrossRef]
12. Huang, Z.; Qian, L.; Yin, Q.; Yu, N.; Liu, T.; Tian, D. Biodegradability studies of poly(butylene succinate) composites filled with sugarcane rind fiber. *Polym. Test.* **2018**, *66*, 319–326. [CrossRef]
13. Puchalski, M.; Szparaga, G.; Biela, T.; Gutowska, A.; Sztajnowski, S.; Krucinska, I. Molecular and Supramolecular Changes in Polybutylene Succinate (PBS) and Polybutylene Succinate Adipate (PBSA) Copolymer during Degradation in Various Environmental Conditions. *Polymers* **2018**, *10*, 251. [CrossRef]
14. Fujimaki, T. Processability and properties of aliphatic polyesters, 'BIONOLLE', synthesized by polycondensation reaction. *Polym. Degrad. Stab.* **1998**, *59*, 209–214. [CrossRef]
15. Cihal, P.; Vopicka, O.; Pilnacek, K.; Poustka, J.; Friess, K.; Hajslova, J.; Dobias, J.; Dole, P. Aroma scalping characteristics of polybutylene succinate based films. *Polym. Test.* **2015**, *46*, 108–115. [CrossRef]
16. Siracusa, V.; Lotti, N.; Munari, A.; Rosa, M.D. Poly(butylene succinate) and poly(butylene succinate-co-adipate) for food packaging applications: Gas barrier properties after stressed treatments. *Polym. Degrad. Stab.* **2015**, *119*, 35–45. [CrossRef]
17. Gigli, M.; Fabbri, M.; Lotti, N.; Gamberini, R.; Rimini, B.; Munari, A. Poly(butylene succinate)-based polyesters for biomedical applications: A review. *Eur. Polym. J.* **2016**, *75*, 431–460. [CrossRef]
18. Cheng, H.H.; Xiong, J.; Xie, Z.N.; Zhu, Y.T.; Liu, Y.M.; Wu, Z.Y.; Yu, J.; Guo, Z.X. Thrombin-Loaded Poly (butylene succinate)-Based Electrospun Membranes for Rapid Hemostatic Application. *Macromol. Mater. Eng.* **2018**, *303*, 1700395. [CrossRef]
19. Costa-Pinto, A.R.; Martins, A.M.; Castelhano-Carlos, M.J.; Correlo, V.M.; Sol, P.C.; Longatto-Filho, A.; Battacharya, M.; Reis, R.L.; Neves, N.M. In vitro degradation and in vivo biocompatibility of chitosan–poly (butylene succinate) fiber mesh scaffolds. *J. Bioact. Compat. Polym.* **2014**, *29*, 137–151. [CrossRef]
20. Wu, D.; Lin, D.; Zhang, J.; Zhou, W.; Zhang, M.; Zhang, Y.; Wang, D.; Lin, B. Selective localization of nanofillers: effect on morphology and crystallization of PLA/PCL blends. *Macromol. Chem. Phys.* **2011**, *212*, 613–626. [CrossRef]
21. Peponi, L.; Sessini, V.; Arrieta, M.P.; Navarro-Baena, I.; Sonseca, A.; Dominici, F.; Gimenez, E.; Torre, L.; Tercjak, A.; López, D. Thermally-activated shape memory effect on biodegradable nanocomposites based on PLA/PCL blend reinforced with hydroxyapatite. *Polym. Degrad. Stab.* **2018**, *151*, 36–51. [CrossRef]
22. Dicker, M.P.M.; Duckworth, P.F.; Baker, A.B.; Francois, G.; Hazzard, M.K.; Weaver, P.M. Green composites: A review of material attributes and complementary applications. *Compos. Part A Appl. Sci. Manuf.* **2014**, *56*, 280–289. [CrossRef]
23. Gurunathan, T.; Mohanty, S.; Nayak, S.K. A review of the recent developments in biocomposites based on natural fibres and their application perspectives. *Compos. Part A Appl. Sci. Manuf.* **2015**, *77*, 1–25. [CrossRef]
24. Lau, K.T.; Hung, P.Y.; Zhu, M.H.; Hui, D. Properties of natural fibre composites for structural engineering applications. *Compos. Part B Eng.* **2018**, *136*, 222–233. [CrossRef]
25. Chun, K.S.; Yeng, C.M.; Hussiensyah, S. Green Coupling Agent for Agro-Waste Based Thermoplastic Composites. *Polym. Compos.* **2018**, *39*, 2441–2450. [CrossRef]

26. Panthapulakkal, S.; Sain, M. Agro-residue reinforced high-density polyethylene composites: Fiber characterization and analysis of composite properties. *Compos. Part A Appl. Sci. Manuf.* **2007**, *38*, 1445–1454. [CrossRef]
27. Vaisanen, T.; Haapala, A.; Lappalainen, R.; Tomppo, L. Utilization of agricultural and forest industry waste and residues in natural fiber-polymer composites: A review. *Waste Manag.* **2016**, *54*, 62–73. [CrossRef] [PubMed]
28. Feng, Y.H.; Li, Y.J.; Xu, B.P.; Zhang, D.W.; Qu, J.P.; He, H.Z. Effect of fiber morphology on rheological properties of plant fiber reinforced poly(butylene succinate) composites. *Compos. Part B Eng.* **2013**, *44*, 193–199. [CrossRef]
29. Terzopoulou, Z.N.; Papageorgiou, G.Z.; Papadopoulou, E.; Athanassiadou, E.; Reinders, M.; Bikiaris, D.N. Development and Study of Fully Biodegradable Composite Materials Based on Poly(butylene succinate) and Hemp Fibers or Hemp Shives. *Polym. Compos.* **2016**, *37*, 407–421. [CrossRef]
30. Lee, J.M.; Ishak, Z.A.M.; Taib, R.M.; Law, T.T.; Thirmizir, M.Z.A. Mechanical, Thermal and Water Absorption Properties of Kenaf-Fiber-Based Polypropylene and Poly(Butylene Succinate) Composites. *J. Polym. Environ.* **2013**, *21*, 293–302. [CrossRef]
31. Tserki, V.; Matzinos, P.; Panayiotou, C. Novel biodegradable composites based on treated lignocellulosic waste flour as filler Part II. Development of biodegradable composites using treated and compatibilized waste flour. *Compos. Part A Appl. Sci. Manuf.* **2006**, *37*, 1231–1238. [CrossRef]
32. Yen, F.S.; Liao, H.T.; Wu, C.S. Characterization and biodegradability of agricultural residue-filled polyester ecocomposites. *Polym. Bull.* **2013**, *70*, 1613–1629. [CrossRef]
33. El Mechtali, F.Z.; Essabir, H.; Nekhlaoui, S.; Bensalah, M.; Jawaid, M.; Bouhfid, R.; Qaiss, A. Mechanical and Thermal Properties of Polypropylene Reinforced with Almond Shells Particles: Impact of Chemical Treatments. *J. Bionic Eng.* **2015**, *12*, 483–494. [CrossRef]
34. Essabir, H.; Nekhlaoui, S.; Malha, M.; Bensalah, M.O.; Arrakhiz, F.Z.; Qaiss, A.; Bouhfid, R. Bio-composites based on polypropylene reinforced with Almond Shells particles: Mechanical and thermal properties. *Mater. Des.* **2013**, *51*, 225–230. [CrossRef]
35. Garcia, A.M.; Garcia, A.I.; Cabezas, M.L.; Reche, A.S. Study of the Influence of the Almond Variety in the Properties of Injected Parts with Biodegradable Almond Shell Based Masterbatches. *Waste Biomass Valoriz.* **2015**, *6*, 363–370. [CrossRef]
36. Quiles-Carrillo, L.; Montanes, N.; Sammon, C.; Balart, R.; Torres-Giner, S. Compatibilization of highly sustainable polylactide/almond shell flour composites by reactive extrusion with maleinized linseed oil. *Ind. Crops Prod.* **2018**, *111*, 878–888. [CrossRef]
37. Garcia, A.V.; Santonja, M.R.; Sanahuja, A.B.; Selva, M.D.G. Characterization and degradation characteristics of poly (epsilon-caprolactone)-based composites reinforced with almond skin residues. *Polym. Degrad. Stab.* **2014**, *108*, 269–279. [CrossRef]
38. Liminana, P.; Garcia-Sanoguera, D.; Quiles-Carrillo, L.; Balart, R.; Montanes, N. Development and characterization of environmentally friendly composites from poly(butylene succinate) (PBS) and almond shell flour with different compatibilizers. *Compos. Part B Eng.* **2018**, *144*, 153–162. [CrossRef]
39. Fu, S.Y.; Feng, X.Q.; Lauke, B.; Mai, Y.W. Effects of particle size, particle/matrix interface adhesion and particle loading on mechanical properties of particulate-polymer composites. *Compos. Part B Eng.* **2008**, *39*, 933–961. [CrossRef]
40. Kim, H.S.; Lee, B.H.; Lee, S.; Kim, H.J.; Dorgan, J. Enhanced interfacial adhesion, mechanical, and thermal properties of natural flour-filled biodegradable polymer bio-composites. *J. Therm. Anal. Calorim.* **2011**, *104*, 331–338. [CrossRef]
41. Li, Y.; Zhang, J.; Cheng, P.J.; Shi, J.J.; Yao, L.; Qiu, Y.P. Helium plasma treatment voltage effect on adhesion of ramie fibers to polybutylene succinate. *Ind. Crops Prod.* **2014**, *61*, 16–22. [CrossRef]
42. Sepe, R.; Bollino, F.; Boccarusso, L.; Caputo, F. Influence of chemical treatments on mechanical properties of hemp fiber reinforced composites. *Compos. Part B Eng.* **2018**, *133*, 210–217. [CrossRef]
43. Shaniba, V.; Sreejith, M.P.; Aparna, K.B.; Jinitha, T.V.; Purushothaman, E. Mechanical and thermal behavior of styrene butadiene rubber composites reinforced with silane-treated peanut shell powder. *Polym. Bull.* **2017**, *74*, 3977–3994. [CrossRef]

44. Phua, Y.J.; Chow, W.S.; Ishak, Z.A.M. Reactive processing of maleic anhydride-grafted poly(butylene succinate) and the compatibilizing effect on poly(butylene succinate) nanocomposites. *Express Polym. Lett.* **2013**, *7*, 340–354. [CrossRef]
45. Zhu, N.Q.; Ye, M.; Shi, D.J.; Chen, M.Q. Reactive compatibilization of biodegradable poly(butylene succinate)/Spirulina microalgae composites. *Macromol. Res.* **2017**, *25*, 165–171. [CrossRef]
46. Chieng, B.W.; Ibrahim, N.A.; Then, Y.Y.; Loo, Y.Y. Epoxidized Vegetable Oils Plasticized Poly(lactic acid) Biocomposites: Mechanical, Thermal and Morphology Properties. *Molecules* **2014**, *19*, 16024–16038. [CrossRef] [PubMed]
47. Orue, A.; Eceiza, A.; Arbelaiz, A. Preparation and characterization of poly(lactic acid) plasticized with vegetable oils and reinforced with sisal fibers. *Ind. Crop. Prod.* **2018**, *112*, 170–180. [CrossRef]
48. Balart, J.F.; Fombuena, V.; Fenollar, O.; Boronat, T.; Sanchez-Nacher, L. Processing and characterization of high environmental efficiency composites based on PLA and hazelnut shell flour (HSF) with biobased plasticizers derived from epoxidized linseed oil (ELO). *Compos. Part B Eng.* **2016**, *86*, 168–177. [CrossRef]
49. Garcia-Garcia, D.; Ferri, J.M.; Montanes, N.; Lopez-Martinez, J.; Balart, R. Plasticization effects of epoxidized vegetable oils on mechanical properties of poly(3-hydroxybutyrate). *Polym. Int.* **2016**, *65*, 1157–1164. [CrossRef]
50. Sarwono, A.; Man, Z.; Bustam, M.A. Blending of Epoxidised Palm Oil with Epoxy Resin: The Effect on Morphology, Thermal and Mechanical Properties. *J. Polym. Environ.* **2012**, *20*, 540–549. [CrossRef]
51. Carbonell-Verdu, A.; Garcia-Garcia, D.; Dominici, F.; Torre, L.; Sanchez-Nacher, L.; Balart, R. PLA films with improved flexibility properties by using maleinized cottonseed oil. *Eur. Polym. J.* **2017**, *91*, 248–259. [CrossRef]
52. Garcia-Garcia, D.; Fenollar, O.; Fombuena, V.; Lopez-Martinez, J.; Balart, R. Improvement of Mechanical Ductile Properties of Poly(3-hydroxybutyrate) by Using Vegetable Oil Derivatives. *Macromol. Mater. Eng.* **2017**, *302*, 1600330. [CrossRef]
53. Ferri, J.M.; Garcia-Garcia, D.; Sanchez-Nacher, L.; Fenollar, O.; Balart, R. The effect of maleinized linseed oil (MLO) on mechanical performance of poly(lactic acid)-thermoplastic starch (PLA-TPS) blends. *Carbohydr. Polym.* **2016**, *147*, 60–68. [CrossRef] [PubMed]
54. Ren, M.; Song, J.; Song, C.; Zhang, H.; Sun, X.; Chen, Q.; Zhang, H.; Mo, Z. Crystallization kinetics and morphology of poly (butylene succinate-co-adipate). *J. Polym. Sci. Part B Polym. Phys.* **2005**, *43*, 323–3241. [CrossRef]
55. Ye, H.-M.; Chen, X.-T.; Liu, P.; Wu, S.-Y.; Jiang, Z.; Xiong, B.; Xu, J. Preparation of Poly(butylene succinate) Crystals with Exceptionally High Melting Point and Crystallinity from Its Inclusion Complex. *Macromolecules* **2017**, *50*, 5425–5433. [CrossRef]
56. Barcík, Š.; Gašparík, M.; Razumov, E.Y. Effect of temperature on the color changes of wood during thermal modification. *Cellul. Chem. Technol.* **2015**, *49*, 789–798.
57. Ostafi, M.-F.; Dinulică, F.; Nicolescu, V.-N. Physical properties and structural features of common walnut (Juglans regia L.) wood: A case-study/Physikalische Eigenschaften und strukturelle Charakteristika des Holzes der Walnuß (Juglans regia L.): Eine Fallstudie. *Die Bodenkultur J. Land Manag. Food Environ.* **2016**, *67*, 105–120. [CrossRef]
58. Luís, R.C.G.; Nisgoski, S.; Klitzke, R.J. Effect of Steaming on the Colorimetric Properties of Eucalyptus saligna Wood. *Floresta e Ambiente* **2018**. [CrossRef]
59. Lopes, J.d.O.; Garcia, R.A.; Latorraca, J.V.d.F.; Nascimento, A.M.d. Color change of teak wood by heat treatment. *Floresta e Ambiente* **2014**, *21*, 521–534. [CrossRef]
60. Yang, H.-S.; Kim, H.-J.; Park, H.-J.; Lee, B.-J.; Hwang, T.-S. Water absorption behavior and mechanical properties of lignocellulosic filler–polyolefin bio-composites. *Compos. Struct.* **2006**, *72*, 429–437. [CrossRef]
61. Xu, X.-L.; Zhang, M.; Qiang, Q.; Song, J.-Q.; He, W.-Q. Study on the performance of the acetylated bamboo fiber/PBS composites by molecular dynamics simulation. *J. Compos. Mater.* **2016**, *50*, 995–1003. [CrossRef]
62. Wu, C.-S.; Hsu, Y.-C.; Liao, H.-T.; Yen, F.-S.; Wang, C.-Y.; Hsu, C.-T. Characterization and Biocompatibility of Chestnut Shell Fiber-Based Composites with Polyester. *J. Appl. Polym. Sci.* **2014**, *131*. [CrossRef]
63. Saeed, U.; Nawaz, M.A.; Al-Turaif, H.A. Wood flour reinforced biodegradable PBS/PLA composites. *J. Compos. Mater.* **2018**, *52*, 2641–2650. [CrossRef]
64. Luo, X.; Li, J.; Feng, J.; Yang, T.; Lin, X. Mechanical and thermal performance of distillers grains filled poly(butylene succinate) composites. *Mater. Des.* **2014**, *57*, 195–200. [CrossRef]

65. Ljungberg, N.; Wesslen, B. The effects of plasticizers on the dynamic mechanical and thermal properties of poly (lactic acid). *J. Appl. Polym. Sci.* **2002**, *86*, 1227–1234. [CrossRef]
66. Quiles-Carrillo, L.; Blanes-Martínez, M.; Montanes, N.; Fenollar, O.; Torres-Giner, S.; Balart, R. Reactive toughening of injection-molded polylactide pieces using maleinized hemp seed oil. *Eur. Polym. J.* **2018**, *98*, 402–410. [CrossRef]
67. Quiles-Carrillo, L.; Montanes, N.; Garcia-Garcia, D.; Carbonell-Verdu, A.; Balart, R.; Torres-Giner, S. Effect of different compatibilizers on injection-molded green composite pieces based on polylactide filled with almond shell flour. *Compos. Part B Eng.* **2018**, *147*, 76–85. [CrossRef]
68. Calabia, B.P.; Ninomiya, F.; Yagi, H.; Oishi, A.; Taguchi, K.; Kunioka, M.; Funabashi, M. Biodegradable poly (butylene succinate) composites reinforced by cotton fiber with silane coupling agent. *Polymers* **2013**, *5*, 128–141. [CrossRef]
69. Frollini, E.; Bartolucci, N.; Sisti, L.; Celli, A. Poly (butylene succinate) reinforced with different lignocellulosic fibers. *Ind. Crop. Prod.* **2013**, *45*, 160–169. [CrossRef]
70. Faulstich de Paiva, J.M.; Frollini, E. Unmodified and modified surface sisal fibers as reinforcement of phenolic and lignophenolic matrices composites: thermal analyses of fibers and composites. *Macromol. Mater. Eng.* **2006**, *291*, 405–417. [CrossRef]
71. Wang, G.; Guo, B.; Xu, J.; Li, R. Rheology, Crystallization Behaviors, and Thermal Stabilities of Poly(butylene succinate)/Pristine Multiwalled Carbon Nanotube Composites Obtained by Melt Compounding. *J. Appl. Polym. Sci.* **2011**, *121*, 59–67. [CrossRef]
72. Dumazert, L.; Rasselet, D.; Pang, B.; Gallard, B.; Kennouche, S.; Lopez-Cuesta, J.-M. Thermal stability and fire reaction of poly(butylene succinate) nanocomposites using natural clays and FR additives. *Polym. Adv. Technol.* **2018**, *29*, 69–83. [CrossRef]
73. Chen, G.X.; Yoon, J.S. Thermal stability of poly(L-lactide)/poly(butylene succinate)/clay nanocomposites. *Polym. Degrad. Stab.* **2005**, *88*, 206–212. [CrossRef]
74. Ferrero, B.; Fombuena, V.; Fenollar, O.; Boronat, T.; Balart, R. Development of natural fiber-reinforced plastics (NFRP) based on biobased polyethylene and waste fibers from Posidonia oceanica seaweed. *Polym. Compos.* **2015**, *36*, 1378–1385. [CrossRef]
75. Fuqua, M.A.; Chevali, V.S.; Ulven, C.A. Lignocellulosic byproducts as filler in polypropylene: Comprehensive study on the effects of compatibilization and loading. *J. Appl. Polym. Sci.* **2013**, *127*, 862–868. [CrossRef]

materials

MDPI

Article

Optimization of Maleinized Linseed Oil Loading as a Biobased Compatibilizer in Poly(Butylene Succinate) Composites with Almond Shell Flour

Patricia Liminana, David Garcia-Sanoguera, Luis Quiles-Carrillo *, Rafael Balart and Nestor Montanes

Technological Institute of Materials (ITM), Universitat Politècnica de València (UPV), Plaza Ferrándiz y Carbonell 1, 03801 Alcoy, Spain; patligre@mcm.upv.es (P.L.); dagarsa@dimm.upv.es (D.G.-S.); rbalart@mcm.upv.es (R.B.); nesmonmu@upvnet.upv.es (N.M.)

* Correspondence: luiquic1@epsa.upv.es; Tel.: +34-966-528-433

Received: 31 January 2019; Accepted: 22 February 2019; Published: 26 February 2019

Abstract: Green composites of poly(butylene succinate) (PBS) were manufactured with almond shell flour (ASF) by reactive compatibilization with maleinized linseed oil *MLO) by extrusion and subsequent injection molding. ASF was kept constant at 30 wt %, while the effect of different MLO loading on mechanical, thermal, thermomechanical, and morphology properties was studied. Uncompatibilized PBS/ASF composites show a remarkable decrease in mechanical properties due to the nonexistent polymer-filler interaction, as evidenced by field emission scanning electron microscopy (FESEM). MLO provides a plasticization effect on PBS/ASF composites but, in addition, acts as a compatibilizer agent since the maleic anhydride groups contained in MLO are likely to react with hydroxyl groups in both PBS end chains and ASF particles. This compatibilizing effect is observed by FESEM with a reduction of the gap between the filler particles and the surrounding PBS matrix. In addition, the T_g of PBS increases from −28 °C to −12 °C with an MLO content of 10 wt %, thus indicating compatibilization. MLO has been validated as an environmentally friendly additive to PBS/ASF composites to give materials with high environmental efficiency.

Keywords: polymer-matrix composites (PMCs); mechanical properties; thermomechanical; electron microscopy; compatibilizers

---

1. Introduction

Nowadays, there is growing interest in the development of biopolymers that could substitute for, or at least compete with, conventional petroleum-based polymers. This need is much more pronounced in the packaging industry due to the high volume of waste it generates. Among different alternatives such as proteins, polysaccharides, bacterial polymers, and so on, biopolyesters (either from petroleum origin or bio-derived) are gaining relevance as they are biodegradable (disintegrable in controlled compost soil conditions). One of these polyesters is poly(butylene succinate) (PBS), which presents interesting possibilities for manufacturing wood plastic composites (WPCs) or natural fiber-reinforced plastics (NFRPs) [1]. In general, the biodegradable synthetic polymers are mainly aliphatic polyesters produced by microbiological and chemical synthesis, natural polymer-based products, and their blends such as poly(lactic acid) (PLA), poly(ε-caprolactone) (PCL), poly(glycolic acid) (PGA), poly(butylene succinate) (PBS), poly(butylene succinate-co-adipate), among others. PBS offers interesting possibilities in the packaging industry due to an excellent combination of flexibility and biodegradability (more correctly, disintegration in controlled compost soil). PBS is produced through the condensation reaction of glycols such as 1,4-butanediol and succinic acid. Currently, PBS is obtained from the petroleum-based route, which involves the use of both 1,4-butanediol and succinic acid from petroleum sources. Nevertheless, it has been proposed as a renewable resource. It is worth remarking that some time ago the U.S. Department of Energy (DoE) classified succinic acid as one of the 12 most promising biobased building blocks for a biorefinery concept. Succinic acid can be obtained from renewable feedstocks such as glucose, sucrose, and glycerol. The main advantage is a lower carbon footprint. New fermentation processes are being developed in order to obtain a cost-effective material [2–4]. With regard to 1,4-butanediol, it can be obtained from the fermentation of dextrose, for example, and biobased BDO represents a valuable building block for a wide variety of engineering materials such as polyurethanes and biobased polymers, with interesting applications in the automotive industry [5]. PBS owns similar properties to some commodities widely used in the food industry. That is why PBS finds its main market in the food packaging industry [6–9]. Although it can fully disintegrate in compost soil, it is still an expensive polymer. For this reason, some research works have focused on manufacturing composite materials with natural fibers [10–12]. Thus, PBS has been used as a matrix for composites with hemp fiber [13], which contributes to lowering the overall cost of the developed material, making this a more environmentally friendly material. The use of wool waste in PBS composites has also been reported [14].

Spain is the second worldwide producer of almonds, just behind the USA. This industry generates a huge amount of waste, mainly shells. Although the use of almond shell waste to remove contaminants by adsorption has been reported [15–17], and they can be a biomass source for energy and fuels [18–20], one of the most attractive uses of almond shells is in the form of powder or flour (ASF), to be used in combination with several polymer matrices to give wood-like composite materials. Some investigations have focused on the partially biodegradable composites of poly(propylene) (PP) with almond shell flour [21,22], and fully biodegradable composites with poly(lactic acid) (PLA) matrices [23].

Almond shells are composed of approximately 38 wt % hemicellulose, 31 wt % cellulose, 28 wt % lignin, and 3 wt % other compounds [24]. This high content of cellulose, lignin, and hemicellulose gives a marked hydrophilic nature to the waste, which is opposite to the typical hydrophobic nature of most polymer matrices. This lack of (or very low) compatibility leads to stress concentration phenomena that, in turn, are responsible for a decrease in overall properties [25–28]. In particular, the mechanical properties of polymer-filled composites, which are directly related to material cohesion, are highly affected by the presence of hydrophilic particles embedded in a hydrophobic matrix. The elongation at break, as well as the impact strength are remarkably diminished [29,30]. To overcome this drawback, or at least to minimize its effects, several technical approaches are widely used such as the use of compatibilizers, surface treatment of particles that provide a more or less intense interaction between the polymer matrix and the embedded particles [31–34]. Plasticizers could act as internal lubricants, thus allowing chain mobility, which enhances processability and improves thermal stability and ductility. In the last decade, a new family of vegetable-oil-derived plasticizers has been proposed [35–37]. Among others, it is worth noting the increasing use of epoxidized oils such as linseed oil (ELO) [31,38], soybean oil (ESBO) [39], palm oil (EPO) [40], epoxidized castor oil (ECO) [41], epoxidized tung oil [42], and so on. Recently, the potential plasticization and compatibilization effects of maleinized oils such as linseed oil (MLO) [23,43], soybean oil (MSBO) [44], cotton seed (MCSO) [45,46], and hemp seed (MHSO) [47] have been reported.

Linseed oil (LO) is composed of about 9–11% saturated fatty acids (5–6% palmitic acid and 4–5% stearic acid) and around 75–90% unsaturated fatty acids (50–55% linolenic acid, 15–20% oleic acid, and 10–15% linoleic acid) [48]. This high unsaturated content allows chemical modification through different paths such as epoxidation, maleinization, hydroxylation, acrylation, etc. Maleinized linseed oil (MLO) has been proven to be a good plasticizer in PLA-based formulations, as reported by Ferri at al. [43]. In particular, they reported a dual effect of MLO on PLA blends with thermoplastic starch (TPS); on the one hand, a plasticization effect allows easy processing, and on the other hand, a compatibilizing effect through the reaction of maleic anhydride with hydroxyl groups in both PLA end chains and starch provides enhanced mechanical properties and thermal stability. It has also been proven to have its in poly(3-hydroxybutyrate) (PHB) formulations [36].

In a previous work with PBS composites and almond shell flour (ASF), maleinized linseed oil was revealed as the best compatibilizer compared to other families such as acrylates, anhydrides, and epoxidized oils [1]. This work is focused on the study of the effect of MLO loading on the final properties of PBS-ASF composites with the aim of improving interface phenomena between the PBS matrix and the embedded ASF particles.

2. Materials and Methods

2.1. Materials

A commercial grade of poly(butylene succinate) (PBS, Bionolle 1020MD, supplied by Showa Denko, Tokyo, Japan) was used as the base material for composites. This commercial grade possesses a melt flow index (MFI) comprised between 20 and 34 g/(10 min) and a density of 1.26 g cm^{-3}.

Almond shell powder/flour (ASF) was purchased from Jesol Materias Primas (Valencia, Spain). The supplied powder was sieved in a vibrational sieve RP09 CISA®(Barcelona, Spain) to obtain a homogeneous particle size of 150 μm.

Maleinized linseed oil (MLO) was supplied by Vandeputte (Mouscron, Belgium) under the trade name of VEOMER LIN. It possesses a viscosity of 10 dPa, measured at 20 °C, and an acid value of 105–130 mg KOH g^{-1}.

2.2. Manufacturing of PBS/ASF/MLO Composites

Initially, PBS and ASF were dried at 50 °C for 24 h to remove moisture since polyesters are highly sensitive to hydrolysis at high temperatures. MLO was heated to 40 °C to reduce its viscosity and enable good pre-mixing with PBS and ASF. Pre-mixing was carried out using ziploc bags with the appropriate compositions of each component (see Table 1). After this pre-mixing stage, the mixtures were compounded in a twin-screw co-rotating extruder from Dupra S.L. (Alicante, Spain) with a screw diameter of 25 mm and a length (L) to diameter (D) ratio (L/D) of 24. The screw speed was set to 40 rpm and the temperature profile was as follows: 120 °C, 125 °C, 130 °C, and 130 °C from the hopper to the die. After extrusion, the compounded materials were pelletized and dried at 50 °C for 24 h before further processing by injection molding in a Meteor 270/75 injection machine from Mateu & Solé (Barcelona, Spain). The temperature profile was set to 110 °C (hopper), 115 °C, 120 °C, and 125 °C (injection nozzle).

Table 1. Composition and labelling of poly(butylene succinate) composites with almond shell flour (ASF) and different maleinized linseed oil (MLO) compatibilizing load.

Reference	PBS wt %	ASF wt %	MLO wt %
PBS	100.0	-	-
PBS + ASF	70.0	30.0	-
PBS + ASF + 2.5MLO	67.5	30.0	2.5
PBS + ASF + 4.5MLO	65.5	30.0	4.5
PBS + ASF + 7.5MLO	62.5	30.0	7.5
PBS + ASF + 10MLO	60.0	30.0	10.0

2.3. Mechanical Characterization

Mechanical properties were obtained from standard tensile, hardness, and impact tests. Tensile tests were carried out in a universal test machine ELIB 50 from S.A.E. Ibertest (Madrid, Spain) following the guidelines of ISO 527-1:2012. A load cell of 5 kN was used and the crosshead speed was set to 10 mm min^{-1}. Hardness was measured in a Shore D durometer model 676-D from J. Bot Instruments (Barcelona, Spain), as indicated by ISO 868:2003. With regard to impact tests, a Charpy's pendulum with an energy of 1 J was used to obtain the impact energy on standard notched samples ("V" type notch with a radius of 0.25 mm), according to ISO 179-1:2010. At least five different samples were used for each test to ensure reproducible results and average values were calculated for each property.

2.4. Morphological Characterization

Fractured samples from impact tests were observed by field emission scanning electron microscopy (FESEM) in a ZEISS ULTRA 55 microscope from Oxford Instruments (Abingdon, UK). All samples were previously subjected to a sputtering process with an Au-Pd alloy to enhance electrical conductivity. The sputtering was carried out in a EMITECH SC7620 sputter coater from Quorum Technologies (Lewes, UK).

2.5. Thermal Characterization

The main thermal transitions were obtained by differential scanning calorimetry (DSC) using a Mettler-Toledo 821 calorimeter (Schwerzenbach, Switzerland). Samples with an average size of 5–7 mg were subjected to a dynamic thermal program consisting on three stages, a first heating from 25 °C to 200 °C, then a controlled cooling from 200 °C to −50 °C and, finally, a second heating process from −50 °C to 300 °C. The scanning rate was 10 °C min^{-1} for all three stages. All DSC runs were conducted in triplicate under a nitrogen atmosphere with a flow rate of 66 mL min^{-1}. Standard sealed aluminum crucibles with a total volume of 40 μL were used. The degree of crystallinity was calculated by the following equation:

$$X_C = \left[\frac{\Delta H_m}{\Delta H_m^0 \times (1-w)}\right] \times 100, \quad (1)$$

where ΔH_m stands for the measured melt enthalpy. $\Delta H^0{}_m$ (J g^{-1}) represents the melt enthalpy of a theoretically fully crystalline PBS polymer, with a value of 110.3 J g^{-1} for PBS [49]. With regard to w, it represents the weight fraction of all added components (except for PBS), including both ASF and MLO. The thermal stability at elevated temperatures was followed by thermogravimetry (TGA) in a TGA/SDTA 851 from Mettler-Toledo. Standard alumina crucibles with a total volume of 70 mL were charged with approx. 6 mg and subsequently subjected to a dynamic heating program from 30 °C to 700 °C at a constant heating rate of 20 °C min^{-1} in air atmosphere.

2.6. Thermomechanical Characterization

Dynamic mechanical thermal analysis (DMTA) was conducted on a DMA1 analyzer from Mettler-Toledo. Single cantilever flexural tests were carried out at a frequency of 1 Hz at a heating rate of 2 °C min^{-1} and a maximum deflection of 10 μm. Rectangular samples with dimensions of 10 × 7 × 1 mm^3 were subjected to a dynamic heating program from −50 °C to 80 °C.

Thermal stability of the developed composite materials was studied by thermomechanical analysis (TMA) in a Q400 TMA analyzer from TA Instruments (New Castle, DE, USA). A constant force of 0.02 N was applied to squared samples of 10 × 10 × 4 mm^3 and subjected to a temperature sweep from −50 °C up to 80 °C at a constant heating rate of 2 °C min^{-1}. The coefficient of linear thermal expansion (CLTE) was calculated both below and above the glass transition temperature (T_g). All thermal runs were done in triplicate.

3. Results

3.1. Effect of MLO Loading on Mechanical Properties of PBS/ASF Composites

Table 2 summarizes the main parameters obtained from mechanical characterization as a function of the MLO content.

Table 2. Summary of mechanical properties of PBS/ASF composites with different MLO compatibilizer loading: tensile properties (tensile modulus—Et, tensile strength—σ_t, and elongation at break—ε_b), Shore D hardness and impact strength from Charpy test.

Reference	E_t (MPa)	σ_t (MPa)	ε_b (%)	Shore D Hardness	Impact Strength (kJ m^{-2})
PBS	417.4 ± 21.1	31.5 ± 0.9	215.6 ± 16.5	60.1 ± 0.5	16.5 ± 0.8
PBS + ASF	787.9 ± 55.8	14.8 ± 0.5	6.3 ± 0.9	71.2 ± 0.3	1.8 ± 0.3
PBS + ASF + 2.5MLO	779.8 ± 33.3	14.3 ± 0.5	17.4 ± 0.3	67.8 ± 0.9	2.5 ± 0.2
PBS + ASF + 4.5MLO	534.6 ± 51.3	13.8 ± 0.3	25.9 ± 1.0	67.2 ± 0.2	3.8 ± 0.5
PBS + ASF + 7.5MLO	423.4 ± 13.1	12.3 ± 0.1	26.1 ± 1.7	64.0 ± 0.4	3.9 ± 0.3
PBS + ASF + 10MLO	269.8 ± 30.7	11.7 ± 1.1	26.6 ± 1.2	62.0 ± 0.7	4.2 ± 0.2

As expected, the addition of ASF into the PBS matrix leads to an increase in stiffness together with a clear decrease in ductility. PBS possesses a σ_t of about 31.5 MPa; this is dramatically reduced by almost half (14.8 MPa) by filling the PBS matrix with 30 wt % ASF. As described previously, the highly hydrophilic nature of ASF does not allow interaction with the highly hydrophobic PBS matrix and this is responsible for poor load transfer from the filler to the matrix. In fact, the filler offers the typical stress concentration phenomenon due to a lack of interaction with the polymeric matrix. This lack of interaction contributes to poor material cohesion (small gaps between the particle filler and the surrounding matrix) and, therefore, the mechanical properties are worse than neat PBS. This stress concentration phenomenon is also evident when observing the extremely high decrease in ε_b, which drops from 215.6% to 6.3% (a decrease of about 97%). In contrast, the material becomes stiffer, as shown by the E_t values. Neat PBS is a flexible polymer with a relatively low modulus of 417.4 MPa; this is almost doubled by the addition of 30 wt % ASF (787.9 MPa). This increase is evident as the modulus represents the ratio of the applied stress (σ) and the obtained elongation (ε) in the linear region. As has been shown in Table 2, the tensile strength decreases (by 50%) but the elongation at break decreases by 97%. Therefore, the σ/ε ratio gives higher values as the decrease in ε is much pronounced than that of σ.

MLO has a dual effect on PBS/ASF composites. As the MLO load increases, both the tensile modulus and the tensile strength decrease due to a plasticizing effect, as reported by Ferri et al. [50] for PLA formulations containing 10–13 wt % MLO. Tensile strength and tensile modulus decrease in all formulations with increasing MLO content. In contrast, ductility is remarkably improved, as can be seen from the elongation at break values, which change from 6.3% (uncompatibilized PBS/ASF composite) to almost 26% for the PBS/ASF composite compatibilized with 4.5 wt % MLO. The elongation at break is not remarkably improved with higher MLO content. Quiles-Carrillo et al. [23] reported similar findings in PLA composites with ASF compatibilized with MLO. As an efficient plasticizer, MLO remarkably improved the ductility of the PLA/ASF composites. Elongation at break increased by 292% and 84% in relation to the uncompatibilized PLA/ASF and to the neat PLA, respectively. Ferri et al. [50] also reported a remarkable increase in elongation at break in PLA formulations with 10–13 wt % MLO. This increase in ductility is related to the lubricant effect of MLO, which increases chain mobility. The plasticizer increases the free volume and, subsequently, chain interactions decrease, leading to improved intermolecular mobility. Chieng et al. [35] also reported similar findings in PLA biocomposites plasticized with epoxidized vegetable oils (EVOs).

Regarding Shore D hardness, its evolution is identical to the tensile modulus. Neat PBS shows a Shore D value of 60.1. The addition of 30 wt % ASF leads to stiffer material with a Shore D value of 71.2. Then, as the MLO content increases, Shore D progressively decreased to values of 62.0 for the PBS/ASF composite with 10 wt % MLO.

The second effect of MLO can be inferred by observing the impact strength values in Table 2. Neat PBS is characterized by good energy absorption, which gives an impact strength of 16.5 kJ m^{-2}. The addition of 30 wt % ASF causes a dramatic decrease in impact strength down to 1.8 kJ m^{-2}. The impact strength is directly related to both ductility and resistance. As previously indicated, both the tensile strength and the elongation at break are remarkably reduced due to poor material cohesion, which does not allow stress transfer and, hence, there is low deformation with low applied stresses. Obviously, the ability to absorb energy is directly linked to the deformation capability, together with the need for high stress to promote deformation. Therefore, ASF addition leads to extremely low impact energy values. The addition of MLO provides a plasticizing effect, as indicated previously; in addition, reactions between MLO and both PBS and ASF could be expected as the absorbed energy increases to almost 4 kJ m^{-2} with 4.5 wt % MLO (which represents a percentage increase of more than 120% with respect to the uncompatibilized PBS/ASF composite). In fact, the maleic anhydride pendant group can react with hydroxyl groups in both PBS (end chains) and ASF particles (cellulose and hemicellulose). This effect has been previously reported by Ernzen et al. with polyamides and maleinized soybean oil [51], and Quiles-Carrillo et al. with polyesters [23,47] with maleinized linseed oil.

3.2. Effect of MLO Loading on Morphology of PBS/ASF Composites

Figure 1 gathers FESEM images corresponding to PBS/ASF composites (uncompatibilized and MLO-compatibilized composites). As can be observed in Figure 1a, polymer-particle adhesion is very poor. It is possible to find a small gap (1–3 μm) between the ASF particle and the surrounding PBS matrix. In addition, this lack of interaction is evident upon a detailed observation of the PBS fracture surface. PBS shows a porous pattern (1 μm) that is the reverse of that of ASF, which is identical to the typical surface of an almond shell at the macroscale [52]. The addition of 2.5 wt % MLO (Figure 1b) provides some interaction as the gap has been reduced. Nevertheless, some ASF particles have been pulled out without any interaction. Once again, it is possible to observe a negative copy of the almond shell microstructure on the PBS matrix in some areas. Compatibilization is more evident with higher MLO loads, as can be seen in Figure 1c, 1d, and 1e with 4.5, 7.5, and 10 wt % MLO, respectively. The typical PBS surface (negative copy of the ASF microstructure) after particle removal is not detected. The gap is almost inexistent and no signs of particle removal can be observed. All these issues indicate the good compatibilizing effect of MLO as it can react with both PBS and ASF.

3.3. Effect of MLO Loading on Thermal Properties of PBS/ASF Composites

Table 3 and Figure 2 show a summary of the main thermal parameters obtained by DSC characterization of PBS/ASF composites. The melt peak temperature remains almost invariable after the addition of ASF and MLO and changes in a very narrow range from 113 °C to 116 °C [53]. A slight decrease in the melt peak temperature can be observed. This could be related to the nucleant effect a lignocellulosic filler exerts on semicrystalline polymers. Accordingly, the overall degree of crystallinity increases after the addition of ASF due to an interfacial phenomenon as the cellulose crystals act as nucleant points for PBS [54,55]. On the other hand, the addition of MLO provides a decrease in crystallinity down to values close to those of neat PBS. This effect is particularly noticeable for high MLO contents in the 7.5–10.0 wt % range. The plasticizing effect of MLO at higher concentrations provides increased chain mobility, which contributes to a slight decrease in the melt peak temperature down to 113 °C. It is worth noting the role of the melt temperature and crystallinity in the overall processing of polymers and their composites [56].

Table 3. Main thermal parameters of neat PBS and PBS/AHF composites with different percentage of MLO obtained by differential canning calorimetry (DSC).

Code	Melt Enthalpy (J g^{-1})	Melt Peak Temperature, T_m (°C)	Xc (%)
PBS	68.9 ± 1.9	115.6 ± 2.3	60.9 ± 1.8
PBS/ASF	61.2 ± 1.7	115.1 ± 1.9	77.4 ± 1.6
PBS/ASF + 2.5MLO	58.6 ± 2.0	114.6 ± 1.7	79.6 ± 1.9
PBS/ASF + 4.5MLO	56.6 ± 2.1	115.2 ± 2.1	77.1 ± 2.0
PBS/ASF + 7.5MLO	47.3 ± 1.9	113.7 ± 1.7	66.9 ± 1.8
PBS/ASF + 10MLO	43.4 ± 1.5	113.9 ± 1.5	64.0 ± 1.4

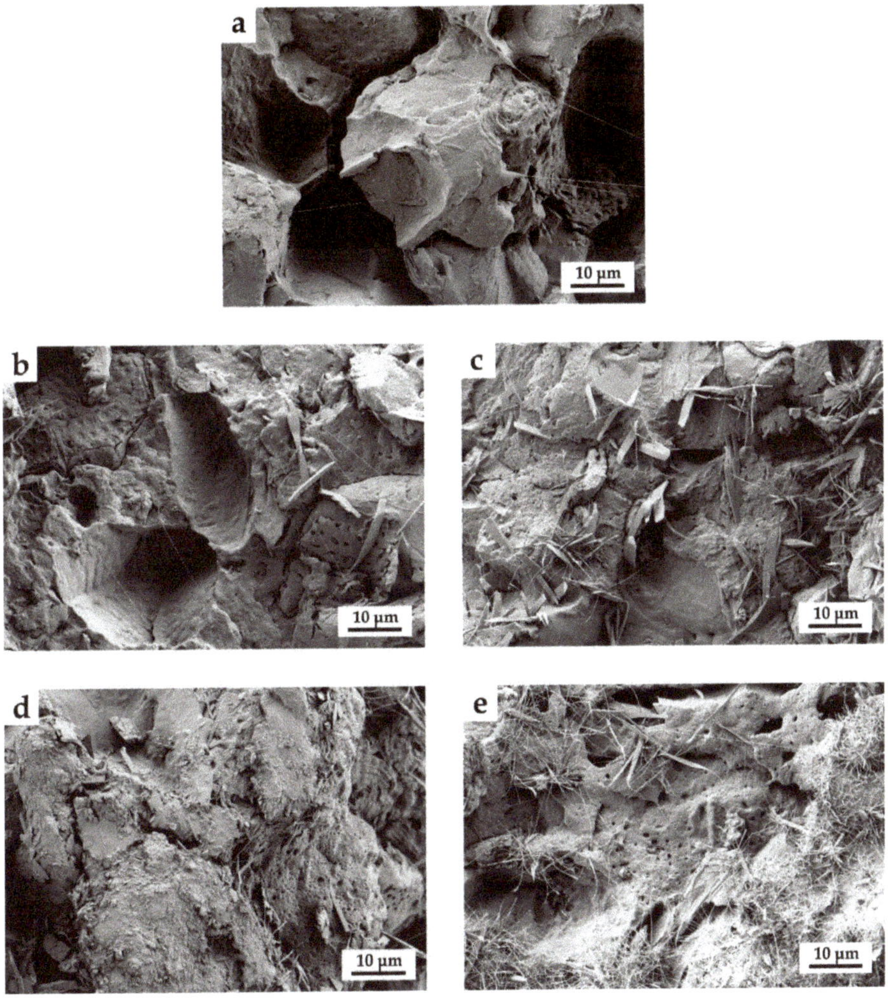

Figure 1. Field emission scanning electron microscopy (FESEM) images of the surface fracture of PBS/ASF composites, uncompatibilized and MLO-compatibilized taken at 1000×: (**a**) PBS + ASF; (**b**) PBS + ASF + 2.5MLO; (**c**) PBS + ASF + 4.5MLO; (**d**) PBS + ASF + 7.5MLO; and (**e**) PBS + ASF + 10MLO. Scale markers of 10 μm.

With regard to thermal degradation at elevated temperatures, Figure 3a shows the TGA thermogram with the weight loss in function of increasing temperature. It has been reported that PBS decomposes by the following mechanisms, which cover typical random chain scission of aliphatic polyesters such as PLA [57], and specific chain scission. Shih et al. reported that the diffusion effect becomes important at elevated temperatures during degradation of PBS [58].

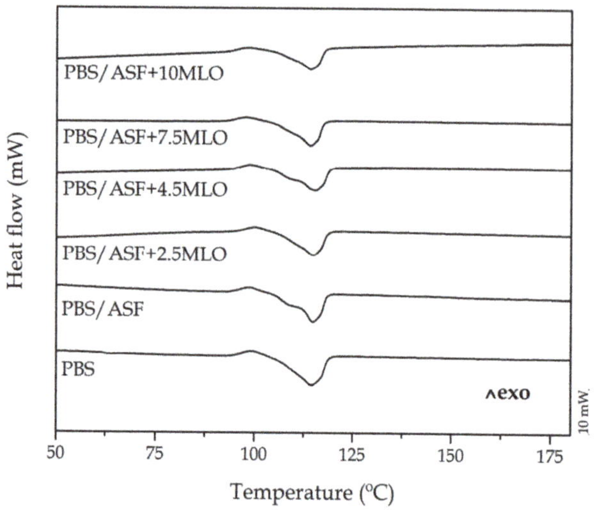

Figure 2. Differential scanning calorimetry (DSC) thermograms of neat PBS, uncompatibilized PBS/ASF composite, and PBS/ASF composites compatibilized with different MLO loading.

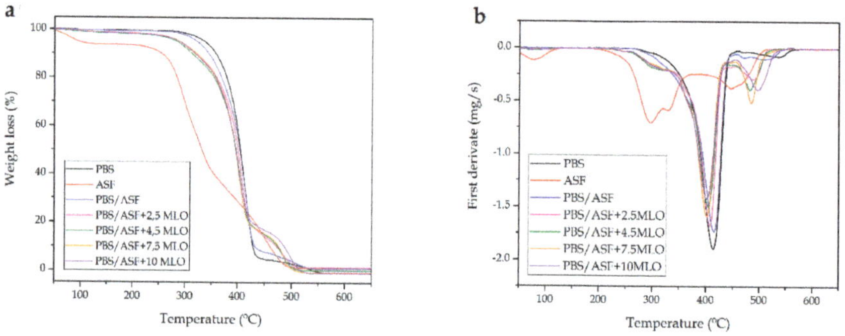

Figure 3. (**a**) Thermogravimetric (TGA) thermograms corresponding to raw ASF, neat PBS and PBS/ASF composites compatibilized with different MLO loading and (**b**) first derivative (DTG) curves.

Addition of 30 wt % ASF into the PBS matrix provides a slight change in the TGA curves as the ASF degradation occurs in different stages. ASF shows the typical degradation profile of a lignocellulosic filler. At about 90–100 °C it is possible to observe a first weight loss that corresponds to the residual moisture contained in the ASF filler [59]. Over this temperature, ASF shows thermal stability up to 210–220 °C. Degradation of cellulose and hemicelluloses starts at about 220 °C and proceeds until 330–340 °C. Decomposition of cellulose and hemicellulose involves complex reactions (dehydration, decarboxylation, etc.) and the breakage of C–H, C–O, and C–C bonds. Lignin is more thermally stable and its degradation occurs in a broad temperature range starting at about 250 °C and ending at 450 °C in a progressive weight loss process. This higher thermal stability of lignin versus cellulose and hemicellulose is due to the presence of aromatic rings. Quiles-Carrillo et al. [60] showed similar degradation behavior in PLA/ASF composites. As has been reported by many authors, the degradation of lignocellulosic materials is a complex issue that mainly involves the degradation of hemicelluloses, cellulose, and lignin [61]. The components that most readily degrade due to a temperature rise are hemicelluloses, which decompose (depolymerize) in the temperature range 180–350 °C [62]. Crystalline areas of cellulose are more thermally stable and start to degrade at 275 °C; the decomposition takes place up to temperatures of 350–370 °C [63]. With regard to lignin, it shows a broader degradation range that starts at about 200 °C and lasts up to 700 °C, showing a typical shoulder at elevated temperatures in TGA thermograms [64]. The TGA curves corresponding to PBS/ASF composites show a slight decrease at the onset degradation temperature since the ASF filler is less thermally stable than PBS [11]. Sanchez-Jimenez at al. studied the thermal decomposition of cellulose and the typical degradation temperature range they reported overlaps with that of PBS [65]. It has been reported that the reactive extrusion of aliphatic polyesters such as PLA with styrene-epoxy acrylic oligomers can provide increased thermal stability due to the branching effect reactive extrusion provides [66]. As has been suggested in this study, MLO provides two overlapping processes: on one hand, a plasticizing effect and, on the other hand, a compatibilization effect due to the reaction of maleic anhydride with hydroxyl groups in PBS and ASF filler. Despite this, the thermal stability at moderate temperatures is not improved, as can be seen in Table 4, with a decreasing values of $T_{5\%}$ with increasing MLO. The same tendency can be observed for the maximum degradation rate temperature (T_{max}) from 415.64. Despite this, it seems that the typical degradation peak of lignin at elevated temperatures is moved to higher temperatures with MLO addition, as can be observed in Figure 3b, which gathers the first derivative TGA curves for all PBS/ASF composites as well as raw ASF and neat PBS.

Table 4. Summary of the main thermal parameters of the degradation process of PBS/ASF with different percentage of MLO.

Reference	* $T_{5\%}$ (°C)	** T_{max} (°C)	Residual Weight (%)
PBS	335.9 ± 1.8	415.64 ± 1.86	0.32 ± 0.12
PBS/ASF	329.6 ± 2.0	414.06 ± 1.98	1.23 ± 0.31
PBS/ASF + 2.5MLO	291.3 ± 2.1	407.18 ± 1.67	1.29 ± 0.32
PBS/ASF + 4.5MLO	288.7 ± 2.1	405.82 ± 2.08	0.68 ± 0.24
PBS/ASF + 7.5MLO	289.6 ± 1.4	400.47 ± 1.87	0.46 ± 0.32
PBS/ASF + 10MLO	289.3 ± 1.8	399.93 ± 1.57	0.42 ± 0.15

* $T_{5\%}$ represents the characteristic temperature for a mass loss of 5%. ** T_{max} represents the characteristic temperature corresponding to the maximum degradation rate.

3.4. Effect of MLO Loading on Thermomechanical Properties of PBS/ASF Composites

Figure 4 shows the evolution of the loss modulus (G″) and the damping factor (tan ∂). The glass transition temperature (T_g) can be estimated (among other methods) by the peak temperature corresponding to the loss modulus (G″). It is important to remark that the T_g value could not be observed by DSC due to the low signal it gives. Neat PBS shows a T_g value of about −29 °C, which is in accordance with other T_g values reported in the literature [67]. The addition of 30 wt % ASF does not promote any change in the T_g value, thus indicating no interaction. In contrast to the expected behavior suggested by mechanical characterization, as the MLO increases the T_g also increases, in a slight way, but detectable by DMTA characterization. The compatibilizing effect of MLO by reaction with both PBS and ASF filler is evidenced by an increase in the T_g value since the reaction of MLO with both polymer matrix and particle filler restricts chain mobility and, hence, the T_g is increased. As reported by Quiles-Carrillo et al. [23] in a previous work, the effect of MLO is evident when comparing the FTIR spectra of uncompatibilized and compatibilized PLA/ASF composites. Comparison of these FTIR spectra indicate that new esters and carboxylic acids are obtained after reactive extrusion with PLA and ASF. They propose the formation of a cellulose-*g*-PLA by the reaction of the multiple anhydride groups in MLO with the hydroxyl groups in both PLA end chains and the lignocellulosic filler. Similar behavior could be expected for this system with a more flexible polymer (PBS) and the same lignocellulosic filler without previous ultraviolet (UV) surface treatment. PBS/ASF composites with 2.5 wt % MLO possess a T_g of −27 °C, which means a slight increase of 2 °C with respect to the PBS (or uncompatibilized PBS/ASF composite). As has been reported, MLO could potentially provide several overlapped effects such as plasticization, chain extension, branching, and even crosslinking. All these phenomena could overlap and, depending on the polymer, one mechanism could stand out over the others [50].

In this system, some plasticization is evident, as suggested by mechanical characterization, but, together with this phenomenon, compatibilization could occur, due to the reaction of maleic anhydride with PBS end chains and the lignocellulosic filler that restricts the Brownian motion of long-chain molecules [68]. On the other hand, it has also been reported that MLO could lead to some polymer crosslinking that could contribute to a slight increase in the T_g. This increase in T_g is clear evidence of the compatibilization effect [69]. As the MLO content increases, a slight increase in T_g is observed as well. Therefore, for PBS/ASF composites compatibilized with 10 wt % MLO, the T_g reaches values of about −23.5 °C.

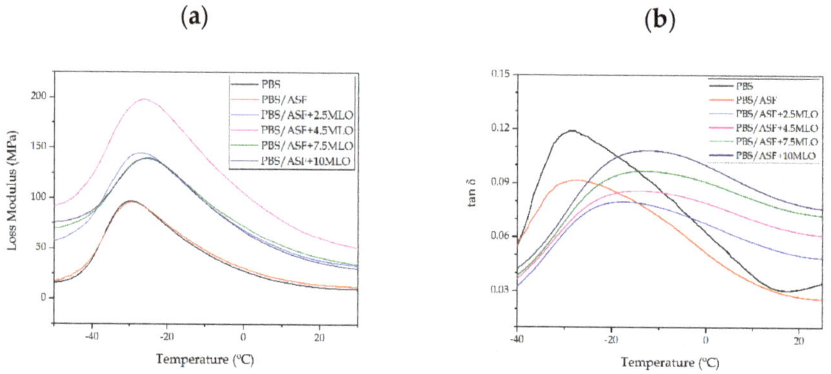

Figure 4. Plot evolution of (**a**) loss modulus (E") and (**b**) damping factor (tan ∂) of neat PBS and PBS/ASF composites with different MLO compatibilizer content.

To obtain a complete characterization of the thermomechanical properties, the change in dimensions as a function of temperature was obtained by thermomechanical analysis (TMA). In particular, the coefficient of linear thermal expansion (CLTE) was calculated both below and above the glass transition temperature, as shown in Table 5. As PBS is extremely flexible, the comparison material has been the uncompatibilized PBS/ASF composite, which shows a CLTE of 59.3 $\mu m\ m^{-1}\ {}^{\circ}C^{-1}$. As previously indicated, the addition of MLO provides an increase in elongation due to a combined effect of plasticization and compatibilization through reaction. This increase in ductility is directly related to an increase in the CLTE. In fact, the CLTE shows a clear increasing trend with the MLO content, up to values of 84.8 $\mu m\ m^{-1}\ {}^{\circ}C^{-1}$ below the T_g. This same tendency can be observed for the CLTE measured above the T_g, as seen in Table 5.

Table 5. Variation of the coefficient of linear thermal expansion (CLTE) of neat PBS and PBS/ASF composites with different MLO compatibilizer content.

Reference	CLTE below T_g ($\mu m\ m^{-1}\ {}^{\circ}C^{-1}$)	CLTE above T_g ($\mu m\ m^{-1}\ {}^{\circ}C^{-1}$)
PBS/ASF	59.3 ± 1.6	128.1 ± 1.8
PBS/ASF + 2.5MLO	65.6 ± 1.6	132.1 ± 2.1
PBS/ASF + 4.5MLO	74.2 ± 2.1	138.7 ± 2.9
PBS/ASF + 7.5MLO	78.9 ± 2.3	142.9 ± 1.7
PBS/ASF + 10MLO	84.8 ± 1.9	144.1 ± 2.4

4. Conclusions

Maleinized linseed oil (MLO) has been validated as a good compatibilizer in green composites of PBS and ASF, with the additional feature of being environmentally friendly. Uncompatibilized PBS/ASF composites show a dramatic decrease in both tensile strength and elongation at break due to the lack of polymer-filler interaction. Accordingly, the impact strength of the uncompatibilized PBS/ASF composite is also reduced. MLO provides a combination of plasticization and compatibilization through the reaction of maleic anhydride pendant groups in the triglyceride structure with the hydroxyl groups contained in both PBS (end chains) and ASF (mainly cellulose and hemicelluloses). This leads to a noticeable increase in elongation at break as well as in the impact strength. MLO reaction with both PBS and ASF is also assessed by an increase in the T_g from −28 °C (neat PBS) up to −12 °C (PBS/ASF composite compatibilized with 10 wt % MLO). The compatibilization effect is also observed by scanning electron microscopy as the gap between the filler and the surrounding matrix almost disappears. Therefore, the herein-developed composites represent an interesting approach to reduce the overall cost of using PBS by using industrial waste coming from the almond industry, thus leading to highly environmentally friendly materials. These PBS/ASF composites can be successfully compatibilized with a vegetable-oil-derived additive, which contributes to the high environmental efficiency of these materials.

Author Contributions: Conceptualization, N.M., D.G.-S. and R.B.; Methodology, Validation, and Formal Analysis, P.L., L.Q.-C., N.M., D.G.-S., and R.B.; Investigation, Resources, Data Curation, and Writing—Original Draft Preparation, P.L. and L.Q.-C.; Writing—Review & Editing, N.M., D.-G.-S. and R.B.; Supervision, N.M. and R.B.; Project Administration, R.B.

Funding: This work was supported by the Ministry of Economy and Competitiveness (MINECO) grant number MAT2017-84909-C2-2-R. L.Q.-C. wants to thank Generalitat Valenciana (GV) for his FPI grant (ACIF/2016/182) and the Spanish Ministry of Education, Culture, and Sports (MECD) for his FPU grant (FPU15/03812).

Conflicts of Interest: The authors declare no conflict of interest.

References

1. Liminana, P.; Garcia-Sanoguera, D.; Quiles-Carrillo, L.; Balart, R.; Montanes, N. Development and characterization of environmentally friendly composites from poly (butylene succinate)(PBS) and almond shell flour with different compatibilizers. *Compos. Part B Eng.* **2018**, *144*, 153–162. [CrossRef]
2. Bechthold, I.; Bretz, K.; Kabasci, S.; Kopitzky, R.; Springer, A. Succinic acid: A new platform chemical for biobased polymers from renewable resources. *Chem. Eng. Technol.* **2008**, *31*, 647–654. [CrossRef]
3. McKinlay, J.B.; Vieille, C.; Zeikus, J.G. Prospects for a bio-based succinate industry. *Appl. Microbiol. Biotechnol.* **2007**, *76*, 727–740. [CrossRef] [PubMed]
4. Bozell, J.J.; Petersen, G.R. Technology development for the production of biobased products from biorefinery carbohydrates—the US Department of Energy's "Top 10" revisited. *Green Chem.* **2010**, *12*, 539–554. [CrossRef]
5. Kim, H.S.; Yang, H.S.; Kim, H.J. Biodegradability and mechanical properties of agro-flour–filled polybutylene succinate biocomposites. *J. Appl. Polym. Sci.* **2005**, *97*, 1513–1521. [CrossRef]
6. Siracusa, V.; Lotti, N.; Munari, A.; Rosa, M.D. Poly(butylene succinate) and poly(butylene succinate-co-adipate) for food packaging applications: Gas barrier properties after stressed treatments. *Polym. Degrad. Stabil.* **2015**, *119*, 35–45. [CrossRef]
7. Vytejckova, S.; Vapenka, L.; Hradecky, J.; Dobias, J.; Hajslova, J.; Loriot, C.; Vannini, L.; Poustka, J. Testing of polybutylene succinate based films for poultry meat packaging. *Polym. Test* **2017**, *60*, 357–364. [CrossRef]
8. Hongsriphan, N.; Sanga, S. Antibacterial food packaging sheets prepared by coating chitosan on corona-treated extruded poly(lactic acid)/poly(butylene succinate) blends. *J. Plast. Film Sheeting* **2018**, *34*, 160–178. [CrossRef]
9. Imre, B.; Pukánszky, B. Compatibilization in bio-based and biodegradable polymer blends. *Eur. Polym. J.* **2013**, *49*, 1215–1233. [CrossRef]
10. Dorez, G.; Taguet, A.; Ferry, L.; Lopez-Cuesta, J.M. Thermal and fire behavior of natural fibers/PBS biocomposites. *Polym. Degrad. Stabil.* **2013**, *98*, 87–95. [CrossRef]
11. Frollini, E.; Bartolucci, N.; Sisti, L.; Celli, A. Poly(butylene succinate) reinforced with different lignocellulosic fibers. *Ind. Crop. Prod.* **2013**, *45*, 160–169. [CrossRef]
12. Kurokawa, N.; Kimura, S.; Hotta, A. Mechanical properties of poly(butylene succinate) composites with aligned cellulose-acetate nanofibers. *J. Appl. Polym. Sci.* **2018**, *135*. [CrossRef]
13. Terzopoulou, Z.N.; Papageorgiou, G.Z.; Papadopoulou, E.; Athanassiadou, E.; Reinders, M.; Bikiaris, D.N. Development and Study of Fully Biodegradable Composite Materials Based on Poly(butylene succinate) and Hemp Fibers or Hemp Shives. *Polym. Compos.* **2016**, *37*, 407–421. [CrossRef]

14. Lee, H.Y.; Choi, D. Influence of Waste Fiber Content on the Thermal and Mechanical Properties of Waste Silk/Waste Wool/PBS Hybrid Biocomposites. *Polym. Korea* **2017**, *41*, 719–726. [CrossRef]
15. Flores-Cano, J.V.; Sanchez-Polo, M.; Messoud, J.; Velo-Gala, I.; Ocampo-Perez, R.; Rivera-Utrilla, J. Overall adsorption rate of metronidazole, dimetridazole and diatrizoate on activated carbons prepared from coffee residues and almond shells. *J. Environ. Manag.* **2016**, *169*, 116–125. [CrossRef] [PubMed]
16. Loffredo, E.; Castellana, G.; Senesi, N. Decontamination of a municipal landfill leachate from endocrine disruptors using a combined sorption/bioremoval approach. *Environ. Sci. Pollut. Res.* **2014**, *21*, 2654–2662. [CrossRef] [PubMed]
17. Loffredo, E.; Castellana, G.; Taskin, E. A Two-Step Approach to Eliminate Pesticides and Estrogens from a Wastewater and Reduce Its Phytotoxicity: Adsorption onto Plant-Derived Materials and Fungal Degradation. *Water Air Soil Pollut.* **2016**, *227*, 1–12. [CrossRef]
18. Adanez-Rubio, I.; Perez-Astray, A.; Mendiara, T.; Teresa Izquierdo, M.; Abad, A.; Gayan, P.; de Diego, L.F.; Garcia-Labiano, F.; Adanez, J. Chemical looping combustion of biomass: CLOU experiments with a Cu-Mn mixed oxide. *Fuel Process. Technol.* **2018**, *172*, 179–186. [CrossRef]
19. Cerone, N.; Zimbardi, F. Gasification of Agroresidues for Syngas Production. *Energies* **2018**, *11*, 1280. [CrossRef]
20. Safari, F.; Javani, N.; Yumurtaci, Z. Hydrogen production via supercritical water gasification of almond shell over algal and agricultural hydrochars as catalysts. *Int. J. Hydrog. Energy* **2018**, *43*, 1071–1080. [CrossRef]
21. Essabir, H.; Nekhlaoui, S.; Malha, M.; Bensalah, M.O.; Arrakhiz, F.Z.; Qaiss, A.; Bouhfid, R. Bio-composites based on polypropylene reinforced with Almond Shells particles: Mechanical and thermal properties. *Mater. Des.* **2013**, *51*, 225–230. [CrossRef]
22. El Mechtali, F.Z.; Essabir, H.; Nekhlaoui, S.; Bensalah, M.; Jawaid, M.; Bouhfid, R.; Qaiss, A. Mechanical and Thermal Properties of Polypropylene Reinforced with Almond Shells Particles: Impact of Chemical Treatments. *J. Bionic Eng.* **2015**, *12*, 483–494. [CrossRef]
23. Quiles-Carrillo, L.; Montanes, N.; Sammon, C.; Balart, R.; Torres-Giner, S. Compatibilization of highly sustainable polylactide/almond shell flour composites by reactive extrusion with maleinized linseed oil. *Ind. Crop. Prod.* **2018**, *111*, 878–888. [CrossRef]
24. Caballero, J.A.; Conesa, J.A.; Font, R.; Marcilla, A. Pyrolysis kinetics of almond shells and olive stones considering their organic fractions. *J. Anal. Appl. Pyrolysis* **1997**, *42*, 159–175. [CrossRef]
25. Faludi, G.; Dora, G.; Imre, B.; Renner, K.; Moczo, J.; Pukanszky, B. PLA/Lignocellulosic Fiber Composites: Particle Characteristics, Interfacial Adhesion, and Failure Mechanism. *J. Appl. Polym. Sci.* **2014**, *131*. [CrossRef]
26. Garcia-Garcia, D.; Carbonell, A.; Samper, M.D.; Garcia-Sanoguera, D.; Balart, R. Green composites based on polypropylene matrix and hydrophobized spend coffee ground (SCG) powder. *Compos. Part B-Eng.* **2015**, *78*, 256–265. [CrossRef]
27. Kabir, M.M.; Wang, H.; Lau, K.T.; Cardona, F. Chemical treatments on plant-based natural fibre reinforced polymer composites: An overview. *Compos. Part B-Eng.* **2012**, *43*, 2883–2892. [CrossRef]
28. Wang, F.; Yang, M.Q.; Zhou, S.J.; Ran, S.Y.; Zhang, J.Q. Effect of fiber volume fraction on the thermal and mechanical behavior of polylactide-based composites incorporating bamboo fibers. *J. Appl. Polym. Sci.* **2018**, *135*, 46148. [CrossRef]
29. Nam, T.H.; Ogihara, S.; Tung, N.H.; Kobayashi, S. Effect of alkali treatment on interfacial and mechanical properties of coir fiber reinforced poly(butylene succinate) biodegradable composites. *Compos. Part B-Eng.* **2011**, *42*, 1648–1656. [CrossRef]
30. Sinha, A.K.; Narang, H.K.; Bhattacharya, S. Mechanical properties of natural fibre polymer composites. *J. Polym. Eng.* **2017**, *37*, 879–895. [CrossRef]
31. Balart, J.F.; Fombuena, V.; Fenollar, O.; Boronat, T.; Sanchez-Nacher, L. Processing and characterization of high environmental efficiency composites based on PLA and hazelnut shell flour (HSF) with biobased plasticizers derived from epoxidized linseed oil (ELO). *Compos. Part B-Eng.* **2016**, *86*, 168–177. [CrossRef]
32. Karmarkar, A.; Chauhan, S.S.; Modak, J.M.; Chanda, M. Mechanical properties of wood-fiber reinforced polypropylene composites: Effect of a novel compatibilizer with isocyanate functional group. *Compos. Part A-Appl. Sci. Manuf.* **2007**, *38*, 227–233. [CrossRef]
33. Labidi, S.; Alqahtani, N.; Alejji, M. *Effect of Compatibilizer on Mechanical and Physical Properties of Green Composites Based on High Density Polyethylene and Date Palm Fiber*; International Conference on Composite Science and Technology: Sorrento, Italy, 2013; pp. 995–997.

34. Pivsa-Art, W.; Chaiyasat, A.; Pivsa-Art, S.; Yamane, H.; Ohara, H. Preparation of Polymer Blends Between Poly(lactic acid) and Poly(butylene adipate-co-terephthalate) and Biodegradable Polymers as Compatibilizers. In *10th Eco-Energy and Materials Science and Engineering Symposium*; Yupapin, P., PivsaArt, S., Ohgaki, H., Eds.; Elsevier Science Bv: Amsterdam, The Netherlands, 2013; Volume 34, pp. 549–554.
35. Chieng, B.W.; Ibrahim, N.A.; Then, Y.Y.; Loo, Y.Y. Epoxidized Vegetable Oils Plasticized Poly(lactic acid) Biocomposites: Mechanical, Thermal and Morphology Properties. *Molecules* **2014**, *19*, 16024–16038. [CrossRef] [PubMed]
36. Garcia-Garcia, D.; Fenollar, O.; Fombuena, V.; Lopez-Martinez, J.; Balart, R. Improvement of Mechanical Ductile Properties of Poly(3-hydroxybutyrate) by Using Vegetable Oil Derivatives. *Macromol. Mater. Eng.* **2017**, *302*. [CrossRef]
37. Narute, P.; Rao, G.R.; Misra, S.; Palanisamy, A. Modification of cottonseed oil for amine cured epoxy resin: Studies on thermo-mechanical, physico-chemical, morphological and antimicrobial properties. *Prog. Org. Coat.* **2015**, *88*, 316–324. [CrossRef]
38. Samper, M.D.; Petrucci, R.; Sanchez-Nacher, L.; Balart, R.; Kenny, J.M. Properties of composite laminates based on basalt fibers with epoxidized vegetable oils. *Mater. Des.* **2015**, *72*, 9–15. [CrossRef]
39. Zhao, Y.Q.; Qu, J.P.; Feng, Y.H.; Wu, Z.H.; Chen, F.Q.; Tang, H.L. Mechanical and thermal properties of epoxidized soybean oil plasticized polybutylene succinate blends. *Polym. Adv. Technol.* **2012**, *23*, 632–638. [CrossRef]
40. Sarwono, A.; Man, Z.; Bustam, M.A. Blending of Epoxidised Palm Oil with Epoxy Resin: The Effect on Morphology, Thermal and Mechanical Properties. *J. Polym. Environ.* **2012**, *20*, 540–549. [CrossRef]
41. Park, S.J.; Jin, F.L.; Lee, J.R. Effect of biodegradable epoxidized castor oil on physicochemical and mechanical properties of epoxy resins. *Macromol. Chem. Phys.* **2004**, *205*, 2048–2054. [CrossRef]
42. Huang, K.; Zhang, P.; Zhang, J.W.; Li, S.H.; Li, M.; Xia, J.L.; Zhou, Y.H. Preparation of biobased epoxies using tung oil fatty acid-derived C21 diacid and C22 triacid and study of epoxy properties. *Green Chem.* **2013**, *15*, 2466–2475. [CrossRef]
43. Ferri, J.M.; Garcia-Garcia, D.; Sanchez-Nacher, L.; Fenollar, O.; Balart, R. The effect of maleinized linseed oil (MLO) on mechanical performance of poly(lactic acid)-thermoplastic starch (PLA-TPS) blends. *Carbohydr. Polym.* **2016**, *147*, 60–68. [CrossRef] [PubMed]
44. Garcia-Campo, M.J.; Quiles-Carrillo, L.; Masia, J.; Reig-Perez, M.J.; Montanes, N.; Balart, R. Environmentally Friendly Compatibilizers from Soybean Oil for Ternary Blends of Poly(lactic acid)-PLA, Poly(epsilon-caprolactone)-PCL and Poly(3-hydroxybutyrate)-PHB. *Materials* **2017**, *10*, 1339. [CrossRef] [PubMed]
45. Carbonell-Verdu, A.; Ferri, J.M.; Dominici, F.; Boronat, T.; Sanchez-Nacher, L.; Balart, R.; Torre, L. Manufacturing and compatibilization of PLA/PBAT binary blends by cottonseed oil-based derivatives. *Express Polym. Lett.* **2018**, *12*, 808–823. [CrossRef]
46. Carbonell-Verdu, A.; Garcia-Garcia, D.; Dominici, F.; Torre, L.; Sanchez-Nacher, L.; Balart, R. PLA films with improved flexibility properties by using maleinized cottonseed oil. *Eur. Polym. J.* **2017**, *91*, 248–259. [CrossRef]
47. Quiles-Carrillo, L.; Blanes-Martinez, M.M.; Montanes, N.; Fenollar, O.; Torres-Giner, S.; Balart, R. Reactive toughening of injection-molded polylactide pieces using maleinized hemp seed oil. *Eur. Polym. J.* **2018**, *98*, 402–410. [CrossRef]
48. Bayrak, A.; Kiralan, M.; Ipek, A.; Arslan, N.; Cosge, B.; Khawar, K.M. Fatty acid compositions of linseed (*Linum usitatissimum* l.) genotypes of different origin cultivated in turkey. *Biotechnol. Biotechnol. Equip.* **2010**, *24*, 1836–1842. [CrossRef]
49. Ren, M.; Song, J.; Song, C.; Zhang, H.; Sun, X.; Chen, Q.; Zhang, H.; Mo, Z. Crystallization kinetics and morphology of poly (butylene succinate-co-adipate). *J. Polym. Sci. Part B: Polym. Phys.* **2005**, *43*, 3231–3241. [CrossRef]
50. Ferri, J.M.; Garcia-Garcia, D.; Montanes, N.; Fenollar, O.; Balart, R. The effect of maleinized linseed oil as biobased plasticizer in poly (lactic acid)-based formulations. *Polym. Int.* **2017**, *66*, 882–891. [CrossRef]
51. Ernzen, J.R.; Bondan, F.; Luvison, C.; Wanke, C.H.; Martins, J.D.N.; Fiorio, R.; Bianchi, O. Structure and properties relationship of melt reacted polyamide 6/malenized soybean oil. *J. Appl. Polym. Sci.* **2016**, *133*. [CrossRef]

52. Bordbar, M. Biosynthesis of Ag/almond shell nanocomposite as a cost-effective and efficient catalyst for degradation of 4-nitrophenol and organic dyes. *RSC Adv.* **2017**, *7*, 180–189. [CrossRef]
53. Yokohara, T.; Yamaguchi, M. Structure and properties for biomass-based polyester blends of PLA and PBS. *Eur. Polym. J.* **2008**, *44*, 677–685. [CrossRef]
54. Luo, X.; Li, J.; Feng, J.; Yang, T.; Lin, X. Mechanical and thermal performance of distillers grains filled poly (butylene succinate) composites. *Mater. Des.* **2014**, *57*, 195–200. [CrossRef]
55. Bendahou, A.; Kaddami, H.; Sautereau, H.; Raihane, M.; Erchiqui, F.; Dufresne, A. Short palm tree fibers polyolefin composites: Effect of filler content and coupling agent on physical properties. *Macromol. Mater. Eng.* **2008**, *293*, 140–148. [CrossRef]
56. Li, J.; Luo, X.; Lin, X. Preparation and characterization of hollow glass microsphere reinforced poly (butylene succinate) composites. *Mater. Des.* **2013**, *46*, 902–909. [CrossRef]
57. Carrasco, F.; Perez-Maqueda, L.A.; Sanchez-Jimenez, P.E.; Perejon, A.; Santana, O.O.; Maspoch, M.L. Enhanced general analytical equation for the kinetics of the thermal degradation of poly(lactic acid) driven by random scission. *Polym. Test* **2013**, *32*, 937–945. [CrossRef]
58. Shih, Y.-F. Thermal Degradation and Kinetic Analysis of Biodegradable PBS/Multiwalled Carbon Nanotube Nanocomposites. *J. Polym. Sci. Part B-Polym. Phys.* **2009**, *47*, 1231–1239. [CrossRef]
59. Faulstich de Paiva, J.M.; Frollini, E. Unmodified and modified surface sisal fibers as reinforcement of phenolic and lignophenolic matrices composites: Thermal analyses of fibers and composites. *Macromol. Mater. Eng.* **2006**, *291*, 405–417. [CrossRef]
60. Quiles-Carrillo, L.; Montanes, N.; Garcia-Garcia, D.; Carbonell-Verdu, A.; Balart, R.; Torres-Giner, S. Effect of different compatibilizers on injection-molded green composite pieces based on polylactide filled with almond shell flour. *Compos. Part B: Eng.* **2018**, *147*, 76–85. [CrossRef]
61. Nabinejad, O.; Sujan, D.; Rahman, M.; Davies, I.J. Determination of filler content for natural filler polymer composite by thermogravimetric analysis. *J. Therm. Anal. Calorim.* **2015**, *122*, 227–233. [CrossRef]
62. Kim, H.-S.; Kim, S.; Kim, H.-J.; Yang, H.-S. Thermal properties of bio-flour-filled polyolefin composites with different compatibilizing agent type and content. *Thermochim. Acta* **2006**, *451*, 181–188. [CrossRef]
63. Poletto, M.; Zattera, A.J.; Forte, M.M.; Santana, R.M. Thermal decomposition of wood: Influence of wood components and cellulose crystallite size. *Bioresour. Technol.* **2012**, *109*, 148–153. [CrossRef] [PubMed]
64. Alemdar, A.; Sain, M. Biocomposites from wheat straw nanofibers: Morphology, thermal and mechanical properties. *Compos. Sci. Technol.* **2008**, *68*, 557–565. [CrossRef]
65. Sánchez-Jiménez, P.E.; Pérez-Maqueda, L.A.; Perejón, A.; Criado, J.M. Generalized master plots as a straightforward approach for determining the kinetic model: The case of cellulose pyrolysis. *Thermochim. Acta* **2013**, *552*, 54–59. [CrossRef]
66. Carrasco, F.; Cailloux, J.; Sanchez-Jimenez, P.E.; Maspoch, M.L. Improvement of the thermal stability of branched poly(lactic acid) obtained by reactive extrusion. *Polym. Degrad. Stabil.* **2014**, *104*, 40–49. [CrossRef]
67. Jin, H.-J.; Lee, B.-Y.; Kim, M.-N.; Yoon, J.-S. Properties and biodegradation of poly (ethylene adipate) and poly (butylene succinate) containing styrene glycol units. *Eur. Polym. J.* **2000**, *36*, 2693–2698. [CrossRef]
68. Ciemniecki, S.L.; Glasser, W.G. *Polymer Blends with Hydroxypropyl Lignin*; ACS Publications: Washington, DC, USA, 1989.
69. Sahoo, S.; Misra, M.; Mohanty, A.K. Effect of compatibilizer and fillers on the properties of injection molded lignin-based hybrid green composites. *J. Appl. Polym. Sci.* **2013**, *127*, 4110–4121. [CrossRef]

materials

MDPI

Article

The Use of Cashew Nut Shell Liquid (CNSL) in PP/HIPS Blends: Morphological, Thermal, Mechanical and Rheological Properties

Mirna Nunes Araújo [1], Leila Lea Yuan Visconte [1,2], Daniel Weingart Barreto [3], Viviane Alves Escócio [1], Ana Lucia Nazareth da Silva [1,2], Ana Maria Furtado de Sousa [4] and Elen Beatriz Acordi Vasques Pacheco [1,2,*]

1 Instituto de Macromoléculas Professora Eloisa Mano, Universidade Federal do Rio de Janeiro, 2.300 Horácio Macedo Av., Technology Center, Building J, 21941-598 Rio de Janeiro, RJ, Brazil; mirnana9@hotmail.com (M.N.A.); lyv@ima.ufrj.br (L.L.Y.V.); vivi75@ima.ufrj.br (V.A.E.); ananazareth@ima.ufrj.br (A.L.N.d.S.)

2 Programa de Engenharia Ambiental (PEA/UFRJ), Universidade Federal do Rio de Janeiro, 149 Athos da Silveira Av., Technology Center, Building J, 21941-909 Rio de Janeiro, RJ, Brazil

3 Escola de Química, Universidade Federal do Rio de Janeiro, 149 Athos da Silveira Av., Technology Center, Building E, 21941-611 Rio de Janeiro, RJ, Brazil; dbarreto@eq.ufrj.br

4 Department of Chemical Processes, Universidade do Estado do Rio de Janeiro, 524 São Francisco Xavier, 20550-900 Rio de Janeiro, RJ, Brazil; ana.furtado.sousa@gmail.com

* Correspondence: elen@ima.ufrj.br; Tel.: +55-21-3938-7224

Received: 14 May 2019; Accepted: 10 June 2019; Published: 13 June 2019

Abstract: Polypropylene (PP) and high impact polystyrene (HIPS) are two polymers that are frequently found in disposable waste. Both of these polymers are restricted from being separated in several ways. An easier way to reuse them in new applications, without the need for separation, would require them to be less immiscible. In this work, cashew nut shell liquid (CNSL), a sub-product of the cashew agroindustry, was added as a third component to PP-HIPS mixtures and its effect as a compatibilizing agent was investigated. Morphological results showed that CNSL acted as an emulsifier by promoting reduction in the domains of the dispersive phase, HIPS, thus stabilizing the blends morphology. Differential scanning calorimetry (DSC) analysis suggests that CNSL is preferably incorporated in the HIPS phase. Its plasticizing effect leads to more flexible materials, but no significant effect could be detected on impact resistance or elongation at break.

Keywords: polymer mixtures; blends; cashew nut shell liquid (CNSL); polypropylene; high impact polystyrene; compatibilization

1. Introduction

The practice of recycling through reprocessing can reduce the volume of waste in landfills, simultaneously generating a new economic activity, saving energy and non-renewable resources [1,2]. The mechanical recycling is based on the conversion of such residues, as post-industrial or post consumption plastic materials into pellets through extrusion reprocessing. This procedure allows the subsequent use of the pellets in the production of various products such as garbage bags, soles, floors, hoses, car components, non-food packaging, etc. [3].

A limitation to mechanical recycling is the heterogeneous composition of the residues, which are formed by different types of plastic materials that are usually incompatible. When mixed, the incompatible materials give rise to products of poor mechanical performance.

The wide consumption of disposable products, as dishes and cups, heavily contributes to the total volume of discarded urban plastic, since they are rapidly discarded after a very short period of usage.

These artifacts are very similar regarding their appearance; unlike polyethylene terephthalate (PET) bottles, which are produced exclusively from PET, disposable dishes and cups can be manufactured either from polypropylene (PP) or from high impact polystyrene (HIPS) [4]. Thus, as these two polymers are incompatible, their mixtures give rise to phase separation and, as a consequence, to materials with poor properties [5]. Thus to be recycled, the previous separation of each post-consumer polymer should be carried out. So, despite the frequency with which they appear in urban garbage, these products are not potentially interesting from technical and economic points of view.

On the other hand, a proposal of processing blends of HIPS and PP without having to proceed to the separation step would allow a more favorable implementation of recycling for these artifacts. However, to succeed in obtaining PP/HIPS mixtures with good properties, a third component must be added in order to overcome the incompatibility between the two polymers. Thus, the use of an appropriate and economically viable compatibilizing agent should be considered. This has been the goal of a number of investigations.

Santana and Manrich [6] evaluated the efficiency of SEBS (styrene-ethylene-co-butylene-styrene copolymer) as a compatibilizing agent in PP/HIPS mixtures of different contents. They found that the incorporation of SEBS leads to more homogeneous morphologies, since the HIPS particles size was reduced and better dispersed throughout the PP matrix. The efficiency of an analogous polymer, SBS (styrene-butadiene-styrene tri-block copolymer) was also evaluated as compatibilizing agent for PP/HIPS by Fernandes et al. [7]. The influence of HIPS on the photodegradation of reprocessed PP was evaluated. It was observed that on increasing HIPS concentration from 10 to 30% *w/w*, the degradation effect on PP decreased. Another report by Fernandes et al. [8] addressed the influence of reprocessing on the degree of degradation by UV radiation. They observed that the resistance to radiation was lowest for virgin PP and its blends, and better resistance was related to those mixtures based on reprocessed PP.

The behavior of HIPS mixtures with low amounts of PP has also been studied. Parres et al. [9] determined thermomechanical and morphological properties of HIPS/PP blends in which the amount of PP varied from 2.5 to 10% *w/w*, in order to simulate the effect of occasional impurities on the properties. The authors observed that on increasing the content of PP in the mixture, mechanical properties were generally lower due to the increasing incompatibility between the polymers.

In this work, the use of cashew nut shell liquid (CNSL), a sub-product from the cashew nut industry, was evaluated as a compatibilizing agent in PP/HIPS blends [10]. As most commercial compatibilizing agents come from fossil resources, the industrial use of CNSL would represent an economic as well as a sustainable alternative.

In the cashew nut industry, the main product is the nut. CNSL is extracted from the husk by solvent extraction at high (180 to 200 °C) or low temperatures. At low temperatures the extraction product is called natural CNSL. At high temperatures, anacardic acid undergoes decarboxilation, giving rise to which is called technical CNSL. Thus, natural and technical CNSL have different compositions, as seen in Table 1 [11].

Table 1. Components of natural and technical CNSL [11].

Component	Natural CNSL (%)	Technical CNSL (%)
Anacardic acid	71–82	0–1.8
Cardanol	1.6–9.2	63–95
Cardol	13.8–20.3	3.8–18.9
2-Methyl-cardol	1.6–3.9	1.2–5.2
Polymeric material	-	0–21.63
Minor components	0–2.2	0–4

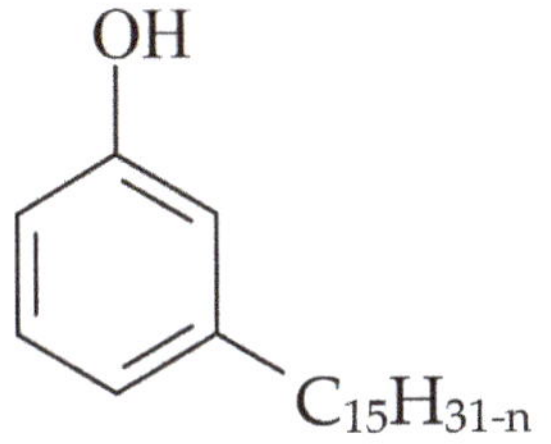

Figure 1. Chemical structure of cardanol; n depends on the number of unsaturations [12].

Technical grade CNSL is a mixture of phenols that combine the aromatic character to a long aliphatic, unsaturated chain. The major constituent is cardanol, which can be present in amounts varying from 68% up to 95% [12]. Its chemical structure is shown in Figure 1. CNSL constituents can be separated and purified, but the percentages obtained are relatively low and the cost of reagents and solvents is relatively high. In addition, contaminations with cardol and polymeric materials may prevent the large-scale production of cardanol [11].

CNSL is a biologically based lipid that is low cost, abundant and comes from a renewable resource. Its peculiar structure gives CNSL the possibility of a wide range of applications, including acting as a substitute for petroleum-based compounds [11,13,14]. The phenolic character allows CNSL to be a non-toxic anti-oxidant and raw material for phenolic and epoxy resins and plasticizers. It is also used as the basis for a number of industrial applications including adhesives, coatings and resins [15,16].

2. Materials and Methods

2.1. Materials

In this work, technical CNSL was provided by Irmãos Fontenele S.A.—Comércio, Indústria e Agricultura, Fortaleza, Ceará, Brazil. Isotactic polypropylene H605 (thermoforming grade), Melting Flow Index (MFI) = 2.1 g/10 min (230 °C, 2.16 kg) and density = 0.905 g/cm^3, was provided by Braskem (Triunfo, Brazil) S.A. High impact polystyrene R 870E (6.2% polybutadiene), MFI = 4g/10 min (200 °C, 5 kg) was provided by Innova S.A. The solvents for the selective extraction were chloroform P.A. from Merck (Rio de Janeiro, Brazil) S.A and acetone P.A. from Sigma-Aldrich Brazil Ltd. (São Paulo, Brazil).

2.2. Preparation and Injection Molding of the Blends

An interpenetrating co-rotating twin screw extruder, model DCT-20 Teck Tril, screw diameter 20 mm and L/D = 36 was used. The temperature profile along the cylinder is seen in Table 2. The screw profile consisted in five KB45 kneading elements at compression zone, assuring the complete fusion of the polymer and leading to a good mixture between the blend components. The composition of mixtures comprised of PP, HIPS and CNSL was varied according to Table 3. The filaments coming out of the extruder passed through a Brabender pelletizer, under the conditions set by the equipment manufacturer.

Table 2. Temperature profile used in the extrusion process.

Zone	1	2	3	4	5	6	7	8	9	Die
Temperature (°C)	90	140	150	150	160	165	170	175	180	180

Table 3. PP/HIPS compositions in the presence of different amounts of CNSL (0; 2.5; 5 phr).

PP/HIPS Ratio (%, *w/w*)	CNSL (phr)	Sample
1:0	0	1:0/0
	2.5	1:0/2.5
	5.0	1:0/5
4:1	0	1:0/0
	2.5	1:0/2.5
	5.0	1:0/5
3:2	0	4:1/0
	2.5	4:1/2.5
	5.0	4:1/5
2:3	0	3:2/0
	2.5	3:2/2.5
	5.0	3:2/5
1:4	0	2:3/0
	2.5	2:3/2.5
	5.0	2:3/5
0:1	0	1:4/0
	2.5	1:4/2.5
	5.0	0:1/5

The blends were injection molded in an Arburg Allrounder, model 270S-400-170, with 30 mm diameter and L/D = 20, into specimens for the analyses. Temperature profile and pressure, shown in Table 4, were selected as to preserve as much as possible the CNSL integrity. The operation procedure followed the ASTM D3641 (2012) standard.

Table 4. Injection parameters used in the processing.

Parameter	Zone				Die
	1	2	3	4	
Temperature (°C)	170	175	180	190	200
Injection pressure (bar)			1000		
Back pressure (bar)			300		

2.3. *Scanning Electron Microscopy (SEM)*

For this test, the samples were immersed in liquid nitrogen and then fractured. As neither PP nor HIPS have any special chemical feature in the chains but just carbon and hydrogen, it was impossible to apply X-ray Dispersive Energy Spectroscopy (EDS) to dye one of the phases. Due to this, the samples related to the blends were immersed in chloroform for 4 h at 23 °C to promote the extraction of the HIPS phase and CNSL and allow a better visualization of phase distribution in the polymer blends. The samples surfaces were then metalized with gold and taken to the microscope (JEOL, model 1065-LV, Tokyo, Japan). Images were obtained from a secondary electron detector with a voltage acceleration of 20 kV.

2.4. Thermal Analysis by Differential Scanning Calorimetry (DSC)

DSC analyses were carried out on the unmixed resins, CNSL and the different mixes, under nitrogen. The equipment was a NETZSCH, model 204 F1 Phoenix (Selb, Germany). Sample weights in the range 7.5 to 8.5 mg were used, as suggested by ASTM E793. The first heating cycle was done at 20 °C/min, from room to 200 °C, kept at that temperature for 2 min, followed by cooling to 0 °C to rest for 5 min. The second heating was performed from 0 °C to 200 °C at a rate of 10 °C/min. The values used in the study were obtained after the second heating.

2.5. Mechanical Tests

The Izod impact resistance was measured following ASTM D256 (2010), in a CEAST Resil Impactor Tester (Pianezza, Italy) at room temperature. To proceed to the test, the sample was fixed vertically and submitted to a 2J hammer released from an angle of 150°.

The flexural modulus was determined by EMIC DL-3000 Universal Machine (São José dos Campos, Brazil), according to ASTM D790 (2015) with loading cell of 1000 kgf. The appropriate rate for the test was set at 1.4 mm/min, according to the aforementioned standard ASTM.

2.6. Rheological Measurements

The dynamic melt rheological measurements were performed using an oscillatory rheometer (AR 2000. TA Instruments with a parallel-plate geometry D = 25 mm). All tests were conducted at 200 °C. The linear viscoelastic zone was assessed by performing strain sweep tests from 0.1% to 100% at 1 Hz. Frequency sweep tests from 0.01 to 600 rad/s were subsequently performed to determine the dynamic properties of the materials at 5% strain under nitrogen atmosphere. The rheological behaviors of the samples were evaluated based on their complex viscosity (η^*) and storage modulus (G') as a function of frequency (ω).

3. Results

3.1. Morphology

Figure 2a–e shows SEM micrographs for neat PP and HIPS and the mixtures of each of these polymers with 5 phr of CNSL. Figure 2e also shows the mixture PP-CNSL 5 phr after CNSL extraction with chloroform. In Figure 2a,c for the neat polymers, PP seems to have a more homogeneous and smoother surface than HIPS, which presents a rougher surface. Such an aspect results from the typical "salami" type morphology of HIPS, since in the production of this material PS is obtained in the presence of polybutadiene (PB). The presence of PB during styrene polymerization leads to grafting and intercrossing of this rubber in such a way that PS domains are involved inside the tenacifier phase, thus giving rise to the "salami" morphology [17].

In Figures 2b and 3d the effect of the addition of CNSL to either HIPS or PP, respectively, can be seen. Cavities and spherical particles appeared in HIPS micrograph, Figure 2b, which can be related to a localized concentration of CNSL. Figure 2d, on the other hand presented a more subtle difference. However, it can be observed that the typically smooth surface started to present slight roughness after the introduction of CNSL. To corroborate the hypotheses the sample was submitted to chloroform extraction for 4h, for the extraction of CNLS.

The result of this extraction is shown in Figure 2e, from which the holes formed due to the chloroform extraction of CNSL can be clearly observed. In the case of HIPS, as this polymer is soluble in chloroform and also in other common solvents for CNSL, the same strategy could not be employed.

In Figure 3 the micrographs refer to PP/HIPS 4:1 without CNSL (Figure 3a) and with 5 phr of CNSL (Figure 3b). Rough regions related to HIPS phase dispersed in the homogeneous PP matrix can be observed. In addition, on comparing the two images, a reduction of HIPS domains brought about by the addition of CNSL to the mixture is suggested. Nevertheless, the presence of CNSL is not explicitly evidenced, except for the slightly rougher appearance of PP phase.

Likewise, micrographs were taken for 1:4 blends without CNSL (Figure 3c) and with 5 phr CNSL (Figure 3d), for comparison. In both cases, the images point to HIPS playing the role of the matrix and PP being the disperse phase. However, there is an explicit dissimilarity between the two morphologies due to the apparent increase in the roughness of the HIPS phase due to the emergence of more prominent notches and recesses not observed in the image related to the blend without CNSL.

In addition, the shape of PP domains has also undergone visible modifications, since the dispersed phase started to show a more circular appearance after CNSL insertion, indicating a more stable morphology than the original one.

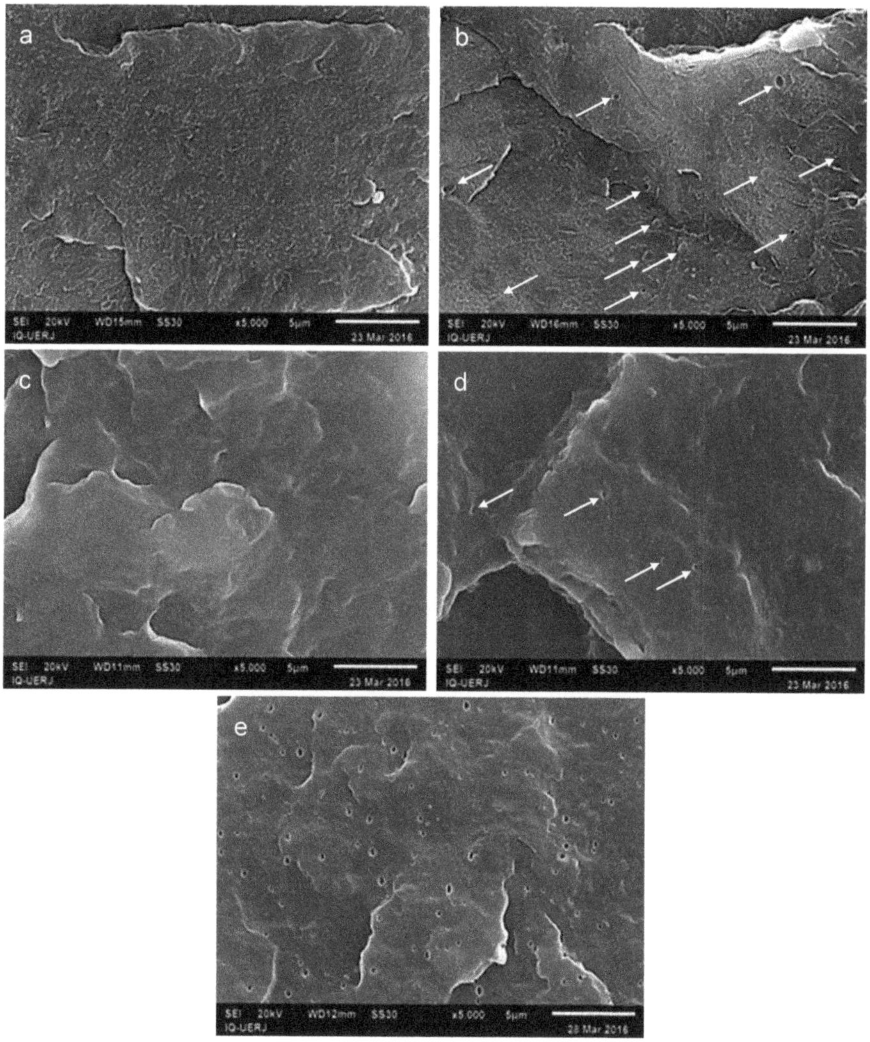

Figure 2. SEM micrographs of neat resins: (**a**) HIPS (without CNSL); (**b**) HIPS with 5 phr CNSL; (**c**) PP (without CNSL); (**d**) PP with 5 phr CNSL; (e) PP with 5 phr of CNSL, after chloroform extraction; 5000× amplification (5 μm scale).

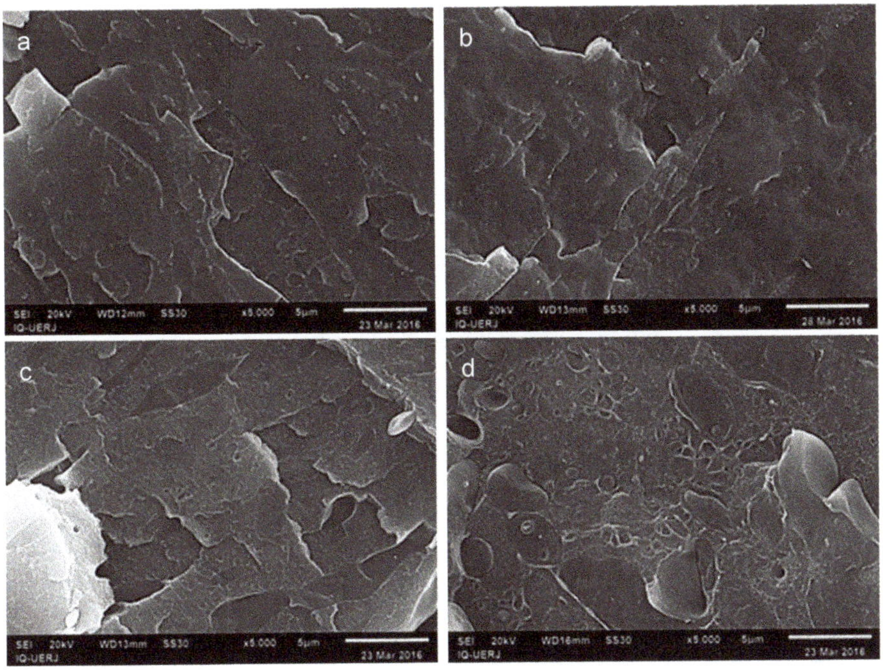

Figure 3. SEM micrographs for the blends PP/ HIPS: (**a**) 4:1 without CNSL; (**b**) 4:1 with 5 phr of CNSL; (**c**) 1:4 without CNSL; (**d**) 1:4 with 5 phr of CNSL. Magnification 5000×.

This result suggests the possibility of forming a more favorable scenario, in which the oil would position itself at the interface of the two phases, thus reducing the tension between them and improving the properties of the blend. After that, all the 12 blends were immersed in chloroform for extraction of the HIPS phase, as well as the CNSL.

Figure 4 shows micrographs for PP/HIPS 4:1 without CNSL, before extraction (Figure 4a), and after extraction (Figure 4b), the samples with 2.5 phr CNSL, after extraction (Figure 4c), and two different regions of sample with 5 phr CNSL, after extraction (Figure 4d,e).

Figure 4a shows that even with an amplification of 2000×, it is not possible to identify the two phases for the blend 4:1/0. When the sample is submitted to chloroform extraction, regions previously occupied by HIPS became observable, Figure 4b, characterized by cavity formation, deep and narrow, throughout the surface. It can be seen from these micrographs that the domains of the dispersed phase present a variety of dimensions, emphasizing the instability of the system. As for the stretched shape of these domains, one can presume that they have arisen during the injection molding process. In the process the melt material is injected under pressure into a cooled mold, which can provide the material with a processing dependent morphology.

Figure 4c–e corresponds to the 4:1 blends with 2.5 and 5 phr of CNSL. The images show morphological aspects which can be attributed to the presence of CNSL, since they do not appear in sample without CNSL. Specifically, in Figure 4d (4:1 blend with 5 phr CNSL, after extraction) aside the cavities coming from the extraction of HIPS, the appearance of concave and rounded depressions was also clearly observed, with the formation of circles in PP matrix. This morphology, nevertheless, cannot be attributed exclusively to the presence of CNSL since this pattern was not seen in the micrograph of neat PP+CNSL, after extraction. Therefore, cavity formation would be a combined effect of both CNSL and HIPS being present in the PP matrix.

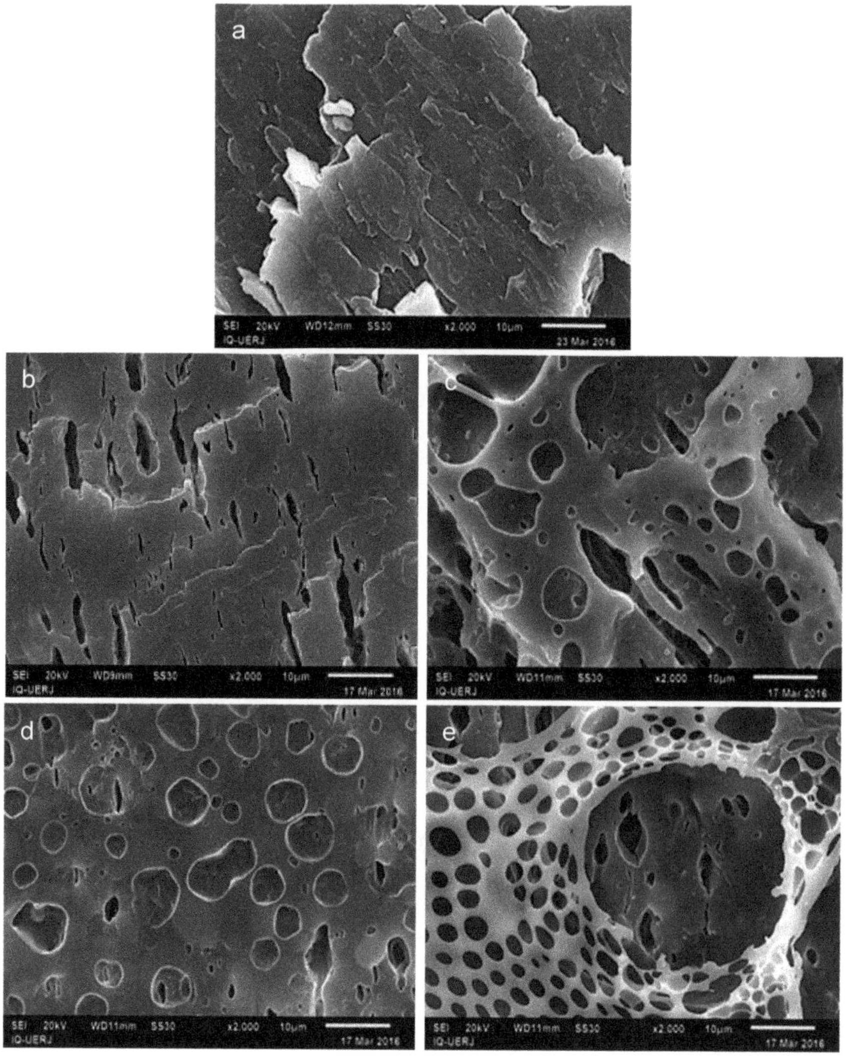

Figure 4. SEM micrographs for PP/HIPS 4:1 without CNSL, before extraction (**a**), and after extraction (**b–e**): without CNSL (**b**), with 2.5 phr CNSL (**c**), and with 5 phr CNSL at different regions of the sample (**d**,**e**). Amplification 2000× (10 μm scale).

However, the same blend 4:1/5, under the analysis of a different spot (Figure 4e), revealed a peculiar morphology type that is different from that showed in Figure 4d. In Figure 4e, a perforated film can be seen on the material surface that has not been observed before. The same pattern appears on a smaller scale in Figure 4c (4:1/2.5), where round and stretched cavities are seen. Such an unusual morphology motivated a deeper investigation that will be described later on.

Blends 3:2 were also evaluated after chloroform extraction of HIPS phase and their SEM images are presented in Figure 5 at two magnifications. On the left side the images were taken at 500× and on the right side, at 2000× magnification. Figure 5a,b shows micrographs of 3:2/0 blend where the evident increase of the HIPS domains are seen, when compared with 4:1/0 blend (Figure 5b) due to the doubled amount of this polymer in the mixture.

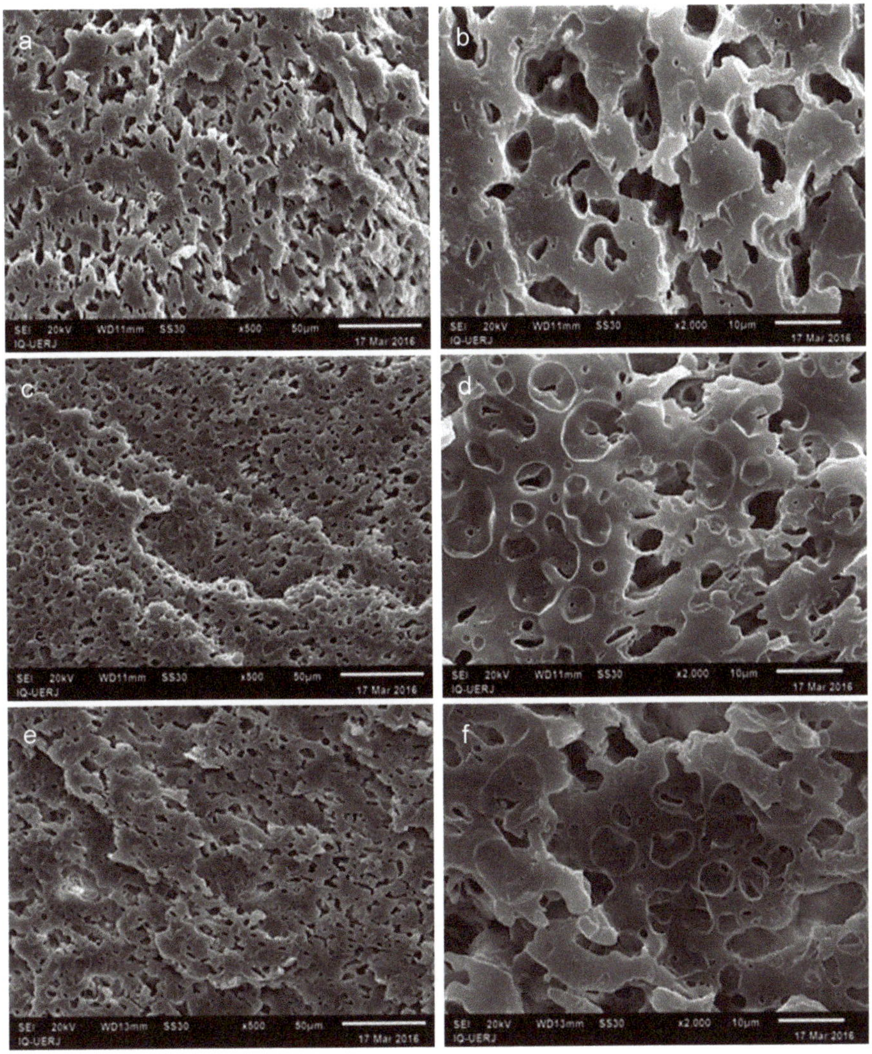

Figure 5. SEM micrographs of 3:2 PP/HIPS blends, after chloroform extraction: (**a**,**b**) without CNSL; (**c**,**d**) with 2.5 phr of CNSL; (**e**,**f**) with 5.0 phr of CNSL; images on left side, magnification of 500×; images on right side magnification of 2000×.

Figure 5c,d, related to 3:2 PP/HIPS blends with 2.5 phr of CNSL, and Figure 5e,f corresponding to the blends with 5phr of CNSL, present a clear reduction in domain size associated to the previous localization of HIPS, in addition to a higher homogeneity of these sizes, suggesting a more stable morphology, as compared with the blend with no CNSL.

Moreover, it was also possible to verify the occurrence of circles in the PP matrix, as seen before in the micrograph of 4:1 blend containing CNSL (Figure 4d), thus corroborating the hypothesis that the formation of these cavities is intrinsically associated to the insertion of CNSL in PP/HIPS mixtures.

Going further with SEM analysis, Figure 6 presents the micrographs for 2:3 blends. In Figure 6a,b, as expected, the number of cavities related to HIPS domains tends to increase as the amount of this polymer in the blends also increases. Nevertheless, in spite of HIPS accounting for 60% of the total mass, it is still possible to verify continuity in the remaining phase (PP). This tendency of PP to play the role of the matrix when in blends with polystyrene (PS) has been widely studied by Omonov et al. [18]. The authors used selective dissolution experiments to quantitatively estimate the total continuity of PS phase in PP/PS compositions. They found that it occurs between 70%/30% and 10%/90% (*w*/*w*), depending on the viscosity ratio of the constituents.

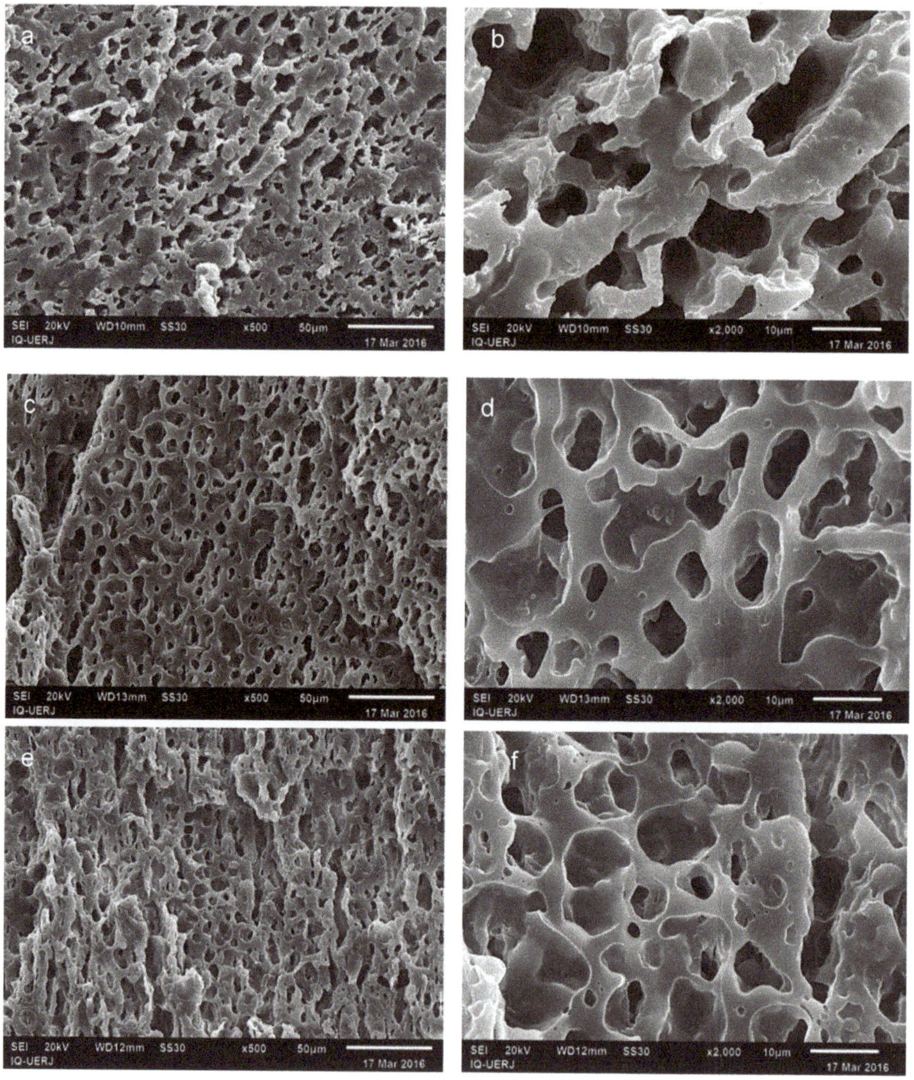

Figure 6. SEM micrographs of 2:3 blends, after chloroform extraction: (**a**,**b**) without CNSL; (**c**,**d**) with 2.5 phr of CNSL; (**e**,**f**) with 5.0 phr of CNSL; images on left side with 500× magnification; images on right side with 2000× magnification.

It is worth noting that the image seen in Figure 6a,b suggests that morphology is very close to a co-continuous one. However, on the addition of CNSL to the mixtures, a reduction of HIPS domains took place (Figure 6c–f), inhibiting the occurrence of a co-continuous system in these compositions.

On the other hand, the reduction in the disperse domains as well as the better homogeneity in the morphology provide better properties to the blends due to the stabilizing effect of CNSL.

Figure 7 shows the micrographs for the 1:4 PP/HIPS mixtures, before and after chloroform extraction, at 2000× magnification. In Figure 7a, PP domains dispersed in the HIPS matrix can be identified, suggesting that in the 1:4 blends without CNSL, a co-continuous morphology has been achieved. If this is so, it can be said that for a PP/HIPS system, in the absence of CNSL and under the imposed processing conditions, a co-continuous morphology can be formed when components ratios vary between 2:3 and 1:4, that is, between 40%/60% and 20%/80% *w*/*w*.

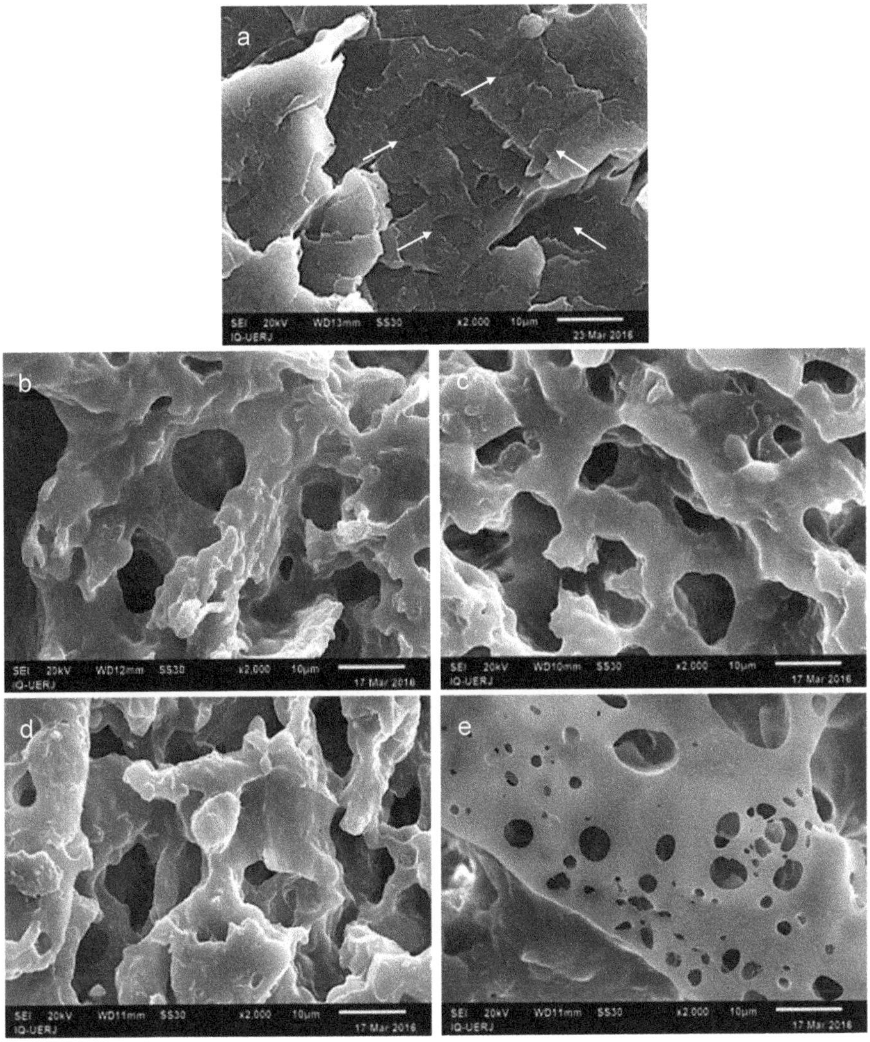

Figure 7. SEM micrographs of 1:4 PP/HIPS blends; before chloroform extraction: (**a**) without CNSL; after chloroform extraction: (**b**) no CNSL, (**c**) with 2.5 phr of CNSL, (**d**) and (**e**) different aspects of blend with 5 phr of CNSL. Magnification of 2000×.

After selective extraction of HIPS (Figure 7b), the 1:4/0 blend presented cavities and deep cracks, in addition to entire regions which have been extracted, an indicative of the effective occurrence continuity of the HIPS phase. However, the remaining PP, also observed in the images, shows the concomitant occurrence of continuity of this phase, most probably partial.

In Figure 7c,d more homogeneous morphologies than that for the blend without CNSL (Figure 7b) are observed. Also, the images presented a visual aspect typical of co-continuous systems, which may suggest a morphological stability much superior to that for the 1:4/0 blend.

Figure 7e shows another region of the 1:4/5 blend where a perforated film is again observed, as in Figure 4e. With the purpose of clarifying the casual formation of this atypical morphology, another selective extraction of the HIPS phase was carried out, this time with acetone instead of chloroform, during the same period of time. Since acetone is not a solvent as good as chloroform for HIPS, the extraction was only partial, leaving HIPS residues on the surface of the sample as seen in Figure 8.

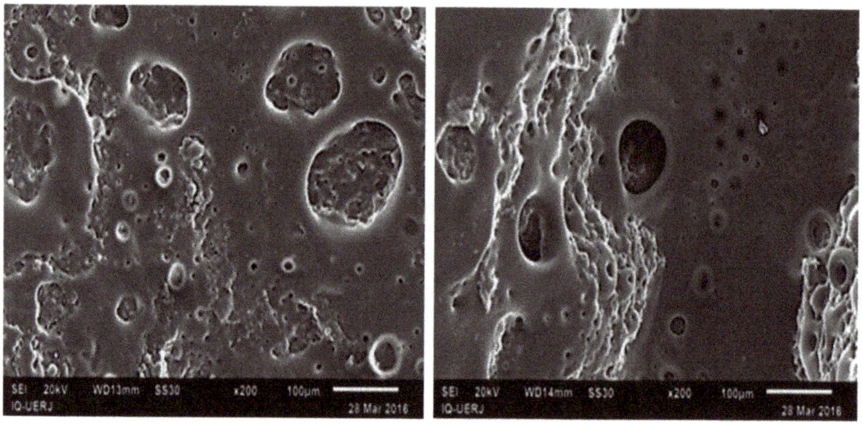

Figure 8. SEM Micrographs of 4:1 PP/HIPS blend with 5 phr of CNSL after acetone extraction of HIPS phase in two different regions of the same sample. Magnification 200×.

On comparing with the atypical morphologies observed after the extraction with chloroform, a similarity in the pattern is found, thus suggesting that the perforated formation would be caused by some remaining amounts of HIPS on the sample surface due to an incomplete HIPS dissolution by chloroform. As this morphology was observed only in those blends with CNSL, one can infer that the oil may have somehow interfered with the HIPS solubility in chloroform.

It is worth noting, as previously seen in Figure 4d which shows the 4:1 blend with 5 phr of CNSL, that a different morphology of HIPS phase was found which suggests that the presence of CNSL in the mixture could be the key for the formation of the peculiar aspect of a perforated film.

3.2. *Thermal Behavior*

From DSC curves polymer transition temperatures as well as heat flow and degree of crystallinity were determined. Glass transition temperatures were taken from the inflection point and a deviation of 2 °C was considered, as suggested by ASTM E1356 (2014). The same procedure was followed for the determination of melting enthalpy according to ASTM E793 (2012).

The degree of crystallinity was determined from the endothermic peak relative to the enthalpy of crystalline melting, calculated by Equation (1) [19].

$$\mathcal{X}_c(\%) = \frac{\Delta H_f}{\Delta H_f^o} \times 100 \quad (1)$$

where, X_c is the degree of crystallinity of the sample; ΔH_f is the variation in the melting enthalpy of the sample, measured by the area of the melting peak in the curve obtained by DSC; ΔH_f^o is the variation in the enthalpy of melting for a 100% crystalline sample (determined by extrapolation).

As polystyrene is amorphous, the final crystallinity of the PP/HIPS blends can be attributed exclusively by the semi-crystalline polypropylene. The theoretical value of melting enthalpy for a 100% crystalline PP, used in the calculations was 207 J/g [20].

Concerning the samples with CNSL, a correction had to be made to compensate for the presence of the added oil. Thus, as suggested by ASTM E793, a standard deviation of 7.8% was used to correct the values of melting enthalpy, from now on referred to as ΔH_f^c. To perform the analysis the protocol described in the experimental part was used.

Figure 9 shows the thermograms related to HIPS with 0, 2.5 and 5.0 phr of CNSL. Looking at the curves it can be seen that a reduction in the glass transition temperatures occurs from 104 ± 2 °C to 96 °C, and then to 89 °C as the addition of CNSL increases from 0 to 5 phr. This can be credited to the role of CNSL as a plasticizer for the HIPS matrix, which contributes to chain mobility, allowing the amorphous phase of the polymer to flow at lower temperatures.

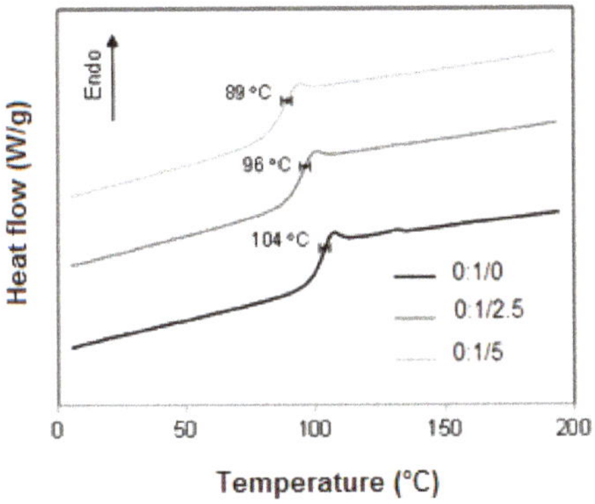

Figure 9. DSC thermograms for HIPS with 0, 2.5 and 5 phr of CNSL.

The thermograms in Figure 10 correspond to the blends PP/HIPS 1:4, with 0, 2.5 and 5 phr of CNSL. Again, the reduction in Tg is observed. However, the crystalline enthalpy of melting did not show any change.

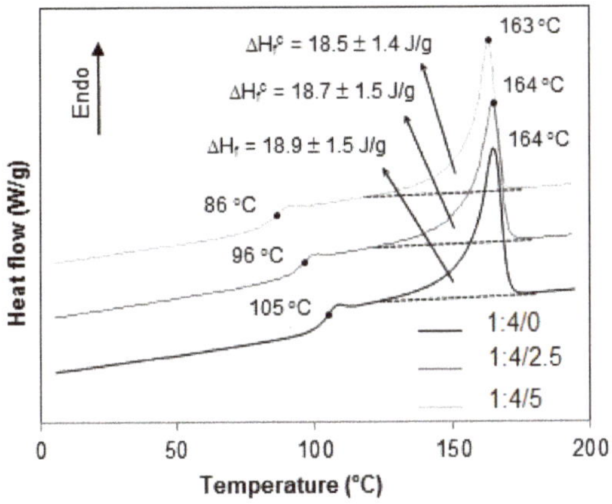

Figure 10. DSC thermograms for 1:4 PP/HIPS blends with 0, 2.5 and 5 phr of CNSL.

However, as the HIPs content decreases, Figures 11–13, the signal associated to T_g^{HIPS} becomes less and less visible till its total disappearance, as seen in Figure 13 for 4:1 blends.

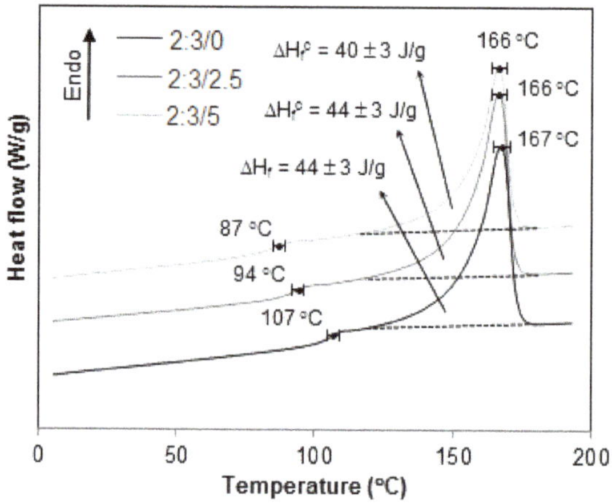

Figure 11. DSC thermograms for 2:3 PP/HIPS blends with 0, 2.5 and 5 phr of CNSL.

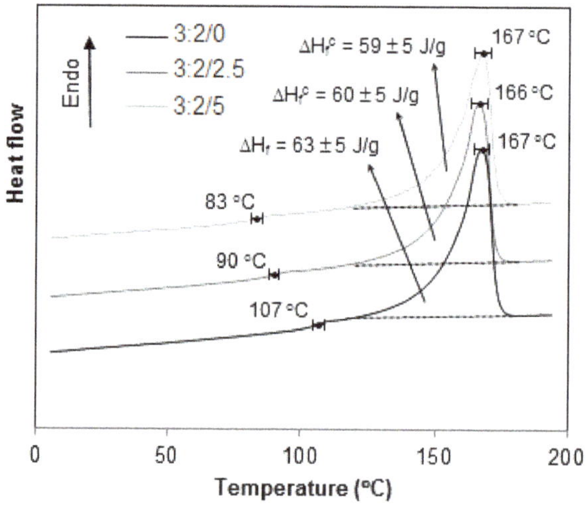

Figure 12. DSC thermograms for 3:2 PP/HIPS blends with 0, 2.5 and 5 phr of CNSL.

The thermograms in Figures 10–14 present the curves for all blends, 1:4, 2:3, 3:2, 4:1 and 1:0, respectively. In general, the same tendency of T_g^{HIPS} decrease as the CNSL content increases is observed. This behavior can be better visualized in Figure 15. In this figure, data for 4:1 blends were dismissed for lack of resolution in T_g^{HIPS} determination. The decay in the values of T_g^{HIPS} may suggests that the oil CNSL was not completely incorporated in PP phase and thus the following possibilities can be raised: 1. CNSL could be distributed throughout the two phases; 2. It could be incorporated only in HIPS phase; 3. It could be placed preferentially in the interface of the two phases. In Figure 15, the 3:2 blends, which have the lowest amount of HIPS, one can see that T_g^{HIPS} decay is more significant as compared with the other blends. This suggests that CNSL may have a higher tendency to be incorporated into the HIPS phase, so that as the amount of HIPS in the blend is reduced, the relative concentration of CNSL increases in relation to the HIPS mass, thus indicating its effect on T_g^{HIPS} reduction even more.

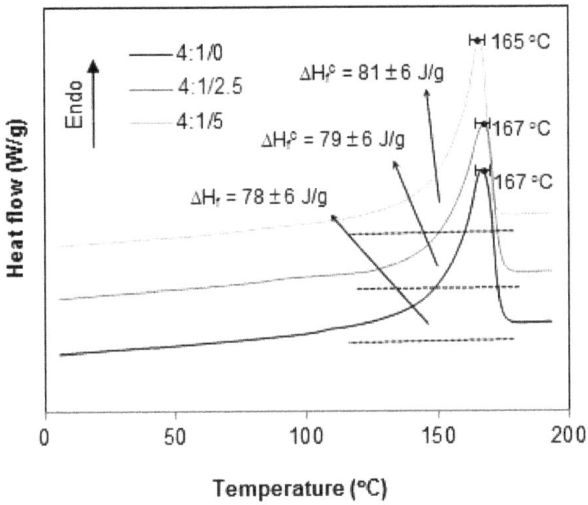

Figure 13. DSC thermograms for 4:1 PP/HIPS blends with 0, 2.5 and 5 phr of CNSL.

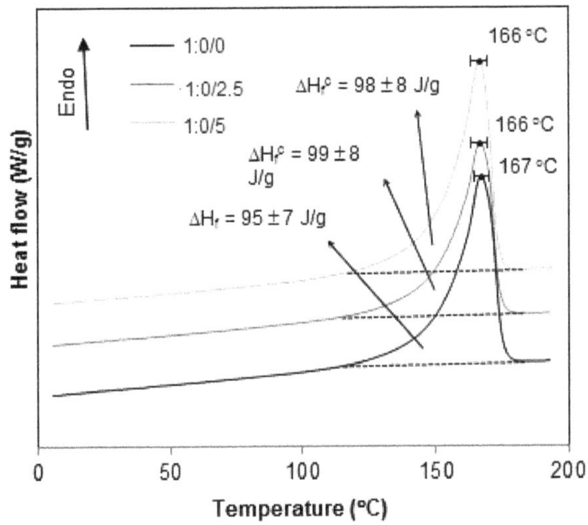

Figure 14. DSC thermograms of PP with 0, 2.5 and 5 phr of CNSL.

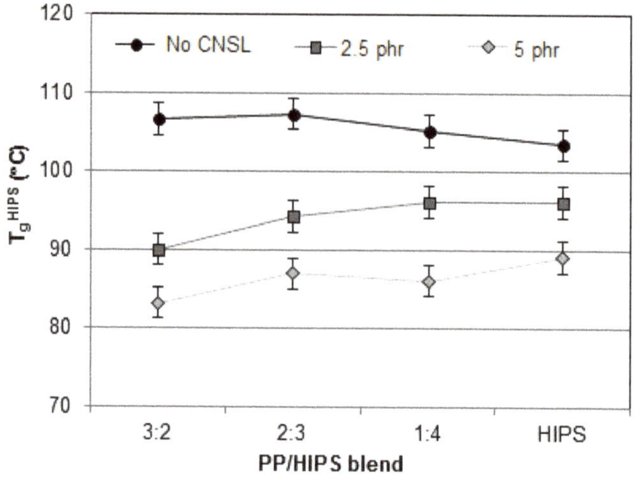

Figure 15. Tg values for HIPS and the 3:2, 2:3 and 1:4 PP/HIPS blends with 0, 2.5 and 5 phr of CNSL.

Concerning T_m^{PP}, the values were determined from the temperature at which the maximum of the crystalline melting peak occurs, as seen in Figures 10–14. Unlike T_g^{HIPS}, no relevant variation in T_m^{PP} has been found considering all samples, thus suggesting that neither the blend with HIPS nor the presence of CNSL had influence on the formation of a PP crystalline phase.

The values of melting enthalpy ΔH_f for PP were obtained from the area under the melting peaks. No significant changes in enthalpy were observed in any sample on the addition of CNSL, as seen in Figures 10–14, only a gradual reduction of these values as the amount of PP was also reduced. This can be better observed in Figure 16, which shows the calculated degree of crystallinity X_c for the different blend compositions. An almost linear relationship between X_c and the contribution of PP to the mixture was found, thus supporting the previous supposition that the presence of HIPS or CNSL does not affect the crystal formation of the PP phase.

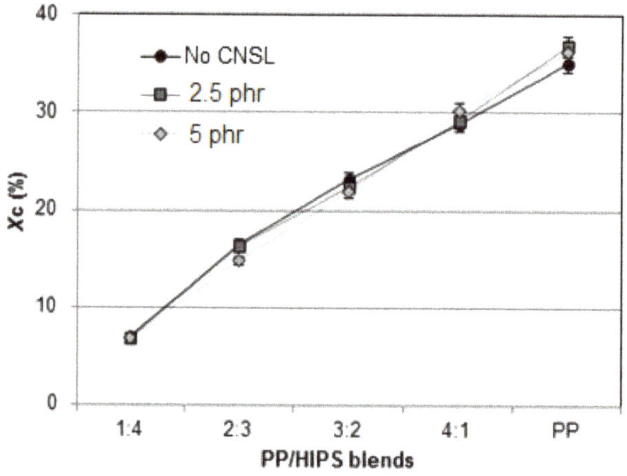

Figure 16. Degree of crystallinity X_c of PP and the PP/HIPS blends with 0, 2.5 and 5 phr of CNSL.

3.3. Mechanical Tests

3.3.1. Impact Strength Analysis

In general, impact strength is denoted as resilience, or the ability a material has to absorb mechanical energy. The values of impact strength for the neat polymers as well as the blends, in the presence or not of CNSL, are presented in Figure 17.

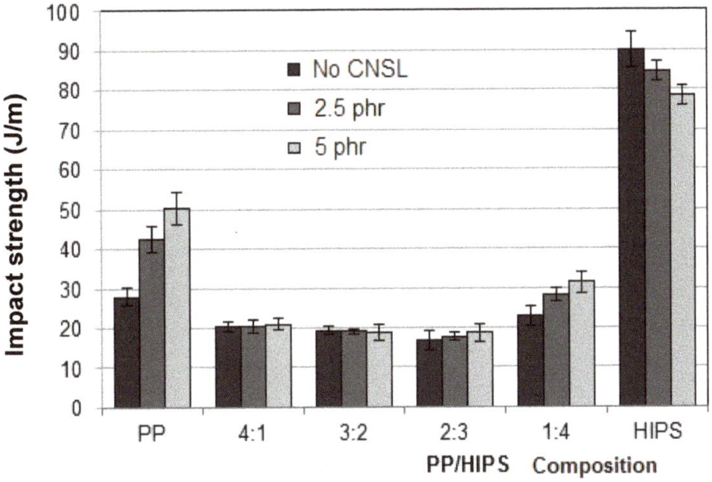

Figure 17. Impact strength results for PP, HIPS and PP/HIPS blends the PP/HIPS blends with 0, 2.5 and 5 phr of CNSL.

The neat polymers presented a visible effect on increasing CNSL addition but in opposite directions. PP experienced a gradual enlargement in impact resistance with the incorporation of a higher amount of CNSL, giving rise to an improvement of 80% in the property for the sample with 5 phr of CNSL, as compared with the sample with no oil. This can be due to a plasticizing effect of CNSL on the PP amorphous phase, thus increasing the efficiency of the sample in absorbing impact energy and improving its toughness property.

The adverse effect of CNSL on HIPS can be the result of a disturbing effect this substance may be imposing to HIPS morphology which would make the transfer of impact energy more difficult throughout the polymer matrix. The morphology formed during HIPS synthesis is such as to maximize the energy transfer. Any factor leading to modification in this morphology will cause, as consequence, a reduction in the impact strength.

Figure 17 also shows the results of impact resistance for the blends. The values for these samples are lower than for the neat PP, much lower than those for neat HIPS and very similar to each other, except for the 1:4 PP/HIPS blends. For the three first sets of blends, the deleterious effects of both PP phase and CNSL are well evidenced. Only the last set, with a larger amount of HIPS, could show a small recovery of the impact strength. As for the effect of PP, Parres et al. [9] showed that low levels of PP are sufficient to cause a drastic reduction of the impact resistance of HIPS, with losses as high as 40% by adding only 5% of PP.

The results can be corroborated by the effect of CNSL on the blends 1:4, previously seen in Figure 8. In the figure the stabilizing effect of CNSL was observed as the morphology of the blends became more homogeneous. Thus, the more uniform distribution of the domains of a phase throughout the other gave rise to a better absorption of the impact energy.

3.3.2. Flexural Analysis

The flexibility of a material can be inferred from the elastic modulus determined in the tensile test, since it is directly related to rigidity. In spite of this, the flexural strength can be directly measured by the flexural application test itself, from which it is possible to obtain more appropriate results for the study of the flexibility of a material, that is, the capacity of the material to yield to a mechanical bending stress.

The results of flexural modulus are presented in Figure 18. Most error bars did not present a significant magnitude and some cannot be detected in the curves.

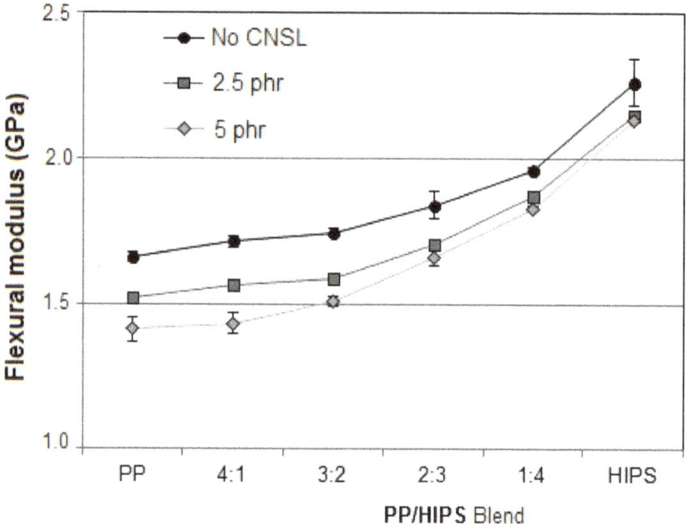

Figure 18. Flexural modulus for neat PP, HIPS and the PP/HIPS blends with 0, 2.5 and 5 phr of CNSL.

From these curves it is observed that increasing flexural modulus is directly related to the HIPS content in the samples, meaning that the higher the ratio between PP and HIPS, the lower the tension needed to promote a specific deformation under the bending stress.

On comparing the results related to the samples with and without CNSL, a reduction in the moduli was found as the oil content increased leading, in all cases, to an increase in material flexibility. This finding agrees with the idea that the CNSL molecules have been inserted in-between the polymer molecules thus creating secondary links, so that these molecules became separated from one another. This way, the cohesive forces among the macromolecules were reduced, thus promoting an increased mobility in the system and improving the flexibility of these materials. This intermolecular effect is characteristic of plasticizer additives [21,22], generally added to blends to increase flexibility, which was observed in the case of CNSL addition.

3.3.3. Dynamic Melt Rheological Measurements

The variation of the complex shear viscosity as a function of the frequency of the neat PP, HIPS and PP/HIPS blends with different CNSL content is shown in Figure 19.

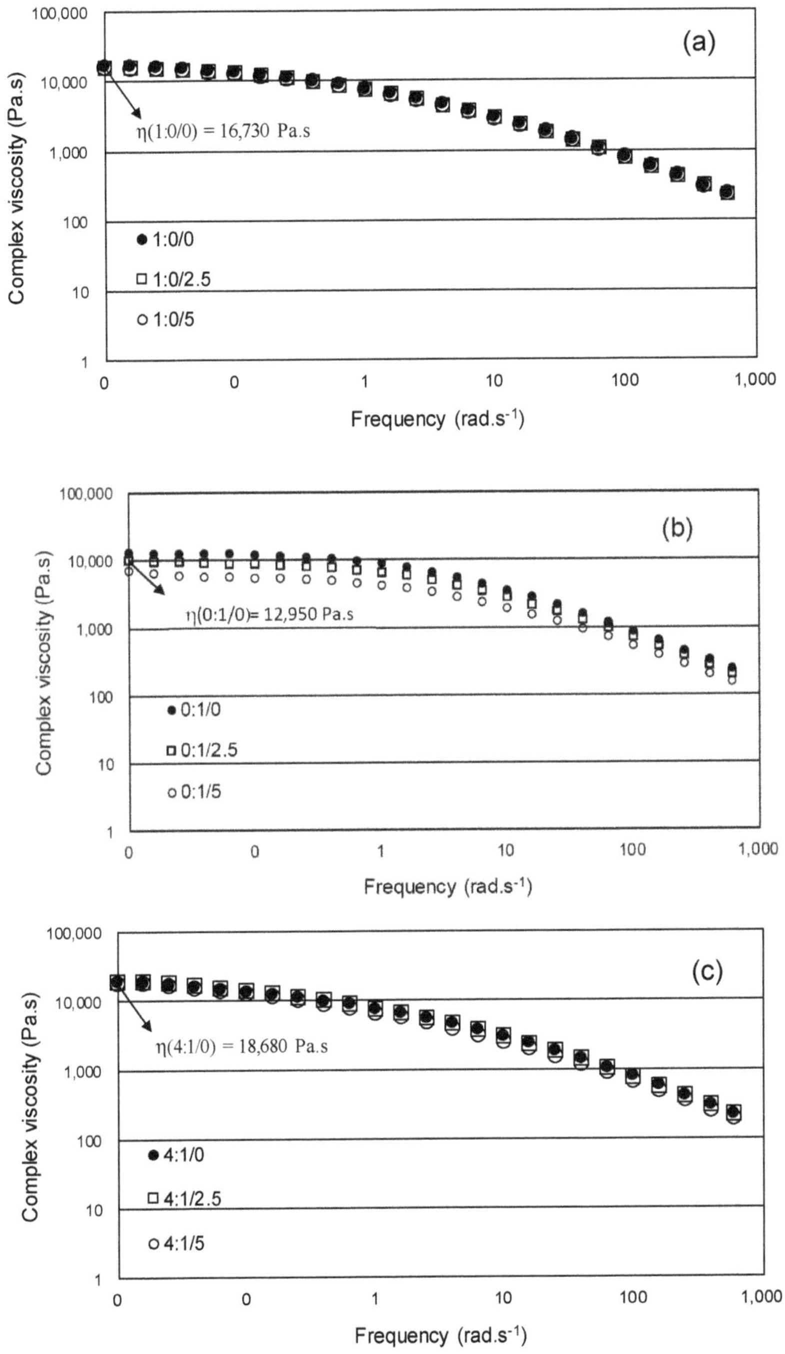

Figure 19. *Cont.*

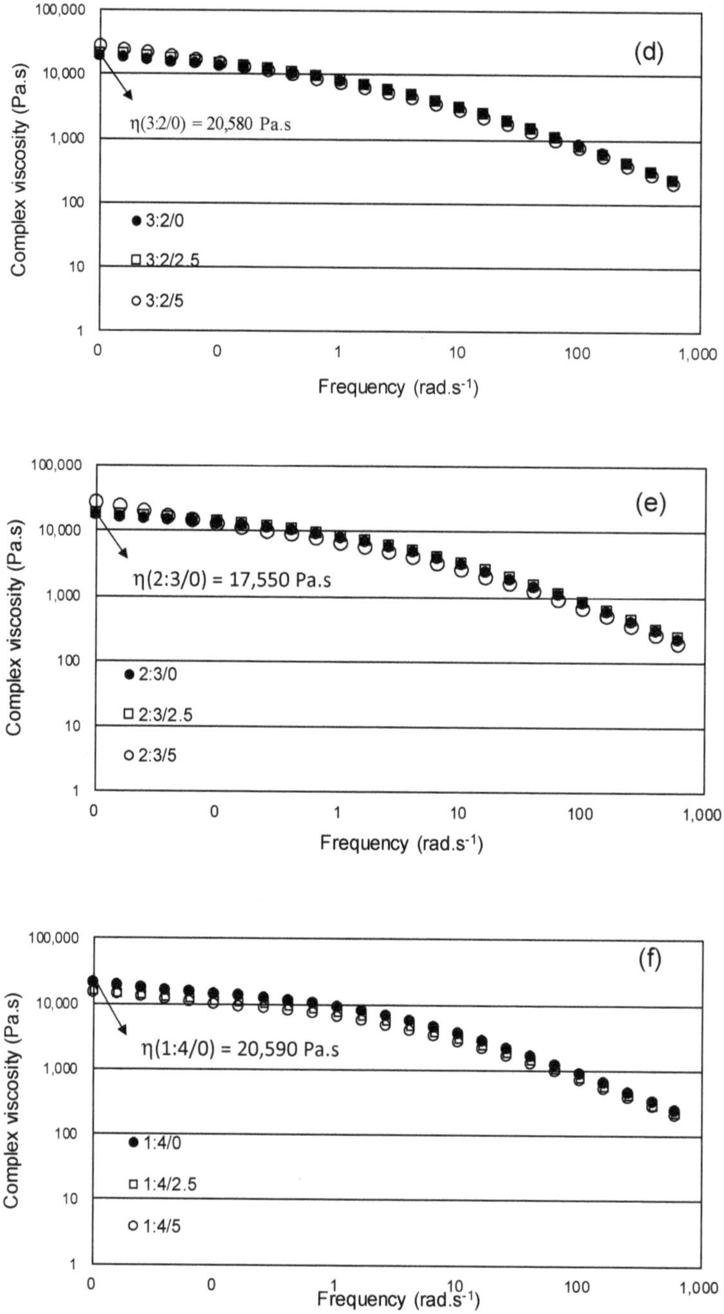

Figure 19. Complex viscosity *versus* frequency curves for: (**a**) PP; (**b**) HIPS and (**c**–**f**) 4:1, 3:2, 2:3 and 1:4 PP/HIPS blends, with 0, 2.5 and 5.0 phr CNSL content.

As seen in Figure 19, the neat PP and all the blends showed a shear-thinning behavior, with an increase in viscosity in the low-frequency region. Figure 19b shows that neat HIPS presents the lowest viscosity values in relation to neat PP and also a Newtonian plateau up to 0.1 rad·s^{-1} and then shows a shear-thinning behavior. It can also be observed that the addition of CNSL oil had a more pronounced effect on the flow behavior of HIPS in relation to PP. As HIPS is added to PP, an increase in viscosity values tends to occur due to the presence of HIPS domains which hinder the PP molecule flow. However, a different flow behavior occurred in 2:3/0 composition, that is, a decrease in viscosity values is observed in relation to other compositions without CNSL. SEM micrographs suggested that a co-continuous morphology occurred, which can explain the viscosity decrease in this blend. When CNSL oil is added to PP/HIPS blends, again it is observed a plasticizer effect, resulting in a more frequency-thinning behavior. As HIPS content increases in the blend composition, the oil tends to present a more pronounced plasticizer effect. It indicates that oil molecules interact more strongly with the HIPS phase, corroborating the SEM analysis. In 1:4/5 PP/HIPS/CNSL composition, the lowest viscosity values are observed, indicating once again that HIPS molecules achieve more mobility in the presence of the oil. This behavior corroborates the DSC results, which showed that the addition of CNSL oil reduces the Tg values of HIPS phase.

The variation of dynamic modulus, G′ and G″, was evaluated. Table 5 shows the values of the crossover point, at which G′ = G″, obtained from G′ and G″ versus the frequency curves. The results were evaluated by the displacement of the G′ x G″ crossover point, which allows the flow behavior of polymeric materials to be predicted [23].

Table 5. Dynamic modulus and frequency values at the G′/G″ crossover point for neat PP, HIPS and blends with 0, 2.5 and 5.0 phr CNSL contents.

PP/HIPS Ratio (%, *w/w*)	CNSL (phr)	Sample	Modulus at Crossover Point G′ = G″ (Pa)	Crossover Point ω_c (rad/s)
1:0	0	1:0/0	20,670	10
	2.5	1:0/2.5	21,080	10
	5.0	1:0/5	20,780	10
4:1	0	4:1/0	21,010	10
	2.5	4:1/2.5	20,730	10
	5.0	4:1/5	17,700	10
3:2	0	3:2/0	22,120	10
	2.5	3:2/2.5	21,900	10
	5.0	3:2/5	19,910	10
2:3	0	2:3/0	21,600	10
	2.5	2:3/2.5	22,940	10
	5.0	2:3/5	18,260	10
1:4	0	1:4/0	25,130	10
	2.5	1:4/2.5	21,620	10
	5.0	1:4/5	18,030	10
0:1	0	0:1/0	25,280	10
	2.5	0:1/2.5	20,250	10
	5.0	0:1/5	17,270	16

Table 5 shows that the addition of HIPS to PP increases the elastic behavior of the blend and the addition of the CNSL oil tends to increase the viscous behavior, indicating its plasticizer effect in the PP/HIPS blends. This effect is more pronounced as HIPS content increases. A different behavior is observed in the 2:3/0 composition, which presents a decrease in the elastic behavior. As mentioned before, this blend showed a co-continuous morphology in SEM analysis. Once again, the rheological data show that the plasticizer effect of the oil is more pronounced in the HIPS phase.

Viscoelastic behavior from rheological measurements can be shown in a different representation called Cole-Cole plots. In this representation, the imaginary viscosity values ($\eta'' = G'/\omega$) is plotted against dynamic viscosity values ($\eta' = G''/\omega$). The plot should be a perfect arc in the absence of higher structures. In this case, the relaxation behavior of the melt will be described by a single relaxation time. Flattening of the arc, the presence of a tail or an increasing correlation indicate the occurrence of a broad relaxation time spectrum, while structural effects result in occurrence of a second arc [23,24].

Figure 20 shows the Cole-Cole plots of the neat PP, HIPS and PP/HIPS blends with different CNSL content.

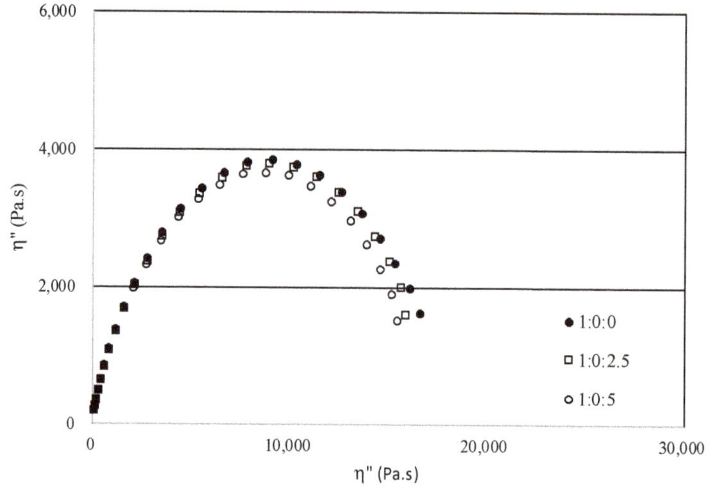

(**a**)

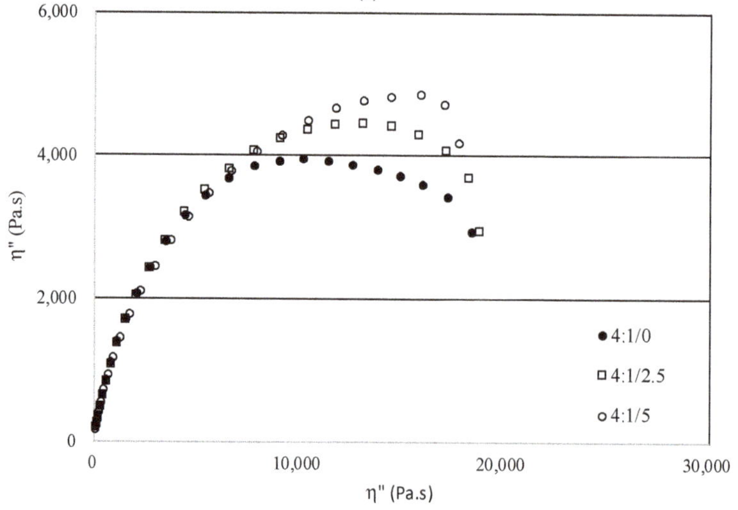

(**b**)

Figure 20. *Cont.*

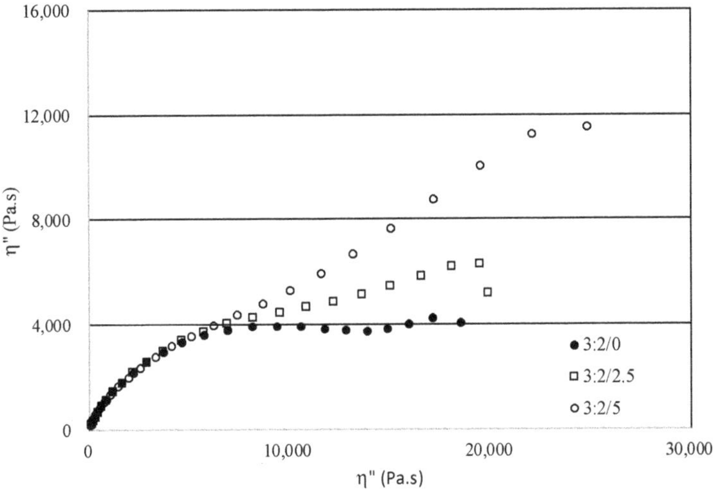

(c)

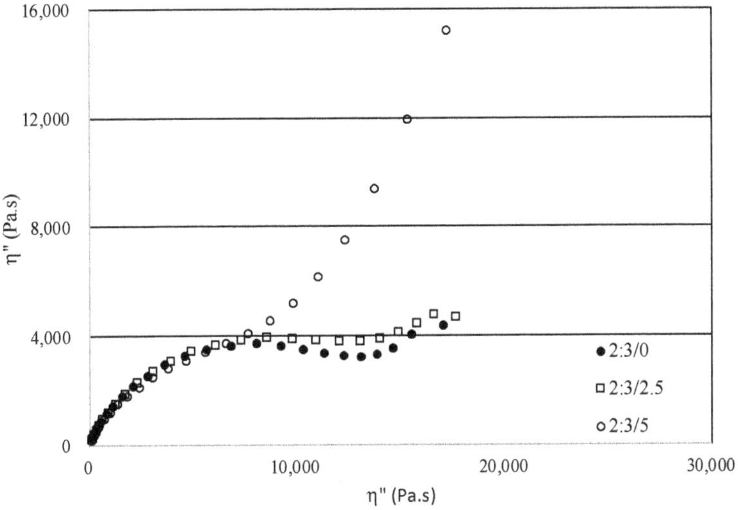

(d)

Figure 20. *Cont.*

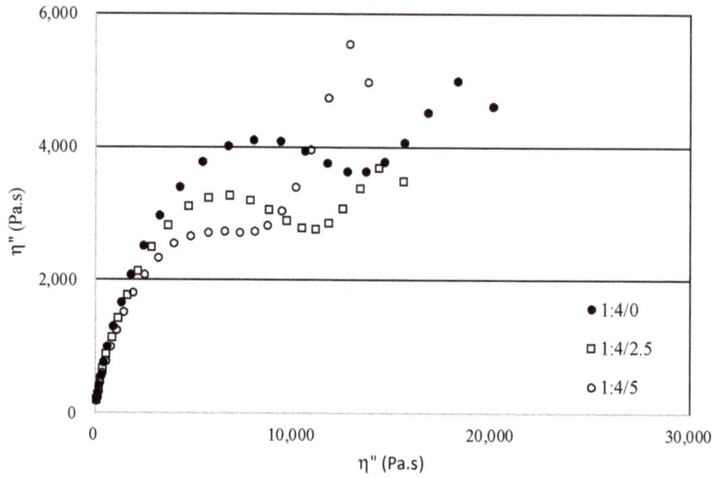

(e)

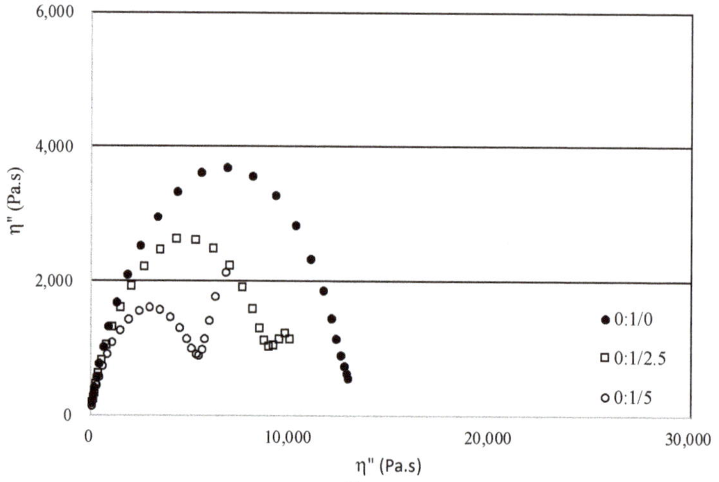

(f)

Figure 20. Cole-Cole plots of PP/HIPS blends (**a**) 1:0; (**b**) 4:1; (**c**) 3:2; (**d**) 2:3; (**e**) 1:4 and (**f**) 0:1, with different CNSL contents.

The Cole-Cole plot is like an arc for neat PP and PP/CNSL compositions, but deviates very strongly from an arc as HIPS and CNSL additive are added to PP, indicating a broad relaxation time due to the different interactions that occur between PP, HIPS and CNSL (when present) molecules. Figure 20f shows a perfect arc for neat HIPS and the appearance of arcs and tails for HIPS/CNSL compositions, indicating the occurrence of structural effects between HIPS and CNSL molecules. These results indicate the more pronounced effect of the CNSL oil on the HIPS phase.

4. Conclusions

The general evaluation of the morphological, mechanical and thermal aspects of the polypropylene (PP)/high impact polystyrene (HIPS) blends was carried out with the objective of investigating the effect of Cashew Nut Shell Liquid (CNSL) addition on these properties. Scanning Electron Microscopy (SEM) micrographies suggest that CNSL has a tendency to be located in the HIPS phase and also in the interface of the domains. This was corroborated by the results of Differential Scanning Calorimetry (DSC).

The addition of different contents of CNSL in PP/HIPS blends showed a general tendency to reduce the size of HIPS domains, thus suggesting a better interaction between the two polymers, providing tension transfer from one phase to the other and favoring stabilization of the morphology. Crystalline melting temperature of PP phase concerning all the blends did not change, thus corroborating the hypothesis that CNSL would be placed preferably in the HIPS phase or in the interphase between the two phases. Mechanical properties are in accordance with the proposed morphology, corroborated by rheological measurements. Cole-Cole plots also indicated that CNSL has a more pronounced effect on the HIPS phase.

Author Contributions: This work was obtained from the development of M.N.A. and the other authors contributed in the orientation, discussion, elaboration and writing of the paper.

Funding: The paper was written from the experimental results of the Master's dissertation that had a scholarship from the State of Rio de Janeiro Research Foundation (FAPERJ).

Acknowledgments: The authors are grateful for the scholarship granted to Mirna Nunes de Araújo of the Foundation for Research Support of the State of Rio de Janeiro (FAPERJ) and National Council for Scientific and Technological Development (CNPq) for financial support.

Conflicts of Interest: The authors declare no conflict of interest.

References

1. Godecke, M.V.; Naime, R.H.; Figueiredo, J.A.S.O. Consumismo e a geração de resíduos sólidos urbanos no Brasil (Consumerism and the generation of solid urban waste in Brazil). *Revista Eletrônica em Gestão Educação e Tecnologia Ambiental* **2012**, *8*, 1700–1712. Available online: http://cascavel.ufsm.br/revistas/ojs-2.2.2/index.php/reget (accessed on 23 April 2019).
2. Huysman, S.; Debaveye, S.; Schaubroeck, T.; De Meester, S.; Ardente, F.; Mathieux, F.; Dewulf, J. The recyclability benefit rate of closed-loop and open-loop systems: A case study on plastic recycling in Flanders. *Resour. Conserv. Recycl.* **2015**, *101*, 53–60. [CrossRef]
3. Ragaert, K.; Delva, L.; Van Geem, K. Mechanical and chemical recycling of solid plastic waste. *Waste Manag.* **2017**, *69*, 24–58. [CrossRef]
4. Theis, V.; Schreiber, D. A inovação e as alternativas de realizar as atividades em P&D: Estudo de caso da Braskem (The innovation and the alternatives to carry out the activities in R&D: Braskem case study). In *Inovação e Aprendizagem Organizacional*; Universidade Feevale: Rio Grande do sul, Brazil, 2013; pp. 141–164. ISBN 978-85-7717-163-7.
5. Mélo, T.J.A.; Carvalho, L.H.; Calumby, R.B.; Brito, K.G.Q.; D'Almeida, J.R.M.; Spieth, E. Mechanical properties and morphology of a PP/HIPS polymer blend compatibilized with SEBS. *Polímeros Ciência e Tecnologia* **2000**, *10*, 82–89. [CrossRef]
6. Santana, R.M.C.; Manrich, S. Morphology and mechanical properties of polypropylene/high-impact polystyrene blends from postconsumer plastic waste. *J. Appl. Polym. Sci.* **2003**, *88*, 2861–2867. [CrossRef]
7. Fernandes, L.L.; Freitas, C.A.; Demarquette, N.R.; Fechine, G.J.M. Photodegradation of thermodegraded polypropylene/high-impact polystyrene blends: Mechanical properties. *J. Appl. Polym. Sci.* **2011**, *120*, 770–779. [CrossRef]
8. Fernandes, L.L.; Freitas, C.A.; Demarquette, N.R.; Fechine, G.J.M. Influence of the type of polypropylene on the photodegradation of blends of polypropylene/high impact polystyrene. *Polímeros* **2012**, *22*, 61–68. [CrossRef]
9. Parres, F.; Balart, R.; López, J.; García, D. Changes in the mechanical and thermal properties of high impact polystyrene (HIPS) in the presence of low polypropylene (PP) contents. *J. Mater. Sci.* **2008**, *43*, 3203–3209. [CrossRef]
10. Paiva, F.F.A.; Garrutti, D.S.; Silva Neto, R.M. *Cashew Industrial Exploitation*, 1st ed.; EMBRAPA Agroindústria Tropical—SEBRAE: Fortaleza, Brazil, 2000; 88p.
11. Mazzetto, S.E.; Lomonaco, D.; Mele, G. Cashew nut oil: Opportunities and challenges in the context of sustainable industrial development. *Química Nova* **2009**, *32*, 732–741. [CrossRef]

12. Gedam, P.H.; Sampathkumaran, P.S. Cashew nut shell liquid: Extraction, chemistry and applications. *Prog. Org. Coat.* **1986**, *14*, 115–157. [CrossRef]
13. Alexander, M.; Thachil, E.T. The effectiveness of cardanol as plasticiser, activator, and antioxidant for natural rubber processing. *Prog. Rubber Plast. Recycl. Technol.* **2010**, *26*, 107–123. [CrossRef]
14. Lomonaco, D.; Cangane, F.Y.; Mazzetto, S.E. Thiophosphate esters of cashew nutshell liquid derivatives as new antioxidants for poly(methyl methacrylate). *J. Therm. Anal. Calorim.* **2011**, *104*, 1177–1183. [CrossRef]
15. Balgude, D.; Sabnis, A.S. CNSL: An environment friendly alternative for the modern coating industry. *J. Coat. Technol. Res.* **2014**, *11*, 169–183. [CrossRef]
16. Oliveira, L.D.M. *Synthesis, Characterization and Functionality of Lubricity Additives Derived from the LCC*; Universidade Federal do Ceará: Fortaleza, Brazil, 2007; Available online: http://www.repositorio.ufc.br/bitstream/riufc/2102/1/2007_dis_Lin_Oliveira.pdf (accessed on 23 April 2019).
17. Grassi, V.G.; Forte, M.M.C.; Dal Pizzol, M.F. Morphologic Aspects and Structure-Properties Relations of High Impact Polystyrene. *Polímeros Ciência e Tecnologia* **2001**, *11*, 158–168. [CrossRef]
18. Omonov, T.S.; Harrats, C.; Moldenaers, P.; Groeninckx, G. Phase continuity detection and phase inversion phenomena in immiscible polypropylene/polystyrene blends with different viscosity ratios. *Polymer* **2007**, *48*, 5917–5927. [CrossRef]
19. Lucas, E.F.; Soares, B.G.; Monteiro, E. *Caracterização de Polímeros. Determinação de Peso Molecular e Análise Térmica (Characterization of Polymers. Determination of Molecular Weight and Thermal Analysis)*, 1st ed.; Editora e-papers: Rio de Janeiro, Brazil, 2001.
20. Wunderlich, B. *Thermal Analysis of Polymeric Materials*; Springer Science + Business Media: Berlim, Germany, 2005.
21. Rosen, S.L. *Fundamental Principles of Polymeric Materials*, 3rd ed.; John Wiley & Sons: New York, NY, USA, 2012.
22. Rahman, M.; Brazel, C.S. The plasticizer market: An assessment of traditional plasticizers and research trends to meet new challenges. *Prog. Polym. Sci.* **2004**, *29*, 1223–1248. [CrossRef]
23. Almeida, J.F.M.; Silva, A.L.N.; Silva, A.H.M.F.T.; Sousa, A.M.F.; Nascimento, C.R.; Bertolino, L.C. Rheological, mechanical and morphological behavior of polylactide/nano-sized calcium carbonate composites. *Polym. Bull.* **2016**, *73*, 3531–3545. [CrossRef]
24. Kiss, A.; Fekete, E.; Pukánszky, B. Aggregation of $CaCO_3$ particles in PP composites: Effect of surface coating. *Compos. Sci. Technol.* **2007**, *67*, 1574–1583. [CrossRef]

materials

MDPI

Article

Origanum majorana L. Essential Oil-Associated Polymeric Nano Dendrimer for Antifungal Activity against *Phytophthora infestans*

Vu Minh Thanh [1,†], Le Minh Bui [2,†], Long Giang Bach [2,3], Ngoc Tung Nguyen [4], Hoa Le Thi [5] and Thai Thanh Hoang Thi [5,*]

1 Institute of Chemistry and Materials, 17 Hoang Sam, Cau Giay, Hanoi 100000, Vietnam; vmthanh222@yahoo.com
2 NTT Hi-Tech Institute, Nguyen Tat Thanh University, 300A Nguyen Tat Thanh, District 4, Ho Chi Minh City 700000, Vietnam; blminh@ntt.edu.vn (L.M.B.); blgiang@ntt.edu.vn (L.G.B.)
3 Center of Excellence for Functional Polymers and NanoEngineering, Nguyen Tat Thanh University, Ho Chi Minh City 700000, Vietnam
4 Center for Research and Technology Transfer (CRETECH), Vietnam Academy of Science and Technology, 18 Hoang Quoc Viet, Cau Giay District, Hanoi 100000, Vietnam; tungnguyen.vast@gmail.com
5 Biomaterials and Nanotechnology Research Group, Faculty of Applied Sciences, Ton Duc Thang University, Ho Chi Minh City 758307, Vietnam; mytom.61303531@gmail.com
* Correspondence: hoangthithaithanh@tdtu.edu.vn
† Co-First author.

Received: 25 March 2019; Accepted: 25 April 2019; Published: 4 May 2019

Abstract: In this study, the introduction of *Origanum majorana L.* essential oil into a polyamidoamine (PAMAM) G4.0 dendrimer was performed for creation of a potential nanocide against *Phytophthora infestans*. The characteristics of marjoram oil and PAMAM G4.0 was analyzed using transmission electron spectroscopy (TEM), nuclear magnetic resonance spectroscopy (^{1}H-NMR) and gas chromatography mass spectrometry (GC-MS). The success of combining marjoram oil with PAMAM G4.0 was evaluated by FT-IR, TGA analysis, and the antifungal activity of this system was also investigated. The results showed that the antifungal activity of oil/PAMAM G4.0 was high and significantly higher than only PAMAM G4.0 or marjoram essential oil. These results indicated that the nanocide oil/PAMAM G4.0 helped strengthen and prolong the antifungal properties of the oil.

Keywords: antifungal activity; dendrimer; *Origanum majorana* L. essential oil; *Phytophthora infestans*

---

1. Introduction

In Vietnam, tomato is ranked tenth in terms of crop value, topping at more than 9.7 billion VND in 2005. However, tomato late blight, caused by *Phytophthora infestans*, is a destructive disease that causes heavy decline in tomato production for many years. According to the General Statistics Office of Vietnam, the late blight caused by *Phytophthora infestans* has drastically hindered the growth of tomato cultivation area in Vietnam from 2009 to 2012, resulting in a halt in production yield [1]. Studies on the distribution as well as the effects of the late blight disease have also been conducted very early. According to the assessment of harm caused by late blight in the suburbs of Hanoi in 1965, the average loss of 30–70%, at the high level, can cause complete loss of productivity. In recent years, the level of the disease is still high. Severe outbreaks of late blight have been observed in many suburbs in Vietnam [1,2].

Recently, fungicides of biological origin have been strongly accentuated as it could overcome the inherent limitations of chemical pesticides. In addition, the approach of using antifungal substances extracted from plants was reported to be a promising agent against fungi. Marjoram (*Origanum majorana* L., Lamiaceae) is a perennial species originating in southern Europe. The plant has been widely cultivated and used in cooking as a spice. More importantly, distillates from marjoram are highly valued for its antimicrobial, antifungal and antioxidant activity. In one study, marjoram oil has been tested against various bacterial and fungal species, demonstrating improved effectiveness against *Beneckea natriegens*, *Erwinia carotovora* and *Moraxella* [3]. In addition, marjoram oil has been studied extensively in the field of plant disease control, especially against *P. infestans.* It is considered to be a fungicidal alternative to chemically derived drugs [4]. Besides marjoram essential oil, *P. infestants* was reported to be inhibited by pathogen-induced proteins in previous studies [5,6]. However, the process for utilizing the pathogen is quite complicated and not suitable for excess products. When it comes to combating microbes, the use of essential oils has various advantages. First, since essential oils are of natural origin, it is completely safe for humans and the environment. Second, since essential oils are mixtures of many different compounds, the use of essential oil as an antimicrobial agent could impair the adapted resistance against drugs of microbes in various ways.

However, essential oil from herbal sources demonstrates many disadvantages including poor stability, solubility, and volatility. One of the approaches to remedy such limitations is the application of nano-encapsulation, which is capable of reducing volatility, improving the stability, water solubility, and efficiency of essential oil-based formulations, while still maintaining therapeutic efficiency of the drug [7–11]. Dendrimer is a spherical, branched, nanoscale material that is more prominent than linear polymers. The dendrimer consists of three parts including core, inner branches, and lateral groups. Among dendrimers, polyamidoamine (PAMAM), a group of branched-chain dendrimers, including amine branching and amide bridging, is the most widely used dendrimer due to its amine functional groups (for even-numbered PAMAMs) and carboxylates (for odd-numbered PAMAMs) that help dissolve dendrimer in polar solvent [12,13]. These groups are also very active so it is easy to react to create new structurally diverse substances [14]. One of the important uses of PAMAM G4.0 (polyamidoamine generation 4) is acting as a carrier system for the transport of biological molecules and drugs for the treatment of cancer diseases [15,16]. Recently, many studies in using dendrimer in biocide has been proved to be much potential [17–20]. For example, Winnicka and co-workers reported the capability of PAMAM dendrimer in increasing the antifungal activity of clotrimazole against different strains of *Candida* [19]. Later, Winnicka continued to design a mixture containing PAMAM G2.0 and ketoconazole that also dramatically enhances the antifungal action of the drug [20]. In another study, Jose and co-workers employed different generation of PAMAM dendrimer (G1.0–G3.0) as agents to improve water solubility of amphotericin B drug, thereafter enhancing its antifungal action [21]. However, there are not much reports in combining extracted essential oil with dendrimer as carrier system.

This study attempted the combination of marjoram essential oil with PAMAM G4.0 to produce the oil/PAMAM G4.0 dendrimer system. The system was then evaluated for preservation efficiency of natural oil functions and tested for antifungal activity against *P. infestans*. The polymer structure and morphology of the PAMAM G4.0 were characterized. TGA analysis and the FT-IR spectrum of PAMAM G4.0 containing this volatile oil were also examined.

2. Materials and Methods

2.1. Materials

Ethylenediamine (EDA) and toluene were purchased from Merck (Darmstadt, Germany). Methyl acrylate (MA) was purchased from Sigma-Aldrich (St. Louis, MO, USA). Methanol was supplied by Fisher Scientific (Houston, TX, USA). Spectra/Por® Dialysis Membrane (MWCO 3.5 kDa) was purchased from Spectrum Laboratories Inc. (Rancho Dominguez, CA, USA). Marjoram essential oil was supplied by NTT Hi-Tech Institute, Nguyen Tat Thanh University, Ho Chi Minh City, Vietnam. *Phytophthora infestants* fungi strain was purchased from Gia Tuong Ltd, Vietnam. Ethanol and acetic acid were purchased from Xilong Chemical, Ltd. (Guangdong, China). All other chemicals were of reagent grade. Distilled water was used in all preparations.

2.2. Methods

2.2.1. Preparation of PAMAM G4.0 Dendrimer

PAMAM dendrimer generation 4.0 (PAMAM G4.0) was synthesized from the EDA core utilizing divergent approach as previously reported in our literature [13], in which the EDA's primary amine groups react with MA's acrylate groups via Michael addition reaction to form half generation PAMAM, followed by the reaction between half generation PAMAM's methyl propionate groups with excess EDA to form full generation PAMAM dendrimer, denoted by Gn.0. Briefly, EDA (20 mL) was added to 150 mL of MA dissolved in methanol. The reaction was in turn kept under stirring for 3 h at 0 °C and then 48 h at room temperature. The removal of impurities and solvent was performed using rotary vacuum evaporator (Strike 300, Lancashire, UK), resulting in the core precursor G0.5. Next, G0.5 was added to EDA solution (130 mL) to obtain PAMA G0.0. The mixture was stirred for 96 h at room temperature, rotated under vacuum using mixed solvent (toluene: methanol is 9:1 v/v). Finally, the resulting mixture was dialyzed by dialysis membrane against methanol to remove excess toluene and EDA and dried under vacuum to remove methanol. This protocol was repeated continuously to obtain PAMAM G4.0.

2.2.2. Characteristics of Dendrimer PAMAM G4.0

The chemical structure of dendrimer PAMAM G4.0 was analyzed using Nuclear Magnetic Resonance spectroscopy (Bruker Advance 500, Bruker Co., Billerica, MA, USA). Deuterated chloroform was used as solvent. For illustration of size and morphology, Transmission Electron Spectroscopy (JEM-1400 TEM; JEOL, Tokyo, Japan) was utilized. A drop of sample solution prepared in distilled water was placed on a carbon–copper grid (300-mesh, Ted Pella Inc., Redding, CA, USA) and air dried in 10 min for the TEM observation.

2.2.3. Evaluation of the Composition of Marjoram Essential Oil

The composition of the marjoram oil was determined by Gas Chromatography Mass Spectrometry (Agilent Technologies, Santa Clara, CA, USA).

2.2.4. Synthesis of PAMAM G4.0 Combining Marjoram Essential Oil

A rotating system was used to drain out all water vapor in the PAMAM G4.0 dendrimer to ensure that PAMAM G4.0 is completely dry. Marjoram essential oil (500 µg, 1000 µg, 2000 µg, and 5000 µg) was quickly added to PAMAM G4.0 according to the ratios (1:1). After being ultrasonicated for 15 min at room temperature, the samples were rotated at 100 °C for 2 min, allowing the oil molecules combine with the dendrimer structure. Samples were stored in refrigerator for later use.

2.2.5. Characterization of Oil/PAMAM G4.0

Marjoram oil, PAMAM G4.0 and the combined system were analyzed with Spectrum Tensor27 FT-IR Spectrometer, using KBr pellets FT-IR Spectrometer in the range of 400–4000 cm^{-1}. All samples were mixed with KBr and pressed into a pellet before the measurement. Thermal gravimetric analysis (TGA) of marjoram essential oil, PAMAM, and oil/PAMAM were carried out using TG Analyzer (Mettler Toledo, Columbus, OH, USA).

2.2.6. Agar diffusion method

The antifungal effect of marjoram oil, PAMAM G4.0 and the system were assessed using agar diffusion method. The PDA (Potato Dextrose Agar) agar plates were prepared by potato infusion at 250.00 g/L, glucose at 20.00 g/L, and agar at 20.00 g/L. After that, fungal solution was prepared and obtained the density of 6.0465×10^8 CFU/mL after 48 h of incubation. The solution was then diluted 2 times, followed by the inoculation of 100 μL diluted fungal solution on the agar surface. Appropriate amount of synthesized oil/PAMAM sample was dissolved in 1 mL of distilled water first to get the desired final concentration, which are 500 ppm, 1000 ppm, 2000 ppm, and 5000 ppm. Next, 100 μL of sample solution was added to each well of 6 mm in diameter. After 48 h, the inhibition zone was measured in diameter. The data were expressed as mean ± SD.

3. Results and Discussion

The preparation of PAMAM dendrimer generation 4.0 (G4.0) was obtained from the initiator EDA core using two-step synthesis in which the odd steps yield half generation PAMAM followed by the even steps which generate full generation PAMAM dendrimer.

3.1. Chemical Structure of Synthesized Dendrimer PAMAM G4.0

The spectra of ^{1}H NMR proton of PAMAM G4.0 is shown in Figure 1. As shown in Figure 1, the protons at 2.351 ppm, 2.55 ppm, 2.77 ppm, 3.04 ppm and 3.23 ppm were assigned to protons in methylene groups of $-CH_2-CO-NH$ (peak d), CH_2-N- (peak b), $-CH_2-NH_2-$ (peak a), $-N-CH_2-CH_2-$ (peak e), and $-CO-NH-CH_2$ (peak c). The presence of all these signals demonstrated that PAMAM G4.0 was successfully synthesized and characterized based on other studies in NMR characterization of fourth-generation PAMAM [13,16].

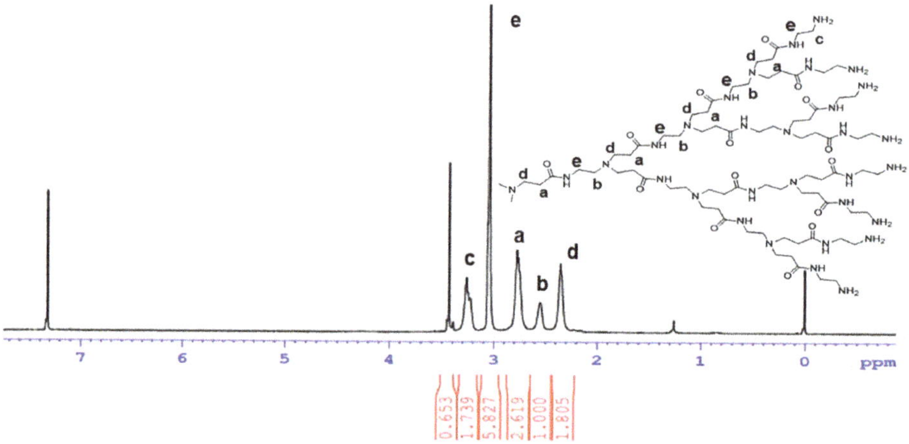

Figure 1. ^{1}H NMR spectrum of PAMAM G4.0.

3.2. Characteristics of PAMAM G4.0 and Oil/PAMAM G4.0

Figure 2 displays the morphology and particle size of PAMAM G4.0 dendrimer. Since the eradication of the fungi depends on the transportation of marjoram essential oil, it is more likely for the dendrimer to penetrate into the spore or mycelium (the size of the spore ranges from 19×10^{-12} to 23×10^{-12} mm) as its size gets tinier [3]. Analysis of TEM results showed that dendrimer PAMAM G4.0 was 20 to 30 nm in size and fairly uniform. There are some aggregates (<70 nm) were shown to occur but in low frequency. After the oil was associated with dendrimer, a significant increment in the diameter of obtained oil/PAMAM G4.0 (40–150 nm) was observed, indicating the successful combination of essential oil with PAMAM G4.0 structure. There are many studies reported that the particle size of less than 500 nm is sufficiently good for the penetration of nanocarriers into fungal cells [22–25]. As a result, G4.0 and oil/PAMAM G4.0 is suitable for associating with oil to fully interact with the fungus's cell membrane as well as easily penetrates into the cell layer, thus destroying fungal cells.

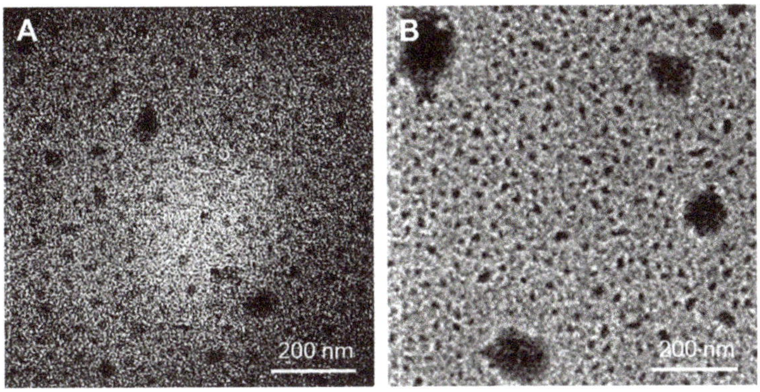

Figure 2. TEM image of polyamidoamine dendrimer PAMAM G4.0 (**A**) and oil/PAMAM G4.0 (**B**).

3.3. Evaluation of the Composition of Marjoram Essential Oil

Figure 3 shows the composition and content of the substances in marjoram essential oil analyzed by GC-MS. From the GC-MS analysis, it is revealed that major components in the oil are 1R-alpha-pinene, sabinene, cymol, cyclohexanol, cineole, alpha-terpinolene, alpha-terpineol, linalyl ester, beta-caryophyllene, alpha-caryophyllene. Chemical structures of these components of marjoram oil are shown in Figure 4. Cineole (28.59%) is the predominant constituent in the oil, followed by alpha-terpinolen (14.75%) and cymol (11.82%). Compared to the analysis results of other studies, the reported composition of the oil is similar. However, component contents are different, possibly due to the difference in cultivation process, the habitat of the plant, the extraction and preservation method [4].

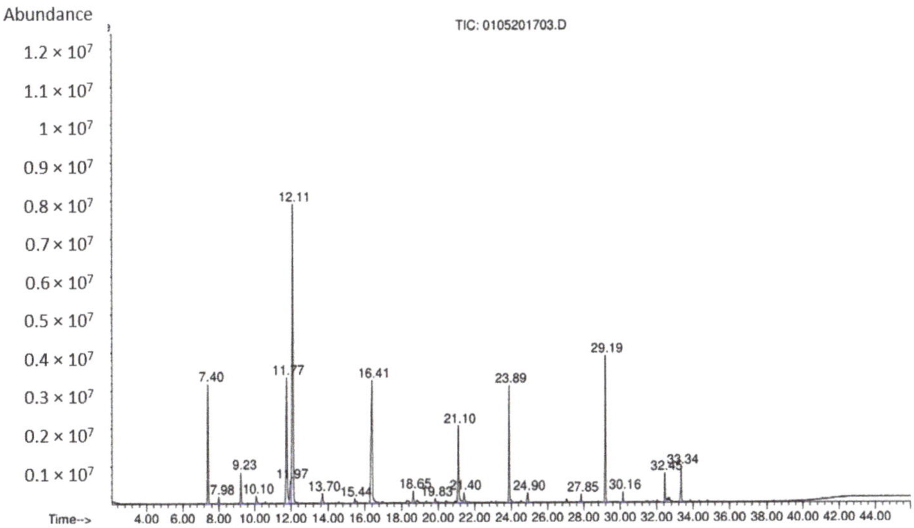

Figure 3. The GC-MS spectrum of marjoram essential oil.

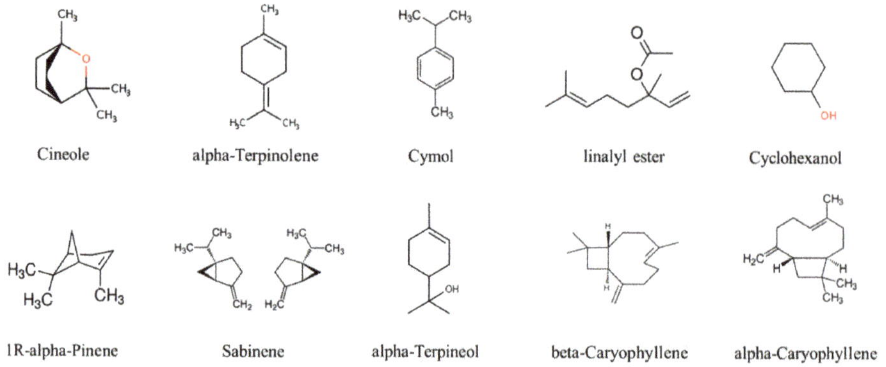

Figure 4. Chemical structures of marjoram oil's components.

3.4. FT-IR Analysis

Figure 5 demonstrates FT-IR spectra of marjoram essential oil, PAMAM G4.0 and oil/PAMAM G4.0. The characteristic peaks of PAMAM G4.0 at positions 1661 cm^{-1} and 1553 cm^{-1} are the signals of the (-CO-NH-) bonds of grades I and II, respectively, which were also observed clearly on the spectrum of oil/PAMAM G4.0. Five characteristic peaks of the essential oil at position 2967 cm^{-1}; 2927 cm^{-1}; 1740 cm^{-1}; 1451 cm^{-1}; and 1376 cm^{-1} have also been observed clearly on the spectrum of oil/PAMAM G4.0. Signaling of the essential oil and dendrimer PAMAM G4.0 were both observed in the spectrum of the synthesized oil/PAMAM G4.0 sample, indicating that the synthesis of dendrimer PAMAM G4.0 associating marjoram oil was successful. Additionally, there is a disappearance of carbonyl peak at 1700 cm^{-1} in oil/PAMAM G4.0 spectrum as compared to essential oil spectrum. C-O absorption band at 1300 cm^{-1} of alcohol functional groups of essential oil's components is not detected as well, indicating the formation of hydrogen bonds between essential oil's components and PAMAM G4.0.

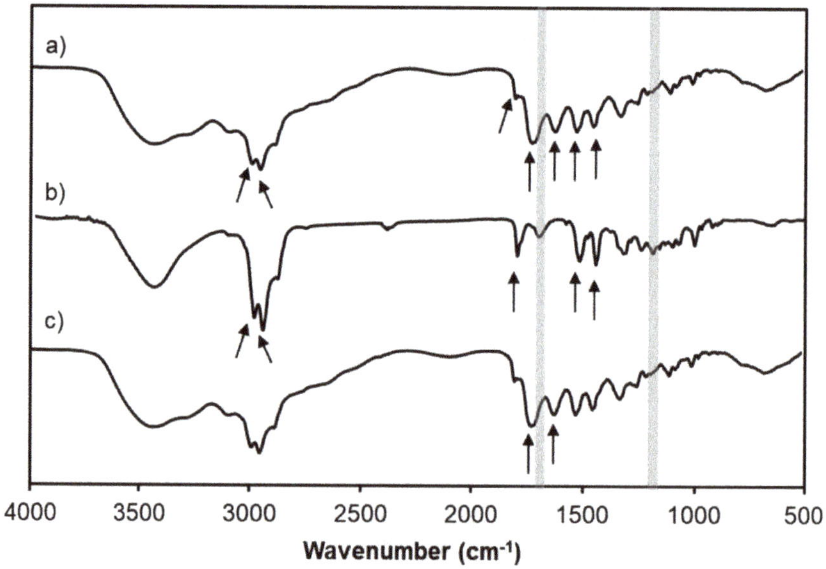

Figure 5. FT-IR spectra of (**a**) oil/PAMAM G4.0; (**b**) Marjoram essential oil and (**c**) PAMAM G4.0.

3.5. TGA Analysis

The actual amount of marjoram essential oil mixed with PAMAM G4.0 was estimated by TGA analysis (Figure 6). The experimental temperature was ramped to 200 °C with heating rate at 3 °C/min and from 200 °C to 600 °C with heating rate at 50 °C /min. The weight loss from room temperature to 100 °C corresponds to the evaporation of marjoram oil. As shown in TGA curves of oil/PAMAM G4.0 with 1000 µg essential oil, the weight loss at 100 °C was around 56%. This means the actual amount of essential oil successfully mixed with 1000 µg PAMAM G4.0 in the final product was about 560 µg. This result indicated the existence of essential oil in oil/PAMAM G4.0 sample.

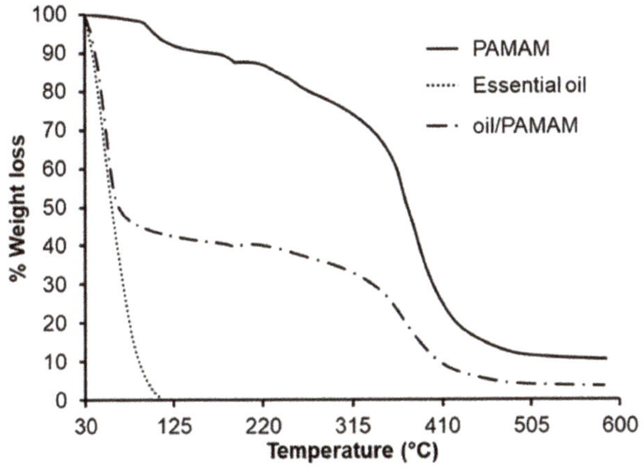

Figure 6. TGA curves of PAMAM G4.0, marjoram essential oil, and oil/PAMAM G4.0.

3.6. Antifungal Experiment

Anti-*Phytophthora infestants* properties of the essential oil, PAMAM G4.0 and the mixture of these two compounds at concentrations of 500, 1000, 2000 and 5000 ppm are displayed and summarized in Figures 7, 8 and Table 1. The combination of PAMAM G4.0 and volatile marjoram exhibits superior antifungal properties compared to the marjoram essential oil and PAMAM G4.0. At the lowest test concentration (500 ppm), PAMAM G4.0 has a diameter of inhibition zone of 6.16 ± 0.34 mm. PAMAM G4.0 anti-fungal properties are due to the fact that their surface containing $-NH_2-$ positively charged particles which could interact and destroy negatively charged microbial membranes. It is also shown that the antifungal property of the essential oil is minimal, as demonstrated by the very small diameter of inhibition zone of 6.33 ± 0.33 mm at 500 ppm concentration and the presence of mycelium around the well. In other words, there are no significant inhibition at both 500 and 1000 ppm of essential oil and PAMAM G4.0, individually. This is contrasted by wells of oil/PAMAM G4.0 at all concentrations, showing recognizable inhibition zones. At higher concentrations of 1000, 2000 and 5000 ppm, the antifungal activity of PAMAM G4.0 and oil, individually, was gradually increased and the strongest activity against *P. infestans* was achieved by the oil/PAMAM G4.0. In other words, oil/PAMAM G4.0 was found to have stronger antifungal ability than that of each individual component. After a series of experiments, it can be concluded that the concentration of oil/PAMAM G4.0 at 500 ppm, which respectively exhibited 10 and 22 times bigger inhibition diameter (9.50 ± 0.50 mm) than that of essential oil and PAMAM G4.0 at the same concentration (500 ppm), provided the optimal antifungal result as compared to other combinations of PAMAM G4.0 and oil. At higher concentration (5000 ppm), the synergistic effect seemed to be lost. This could be explained by the reason that not all marjoram essential oil used was successfully mixed with PAMAM G4.0, as revealed in TGA analysis. Due to steric hindrance effect, reduced ability of essential oil to be mixed and combined with PAMAM as the concentration of PAMAM increased. In addition, the diffusion of mixed essential oil from the mixture oil/PAMAM G4.0 out of the well to the agar surface in agar diffusion test could also be restricted with the increasing of PAMAM molecules concentration. Taken together, thanks to the synergy between PAMAM G4.0 and oil, not only the required concentration for each component was reduced significantly, it was also possible to maintain the sufficient result, which is important for reduction of environmental pollution risk.

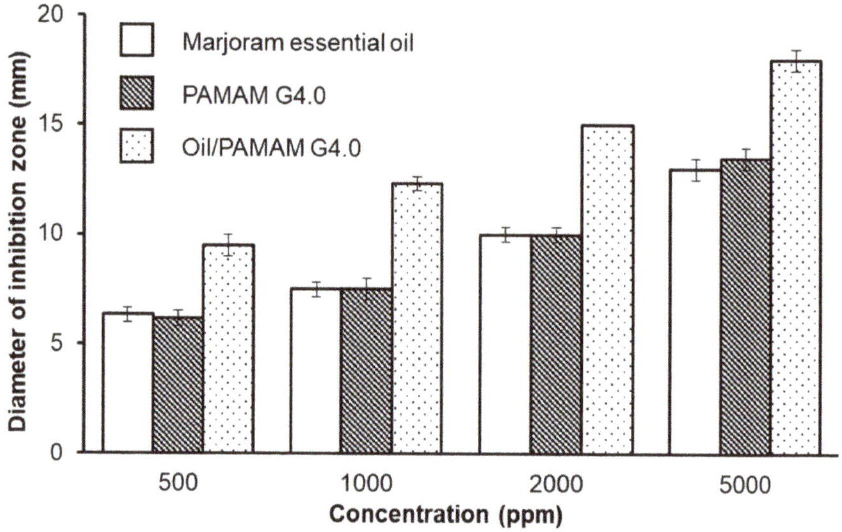

Figure 7. The graph concluding average inhibition zone (d. mm) against *Phytophthora infestants* of marjoram essential oil, PAMAM G4.0 and oil/PAMAM G4.0 based on different concentrations: 500, 1000, 2000 and 5000 ppm.

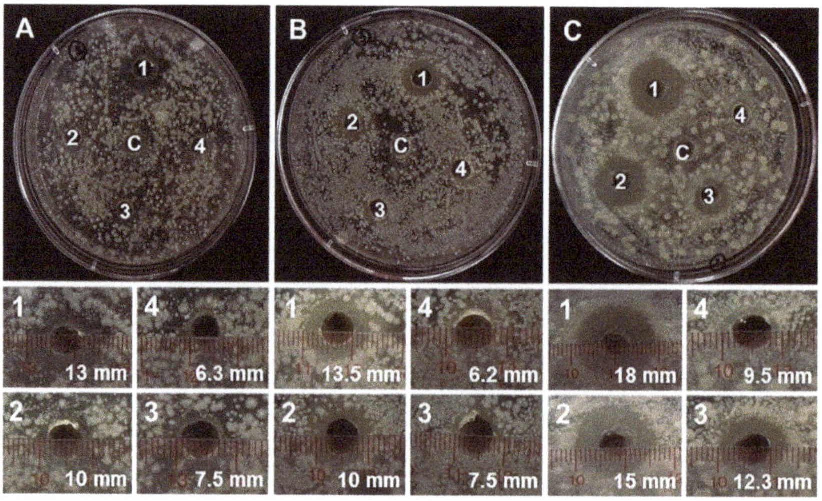

Figure 8. Diameters of inhibition zone of (**A**) Marjoram essential oil; (**B**) dendrimer PAMAM G4.0; (**C**) oil/PAMAM G4.0 after 48 h with different concentrations: (1) 5000 ppm; (2) 2000 ppm; (3) 1000 ppm: (4) 500 ppm.

Table 1. Average diameters of inhibition zone against *Phytophthora infestants* of Marjoram essential oil, PAMAM G4.0 and oil/PAMAM G4.0 based on different concentrations: 500, 1000, 2000 and 5000 ppm.

Concentration (ppm)		Diameter of Inhibition Zone (d. mm)
Marjoram Essential Oil	PAMAM G4.0	
500	-	6.33 ± 0.33
1000	-	7.50 ± 0.33
2000	-	10.00 ± 0.33
5000	-	12.50 ± 0.50
-	500	6.16 ± 0.34
-	1000	7.50 ± 0.50
-	2000	10.00 ± 0.33
-	5000	12.00 ± 0.50
500	500	9.50 ± 0.50
1000	1000	12.33 ± 0.33
2000	2000	15.00 ± 0.00
5000	5000	18.00 ± 0.50

As revealed in GC-MS result, marjoram essential oil comprised several components with very different chemical structures, which are mostly hydrophobic and some containing hydrophilic functional group in the structure. We also interpreted that there must be a hydrogen bonding between hydrophilic OH or C=OH groups of some components in the essential oil and NH_2 groups on the surface of PAMAM G4.0. The rest hydrophobic components tended to be entrapped within the large internal cavity of dendrimer, which is a common tendency of hydrophobic substances when being mixed with PAMAM, thus reducing the essential oil volatility. This phenomenon has been reported in many previous studies [16,26]. With the particle size of less than 500 nm, oil/PAMAM G4.0 (average diameter around 40–120 nm) is considered as a sufficiently good material for the penetration of nanocarriers into fungal cells [22–25]. In addition, PAMAM G4.0 with large number of positively charged surface $-NH_2-$ groups facilitates the interaction and destruction of microbial negatively charged membrane. As a result, oil/PAMAM G4.0 fully interacts with the fungus's cell membrane as well as easily penetrates into the cell layer, thus destroying fungal cells.

The synergistic effect in antifungal properties between the essential oils and oil/PAMAM G4.0 could be explained by the reduction of the oil volatility when it is mixed with PAMAM G4.0 structure, therefore improving the stability and maintaining the antifungal effectiveness better than essential oil in its free form. Moreover, the extremely poor water solubility of essential oil could also be improved by associating with PAMAM dendrimer, which perhaps due to the hydrogen bonds and hydrophobic interactions between the essential oils and functional groups of PAMAM G4.0. There are numerous literatures that successfully demonstrated the use of dendrimer to enhance the solubilization of hydrophobic molecules [27]. Many of such studies revealed the capability of PAMAM dendrimer in antifungal activity improvement as consequence of enhanced solubility of poorly soluble substances [28,29]. For instance, it was reported that the mixture of PAMAM G2.0 and ketoconazole, which is hydrophobic, was 16 times more potent against *Candida* in comparison with free drug [19].

4. Conclusions

In this study, TEM and ^{1}H-NMR technique were employed to characterize the PAMAM G4.0 material and the marjoram essential oil was analyzed for chemical composition using GC-MS. The two samples and the oil/PAMAM G4.0 were jointly analyzed by FT-IR, showing the successful associating of oil and PAMAM G4.0 dendrimer. The resulting PAMAM G4.0@oil was shown to be more effective in terms of antifungal activity compared with PAMAM G4.0 and marjoram volatile oil with same concentration. The enhanced anti-microbial property of the oil/PAMAM G4.0 could be due to the restricted evaporation of the essential oil, caused by the encapsulation. These results suggest that the association of marjoram oil with PAMAM G4.0s is promising in combating late blight in tomatoes and that later studies should focus on determining the suitable concentration for optimal inhibition and conducting field trials.

Author Contributions: Investigation, L.G.B., T.L.H., N.T.N.; Writing-Original Draft Preparation, V.M.T.; Supervision, T.T.H.T. and L.M.B.

Funding: This research received no external funding.

Conflicts of Interest: The authors declare no conflicts of interest.

References

1. Genova, C.; Weinberger, K.; Chanthasombath, T.; Inthalungdsee, B.; Sanatem, K.; Somsak, K. *Postharvest Loss in the Supply Chain for Vegetables: The Case of Tomato, Yardlong Bean, Cucumber and Chili in Lao PDR*; AVRDC: Shanhau, Taiwan, 2006.
2. Thanh, D.; Tarn, L.; Hanh, N.; Tuyen, N.; Srinivasan, B.; Lee, S.-Y.; Park, K.-S. Biological Control of Soilborne Diseases on Tomato, Potato and Black Pepper by Selected PGPR in the Greenhouse and Field in Vietnam. *Plant Pathol. J.* **2009**, *25*, 263–269. [CrossRef]
3. Deans, S.G.; Svoboda, K.P. The antimicrobial properties of marjoram (Origanum majorana L.) Volatile Oil. *Flavour Fragr. J.* **1990**, *5*, 187–190. [CrossRef]
4. Busatta, C.; Vidal, R.; Popiolski, A.; Mossi, A.; Dariva, C.; Rodrigues, M.; Corazza, F.; Corazza, M.; Oliveira, J.V.; Cansian, R.; et al. Application of Origanum majorana L. essential oil as an antimicrobial agent in sausage. *Food Microbiol.* **2008**, *25*, 207–211. [CrossRef] [PubMed]
5. Woloshuk, C.P. Pathogen-Induced Proteins with Inhibitory Activity toward Phytophthora infestans. *Plant Cell* **1991**, *3*, 619–628.
6. Van Der Vossen, E.; Sikkema, A.; Hekkert, B.T.L.; Gros, J.; Stevens, P.; Muskens, M.; Wouters, D.; Pereira, A.; Stiekema, W.; Allefs, S. An ancient R gene from the wild potato species Solanum bulbocastanum confers broad-spectrum resistance to Phytophthora infestans in cultivated potato and tomato. *Plant J.* **2003**, *36*, 867–882.
7. Bilia, A.R.; Guccione, C.; Isacchi, B.; Righeschi, C.; Firenzuoli, F.; Bergonzi, M.C. Essential Oils Loaded in Nanosystems: A Developing Strategy for a Successful Therapeutic Approach. *Evid.-Based Complement. Altern. Med.* **2014**, *2014*, 1–14.
8. Nguyen, D.H.; Joung, Y.K.; Choi, J.H.; Moon, H.T.; Park, K.D. Targeting ligand-functionalized and redox-sensitive heparin-Pluronic nanogels for intracellular protein delivery. *Biomed. Mater.* **2011**, *6*, 55004. [CrossRef]
9. Dai, L.; Zhang, B.; Li, J.; Liu, J.; Luo, Z.; Cai, K. Redox-Responsive Nanocarrier Based on Heparin End-Capped Mesoporous Silica Nanoparticles for Targeted Tumor Therapy in Vitro and in Vivo. *Langmuir* **2014**, *30*, 7867–7877. [CrossRef] [PubMed]
10. Nguyen, D.H.; Lee, J.S.; Choi, J.H.; Park, K.M.; Lee, Y.; Park, K.D. Hierarchical self-assembly of magnetic nanoclusters for theranostics: Tunable size, enhanced magnetic resonance imagability, and controlled and targeted drug delivery. *Acta Biomater.* **2016**, *35*, 109–117. [CrossRef] [PubMed]
11. Yallapu, M.M.; Gupta, B.K.; Jaggi, M.; Chauhan, S.C. Fabrication of curcumin encapsulated PLGA nanoparticles for improved therapeutic effects in metastatic cancer cells. *J. Colloid Interface Sci.* **2010**, *351*, 19–29. [CrossRef]
12. Nguyen, T.L.; Nguyen, T.H.; Nguyen, C.K.; Nguyen, D.H. Redox and pH Responsive Poly (Amidoamine) Dendrimer-Heparin Conjugates via Disulfide Linkages for Letrozole Delivery. *BioMed Int.* **2017**, *2017*, 1–7. [CrossRef]
13. Thanh, V.M.; Nguyen, T.H.; Tran, T.V.; Ngoc, U.-T.P.; Ho, M.N.; Nguyen, T.T.; Chau, Y.N.T.; Le, V.T.; Tran, N.Q.; Nguyen, C.K.; et al. Low systemic toxicity nanocarriers fabricated from heparin-mPEG and PAMAM dendrimers for controlled drug release. *Mater. Sci. Eng. C* **2018**, *82*, 291–298. [CrossRef] [PubMed]
14. Fréchet, J.M.J.; Tomalia, D.A. *Dendrimers and Other Dendritic Polymers: Frechet/Dendrimers*; John Wiley & Sons, Ltd.: Chichester, UK, 2001.
15. Nguyen, T.T.C.; Nguyen, C.K.; Nguyen, T.H.; Tran, N.Q. Highly lipophilic pluronics-conjugated polyamidoamine dendrimer nanocarriers as potential delivery system for hydrophobic drugs. *Mater. Sci. Eng. C* **2017**, *70*, 992–999. [CrossRef] [PubMed]

16. Tran, N.Q.; Nguyen, C.K. Biocompatible nanomaterials based on dendrimers, hydrogels and hydrogel nanocomposites for use in biomedicine. *Adv. Sci. Nanosci. Nanotechnol.* **2017**, *8*, 15001.
17. Chen, C.Z.; Cooper, S.L. Interactions between dendrimer biocides and bacterial membranes. *Biomaterials* **2002**, *23*, 3359–3368. [CrossRef]
18. Chen, C.Z.; Beck-Tan, N.C.; Dhurjati, P.; Van Dyk, T.K.; LaRossa, R.A.; Cooper, S.L. Quaternary Ammonium Functionalized Poly(propylene imine) Dendrimers as Effective Antimicrobials: Structure–Activity Studies. *Biomacromolecules* **2000**, *1*, 473–480. [CrossRef] [PubMed]
19. Winnicka, K.; Sosnowska, K.; Wieczorek, P.; Sacha, P.T.; Tryniszewska, E. Poly(amidoamine) Dendrimers Increase Antifungal Activity of Clotrimazole. *Biol. Pharm. Bull.* **2011**, *34*, 1129–1133. [CrossRef] [PubMed]
20. Winnicka, K.; Wroblewska, M.; Wieczorek, P.; Sacha, P.T.; Tryniszewska, E. Hydrogel of Ketoconazole and PAMAM Dendrimers: Formulation and Antifungal Activity. *Molecules* **2012**, *17*, 4612–4624. [CrossRef]
21. Jose, J.; Charyulu, R.N. Prolonged drug delivery system of an antifungal drug by association with polyamidoamine dendrimers. *Int. J. Pharm. Investig.* **2016**, *6*, 123–127. [CrossRef] [PubMed]
22. Paomephan, P.; Assavanig, A.; Chaturongakul, S.; Cady, N.C.; Bergkvist, M.; Niamsiri, N. Insight into the antibacterial property of chitosan nanoparticles against Escherichia coli and Salmonella Typhimurium and their application as vegetable wash disinfectant. *Food Control* **2018**, *86*, 294–301. [CrossRef]
23. Shankar, S.; Pangeni, R.; Park, J.W.; Rhim, J.-W. Preparation of sulfur nanoparticles and their antibacterial activity and cytotoxic effect. *Mater. Sci. Eng. C* **2018**, *92*, 508–517. [CrossRef]
24. Agnihotri, S.A.; Mallikarjuna, N.N.; Aminabhavi, T.M. Recent advances on chitosan-based micro- and nanoparticles in drug delivery. *J. Control. Release* **2004**, *100*, 5–28. [CrossRef] [PubMed]
25. Sharma, S. Enhanced antibacterial efficacy of silver nanoparticles immobilized in a chitosan nanocarrier. *Int. J. Boil. Macromol.* **2017**, *104*, 1740–1745. [CrossRef]
26. Teow, H.M.; Zhou, Z.; Najlah, M.; Yusof, S.R.; Abbott, N.J.; D'Emanuele, A. Delivery of paclitaxel across cellular barriers using a dendrimer-based nanocarrier. *Int. J. Pharm.* **2013**, *441*, 701–711. [CrossRef] [PubMed]
27. Choudhary, S.; Gupta, L.; Rani, S.; Dave, K.; Gupta, U.; Gupta, D.U. Impact of Dendrimers on Solubility of Hydrophobic Drug Molecules. *Front. Pharmacol.* **2017**, *8*, 261. [CrossRef]
28. Voltan, A.R.; Quindós, G.; Alarcón, K.P.M.; Fusco-Almeida, A.M.; Mendes-Giannini, M.J.S.; Chorilli, M.; Mendes-Giannini, M.J. Fungal diseases: Could nanostructured drug delivery systems be a novel paradigm for therapy? *Int. J. Nanomed.* **2016**, *11*, 3715–3730. [CrossRef]
29. Bondaryk, M.; Staniszewska, M.; Zielińska, P.; Urbańczyk-Lipkowska, Z. Natural antimicrobial peptides as inspiration for design of a new generation antifungal compounds. *J. Fungi* **2017**, *3*, 46. [CrossRef]

MDPI
St. Alban-Anlage 66
4052 Basel
Switzerland
Tel. +41 61 683 77 34
Fax +41 61 302 89 18
www.mdpi.com

Materials Editorial Office
E-mail: materials@mdpi.com
www.mdpi.com/journal/materials

www.ingramcontent.com/pod-product-compliance
Lightning Source LLC
LaVergne TN
LVHW070147100526
838202LV00015B/1905